D1202076

X-Ray Analysis and the Structure of Organic Molecules

Distribution
VCH, P.O. Box 101161, D-69451 Weinheim (Federal Republic of Germany)
Switzerland: VCH, P.O. Box, CH-4020 Basel (Switzerland)
United Kingdom and Ireland: VCH (UK) Ltd., 8 Wellington Court, Wellington Street, Cambridge CB1 1HZ (England)
USA and Canada: VCH, Suite 909, 220 East 23rd Street, New York, NY 10010-4606 (USA)

ISBN 3-906390-14-4

Jack D. Dunitz

X-Ray Analysis and the Structure of Organic Molecules

(2nd Corrected Reprint)

Verlag Helvetica Chimica Acta, Basel

Weinheim · New York · Basel
Cambridge · Tokyo

Prof. Dr. Jack D. Dunitz
Laboratorium für Organische Chemie
ETH-Zentrum
Universitätstrasse 16
CH-8092 Zürich

Published jointly by
VHCA, Verlag Helvetica Chimica Acta, Basel (Switzerland)
VCH Verlagsgesellschaft mbH, Weinheim (Federal Republic of Germany)
VCH Publishers, Inc., New York, NY (USA)

Editorial Director: Dr. M. Volkan Kisakürek
Production Manager: Jakob Schüpfer
Cover Design: Bruckmann & Partner, Basel

Library of Congress Card No. applied for.

A CIP catalogue record for this book is available from the British Library.

Die Deutsche Bibliothek - CIP-Einheitsaufnahme
Dunitz, Jack D.:
X-ray analysis and the structure of organic molecules / by Jack D. Dunitz. - 2., corr. reprint. - Basel : VHCA ; Weinheim ; New York : VCH, 1995
ISBN 3-906390-14-4

Printed on acid-free paper.

Printing: Birkhäuser+GBC AG, CH-4153 Reinach BL
Printed in Switzerland

TO BARBARA

Preface

This book has developed out of a series of lectures given at Cornell University in September and October 1976 when I held the George Fisher Baker Non-Resident Lectureship in Chemistry. I am most grateful to the Baker Lecture Committee for the invitation to act as Baker Lecturer. Thanks to the friendship and hospitality of my hosts, who did everything possible to make me feel at home at Cornell, the weeks I spent there were a rich and happy experience that I shall cherish to the end of my days.

The task of producing a finished manuscript has been greatly eased by the help I have had from several friends and colleagues. Parts of the original manuscript were read by Kevin Brown, Hans-Beat Bürgi, Erik Bye, Derek Chadwick, Phillip Coppens, Volker Gramlich, Christoph Kratky, Richard Rosenfield, Bernd Schweizer, Paul Seiler, and Fritz Winkler; I am grateful to them all for their constructive comments and criticisms. I owe special thanks to Kenneth Trueblood of the University of California at Los Angeles for reading virtually the entire first version of the manuscript, for drawing my attention to countless errors, ambiguities, and stylistic inelegancies in it, and for suggesting numerous ways of improving the quality and intelligibility of the text. It was most fortunate for me that Ken chose to spend a sabbatical year in my laboratory in Zurich, where I could benefit from almost daily discussions with him. I am also indebted to Max Dobler for providing me with X-ray photographs and diagrams for several figures and for preparing the computer programs that are listed in the two appendices.

Chapter 10 contains previously unpublished material that was developed in the course of a long-standing collaboration with Hans-Beat Bürgi, and I am very grateful to him for his generous permission to include this material here under my own name.

I must also acknowledge my indebtedness to the many other researchers who have contributed, knowingly or unknowingly, to the subject matter of this book. Where possible, I have indicated my sources of information by literature references. These are fairly complete as far as factual knowledge is concerned, but they are hopelessly inadequate in assigning credit for ideas. More often than not, the origins of ideas are forgotten, if they are ever consciously known. Like many scientists, I have the habit of absorbing fragments of other people's ideas—from lectures, from books, from conversations—and nurturing them as my own. In my opinion, science lives on this kind of cross-fertilization, so I hope I need not feel too guilty about such borrowings.

Many readers will not have to be told that the quotations at the beginning of each chapter are from Lewis Carroll, but for readers who were not brought up among English books I had better explain that Lewis Carroll was the pseudonym of a nineteenth-century mathematician, Charles Lutwidge Dodgson (1832-1898), one of the founders of mathematical logic and the author of *Alice's Adventures in Wonderland* and *Through the Looking Glass.*

Finally, I must record my appreciation to my secretary, Lucie a Marca, for her skill in typing the manuscript and for her patience in coping with its innumerable modifications, emendations, and revisions.

Zurich JACK DUNITZ

Preface to Corrected Reprint, 1995

The first edition of this book contained errors and misprints. I am obliged to many friends and colleagues for their careful reading of the text and for informing me about the errors they detected. Most are harmless, but some might be sources of confusion. Now that the first edition is out of print, Dr. *M. Volkan Kısakürek* has kindly offered to produce a corrected reprint under the auspices of *Helvetica Chimica Acta Publishers*, and I am most grateful to him for this initiative.

Zurich, June 1995 JACK DUNITZ

Contents

X-RAY ANALYSIS AND THE STRUCTURE OF ORGANIC MOLECULES

Introduction

"Where shall I begin, please your Majesty?" he asked.
"Begin at the beginning," the King said gravely,
"and go on till you come to the end: then stop."

When I first thought of writing this book, I imagined a sequel to J. M. Robertson's *Organic Crystals and Molecules* (Cornell University Press, Ithaca, N.Y., 1953)—an account of X-ray structure analysis with applications to organic chemistry, but from the perspective of the mid 1970s rather than the early 1950s. Robertson's book had influenced a whole generation of chemical crystallographers, but, with all its merits, it had become somewhat outdated. Important developments had occurred during the intervening quarter century, and it seemed appropriate that I, as one of his former students and collaborators, should try to produce a more up-to-date version, incorporating these developments but preserving the overall plan of his book.

I soon found that my plan was illusory. Robertson's book was divided into two parts, the first dealing mainly with the nature of crystals and with methods of crystal structure analysis, the second with the results obtained from such analyses of organic compounds. Apart from the development of direct methods, the advances since the early 1950s have been mainly technical—the development of automatic diffractometers to measure diffraction patterns of crystals and of electronic computers to carry out the necessary calculations—but the effects of these advances have been overwhelming. In 1951 the structures of about 100 organic crystals had been established, and Robertson could discuss nearly all of them in some detail. Today, the figure stands at about 16,000, so that it is quite out of the question for me to try to emulate Robertson's comprehensive coverage of the results.

This increase in the power and rapidity of crystal structure analysis has radically altered the relationship between crystallography and chemistry. By the early 1950s, results of crystal structure analysis had already begun to have an impact on chemistry, but this was barely noticed by most practicing chemists. In problems of molecular structure determination—e.g., of complex natural products—crystal structure analysis was called on only as a last resort, when all other methods had failed. Usually it failed too, although there were notable exceptions, when favorable circumstances happened to coincide with outstanding skill and perseverance on the part of the investigator.

During the last quarter century, however, crystal analysis has become the main source of information on the structure of complex organic molecules. Today the research chemist can hardly avoid confrontation with the crystallographic literature, and an increasing number of chemists are becoming convinced of the desirability of having X-ray diffraction equipment in their own laboratories and carrying out crystal analyses for themselves. It is therefore important that chemists acquire at least a basic knowledge of the theory and practice of X-ray analysis, enough to provide the background necessary to form, say, a reasonably critical judgment of the results reported in a crystallographic paper.

Like most other branches of scientific investigation, crystal structure analysis has developed a jargon of its own, which sometimes makes it difficult for outsiders to know what the initiates are talking about. Lacking the facility for critical appraisal, many chemists regard the results of X-ray analysis as sacrosanct, as representing the ultimate test of a proposed molecular structure. Actually, there have been several cases where results obtained by X-ray analysis have had to be revised. Indeed, it seems fair to say that there are no absolutely rigorous criteria for establishing the unequivocal correctness of a crystal structure analysis, especially if the crystal lacks a center of symmetry. However, this difficulty should not encourage us to adopt a solipsistic attitude; most published results obtained by X-ray analysis are almost certainly correct in their essential features. Still, it is well to be wary, and any "unusual results," however exciting, need especially careful appraisal.

Largely because of the advances in electronic circuitry and computer technology, most of the steps in a routine X-ray analysis can be carried out now in an efficient, more-or-less automatic procedure by relatively inexperienced persons. Nevertheless, it is worth stressing that, contrary to a widely held belief, a fully automatic, hundred-percent reliable method of measuring the X-ray diffraction pattern of a crystal and transforming this pattern into an image of the structure does not exist. Although automatic data measurement and structure solution and refinement will often be successful, it will fail in some cases and lead to more-or-less obviously incorrect results in others. The flaw may arise from an intrinsic property of the crystal (twinning, disorder, space group ambiguity), or it may be caused simply by fallibility of the equipment or the operator. In any case chemists should not believe everything that purports to be a result of crystal structure analysis, and they should know enough about the basic principles to enable them to indulge their scepticism in a reasonably informed manner.

In the first part of this book, I try to sketch these principles and draw attention to the potentialities and limitations of X-ray analysis as a method for obtaining information about crystal and molecular structure. The treatment is not intended to be complete but rather to provide the background necessary for describing recent developments in such a way as to make them comprehensible to nonspecialists.

Chapter 1 deals with the basic theory underlying the diffraction of X-rays by ideal and imperfect crystals. The Fourier transform relationship between diffraction pattern and distribution of scattering matter is introduced at the beginning; this makes it possible to discuss scattering by systems of varying complexity from a unified viewpoint. I hope that the mathematical formalism will not frighten away too many readers. Those who are unfamiliar with Fourier transforms will probably need some help from outside sources, but those with even a rudimentary knowledge should manage to wade through the initial difficulties.

Chapter 2 is a brief account of the internal symmetry of crystals. The theory of space groups (in two and three dimensions) is a closed subject, which was completely worked out before the beginning of this century. A complete tabulation of space groups is given in Volume I of *International Tables for X-ray Crystallography* (Kynoch Press, Birmingham, Eng., 3rd ed., 1969), and the main aim of Chapter 2 is to enable the reader to use these tables with some degree of confidence. The assignment of a crystal to a given space group is usually one of the first steps in a structure analysis, but it is not always straightforward; some of the difficulties are described.

In Chapter 3 we come to the heart of the matter—crystal structure analysis or methods of solving the "phase" problem. There are two main lines of attack. One is based on the presence of a "heavy atom" that contributes more than its fair share to the overall diffraction pattern; the other is based on so-called "direct methods." Direct methods have something of a hit-or-miss character, but, backed by high-speed computing facilities, they have come into general use during the last decade or so (after a period in which their potentialities were a matter of some controversy). Since they are barely mentioned in most introductory texts (a notable exception is M. M. Woolfson, *An Introduction to X-ray Crystallography,* Cambridge University Press, 1970, pp. 294–322), I thought it worthwhile to describe the underlying assumptions in more detail than usual in a book addressed mainly to nonspecialists. Another feature of

Chapter 3 is the extended discussion of the use of anomalous scattering for determining the absolute spatial orientation of chiral crystals and, hence, the absolute configuration of their constituent molecules. This seems to be at present the only reliable, generally applicable method for determining absolute configuration, and as such it should be of prime significance for chemists. Since most books on stereochemistry either avoid the subject or dismiss it in a few lines, I hope readers will agree that the attention devoted to it here is justified.

More computing dollars are probably spent in refining crystal structures than in actually solving them (although I have no figures to support this assertion). Refinement here means the adjustment of the structural parameters (describing atomic positions and mean-square vibrational amplitudes) to yield optimal agreement with the experimental data. Refinement is usually done by least-squares analysis, involving a multiple nonlinear regression of several hundred variables to several thousand observations; because of the nonlinearity, several iterations may be required to reach convergence. Chapter 4 contains a short introduction to least-squares analysis in general and to its use in crystal structure analysis in particular. The problem of appropriate weighting of the observational data is taken up (in a manner that I hope does not appear too frivolous) but left pendent.

Chapter 5 deals with the presentation and treatment of the results of crystal structure analyses. It offers a short introduction to the complexities of calculations in oblique coordinate systems and of transforming vector and tensor components from one coordinate system to another. It also discusses the possibilities and limitations of deriving information about the rigid-body motions (and maybe some non-rigid-body motions as well) of molecules in crystals by analyzing the anisotropic temperature factors.

The all-important experimental aspects of X-ray analysis are dealt with briefly in Chapter 6. The treatment is too superficial to serve as a do-it-yourself manual for the budding experimentalist and is intended only to show that crystallography is still an experimental science where the quality of the results depends on the quality of the raw experimental data, as well as on the way these are handled. The measured intensities of the diffracted X-ray beams are always affected, to a greater or lesser degree, by "extinction," and this chapter therefore includes a short description of extinction phenomena and their interpretation in terms of current models, which cannot be regarded as completely satisfactory.

The second part of the book describes some of the ways in which the results of crystal structure analyses have influenced chemistry, particularly organic chemistry. For reasons already mentioned, a comprehensive treatment of this area is not possible, and in Chapters 7 and 8 I concentrate on topics of personal interest, excluding other, equally important ones. I also include in this part two chapters dealing with specialized aspects of molecular structure: Chapter 9 discusses geometric constraints in cyclic molecules, and Chapter 10 deals with symmetry properties of potential energy surfaces involving cyclic coordinates.

In writing the book I have tried to preserve something of the informality of the lectures in which much of the material was originally presented. The lecturer offers his audience the chance to accompany him on a stroll through a landscape or a city he knows well, pausing from time to time to admire a particularly pretty vista or a noteworthy building. He does not need to go into details and can afford to be selective; this means that he can pass over aspects that do not interest him or that are too difficult for him. The main disadvantage of this kind of account is that it tends to be unbalanced. Too much attention may be paid to minor topics, while important areas may be passed over in a couple of sentences. The lecturer can afford to do this, although he may try to salve his conscience by providing lists of recommended reading matter on topics he has neglected or given insufficient attention to.

The readers of a book are entitled to expect something quite different. Instead of a guided-tour lecture, they may expect a comprehensive, detailed guidebook, with descriptions of all plants and animals they may encounter, heights of all mountains with detailed routes for their ascent, lists of all pictures in the various museums, and reliable, up-to-date maps to help them find their way in unfamiliar surroundings. Unfortunately, I have not been able to provide such a guide to modern chemical crystallography. Although I have tried to fill in the most blatant omissions in the original presentation, the book is still far too uneven and incomplete, too much colored by my own interests and by the interests I attributed, rightly or wrongly, to my audience at Cornell, an audience composed mainly of chemists. Indeed, I like to think that this book is written primarily for chemists, not to enable them to do crystal structure analyses, but to help them read the crystallographic literature, appraise it in a reasonably informed manner, and extract structural information according to their particular interests. At the same time, I hope the book will be of value to scientists actually engaged or about to become engaged in X-ray crystal structure analysis.

PART ONE

CRYSTAL STRUCTURE ANALYSIS

1. Diffraction of X-Rays by Crystals

Alice thought to herself, "I don't see how he can *ever* finish, if he doesn't begin."

Scattering

When electromagnetic radiation passes through a crystal (or matter in general), the electrons are perturbed by the rapidly oscillating electric field and are set into oscillation about their nuclei, with a frequency identical to that of the incident radiation. According to electromagnetic theory an oscillating dipole acts as a source of an electromagnetic wave. Thus, each electron in the medium acts as a source of radiation that travels outward with a spherical wave front. In other words, the incident radiation is scattered by the medium without alteration in its frequency. This is coherent scattering. Other kinds of scattering involving frequency changes also occur but are not of interest here.

Superposition of Waves

Imagine that an oscillatory disturbance occurs at some point we choose as an origin. Let the electric field here be $E_0 \cos 2\pi\nu t$. The disturbance spreads out with velocity c in all directions to produce a spherical wave front. At a sufficiently large distance from the origin, any portion of this wave front approximates a plane. Consider the propagation of our wave to some point at radial distance R from the source. At this point, at time t, the phase of the wave is that of the wave that was emitted from the source at time $t-R/c$ and is therefore

$$2\pi\nu(t-R/c) \qquad \text{or} \qquad 2\pi(\nu t-R/\lambda)$$

where λ is the wavelength ($\lambda = c/\nu$). Notice that an *increase* in distance by an amount ΔR corresponds (at a given time t) to a decrease in phase of $2\pi\Delta R/\lambda$. The energy carried by the wave (its intensity) is proportional to the square of its amplitude. The amplitude at a distance R from the source is therefore inversely proportional to R since the energy E_0^2 is distributed over the surface of a sphere. For large R an increase in path length by ΔR leaves the amplitude essentially constant.

The superposition of two waves of the same frequency but differing in phase is expressed by

$$E_1 \cos 2\pi(\nu t+\delta_1) + E_2 \cos 2\pi(\nu t+\delta_2) = E \cos 2\pi(\nu t+\delta)$$

where

$$E \cos 2\pi\delta = E_1 \cos 2\pi\delta_1 + E_2 \cos 2\pi\delta_2$$
$$E \sin 2\pi\delta = E_1 \sin 2\pi\delta_1 + E_2 \sin 2\pi\delta_2$$

The calculation can be represented graphically in simple form (Fig. 1.1). The resultant intensity E^2 is equal to

$$(E_1 \cos 2\pi\delta_1 + E_2 \cos 2\pi\delta_2)^2 + (E_1 \sin 2\pi\delta_1 + E_2 \sin 2\pi\delta_2)^2 = E_1^2 + E_2^2 + 2E_1E_2 \cos 2\pi(\delta_2 - \delta_1)$$

and hence depends on the phase difference between the individual waves. It differs from the sum of the separate intensities unless $\delta_2 - \delta_1 = \pm 1/4$.

As Fig. 1.1 suggests, it is usually more convenient to carry out such calculations using complex exponential notation. The above superposition would then be written

$$E_1 \exp [2\pi i(\nu t+\delta_1)] + E_2 \exp [2\pi i(\nu t+\delta_2)] = [E_1 \exp (2\pi i\delta_1) + E_2 \exp (2\pi i\delta_2)] \exp (2\pi i\nu t)$$

The phase constants are incorporated with the real amplitudes to yield complex amplitudes, and each wave can be visualized as a vector of length $|E_i|$ inclined at an angle δ_i to the real axis; the result of superimposing two or more waves is obtained by the rules of vector addition. The resultant intensity is obtained by multiplication by the complex conjugate quantity.

It is clear that the result of combining several waves of the same frequency depends on the amplitudes of the component waves *and* on the phase differences between them. The constant frequency is not of interest here and can be left out of the calculation.

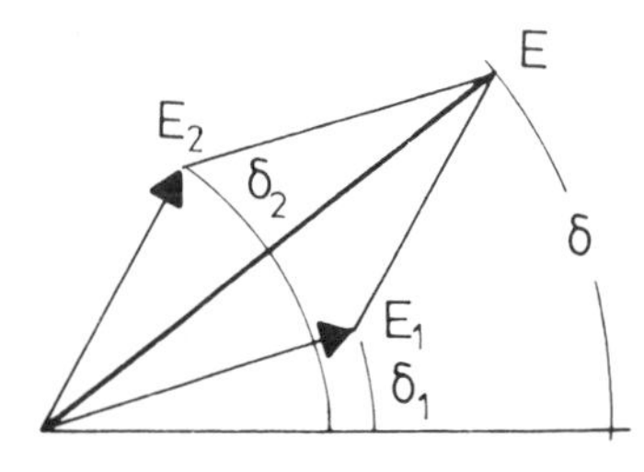

Figure 1.1. Combination of two waves of the same frequency, with amplitudes E_1 and E_2 and phase angles δ_1 and δ_2.

Interference

Waves scattered at different points interfere with one another—i.e., in certain directions the scattered wavelets reinforce one another, in other directions they destroy one another. Interference phenomena are a characteristic of waves, and their occurrence is sufficient to justify our description of the phenomena in terms of a wave model. They arise as a result of differences in path length (distance from source to secondary scatterer plus distance from secondary scatterer to observer) between waves scattered at different points. These path length differences obviously depend on the angle between the incident and scattered beams.

Consider a small volume element $d\mathbf{r}$ containing $\rho(\mathbf{r})d\mathbf{r}$ electrons at vector distance $\mathbf{r}$ from an arbitrary origin, and let the directions of incident and scattered waves be indicated by unit vectors $\mathbf{s}_0$ and $\mathbf{s}$, respectively. Comparing the paths of the waves scattered at the origin and at $\mathbf{r}$ we see (Fig. 1.2) that the latter is longer by $\mathbf{r}\cdot(\mathbf{s}_0-\mathbf{s})$. Since an increase in path length by Δ corresponds to a decrease in phase angle of $2\pi\Delta/\lambda$, the phase of the wave scattered at $\mathbf{r}$ is $2\pi\mathbf{r}\cdot(\mathbf{s}-\mathbf{s}_0)/\lambda$ with respect to the origin.

The complex amplitude of the wave scattered at $\mathbf{r}$ is then proportional to

$$\rho(\mathbf{r}) \exp\,[2\pi i\mathbf{r}\cdot(\mathbf{s}-\mathbf{s}_0)/\lambda]\; d\mathbf{r}$$

which we abbreviate to

$$\rho(\mathbf{r}) \exp\,(2\pi i\mathbf{r}\cdot\mathbf{R})\; d\mathbf{r}$$

where

$$\mathbf{R} = (\mathbf{s}-\mathbf{s}_0)/\lambda \tag{1.1}$$

If $\rho(\mathbf{r})$ describes some continuous electron-density distribution, the complex amplitude of the superposition of all the wavelets scattered by the distribution is obtained by integration

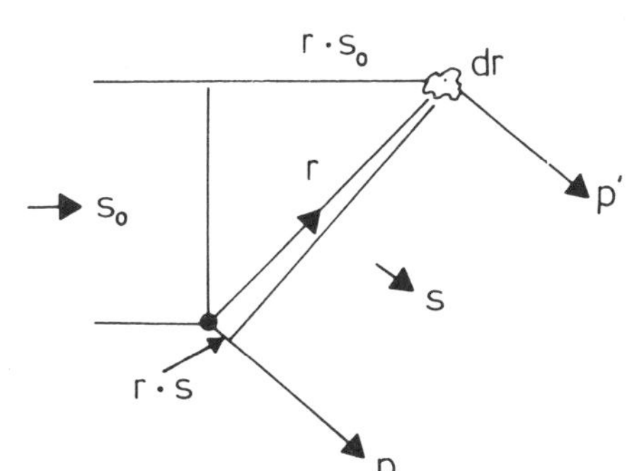

Figure 1.2. The path P' is $\mathbf{r}\cdot(\mathbf{s}_0-\mathbf{s})$ longer than the path P, corresponding to a decrease in phase of $2\pi\mathbf{r}\cdot(\mathbf{s}_0-\mathbf{s})/\lambda$.

$$F(\mathbf{R}) = \int \rho(\mathbf{r}) \exp(2\pi i \mathbf{r}\cdot\mathbf{R})\, d\mathbf{r} \tag{1.2}$$

The vector **R**, frequently called the scattering vector, points along the bisector of the angle between **s** and $-\mathbf{s}_0$ (Fig. 1.3) and plays an important part in scattering theory. The angle between $\mathbf{s}_0$ and **s**, the scattering angle, is usually taken as 2θ, in which case

$$|\mathbf{R}| = 2 \sin \theta/\lambda \tag{1.3}$$

Note that **R** has the dimension of reciprocal length.

If the wavelength λ is about the same order of magnitude as the distances between scattering points, then interference effects will occur at measurable scattering angles. That is why we need X-rays ($\lambda \sim 10^{-8}$ cm ~ 1 Å) to study the structure of molecules.

The quantity $F(\mathbf{R})$ depends on the electron density distribution and on the scattering vector **R** and is called the structure factor; its absolute value is known as the structure amplitude of the scattering in the direction defined by **R**. To get rid of the proportionality factor, $|F(\mathbf{R})|$ is usually expressed as a ratio, the resultant amplitude of the radiation scattered by the distribution $\rho(\mathbf{r})$ to that scattered by a single free electron at the origin.

The above derivation assumes that phase differences between wavelets scattered at different points depend only on path differences. In other words, it is assumed that there is no intrinsic phase change associated with the scattering process, or if there is such a change, it is the same for all the electrons. This assumption is not quite correct, as we shall see in Chapter 3. When the assumption breaks down, we refer to anomalous scattering, otherwise to normal scattering. For the time being, we restrict ourselves to normal scattering.

Another serious shortcoming of this simplified treatment is that energy conservation is not maintained. The intensity of the incident beam is attenuated as it travels through the sample and is scattered at successive volume elements. This loss in intensity, as well as any intensity loss caused by absorption, has been neglected. In other words, Eq. 1.2 can be expected to hold only for infinitesimally small samples where energy losses caused by scattering and absorption are negligible.

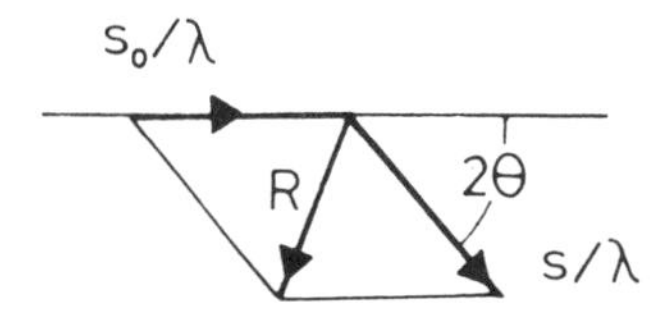

Figure 1.3. Geometrical interpretation of $\mathbf{R} = (\mathbf{s}-\mathbf{s}_0)/\lambda$. Since $\mathbf{s}_0$ and **s** are unit vectors, $|\mathbf{R}| = 2 \sin \theta/\lambda$.

We now discuss Eq. 1.2 in some detail.

1. In general, $F(\mathbf{R})$ is a complex quantity.

2. The intensity of scattering $I(\mathbf{R})$ is given by $F(\mathbf{R})\,F^*(\mathbf{R})$. In principle, the intensity and hence $|F(\mathbf{R})|$ can be measured experimentally as a function of $\mathbf{R}$. However, the phase angle associated with $F(\mathbf{R})$ is not an observable quantity.

3. Provided $\rho(\mathbf{r})$ is real, then $F(-\mathbf{R}) = F^*(\mathbf{R})$, and the scattered intensity is unaffected by reversing the directions of incident and scattered beams. Likewise, the same intensity pattern is produced by a given distribution $\rho(\mathbf{r})$ and by the distribution $\rho(-\mathbf{r})$ obtained by inversion through the origin.

4. If the distribution $\rho(\mathbf{r})$ is centrosymmetric—i.e., if $\rho(\mathbf{r}) = \rho(-\mathbf{r})$, then $F(\mathbf{R})$ is a real quantity. There is still an ambiguity concerning its sign.

Fourier Transforms

Eq. 1.2,

$$F(\mathbf{R}) = \int \rho(\mathbf{r}) \exp\,(2\pi i\mathbf{R}\cdot\mathbf{r})\, d\mathbf{r} \tag{1.2}$$

is an example of a Fourier transformation. The structure factor $F(\mathbf{R})$ is called the Fourier transform of the scattering density $\rho(\mathbf{r})$, and $\rho(\mathbf{r})$ can be expressed in turn as the inverse Fourier transform of $F(\mathbf{R})$,

$$\rho(\mathbf{r}) = \int F(\mathbf{R}) \exp\,(-2\pi i\mathbf{R}\cdot\mathbf{r})\, d\mathbf{R}, \tag{1.4}$$

a remarkable reciprocal relationship.

Just as a periodic function can be represented as a Fourier series, so an arbitrary non-periodic function (subject to certain conditions that do not concern us here) can be represented by integrals of the type shown.[1] By analogy with the Fourier series representation of a periodic function, the Fourier transform can be regarded as the "frequency spectrum" of a nonperiodic function, where the frequencies are no longer restricted to integers. In this connection, note that the dimensions of the variables $\mathbf{r}$ and $\mathbf{R}$ are reciprocal. If $\mathbf{r}$ is a variable distance, then $\mathbf{R}$ has the dimen-

[1]The theory of Fourier integrals and Fourier series can be found in many standard mathematical texts. See, for example, R. Courant, *Differential and Integral Calculus,* Vol. 2, Blackie and Son, London and Glasgow, 1936, pp. 318–323. Applications to spectral analysis plus a compilation of Fourier transforms of many functions can be found in G. M. Jenkins and D. G. Watts, *Spectral Analysis and its Applications,* Holden-Day, San Francisco, 1968.

sions of reciprocal length. We are thus led to talk about two spaces—a direct space and a reciprocal space.

Familiarity with the properties of Fourier transforms is a great help in relating qualitative and quantitative features of the scattering produced by an object to the distribution of scattering density within the object—its "structure" as revealed by the scattering pattern. If we know the structure, we can always calculate the scattering pattern. However, the reverse calculation is more difficult because the scattering pattern provides information only about the absolute value of $F(\mathbf{R})$ but not about its phase. This means that from the scattering pattern, we can calculate the structure only if we can somehow find the missing phase information. From qualitative features of the structure or scattering pattern, we can sometimes draw qualitative inferences about the other.

For future use, we list some simple theorems concerning Fourier transforms. We write

$$F(\mathbf{R}) = \int f(\mathbf{r}) \exp (2\pi i \mathbf{R} \cdot \mathbf{r})\, d\mathbf{r} = T[f(\mathbf{r})]$$

$$f(\mathbf{r}) = \int F(\mathbf{R}) \exp (-2\pi i \mathbf{R} \cdot \mathbf{r})\, d\mathbf{R} = T^{-1}[F(\mathbf{R})]$$

where upper and lower cases of the same letter refer to a pair of Fourier transforms.

1. The transform of the sum of several functions $f_1(\mathbf{r}), f_2(\mathbf{r}) \ldots$ in the same space is the sum of their transforms:

$$F(\mathbf{R}) = T[f_1(\mathbf{r})+f_2(\mathbf{r})+ \ldots +f_n(\mathbf{r})] = F_1(\mathbf{R}) + F_2(\mathbf{R})+ \ldots +F_n(\mathbf{R})$$

2. Direct and inverse transformation applied successively leave $f(\mathbf{r})$ unaltered:

$$T^{-1}T[f(\mathbf{r})] = T^{-1}[F(\mathbf{R})] = f(\mathbf{r})$$

3. Direct and inverse transforms of $f(\mathbf{r})$ differ by inversion through the origin of reciprocal space:

$$T^{-1}[f(\mathbf{r})] = \int f(\mathbf{r}) \exp (-2\pi i \mathbf{R} \cdot \mathbf{r})\, d\mathbf{r}$$

$$= \int f(\mathbf{r}) \exp (2\pi i [-\mathbf{R}] \cdot \mathbf{r})\, d\mathbf{r}$$

$$= F(-\mathbf{R})$$

Thus, direct or inverse transformation applied twice in succession inverts $f(\mathbf{r})$ through the origin:

$$TT[f(\mathbf{r})] = T^{-1}T^{-1}[f(\mathbf{r})] = f(-\mathbf{r})$$

4. Direct and inverse transforms of a real function are complex conjugates: $T[f(\mathbf{r})] = F(\mathbf{R})$, $T^{-1}[f(\mathbf{r})] = F^*(\mathbf{R}) = F(-\mathbf{R})$. Thus, if $f(\mathbf{r})$ is real, the absolute value of $F(\mathbf{R})$ is symmetric and its phase angle is antisymmetric with respect to inversion through the origin of reciprocal space: $|F(\mathbf{R})| = |F(-\mathbf{R})|$. Another corollary is that if $f(\mathbf{r})$ is real and centrosymmetric, $F(\mathbf{R})$ is also real and centrosymmetric.

5. Displacement of $f(\mathbf{r})$ by a translation $\mathbf{r}'$ corresponds to multiplication of its transform by a phase factor $\exp(2\pi i\mathbf{R}\cdot\mathbf{r}')$:

$$\begin{aligned} T[f(\mathbf{r}-\mathbf{r}')] &= \int f(\mathbf{r}-\mathbf{r}')\exp(2\pi i\mathbf{R}\cdot\mathbf{r})\,d\mathbf{r} \\ &= \exp(2\pi i\mathbf{R}\cdot\mathbf{r}')\int f(\mathbf{r}-\mathbf{r}')\exp[2\pi i\mathbf{R}\cdot(\mathbf{r}-\mathbf{r}')]\,d\mathbf{r} \\ &= \exp(2\pi i\mathbf{R}\cdot\mathbf{r}')\,F(\mathbf{R}) \end{aligned}$$

> "What *is* the use of repeating all that stuff," the Mock Turtle interrupted, "if you don't explain it as you go on? It's by far the most confusing thing I ever heard!"

Given a density function, we can always compute the structure factor (Fourier transform) by integration. However, most density functions of chemical interest can be regarded as sums of simpler functions corresponding to atoms, molecules, and so forth. We can often therefore replace the integral of Eq. 1.2 by a sum of integrals over simpler density functions, each multiplied by an appropriate phase factor. We shall begin by evaluating the structure factors of simple distributions and see how they can be built into more complex ones. This leads us to consider the scattering produced by: (a) a single point atom, (b) sets of point atoms, (c) periodic arrays, (d) atoms, (e) molecules, and (f) crystals of various degrees of order.

Scattering by a Point Atom

The scattering from a single point atom is independent of the scattering angle since by definition no path differences are involved. If we want to use Eq. 1.2, we have to define a function $\delta(\mathbf{r})$ with the somewhat unusual properties that

$$\delta(\mathbf{r}) = 0 \quad \text{except when} \quad \mathbf{r} = 0$$

$$\int \delta(\mathbf{r})d\mathbf{r} = c$$

Functions with these properties can be concocted as limiting cases of standard functions, e.g., of a Gaussian function

$$\frac{a}{\sqrt{\pi}} \mathop{\mathscr{L}}_{a\to\infty} \exp(-a^2r^2)$$

and are known as δ-functions. For a δ-function situated at vector distance $\mathbf{r}'$ from the origin, Eq. 1.2 becomes

$$F(\mathbf{R}) = \int \delta(\mathbf{r}-\mathbf{r}') \exp(2\pi i\mathbf{R}\cdot\mathbf{r})\, d\mathbf{r}$$

$$= c \exp(2\pi i\mathbf{R}\cdot\mathbf{r}')$$

since the product within the integral is zero except for $\mathbf{r} = \mathbf{r}'$. If the point atom is placed at the origin, then $F(\mathbf{R}) = c$, the amount of scattering density concentrated at the point in question, and is independent of $\mathbf{R}$ and hence, by Eq. 1.3, is independent of scattering angle.

Point atoms have no physical significance as far as X-ray diffraction is concerned because the electrons that do the scattering cannot be concentrated at a point. However, in crystal structure analysis it is often useful to analyze not the observed diffraction pattern but the pattern that *would be produced* by an assemblage of point atoms occupying the same positions in space as the actual atomic centers. We shall see examples of this when we discuss direct methods. In neutron diffraction, where the scattering is primarily due to the atomic nuclei, the nuclei can be regarded as effective point scatterers because their dimensions are negligible compared with the neutron wavelength.

Scattering by an Assemblage of Point Atoms

For a set of atoms at vector distances $\mathbf{r}_1$, $\mathbf{r}_2$, $\mathbf{r}_3 \ldots$ from the origin

$$F(\mathbf{R}) = \sum c_i \exp(2\pi i\mathbf{R}\cdot\mathbf{r}_i)$$

where the coefficients c_i depend only on the relative "weights" of the atoms and not on the scattering angle. If the atomic distribution is centrosymmetric, then we can group the atoms in pairs. Taking the origin at the center of symmetry, each pair yields a contribution

$$c_i[\exp(2\pi i\mathbf{R}\cdot\mathbf{r}_i) + \exp(-2\pi i\mathbf{R}\cdot\mathbf{r}_i)] = 2c_i \cos 2\pi\mathbf{R}\cdot\mathbf{r}_i$$

so that the summation reduces to

$$F(\mathbf{R}) = 2\sum c_i \cos 2\pi\mathbf{R}\cdot\mathbf{r}_i$$

Scattering by a Periodic Array: Diffraction

A lattice is a set of translations that act on a unit of pattern, repeating it over and over to produce a periodic array. We begin by taking the unit of pattern to be a single point atom that is repeated at

$$\mathbf{r}(n_1,n_2,n_3) = n_1\mathbf{a}_1 + n_2\mathbf{a}_2 + n_3\mathbf{a}_3 \tag{1.5}$$

where n_1, n_2, n_3 are integers, and $\mathbf{a}_1$, $\mathbf{a}_2$, $\mathbf{a}_3$ are the basis vectors or periodicities defining the lattice (Fig. 1.4). The parallelepiped defined by $\mathbf{a}_1$, $\mathbf{a}_2$, $\mathbf{a}_3$ is known as the unit cell; its volume $V = \mathbf{a}_1 \cdot \mathbf{a}_2 \times \mathbf{a}_3$. This is a good place to introduce the idea of the "reciprocal lattice." For any three noncoplanar vectors $\mathbf{a}_1$, $\mathbf{a}_2$, $\mathbf{a}_3$ we can define a set of vectors $\mathbf{b}_1$, $\mathbf{b}_2$, $\mathbf{b}_3$ reciprocal to the first set in the sense that

$$\begin{aligned} \mathbf{a}_j\cdot\mathbf{b}_j &= 1 \\ \mathbf{a}_j\cdot\mathbf{b}_k &= 0 \qquad j \neq k \end{aligned}$$

It is easy to see that the vectors,

$$\begin{aligned} \mathbf{b}_1 &= \frac{\mathbf{a}_2 \times \mathbf{a}_3}{\mathbf{a}_1\cdot\mathbf{a}_2 \times \mathbf{a}_3} & \mathbf{b}_2 &= \frac{\mathbf{a}_3 \times \mathbf{a}_1}{\mathbf{a}_2\cdot\mathbf{a}_3 \times \mathbf{a}_1} & \mathbf{b}_3 &= \frac{\mathbf{a}_1 \times \mathbf{a}_2}{\mathbf{a}_3\cdot\mathbf{a}_1 \times \mathbf{a}_2} \\ &= \mathbf{a}_2 \times \mathbf{a}_3/V & &= \mathbf{a}_3 \times \mathbf{a}_1/V & &= \mathbf{a}_1 \times \mathbf{a}_2/V \end{aligned} \tag{1.6}$$

have just these properties. If $\mathbf{a}_1$, $\mathbf{a}_2$, $\mathbf{a}_3$ are basis vectors with the dimensions of length, then $\mathbf{b}_1$, $\mathbf{b}_2$, $\mathbf{b}_3$ are reciprocal basis vectors with the dimensions of reciprocal length. Since the scattering vector $\mathbf{R} = (\mathbf{s}-\mathbf{s}_o)/\lambda$ also has the dimension of reciprocal length, it can conveniently be expressed in terms of its components on the reciprocal basis vectors:

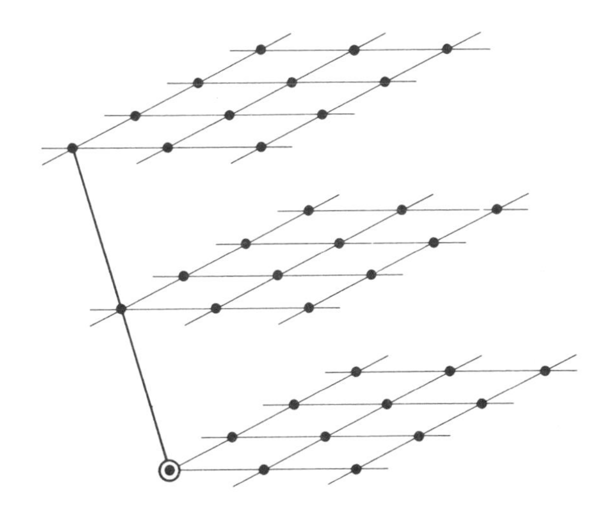

Figure 1.4. Representative point at the origin is repeated by the three lattice translations to produce a periodic structure.

$$\mathbf{R}(h_1, h_2, h_3) = h_1\mathbf{b}_1 + h_2\mathbf{b}_2 + h_3\mathbf{b}_3 \tag{1.7}$$

This has the advantage that scalar quantities such as $\mathbf{R} \cdot \mathbf{r}$ can be written

$$\mathbf{R}(h_1, h_2, h_3) \cdot \mathbf{r}(n_1,n_2,n_3) = h_1n_1 + h_2n_2 + h_3n_3 \tag{1.8}$$

The structure factor of a lattice of point atoms is then

$$\begin{aligned}\mathscr{F}(\mathbf{R}) &= c \sum_{n_1,n_2,n_3} \exp\,(2\pi i\mathbf{R}\cdot\mathbf{r}) \\ &= c \sum_{n_1,n_2,n_3} \exp\,[2\pi i(h_1n_1+h_2n_2+h_3n_3)]\end{aligned} \tag{1.9}$$

The individual contributions of the sum are in phase only if

$$h_1n_1 + h_2n_2 + h_3n_3 = \text{integer} \tag{1.10}$$

in which case

$$\mathscr{F}(\mathbf{R}) = cN \tag{1.11}$$

where N is the total number of lattice points in the sum. Since n_1, n_2, n_3 are integers, Eq. 1.10 implies that h_1, h_2, h_3 must also be integers. In other words, constructive interference (diffraction) from a lattice can occur only in certain directions—i.e., those for which the vector $\mathbf{R} = (\mathbf{s}-\mathbf{s}_o)/\lambda$ has integral components on the reciprocal basis vectors. The termini of all such vectors $\mathbf{H}(h_1, h_2, h_3)$ with integral components form what is known as the reciprocal lattice, and the individual termini are called reciprocal lattice (R. L.) points. Diffraction occurs only when the scattering vector $\mathbf{R}$ coincides with a R. L. vector $\mathbf{H}$. The reciprocal lattice (in reciprocal space) can be regarded as the Fourier transform of the direct lattice (in real space) and vice-versa.

$$\mathbf{R} = \frac{\mathbf{s}-\mathbf{s}_o}{\lambda} = \mathbf{H} = h_1\mathbf{b}_1 + h_2\mathbf{b}_2 + h_3\mathbf{b}_3, \quad h_1, h_2, h_3 \text{ integers} \tag{1.12}$$

can be expressed in a different form by forming the scalar products of both sides with $\mathbf{a}_1$, $\mathbf{a}_2$, $\mathbf{a}_3$, respectively, to obtain what are known as the Laue conditions:

$$\begin{aligned}\frac{\mathbf{s}-\mathbf{s}_o}{\lambda} \cdot \mathbf{a}_1 &= h_1 \\ \frac{\mathbf{s}-\mathbf{s}_o}{\lambda} \cdot \mathbf{a}_2 &= h_2 \\ \frac{\mathbf{s}-\mathbf{s}_o}{\lambda} \cdot \mathbf{a}_3 &= h_3\end{aligned} \tag{1.13}$$

Eqs. 1.13, which state that the path difference corresponding to each lattice translation must equal an integral number of wavelengths, have to be satisfied simultaneously for diffraction to occur.

The diffraction condition can be expressed in yet another way. Dividing the Laue conditions by h_1, h_2, and h_3, respectively, and subtracting one from another, we obtain, for example,

$$\mathbf{R} \cdot \left(\frac{\mathbf{a}_1}{h_1} - \frac{\mathbf{a}_2}{h_2}\right) = 0$$

This says that the R. L. vector $\mathbf{H}$ is perpendicular to $\mathbf{a}_1/h_1 - \mathbf{a}_2/h_2$. Similarly, $\mathbf{H}$ is perpendicular to $\mathbf{a}_1/h_1 - \mathbf{a}_3/h_3$ and hence to the plane containing $\mathbf{a}_1/h_1$, $\mathbf{a}_2/h_2$ and $\mathbf{a}_3/h_3$. Crystallographers refer to such a plane (Fig. 1.5) or rather to the sequence of parallel equispaced planes as having Miller indices $(h_1h_2h_3)$. The spacing $d(h_1h_2h_3)$—the perpendicular distance between any two adjacent planes in the sequence—equals the projection of $\mathbf{a}_1/h_1$, say, on $\mathbf{H}$

$$d(h_1h_2h_3) = \frac{\mathbf{a}_1}{h_1} \cdot \frac{\mathbf{H}}{|\mathbf{H}|}$$

but $\mathbf{a}_1 \cdot \mathbf{H} = h_1$ and $|\mathbf{H}| = |\mathbf{R}| = 2 \sin \theta/\lambda$ (Eq. 1.3) so

$$\frac{1}{d(h_1h_2h_3)} = \frac{2 \sin \theta}{\lambda} \tag{1.14}$$

which is Bragg's Law. If the Laue conditions for diffraction from a lattice are satisfied, the scattering vector $\mathbf{R}$ is perpendicular to the crystal plane $(h_1h_2h_3)$. Since by definition this vector points along the bisector of the incident and diffracted beams, the latter can be regarded as having been produced by reflection of the incident beam from the plane (Fig. 1.5), but this occurs only at a particular angle of incidence given by Bragg's Law.

It is helpful to think of diffraction from a crystal in terms of reflection from planes, especially when the crystal morphology is sufficiently well developed that the external crystal faces can be recognized and identified. It is even more helpful conceptually to replace the set of planes by the set of R. L. vectors $\mathbf{H}$, each normal to a given plane and of magnitude inversely proportional to the spacing of the given plane. The diffraction condition can then be visualized in an elegant manner with the help of a construction due to Ewald.

From the center of a sphere of radius $1/\lambda$ draw a line in the direction of the incident beam to cut the sphere at a point that is taken as the origin of the reciprocal lattice. The diffraction condition is then satisfied

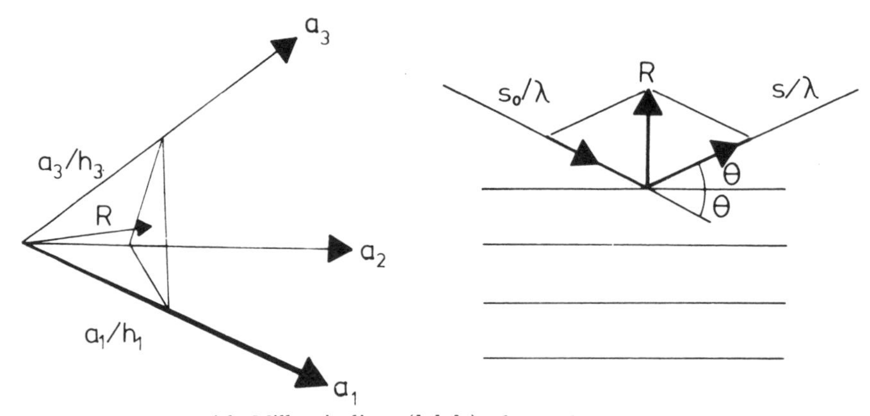

Figure 1.5. Plane with Miller indices $(h_1h_2h_3)$—here (2 3 2). The lattice translations produce a sequence of parallel, equispaced planes. $\mathbf{R} = (\mathbf{s}-\mathbf{s}_0)/\lambda$ is perpendicular to the planes; hence the diffracted beam along **s** can be regarded as the reflection of the incident beam along $\mathbf{s}_0$.

if an R. L. point, the terminus of an R. L. vector **H**, is situated on the surface of the sphere (Fig. 1.6). For a fixed radius $(1/\lambda)$ and orientation of the reciprocal lattice, this will not generally be the case—i.e., diffraction will not occur. However, as the crystal is rotated, the reciprocal lattice is rotated with it, so that all reciprocal lattice points within the distance $2/\lambda$ from the origin can be brought into coincidence with the surface of the sphere. The scattering angle then depends on the length $(2 \sin \theta/\lambda)$ of the particular **H** vector that happens to satisfy this condition.

The distribution of spots on an X-ray photograph of a crystal is a more or less distorted picture of a part of its reciprocal lattice. The intensities of the spots depend on the nature of the unit of pattern, but their positions depend only on the X-ray wavelength, the experimental geometry, and the reciprocal lattice. Methods of "indexing" X-ray photographs—i.e., of assigning the correct triplet of integers $(h_1h_2h_3)$ to each spot—are described briefly in Chapter 6 and in more detail in several books.[2]

In a crystal the lattice operates on a unit of pattern that is more complex than a point atom. However, this increase in complexity has no effect on the diffraction conditions, which depend only on the lattice translations—i.e., on the size and shape of the unit cell and not on the nature of its contents. The intensities of the diffracted beams do, however, depend on the contents of the unit cell, and when we consider diffraction from a crystal, we can simply replace the constant c in

[2]See, for example, M. J. Buerger, *X-ray Crystallography*, Wiley, New York, 1942.

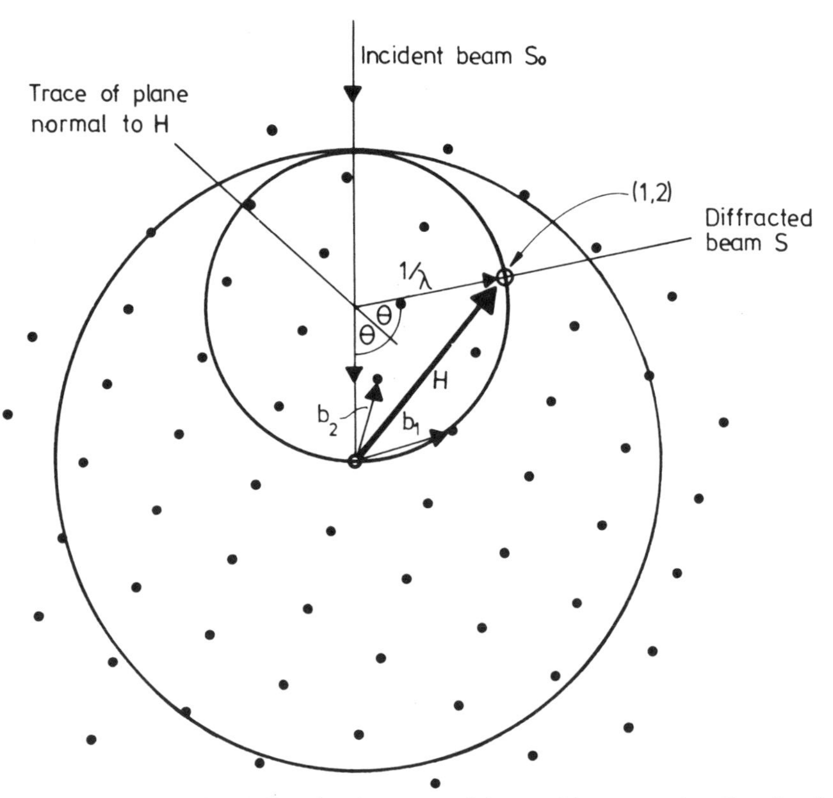

Figure 1.6. Condition for diffraction in terms of the Ewald construction (for simplicity in two dimensions). The reciprocal lattice point **H**(1, 2) lies on the diffraction sphere. Since the vector **H**(1, 2) is perpendicular to the plane with Miller indices (1, 2), the diffracted beam in the direction **s** may be thought of as being reflected from this plane. As the reciprocal lattice is rotated around the origin, all reciprocal lattice points within distance $2/\lambda$ from the origin are brought into coincidence with the surface of the diffraction sphere.

Eq. 1.11 by the structure factor $F(\mathbf{R})$ of the unit of pattern:

$$\mathcal{F}(\mathbf{R}) = NF(\mathbf{R}) \tag{1.15}$$

At the reciprocal lattice points, $F(\mathbf{R})$ is multiplied by N, the number of unit cells in the finite crystal, usually a very large number; at points that do not satisfy the diffraction conditions, $F(\mathbf{R})$ is multiplied by a much smaller number, of the order of $N^{\frac{1}{2}}$. The corresponding intensities are thus proportional to N^2 when the in-phase relationship between N unit cells is satisfied, so that the crystal acts as an amplifier that allows us to measure $F(\mathbf{R})$ at the reciprocal lattice points but not, in general, between them (Fig. 1.7).

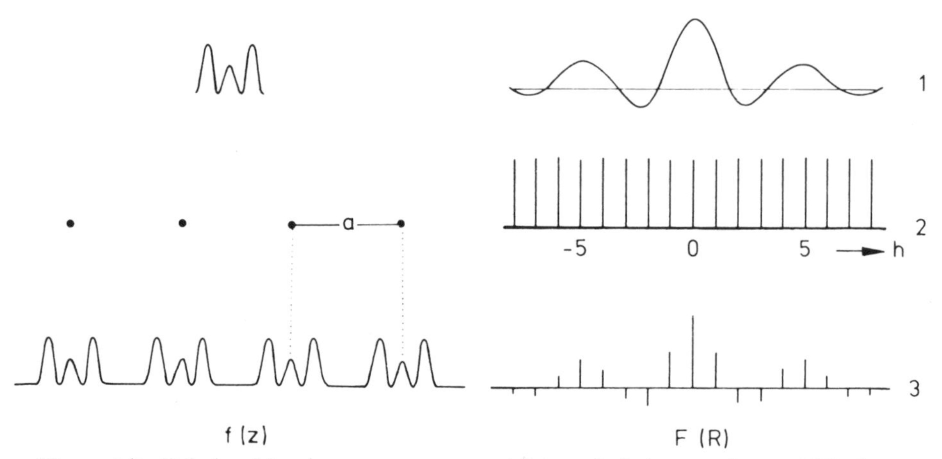

Figure 1.7. Relationships between structures $f(z)$ and their transforms $F(R)$ (in one dimension for simplicity). (1) Unit of pattern. (2) Portion of a lattice of point atoms (to be extended far beyond the confines of the paper). (3) Repetition of unit of pattern by the lattice.

Further Properties of Fourier Transforms: Convolution

The operation that converts a single unit of pattern into many identical copies arranged in a lattice is conveniently discussed as a special case of a more general operation known as convolution. The convolution or folding (Faltung) of two functions $f(\mathbf{x})$ and $g(\mathbf{x})$ is defined as

$$c(\mathbf{x}) = \int f(\mathbf{x}')\, g(\mathbf{x}-\mathbf{x}')\, d\mathbf{x}' \qquad (1.16)$$

Although f and g appear to have nonequivalent roles here, they are really equivalent. To see this, simply change the variable by writing $\mathbf{x} - \mathbf{x}' = \mathbf{x}''$ to obtain

$$c(\mathbf{x}) = \int f(\mathbf{x}-\mathbf{x}'')\, g(\mathbf{x}'')\, d\mathbf{x}''$$

The convolution operation is a useful way to construct complicated functions from simple ones; some examples are shown in Fig. 1.8. It is important in diffraction theory because the Fourier transform of the convolution is the *product* of the Fourier transforms of the separate functions (convolution theorem):

$$C(\mathbf{X}) = \int c(\mathbf{x}) \exp\,(2\pi i\mathbf{X}\cdot\mathbf{x})\, d\mathbf{x}$$

$$= \iint f(\mathbf{x}')\, g(\mathbf{x}-\mathbf{x}') \exp\,(2\pi i\mathbf{X}\cdot\mathbf{x})\, d\mathbf{x}\, d\mathbf{x}'$$

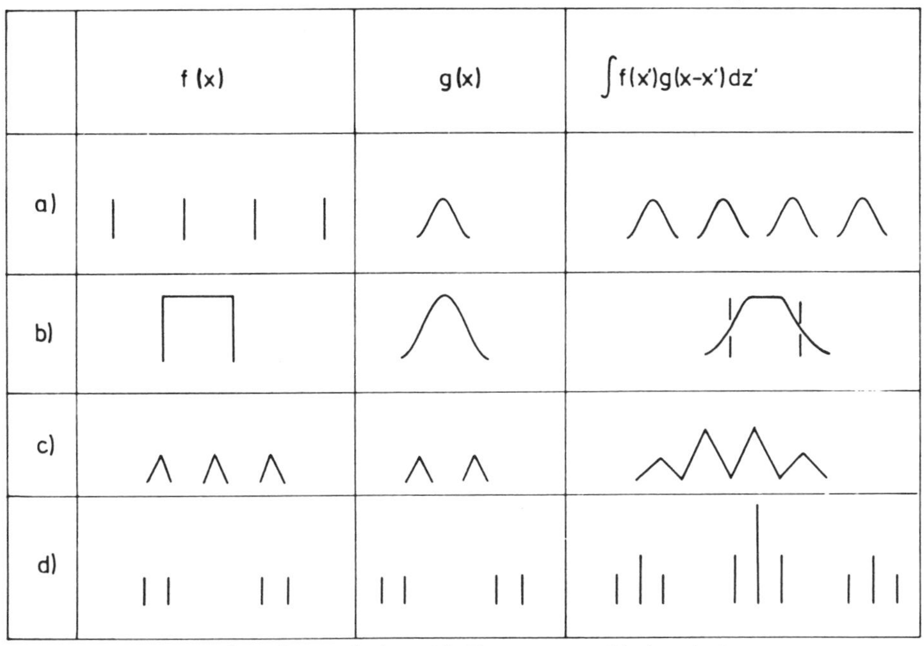

Figure 1.8. Examples of convolution; (d) illustrates a self-convolution.

$$= \int f(\mathbf{x}') \exp (2\pi i\mathbf{X}\cdot\mathbf{x}')\, d\mathbf{x}' \int g(\mathbf{x}-\mathbf{x}') \exp [2\pi i\mathbf{X}\cdot(\mathbf{x}-\mathbf{x}')]\, d\mathbf{x}$$
$$= F(\mathbf{X})\, G(\mathbf{X}) \tag{1.17}$$

If one of the functions, say f, is a δ-function located at $\mathbf{x}_0$, the convolution with g is

$$c(\mathbf{x}) = \int \delta(\mathbf{x}'-\mathbf{x}_0)\, g(\mathbf{x}-\mathbf{x}')\, d\mathbf{x}'$$
$$= g(\mathbf{x}-\mathbf{x}_0)$$

That is, the convolution of $g(\mathbf{x})$ with $\delta(\mathbf{x}-\mathbf{x}_0)$ merely shifts the origin of g to $\mathbf{x}_0$. Convolution of $g(\mathbf{x})$ with a sum of δ-functions arranged on a lattice then converts $g(\mathbf{x})$ into a set of copies arranged on the same lattice. The Fourier transform of the whole pattern is the transform of the unit of pattern multiplied by the transform of the lattice, as illustrated in Fig. 1.7. Since the transform of the lattice is essentially zero between the R. L. points, we could also say that the transform of the unit of pattern is sampled at these points in reciprocal space.

The convolution theorem can also be applied in its converse sense. If a function C can be expressed as the product of two other functions,

$$C(\mathbf{X}) = F(\mathbf{X})\, G(\mathbf{X})$$

then the Fourier transform of C is the convolution of the transforms of F and G. For example, a crystal of finite size can be regarded as the product of an infinite crystal with a step function, one that has the value unity over the size of the crystal specimen and zero elsewhere. The transform of the finite crystal is then the convolution of the weighted reciprocal lattice (the transform of the infinite crystal) with the transform of the step function, which can easily be derived. In one dimension, the step function can be written:

$$s(x) = 1 \qquad |x| \leq l/2$$

$$s(x) = 0 \qquad |x| > l/2$$

where the origin is chosen at the midpoint of the step of length l to make $s(x)$ an even function. The Fourier transform is then

$$S(X) = 2\int_0^{l/2} \cos 2\pi Xx \, dx$$

$$= \frac{2 \sin \pi Xl}{2\pi X} \tag{1.18}$$

$$= lS'(X) \quad \text{where } S'(X) = \frac{\sin \pi Xl}{\pi Xl}$$

shown in Fig. 1.9. The transform $S(X)$ is a sharp peak of height l (the integral over the step function) that drops to zero at $X = 1/l$ and runs through a sequence of ripples of alternating sign and gradually diminishing amplitude, the subsequent zeroes occurring at integral multiples

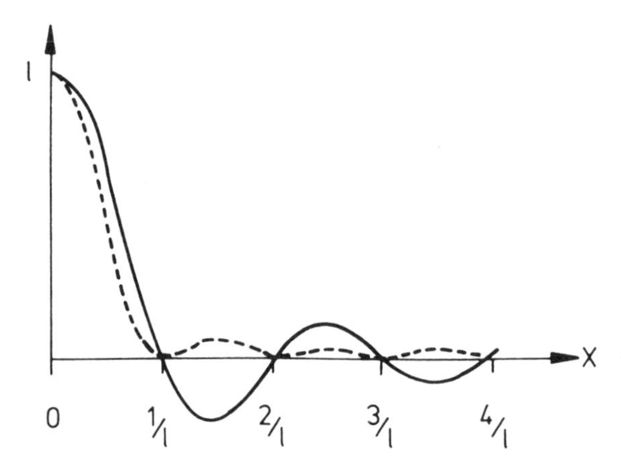

Figure 1.9. $S(X) = \sin \pi Xl/\pi X$ as a function of X. This is the transform of a step function of length l. The square of the transform is also shown (dashed line).

of $1/l$. If $S(X)$ is interpreted as a structure factor, then at $X = 0$ all scattering contributions from the step function are in phase, but with increasing scattering angle $[2 \sin^{-1} (\lambda X/2)]$, phase differences between different parts of the step come into play to produce destructive interference at $X = n/l$ (l integral), with subsidiary maxima and minima at intermediate values of X. Fig. 1.9 also shows the behavior of $S^2(X)$, which would be proportional to the distribution of intensity scattered by the step function. This intensity is effectively concentrated in the region $X < 1/l$, the height of the first ripple being only $(2/3\pi)^2 \sim 0.045$ the height at the origin.

This illustrates a general feature of the mutual relationship between a pair of Fourier transforms. As one function, $s(x)$, becomes more concentrated in its space (i.e., as l becomes smaller), its transform $S(X)$ becomes less concentrated in the reciprocal space. If $s(x)$ becomes infinitely narrow (though remaining integrable, e.g., a δ-function) $S(X)$ becomes a constant, independent of X. Likewise, if $s(x)$ becomes infinitely wide, then $S(X)$ is concentrated into an infinitesimally narrow region near $X = 0$.[3]

The diffracted spectra from a finite crystal do not occur at infinitely sharp scattering angles because the R. L. points are spread over a finite volume of reciprocal space by convolution with $S(X)$. As the crystal becomes smaller, the spectra become broader.

As an example, we estimate the width of a diffracted peak from a crystallite of linear dimension equal to 1μ (10^4 Å) using a perfectly parallel beam of X-rays of wavelength 1.542 Å ($CuK\alpha$ radiation). By "width" we mean the angular distance from the peak center to the scattering angle (2θ) where the scattered intensity is half the peak intensity. This occurs when $S(X) = l/\sqrt{2}$, or when $X = 0.44/l$. Substituting $l = 10^4$ Å, we obtain

$$X = 4.4 \times 10^{-5}\ \text{Å}^{-1} = 2 \sin\theta/1.542\ \text{Å}$$

whence $$2\theta = 6.8 \times 10^{-5}\ \text{radian} \sim 0.004^\circ$$

Only when crystallite dimensions are of the order of a few hundred Ångstroms is there any perceptible broadening of the reflections due to finite crystal size. For larger crystallites, more important factors in determining the actual reflection width are the finite area of the crystal shadowed by the X-ray beam and the angular divergence of the beam.

[3]This kind of relationship is well known to students of quantum mechanics, where the space distribution $\Psi(x)$ and momentum distribution $\varphi(p_x)$ of a particle are related by the wave function as a pair of Fourier transforms; hence the uncertainty relationship.

Scattering by Atoms

The radiation normally used in X-ray diffraction has a wavelength in the range 0.5–2.0 Å and is quite capable of resolving individual atoms in molecules. The electron density in a molecule can therefore be regarded in good approximation as a simple superposition of the density peaks associated with the individual atoms, and we can write Eq. 1.2 as:

$$F(\mathbf{R}) = \int \sum_i \rho_i(\mathbf{r}-\mathbf{r}_i) \exp(2\pi i\mathbf{R}\cdot\mathbf{r})\, d\mathbf{r}$$

$$= \int \sum_i \rho_i(\mathbf{r}-\mathbf{r}_i) \exp[2\pi i\mathbf{R}\cdot(\mathbf{r}-\mathbf{r}_i)] \exp(2\pi i\mathbf{R}\cdot\mathbf{r}_i)\, d\mathbf{r}$$

$$= \sum f_i(\mathbf{R}) \exp(2\pi i\mathbf{R}\cdot\mathbf{r}_i)$$

where $f_i(\mathbf{R}) = \int \rho_i(\mathbf{r}) \exp(2\pi i\mathbf{R}\cdot\mathbf{r})\, d\mathbf{r}$

Note that $f_i(0) = \int \rho_i(\mathbf{r})\, d\mathbf{r} = Z_i$, the total number of electrons in the atom i.

The quantity $f_i(\mathbf{R})$ is the Fourier transform of the electron density distribution $\rho_i(\mathbf{r})$ of the ith atom. It represents the scattering power of that atom and is independent of the atomic position $\mathbf{r}_i$. For an isolated atom, $\rho(\mathbf{r})$ has spherical symmetry, and hence $f(\mathbf{r})$ is also isotropic. The integration is carried out over spherical density shells (Fig. 1.10):

$$f(R) = 4\pi \int_0^{\pi/2} \int_0^{\infty} r^2 \rho(r) \cos(2\pi Rr\cos\alpha) \sin\alpha\, d\alpha\, dr$$

$$= 4\pi \int_0^{\infty} r^2 \rho(r) \frac{\sin 2\pi Rr}{2\pi Rr}\, dr \tag{1.19}$$

Atomic scattering factors (form factors) of isolated neutral atoms and

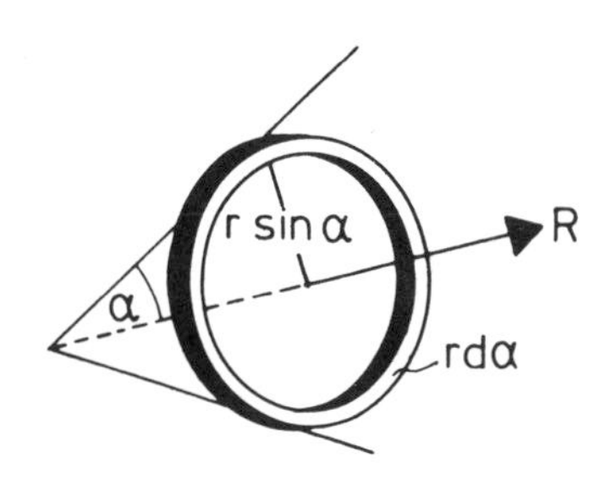

Figure 1.10. Volume element for a spherical shell of radius r and constant density $\rho(r)$ is $2\pi r^2 \sin\alpha\, dr\, d\alpha$. Because of spherical symmetry the vector $\mathbf{R}$ can be chosen to lie in any direction. We take it along the axis of the cone of half-angle α so that $\mathbf{R}\cdot\mathbf{r} = Rr\cos\alpha$.

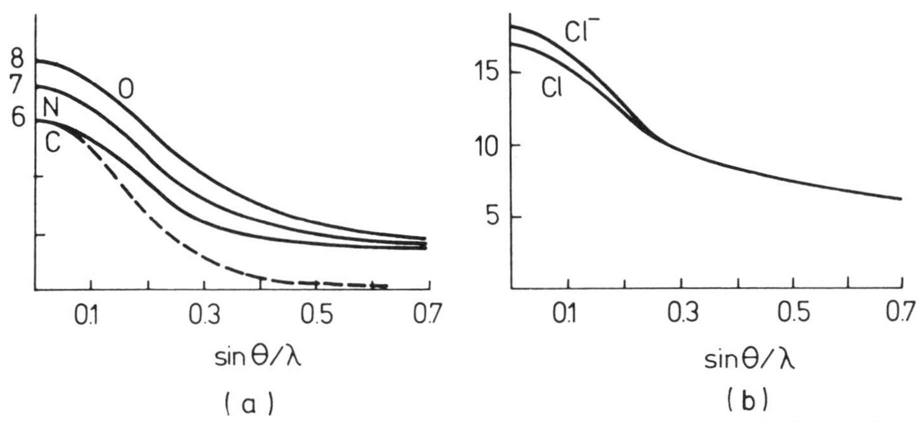

Figure 1.11. Form factors for isolated atoms at rest. (a) C, N, and O. (b) Cl and Cl^-. In (a) the scattering factor is also shown for a carbon atom with mean-square vibration amplitude of 0.1 $Å^2$.

ions have been calculated by many authors using various quantum mechanical models[4] to obtain the approximate electron densities, $\rho(r)$. Some typical results are shown in Fig. 1.11. At zero scattering angle, the electrons scatter in phase, so $f(0)$ equals the total number of electrons in the atom or ion. As the scattering angle increases, interference occurs between different regions of the electron cloud, and the net scattering power of the atom falls off, but even at large scattering angles the contributions of inner core electrons are still manifest.

Of course, the atoms in molecular crystals are not isolated and they are not at rest. Thermal motion makes the electron cloud more diffuse and thus reduces the scattering power at higher angles, as discussed in the next section. The redistribution of the electron density on molecule formation is a matter of great interest for chemists, and it can be studied in principle by comparing the measured electron density distribution with that obtained by simple superposition of the isolated atom densities. We return to this topic in Chapter 8.

Effect of Atomic Vibrations

The atoms in a crystal are not at rest but vibrate about their mean positions. The effect is to make the time-averaged, space-averaged electron density peaks more diffuse, leading to a more rapid decline of the atomic scattering factors than would occur for stationary atoms (see

[4]For a fairly complete compilation from various sources see *International Tables for X-Ray Crystallography,* Vol. III, Kynoch Press, Birmingham, 1962.

Fig. 1.11). Since the mean vibration amplitudes tend to increase with temperature, the reduction of scattering power at large scattering angles becomes more pronounced at high temperatures (*temperature factor*). As first shown by Debye,[5] this reduction in scattering power can be approximated by multiplying $f(\mathbf{R})$ by the exponential function $\exp(-B \sin^2 \theta/\lambda^2)$, where $B = 8\pi^2\langle u^2\rangle$ and $\langle u^2\rangle$ is the mean-square vibrational amplitude; hence it is also known as the *Debye factor*. Debye's treatment was later modified slightly by Waller,[6] so the multiplicative factor is also known as the *Debye-Waller factor*.

Debye's paper, published only a few months after the discovery of X-ray diffraction by crystals, is remarkable for the physical intuition it showed at a time when almost nothing was known about the structure of solids at the atomic level. It was later described by Ewald, another of the founding fathers and the one to whom we are indebted for the early history of the subject, as follows: "A further factor which was soon added by Debye in papers at the time, unequalled in daring, to tackle a question of seemingly unsurmountable difficulty: the temperature factor. The temperature displacements of the atoms in a lattice are of the order of magnitude of the atomic distances and therefore may produce any changes in phase whatsoever in the individual scattered waves. That an interference effect results in spite of this disorder is due (1) to the relative scarcity of large deviations from regularity and (2) to the greater efficiency of producing scattered intensity whenever a regular array holds good for a time in an element of volume. The result is a factor of exponential form whose exponent contains besides the temperature the order of the interference only."[7]

A simplified derivation (in one dimension) that retains some of the essence of Debye's treatment is to consider a row of atoms that would be exactly related by a translation a if the atoms were at rest. We imagine, however, that at a given instant each atom is randomly displaced from its ideal position by an amount u_n, which is small relative to the row translation. The structure factor for the instantaneous arrangement is, for the one-dimensional case,

$$F(X) = f(X) \sum_n \exp\left[2\pi i X(na+u_n)\right] \qquad N \text{ terms}$$

[5]P. Debye, *Verh. Deut. Physik. Ges.* **15,** 738 (1913).

[6]I. Waller, *Z. Physik.* **17,** 398 (1923).

[7]P. P. Ewald, *Curr. Sci*, 11 (1937). Reprinted in *Early Papers on the Diffraction of X-rays by Crystals,* J. M. Bijvoet, W. G. Burgers, and G. Hägg, Eds., Oosthoek, Utrecht, 1969; published for the International Union of Crystallography.

where $f(X)$ is the form factor of the atoms at rest. When $X = m/a$ (m integral)—i.e., when the diffraction condition is satisfied,

$$F(X) = f(X) \sum \exp (2\pi i X u_n)$$

$$F(X)F^*(X) = f^2(X)[N + 2 \sum_{n \neq m} \cos 2\pi X(u_n - u_m)]$$

The expectation value of $F^2(X)$ is obtained by replacing each $u_n - u_m$ term by its expectation value, which is $\sqrt{2}$ times the root-mean-square displacement. The sum contains $N(N-1)/2$ terms, and N is supposed to be very large, so

$$\langle F^2 \rangle = f^2 N^2 \cos \{2\pi X \sqrt{2} \langle u^2 \rangle^{\frac{1}{2}}\}$$

For small x,

$$\cos x = 1 - \frac{x^2}{2} = \exp (-x^2/2)$$

Now $\langle u^2 \rangle^{\frac{1}{2}}$ is small compared with a, so if $X = m/a$ is not too large, we have approximately

$$\langle F^2 \rangle = f^2 N^2 \exp (-4\pi^2 X^2 \langle u^2 \rangle)$$

or

$$\langle F^2 \rangle^{\frac{1}{2}} = fN \exp (-8\pi^2 \langle u^2 \rangle \sin^2 \theta / \lambda^2) \tag{1.20}$$

For the three-dimensional case we only have to interpret $\langle u^2 \rangle$ as the mean-square vibrational amplitude in the direction parallel to the scattering vector $\mathbf{H}$, i.e., perpendicular to the $(h_1 h_2 h_3)$ plane.

The effect of vibration on the time-averaged, space-averaged electron density can be simulated by replacing the stationary atom at $\mathbf{r}_i$ by its convolution with a probability distribution function centered at $\mathbf{r}_i$. Usually we assume that the vibration is harmonic and hence that the probability distribution is Gaussian. This has a particular mathematical convenience because the Gaussian function has the special property that its Fourier transform is also a Gaussian. For the one-dimensional case

$$g(x) = \left(\frac{a}{\pi}\right)^{\frac{1}{2}} \exp (-ax^2) \tag{1.21}$$

and

$$G(X) = 2 \left(\frac{a}{\pi}\right)^{\frac{1}{2}} \int_0^\infty \exp (-ax^2) \cos 2\pi X x \, dx$$

$$= \exp (-\pi^2 X^2 / a) \tag{1.22}$$

Good mathematicians can work out the integration themselves; others can refer to any compilation of standard integrals.[8]

For the two-dimensional case,

$$g(r) = \left(\frac{a}{\pi}\right) \exp(-ar^2) = \left(\frac{a}{\pi}\right) \exp[-a(x^2+y^2)] \tag{1.23}$$

This circularly symmetric Gaussian has another remarkable property—its projection on any radial line is also Gaussian, e.g.,

$$g'(x) = \int_{-\infty}^{\infty} g(x,y)dy = \left(\frac{a}{\pi}\right)^{\frac{1}{2}} \exp(-ax^2) \tag{1.24}$$

Now we can compute the transform of $g(r)$ along any radial line in reciprocal space:

$$G(X, Y) = \iint g(x,y) \exp[2\pi i(Xx+Yy)]\, dx\, dy$$

$$\begin{aligned} G(X, 0) &= \iint g(x, y) \exp(2\pi iXx)\, dx\, dy \\ &= \int g'(x) \exp(2\pi iXx)\, dx \\ &= \exp(-\pi^2X^2/a) \end{aligned} \tag{1.25}$$

the same result as in Eq. 1.22. Since $g(x, y)$ has circular symmetry, its transform along any radial vector **R** is the same as along X.

By similar arguments we can derive the Fourier transform of the three-dimensional Gaussian function with spherical symmetry

$$g(r) = \left(\frac{a}{\pi}\right)^{3/2} \exp(-ar^2) \tag{1.26}$$

$$G(R) = \exp(-\pi^2R^2/a) \tag{1.27}$$

Since $G(R)$ could also have been obtained from Eq. 1.19 as

$$G(R) = 4\pi \left(\frac{a}{\pi}\right)^{3/2} \int_0^{\infty} r^2 \exp(-ar^2)\, \frac{\sin 2\pi Rr}{2\pi Rr}\, dr$$

we have actually evaluated the integral

[8]For example, *Handbook of Tables for Mathematics,* R. C. Weast and S. M. Selby, Eds., 3rd ed., Chemical Rubber Co., Cleveland, Ohio, 1967.

$$\int_0^\infty r \exp(-ar^2) \sin 2\pi Rr \, dr = \left(\frac{R}{2}\right)\left(\frac{\pi}{a}\right)^{3/2} \exp(-\pi^2R^2/a)$$

in a roundabout way.

If the one-dimensional Gaussian function is rewritten as a normal probability distribution by replacing a in Eq. 1.21 by $(1/2U^2)$,

$$g(r) = \frac{1}{U(2\pi)^{\frac{1}{2}}} \exp(-r^2/2U^2) \tag{1.28}$$

where U^2, the mean-square deviation, can be interpreted in the harmonic oscillator approximation as the mean-square vibration amplitude of an atom about its equilibrium position. The corresponding form of the transform is:

$$\begin{aligned} G(R) &= \exp(-2\pi^2U^2R^2) \\ &= \exp(-8\pi^2U^2\sin^2\theta/\lambda^2) \\ &= \exp(-B\sin^2\theta/\lambda^2) \text{ with } B = 8\pi^2U^2 \end{aligned} \tag{1.29}$$

which is the Debye-Waller temperature factor. From Eq. 1.25 and 1.27, the same result is obtained for isotropic vibrations in two and three dimensions. Since the effect of vibration is to convolute the electron density of the stationary atom with the probability distribution function, the transform of the vibrating atom is the product of $f^0(\mathbf{R})$, the atomic form factor, with $G(\mathbf{R})$, the temperature factor,

$$f(\mathbf{R}) = f^0(\mathbf{R})\, G(\mathbf{R}) \tag{1.30}$$

In general, atomic vibrations in crystals are anisotropic—i.e., the mean-square amplitude varies with direction, which leads to a dependence of the corresponding temperature factor on the direction of the scattering vector **R**. For an anisotropic vibration in three dimensions we have to replace Eq. 1.28 by

$$g(x_1x_2x_3) = \frac{(U_1U_2U_3)^{-1}}{(2\pi)^{3/2}} \exp(-[(x_1^2/2U_1^2)+(x_2^2/2U_2^2)+(x_3^2/2U_3^2)]) \tag{1.31}$$

The probability distribution is now an ellipsoid, often called the vibration ellipsoid or thermal ellipsoid, with its principal axes along x_1, x_2, x_3, and U_1^2, U_2^2, U_3^2 are the mean-square amplitudes of vibration along these directions. In general, the principal axes do not coincide with crystal axes (although restrictions can be imposed by crystallographic symmetry). The transform of Eq. 1.31, the anisotropic temperature factor, analogous to Eq. 1.29 is

$$G(\mathbf{R}) = G(R_1R_2R_3) = \exp\,[-2\pi^2(U_1^2R_1^2+U_2^2R_2^2+U_3^2R_3^2)] = \exp\,(-T) \tag{1.32}$$

where R_1, R_2, R_3 are the components of $\mathbf{R}$ along the principal axes of the vibration ellipsoid. By a suitable axis transformation (Chapter 5), T can be expressed in terms of components along the reciprocal lattice base vectors:

$$\begin{aligned} T &= 2\pi^2(U_{11}h_1^2b_1^2+U_{22}h_2^2b_2^2+U_{33}h_3^2b_3^2+2U_{12}h_1h_2b_1b_2+2U_{13}h_1h_3b_1b_3 \\ &\quad +2U_{23}h_2h_3b_2b_3) \\ &= 2\pi^2 \sum_i \sum_j U_{ij}h_ih_jb_ib_j \end{aligned} \tag{1.33}$$

The mean-square amplitude of vibration in the direction of the unit vector $\boldsymbol{l}$ with components l_1, l_2, l_3 on the same set of reciprocal lattice base vectors is given by the quadratic form

$$\begin{aligned} U^2(l_1l_2l_3) &= U_{11}l_1^2+U_{22}l_2^2+U_{33}l_3^2+2U_{12}l_1l_2+2U_{13}l_1l_3+2U_{23}l_2l_3 \\ &= \sum_i \sum_j U_{ij}l_il_j \end{aligned} \tag{1.34}$$

It may not be immediately obvious that Eq. 1.33 is equivalent to the exponent in Eq. 1.29 when U^2 is isotropic. From Eq. 1.3 the quantity $4\sin^2\theta/\lambda^2$ is the square of the scattering vector $\mathbf{R}$, and from Eq. 1.7 this can also be expressed as:

$$\begin{aligned} |\mathbf{R}|^2 &= |\mathbf{R}\cdot\mathbf{R}| \\ &= h_1^2b_1^2+h_2^2b_2^2+h_3^2b_3^2+2h_1h_2b_1b_2\cos(b_1b_2)+2h_1h_3b_1b_3\cos(b_1b_3)+ \\ &\quad 2h_2h_3b_2b_3\cos(b_2b_3) \end{aligned}$$

Substitution of the latter expression for $4\sin^2\theta/\lambda^2$ in Eq. 1.29 and comparison of the result with Eq. 1.33 shows that for an isotropic temperature factor,

$$U_{ii} = U^2$$

and

$$U_{ij} = U^2\cos(b_ib_j)$$

To see through the cumbersome algebraic expressions involved in the preceding discussion, the reader should study Fig. 1.12, which conveys the essence of the matter in a simple two-dimensional example. The larger the mean-square vibration amplitude is in any direction, the greater the fall-off in scattering power in that direction.

In the early days of crystal structure analysis (before electronic com-

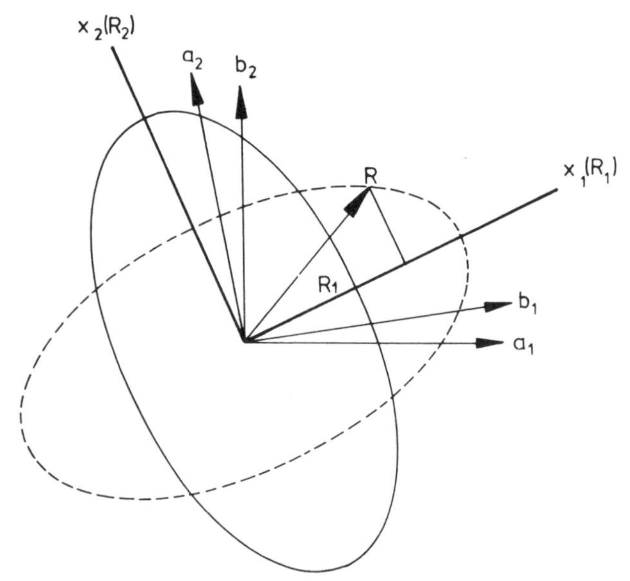

Figure 1.12. Full curve: equiprobability contour line of anisotropic vibration ellipsoid

$$g(x_1x_2) = \frac{1}{2\pi U_1 U_2} \exp\left(-[(x_1^2/2U_1^2)+(x_2^2/2U_2^2)]\right)$$

Dashed curve: equiprobability contour line of anisotropic temperature factor

$$G(\mathbf{R}) = \exp\left[-2\pi^2(U_1^2R_1^2+U_2^2R_2^2)\right]$$

If **R** is expressed in terms of its components along the reciprocal lattice vectors $\mathbf{b}_1$, $\mathbf{b}_2$, the exponent contains cross terms since the equation of the ellipse is no longer in standard form.

uters), it was normal to introduce an overall isotropic temperature factor applicable to all the atoms. Now a crystal structure analysis is not considered complete unless temperature factor parameters are determined for the individual atoms; moreover, significantly improved agreement between calculated and observed structure amplitudes can usually be obtained by using anisotropic rather than isotropic temperature factors.

Transform of a Molecule

If the atomic positions are $\mathbf{r}_i$ referred to some origin, the transform of a group of atoms (the molecular transform) is

$$\begin{aligned} F(\mathbf{R}) &= \sum_i f_i(\mathbf{R}) \exp(2\pi i\mathbf{R}\cdot\mathbf{r}_i) \\ &= A(\mathbf{R}) + iB(\mathbf{R}) \end{aligned} \tag{1.35}$$

where
$$A(\mathbf{R}) = \sum_i f_i(\mathbf{R}) \cos 2\pi\mathbf{R}\cdot\mathbf{r}_i$$
$$B(\mathbf{R}) = \sum_i f_i(\mathbf{R}) \sin 2\pi\mathbf{R}\cdot\mathbf{r}_i$$

Since no repetition is involved, the vector **R** is not restricted here to have integral components on the reciprocal basis vectors.

The contribution of each atom is $f_i(\mathbf{R})$, the atomic form factor, times a phase factor that depends on the displacement from the origin. The quantity $F(\mathbf{R})$ can be represented graphically as the resultant of a number of vectors, each of length $f_i(\mathbf{R})$ and inclined to the real axis at an angle $\varphi_i = 2\pi\mathbf{R}\cdot\mathbf{r}_i$ (Fig. 1.13). If the molecule is centrosymmetric and the origin is taken at the center of symmetry, then $B(\mathbf{R}) = 0$ since the sine terms will cancel in pairs ($\sin \varphi = -\sin(-\varphi)$); $F(\mathbf{R})$ is then real. If the molecule is noncentrosymmetric, the real and imaginary parts have no intrinsic significance because they depend on the arbitrary choice of origin. For given **R**, a shift to a new origin adds a constant φ_o to all the phase angles φ_i—i.e., it simply rotates the resultant $F(\mathbf{R})$ by an angle φ_o, leaving its magnitude unchanged. The general decrease in the magnitude of $F(\mathbf{R})$ with increasing scattering angle follows the decrease in the atomic form factors, which are here understood to include the temperature factor, as in Eq. 1.30.

The molecular scattering amplitude is centrosymmetric with respect to the origin of reciprocal space whether the molecule is centrosymmetric or not. Any other symmetry present in the molecule, e.g., mirror planes, rotation axes, and so forth, leads to a corresponding symmetry in $|F(\mathbf{R})|$. Other kinds of regularity—e.g., regular repetition of an atom

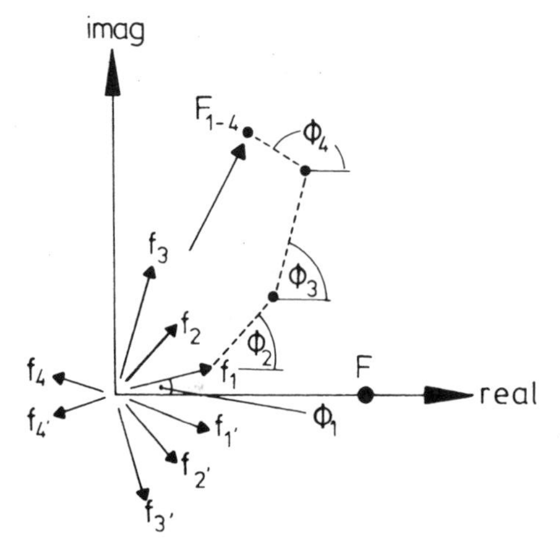

Figure 1.13. Addition of vectors f_1, f_2, f_3, f_4 (complex numbers) to give resultant F_{1-4}. Further addition of $f_{1'}, f_{2'}, f_{3'}, f_{4'}$ from atoms related by center of symmetry produces resultant F on the real axis.

or group of atoms within the molecule—also lead to characteristic features in the molecular transform.

Fig. 1.14 shows a section of the transform of a regular hexagonal arrangement of point atoms, as in the carbon skeleton of benzene. For comparison, the corresponding section for a group of 10 atoms arranged as in the carbon skeleton of naphthalene (which can be regarded approximately as two benzene rings related by a translation) is also shown. The major peaks and troughs in the naphthalene transform are arranged in the same pattern as the peaks of the benzene transform, but some of the signs have been inverted. This is because the translation operation in real space has the effect of multiplying the benzene transform by a set of cosine fringes.

As mentioned earlier (e.g., Fig. 1.7) the transform of an array of molecules regularly arranged on a lattice is the product of the molecular transform with the transform of the lattice; the molecular transform is "sampled" at the reciprocal lattice points. In general, the unit of pattern, the unit acted on by the lattice translations, consists not of a single

Figure 1.14. Planar arrays of point atoms arranged as in the carbon skeletons of benzene (left) and naphthalene (right) and parallel sections through their respective Fourier transforms. Contours are drawn at equal intervals with the zero level dotted and negative levels dashed. The origin is at the central peak in both transforms. From Thiessen and Busing, *Acta Crystallogr.* **A30,** 814 (1974).

molecule but of several molecules, related usually by certain symmetry operations, those included in the space group of the crystal (see Chapter 2). In this case, it is the transform of the entire unit of pattern that is sampled at the reciprocal lattice points.

Nevertheless, certain features of the molecular transform may still be recognizable in the sampled transform of the unit cell contents. For example, the orientation of a large planar group of atoms would be recognizable by the appearance of a column of strong reflections (along the plane normal) emanating from the origin of the reciprocal lattice. Similarly, for benzene derivatives it may be possible to recognize that certain groups of strong reflections are arranged in reciprocal space roughly corresponding to the main peaks of the benzene transform.

In the early days of X-ray analysis this kind of information was often used to solve crystal structures. Lonsdale's analysis of hexamethylbenzene[9] was one of the first and is still one of the best examples. Now one tends to rely on more or less automated computer processing of the experimental intensity data. This is probably much more efficient but not as much fun.

Anyone interested in the visual presentation of Fourier transforms obtained by diffraction of light by suitable masks should study the beautiful collections of photographs that are available.[10] Optical transforms, apart from their aesthetic merits, provide a useful way of visualizing the basic properties of Fourier transformation.

Disorder

Actual crystals deviate in a number of ways from the idealized picture of a perfectly regular array in which equivalent atoms in different unit cells are related exactly by a lattice translation. The temperature factor is one expression of what we might call dynamic disorder—the atoms are not stationary but vibrate about their equilibrium positions so that the instantaneous separation between "equivalent" atoms in different unit cells does not correspond exactly to a lattice translation. As we have seen, the thermal motion leads to a faster decrease in scattering power at higher scattering angles.

Where does the missing intensity go? If the vibrations of individual atoms were completely random—i.e., if each atom were to vibrate about

[9] K. Lonsdale, *Proc. Roy. Soc., London* **A123,** 494 (1929).

[10] C. A. Taylor and H. Lipson, *Optical Transforms,* Bell, London, 1964. G. Harburn, C. A. Taylor, and T. R. Welberry, *Atlas of Optical Transforms,* Bell, London, 1975.

its equilibrium position independently of its neighbors, the decrease in scattering power at the reciprocal lattice points would occur at the expense of a more-or-less uniform increase in scattered intensity between the reciprocal lattice points. In other words, the general background intensity would increase slightly.

However, this picture is too simple. The thermal vibrations in a crystal behave like traveling waves, with wavelengths ranging from the macroscopic dimensions of the crystal to the unit cell dimensions. The instantaneous effect of each wave is to disturb the lattice by a modulation with a definite wavelength Λ. It can be shown[11] that the Fourier transform of the disturbed lattice consists of the reciprocal lattice points of the undisturbed lattice plus satellite points at distances inversely proportional to Λ. The intensity of the satellites depends on the amplitude of the disturbing wave. Vibrations with large Λ generally have lower energy and hence larger amplitude than those with small Λ, and the corresponding, more intense satellites will be crowded into a small region surrounding the main reciprocal lattice points. These are then surrounded by a halo of diffuse intensity. In other words, the background tends to increase more in regions close to the reciprocal lattice points than in regions distant from them.

When the incident radiation interacts with the lattice vibrations, it exchanges momentum and energy with the latter, raising or lowering its energy by one phonon (quantum of lattice vibrational energy) or in some circumstances by two or more phonons. The corresponding change of wavelength is small for X-rays because the energy of phonons (10^{-2} eV) is only a tiny fraction of the X-ray energy ($\sim 10^4$ eV). For thermal neutrons the wavelength changes that occur in such scattering processes are much larger, and inelastic neutron scattering can yield important information about the dynamic properties of solids and liquids.[12]

As far as molecular crystals are concerned, the normal vibrations can usually be separated into two classes, molecular and lattice vibrations. Molecular vibrations involve distortions of covalent bonds and generally have much higher frequencies and correspondingly smaller amplitudes than the lattice vibrations, which involve only distortions of the relatively weak intermolecular bonds. Each molecule can therefore usually be regarded as behaving as a more-or-less rigid grouping that is displaced

[11]J. L. Amóros and M. Amóros, *Molecular Crystals: Their Transforms and Diffuse Scattering,* Wiley, New York, 1968, pp. 327–333.

[12]B. T. M. Willis, Ed., *Chemical Applications of Thermal Neutron Scattering,* Oxford University Press, 1973.

from its equilbrium position and orientation as a whole. Exceptions can occur, especially for low-frequency torsional vibrations of molecules.

In addition to the disorder caused by thermal vibrations, most crystals also manifest other kinds of deviations from the idealized structure that would be obtained by exact repetition of identical unit cells. Such imperfections or defects are often associated with impurities or vacancies, and they may be localized round distinct centers (point defects) or propagate in one or two dimensions (line and plane defects).

Crystallographers often like to think of imperfect crystals as being composed of many blocks, each perfect but slightly disoriented with respect to its neighbors (mosaic block model). When the crystal diffracts the incident X-ray beam, exact phase relationships hold only between the unit cells contained within the confines of a single block. The discontinuities between the blocks interrupt the coherence of the scattering process so that Eq. 1.15 applies for each block separately rather than for the crystal as a whole. The diffracted intensity is then the sum of the intensities contributed by the various blocks. If the blocks are very small (of the order of a few hundred Å) or if they are strongly disoriented with respect to each other, the diffracted spectra may show appreciable "mosaic spread"—i.e., they do not occur at a sharp angle of incidence but over an appreciable range of angles.

As far as the thermal vibrations are concerned, the time-averaged separations between equivalent atoms in different unit cells correspond to lattice translations even if the instantaneous separations do not. Many crystals show another kind of disorder, one in which the average contents of the unit cells are identical only when the averaging is carried out over the different unit cells, i.e., over space as well as over time. We can refer to this as static disorder, as opposed to the dynamic disorder associated with the thermal vibrations. The structure factor depends on the contents of the space-averaged, time-averaged unit cell, obtained by superimposing the contents of all the cells in the crystal, and it is only this averaged content (which may not be present in a single cell!) that can be established by analyzing the X-ray diffraction pattern.

Substitutional disorder, where a given site is occupied at random in different unit cells by two or more types of atoms, is more common in minerals and alloys than in organic structures. However, a few cases are known—e.g., 2-amino-4-methyl-6-chloropyrimidine, where methyl groups and chlorine atoms are randomly distributed throughout the crystal in two chemically equivalent but crystallographically unequivalent positions.[13] In this kind of situation, the intensities of the Bragg reflections depend on the content of the averaged unit cell, but there is also a contribution to non-Bragg scattering (incoherent scattering) that depends on the difference between the scattering powers of the atoms involved.

Consider the simple case where unit cells are assumed to be in perfect register but occupied at random by two kinds of atoms, A and B, in relative proportions, p_A and p_B, with $p_A + p_B = 1$. The mean atomic scattering power is $\langle f \rangle$ where

$$\langle f \rangle = pf_A + (1-p)f_B$$

The structure factor of the mixed crystal is

$$F(\mathbf{R}) = \sum_{j=1}^{N} (\langle f \rangle + \Delta f_j) \exp(2\pi i \mathbf{R} \cdot \mathbf{r}_j)$$

where the summation extends over all N unit cells in the crystal. The quantities Δf_j are either $f_A - \langle f \rangle$ or $f_B - \langle f \rangle$, depending whether the jth unit cell is occupied by an A or a B atom. For the scattered intensity we have

$$\begin{aligned} F(\mathbf{R})F^*(\mathbf{R}) &= \sum_{j=1}^{N}\sum_{k=1}^{N} (\langle f \rangle + \Delta f_j)(\langle f \rangle + \Delta f_k) \exp[2\pi i \mathbf{R}\cdot(\mathbf{r}_j - \mathbf{r}_k)] \\ &= \langle f \rangle^2 \sum\sum \exp[2\pi i \mathbf{R}\cdot(\mathbf{r}_j - \mathbf{r}_k)] \\ &\quad + \langle f \rangle \sum\sum (\Delta f_j + \Delta f_k) \exp[2\pi i \mathbf{R}\cdot(\mathbf{r}_j - \mathbf{r}_k)] \\ &\quad + \sum\sum \Delta f_j \Delta f_k \exp[2\pi i \mathbf{R}\cdot(\mathbf{r}_j - \mathbf{r}_k)] \end{aligned}$$

The first term equals $N^2\langle f \rangle^2$ when $\mathbf{R} = \mathbf{H}$, a reciprocal lattice vector, and corresponds to the Bragg scattering from a crystal built from atoms with scattering power $\langle f \rangle$. The second term is zero since summations of the type

$$\sum_{j=1}^{N} \Delta f_j = N[p(f_A - \langle f \rangle) + (1-p)(f_B - \langle f \rangle)]$$

equal zero, and the third also disappears, except for the contributions with $j = k$, which amount to

[13] C. J. B. Clews and W. Cochran, *Acta Crystallogr.* **1**, 4 (1948).

$$\sum_{j=1}^{N} (\Delta f_j)^2 = N[p(f_A-\langle f\rangle)^2 + (1-p)(f_B - \langle f\rangle)^2]$$
$$= N\{pf_A^2 + (1-p)f_B^2 + \langle f\rangle^2 - 2\langle f\rangle[pf_A + (1-p)f_B]\}$$
$$= N(\langle f^2\rangle - \langle f\rangle^2) \qquad (1.36)$$

Note that this term is independent of the scattering vector **R** and yields a continuous, diffuse background (incoherent scattering) on which the Bragg reflections are superimposed. A little algebra shows that the right side of Eq. 1.36 can be rewritten as $Np(1-p)(f_A-f_B)^2$; the closer p approaches $\frac{1}{2}$ and the greater the difference between the individual scattering powers f_A and f_B, the more intense this diffuse background will be.

For molecular crystals, static disorder can occur when two or more different orientations or conformations of a molecule have almost the same space-filling requirements—e.g., the crystal structure of azulene (Fig. 1.15).[14] In such cases it is still true that the packing arrangement of minimum potential energy is perfectly ordered. It could be built, for example, by regular repetition of molecules in a given orientation or by regular alternation of, say, two possible orientations, as indicated in Fig. 1.16a,b. However, depending on the ability of the intermolecular interactions to discriminate between correct and incorrect orientations, mistakes in the packing arrangement can occur, as in Fig. 1.16c,d. The loss of packing energy, if small, will be compensated by an increase in the configurational entropy of the crystal. In the limit, we would have a completely random sequence of the two orientations. On the other hand, if mistakes were infrequent, the crystal would have rather large ordered domains (Fig. 1.16e) and would give the same X-ray diffraction pattern as a twinned crystal. Indeed, the presence of such domains can sometimes be detected with an ordinary microscope using polarized light. The limiting cases of random or domain disorder can be distinguished, at least in principle, from the diffracted intensities once the averaged structure is known. Let $F_A(\mathbf{H})$ and $F_B(\mathbf{H})$ be the structure factors for unit cells containing the two possible orientations A and B, and let p be the probability that the orientation A occurs in a given unit cell. For the random structure, the structure factor of an averaged unit cell is

$$pF_A + (1-p)\,F_B$$

[14] J. M. Robertson, H. M. M. Shearer, G. A. Sim, and D. G. Watson, *Acta Crystallogr.* **15**, 1 (1962).

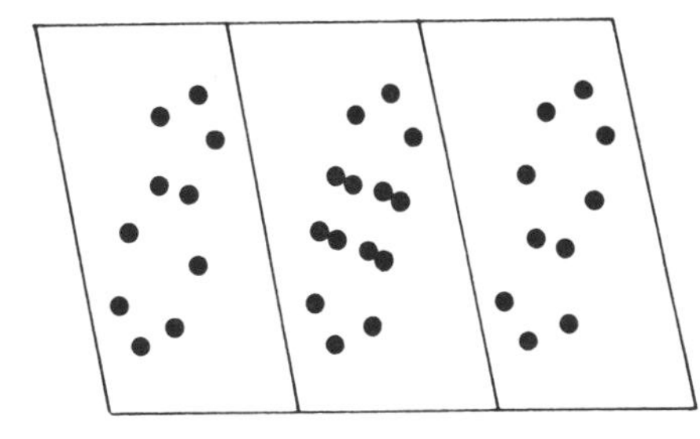

Figure 1.15. Azulene. Left and right: two molecular orientations in projection down a crystal axis; center: superposition. Each molecular site contains only one of the two orientations shown in the superposition, but the sequence is random.

and the intensity is proportional to the square of this quantity. For the domain structure, the net intensity is the sum of the intensities of the two kinds of domain, which is proportional to

$$pF_A^2 + (1-p)\, F_B^2$$

If F_A and F_B differ enough for certain reflections, careful comparison of the corresponding observed intensities may show better agreement for one model than for the other.

The presence of static disorder in crystals can sometimes be recognized from the occurrence of diffuse scattering, either as haloes or streaks, around the intense Bragg reflections. However, the diffuse scattering associated with disorder may be no more pronounced than that given by the ever-present thermal vibrations. Just as for the fluctuating thermal disorder, the presence of static disorder does not necessarily destroy the coherence relationship between different unit cells in the crystal or mosaic block to which they belong, and "normal" X-ray diffraction patterns may be produced from disordered crystals. Actually, orientational disorder is sometimes hard to detect, and it may be present, though undetected, in many crystals described as having perfectly ordered structures.

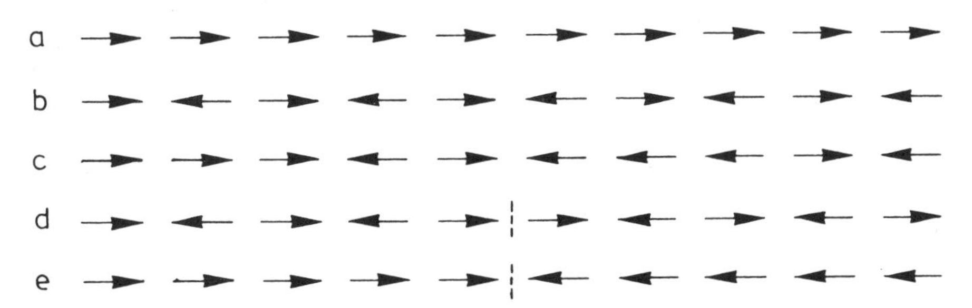

Figure 1.16. Representation of ordered and disordered rows of molecules in two possible orientations. (a) Ordered *A*. (b) Superstructure; ordered alternation of *A*, *B*. (c) Disordered *A*, *B*. (d) Superstructure with disorder. (e) Domains. The space-filling requirements of the two orientations are assumed to be very similar, as in azulene (Fig. 1.15).

Diffuse reflections and streaks are characteristic of the diffraction patterns of crystals that contain stacking faults, particularly those that belong to so-called OD (order–disorder) families.[15] The typical feature of such crystals is that they are built from layers, such that pairs of successive layers can be formed in two or more geometrically and hence energetically equivalent ways.

Hexagonally close-packed layers of spheres provide the simplest example of an OD structure. Given one such layer, the next layer can be placed with its spheres above one of two possible sets of interstices, marked 1 and 2 in Fig. 1.17—i.e., the new layer may be related to the first by either of two displacements:

$$\mathbf{S}_1 = \frac{2}{3}\mathbf{a} + \frac{1}{3}\mathbf{b} + \mathbf{e}$$

$$\mathbf{S}_2 = \frac{1}{3}\mathbf{a} + \frac{2}{3}\mathbf{b} + \mathbf{e}$$

that are entirely equivalent as far as the first layer is concerned; the same holds for any pair of successive layers. The sequence of layers is then defined by specifying the sequence of stacking vectors. The sequence $\mathbf{S}_1, \mathbf{S}_1, \mathbf{S}_1, \mathbf{S}_1, \ldots$ corresponds to cubic closest packing, $\mathbf{S}_1, \mathbf{S}_2, \mathbf{S}_1, \mathbf{S}_2, \ldots$ to hexagonal closest packing, but more complex sequences are possible. As far as the packing energy is concerned, all interactions between successive layers are equivalent, but interactions between next nearest and more distant layers depend on the stacking sequence. Since the interaction energy must decrease fairly rapidly with increasing distance, mistakes in the stacking sequence may occur.

If the stacking sequence were completely random, the average unit cell would be that obtained by superimposing the contents of all the layers. Its periodicity in the **e** direction would be the interlayer separation, and its area in the layer plane would be only one-third of the area of the layer cell. All sites of the type 0, 1, 2 (Fig.1.17) would seem to be occupied although in any given layer, sites of only one type would actually be occupied. The diffraction pattern would be that corresponding to this averaged cell, but diffuse scattering could also be expected to arise from any residual preference for a particular type of stacking sequence, interrupted, of course, by frequent mistakes.

The characteristic feature of the diffraction pattern of an OD disordered crystal is then the presence of both sharp and diffuse reflec-

[15]K. Dornberger-Schiff, *Lehrgang über OD-Strukturen,* Akademie-Verlag, Berlin, 1966; *Acta Crystallogr.* **9,** 593 (1956).

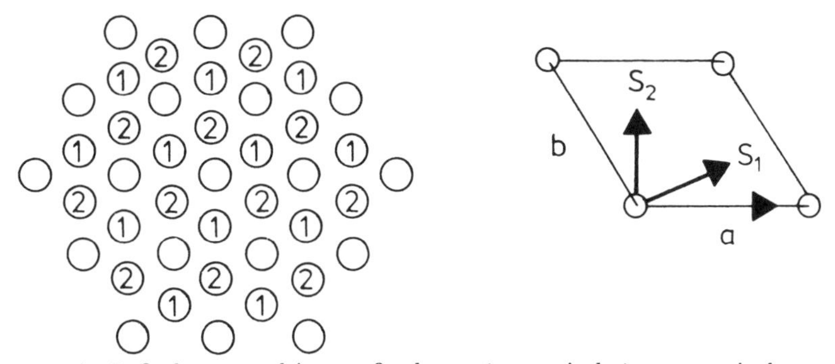

Figure 1.17. Left: hexagonal layer of spheres (open circles); two equivalent positions for an adjacent layer separated by **e** (numbered circles). Right: relationship between the two possible displacements of adjacent layers

$$\mathbf{S}_1 = \frac{2}{3}\mathbf{a} + \frac{1}{3}\mathbf{b} + \mathbf{e}$$

$$\mathbf{S}_2 = \frac{1}{3}\mathbf{a} + \frac{2}{3}\mathbf{b} + \mathbf{e}$$

The scale is double that of the left diagram.

tions, the former arising from the "superposition" cell. In addition, the diffraction pattern may show systematic absences that do not correspond to any of the 230 crystallographic space groups possible for three-dimensionally periodic structures (Chapter 2); this is hardly surprising since a disordered structure is nonperiodic and cannot really be described in terms of a repeating unit cell.

The Crystal as a Fourier Series

Any periodic function can be expressed as a Fourier series, as a sum of waves, each of which fits exactly into the given periodicity. For a one-dimensional function with periodicity a we have

$$\begin{aligned}\rho(x) &= \sum_h |A_h| \cos 2\pi(hx/a - \alpha_h) \quad h = \text{integer} \\ &= \sum_h C_h \cos 2\pi hx/a + S_h \sin 2\pi hx/a \qquad (1.37)\end{aligned}$$

where $\qquad C_h = |A_h| \cos 2\pi\alpha_h, \quad S_h = |A_h| \sin 2\pi\alpha_h$

The important things to note are (Fig. 1.18):

1. The wavelengths of the component Fourier waves are a/h, that is, they are submultiples of the unit period; each component wave has to "fit" *exactly* into the periodicity of the regular repeating pattern.

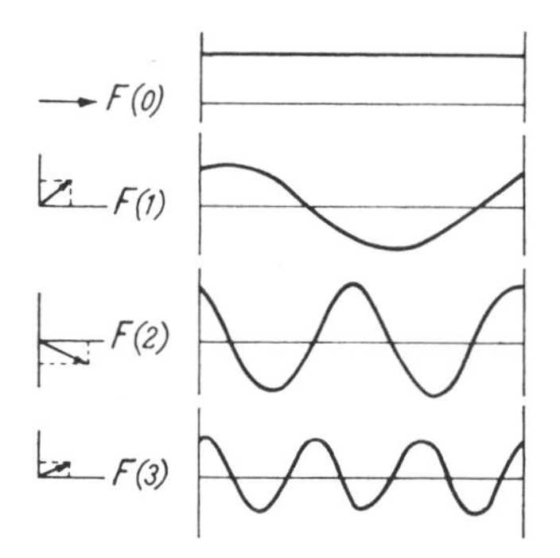

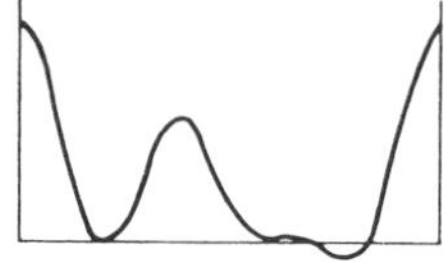

Figure 1.18. Summation of a Fourier series. The periodic function on the right can be broken down into the four Fourier waves illustrated. The amplitude and phase of each Fourier wave can be represented by a vector. From *Chimia* **17**, 170 (1963).

2. The amplitude and phase of each Fourier wave can be represented as a vector with length $|A_h|$ and phase angle α_h.

If the function $\rho(x)$ is given, the coefficients C_h and S_h, and hence the amplitude $|A_h|$ and phase angle α_h of every Fourier wave, can be calculated by integration

$$C_h = \frac{1}{a}\int_0^a \rho(x)\cos 2\pi hx/a\, dx$$

$$S_h = \frac{1}{a}\int_0^a \rho(x)\sin 2\pi hx/a\, dx$$

$$A_h = \frac{1}{a}\int_0^a \rho(x)\exp(2\pi ihx/a)\, dx$$

We can therefore reconstruct the function $\rho(x)$ by adding the individual Fourier waves with the correct amplitudes and phases.

Since the electron density in a crystal is periodic in three dimensions, it can be expressed as a three-dimensional Fourier series

$$\rho(xyz) = \sum_h\sum_k\sum_l |A(hkl)| \cos 2\pi(hx/a+ky/b+lz/c-\alpha(hkl)) \quad (1.38)$$

$$A(hkl) = \frac{1}{V}\int_0^a\int_0^b\int_0^c \rho(xyz) \exp[2\pi i(hx/a+ky/b+lz/c)]\, dx\, dy\, dz = F(hkl)/V \quad (1.39)$$

and the same considerations apply. If the function $\rho(x,y,z)$ were known, the amplitudes $|A(hkl)|$ and phase angles $\alpha(hkl)$ of the component waves could be calculated and the function reconstructed by addition, as in the one-dimensional case. To visualize the individual density waves in three dimensions, recall that the equations

$$hx/a+ky/b+lz/c = 0, 1, 2, \ldots \text{etc.}$$

describe a sequence of equally spaced, parallel planes, the first passing through the origin, the second making intercepts a/h, b/k, c/l along the coordinate axes, etc. This is just the sequence of planes with Miller indices (***hkl***) (Fig. 1.5). For $\alpha(hkl)=0$, these planes correspond to successive maxima of the (***hkl***) wave, for $\alpha(hkl)=1/2$ to successive minima. For intermediate phase angles the maxima and minima are displaced by appropriate distances along the wave normal. The wavelength of the (***hkl***) wave is obviously $d(hkl)$, the spacing of the (***hkl***) set of planes (Fig. 1.5).

Since $\rho(xyz)$ represents the electron density in the crystal, we can think of each of its Fourier components as a standing electron-density wave. As pointed out earlier, diffraction from a crystal is equivalent to reflection of the incident beam from each set of planes with integral Miller indices. However, each such set of planes can now be identified with consecutive crests of an electron-density wave with amplitude $|A(hkl)|$. Thus each wave must be able to reflect the X-ray beam at a definite angle of incidence corresponding to Bragg's Law (Eq. 1.14). We shall see that this is the case, and moreover, a very important result, that the intensity of the reflected beam is proportional to $|A(hkl)|^2$. Each density wave can be rewritten as

$$\rho'(\xi) = |A(hkl)| \cos 2\pi(\xi/d(hkl) - \alpha(hkl))$$

where ξ is a distance measured along the wave normal from the origin of the x, y, z coordinate system. For reflection at an angle θ the path difference between a plane passing through the origin and a parallel plane at ξ is $2\xi \sin\theta$ (Fig. 1.19), corresponding to a phase difference $4\pi\xi \sin\theta/\lambda$. By analogy to Eq. 1.2 we have

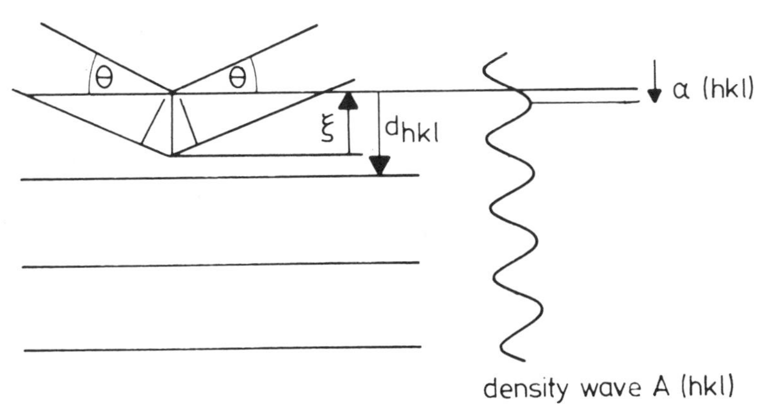

Figure 1.19. The *hkl* density wave runs parallel to the normal to the set of planes with Miller indices (*hkl*); its wavelength equals the spacing of the planes.

$$F(\theta) = \int_0^d \rho'(\xi) \exp(4\pi i\xi \sin\theta/\lambda)\, d\xi$$

$$= \int_0^d |A(hkl)|\, [\cos 2\pi(\xi/d(hkl) - \alpha(hkl))] \exp(4\pi i\xi \sin\theta/\lambda)\, d\xi$$

The integral is zero unless

$$4\pi\xi \sin\theta/\lambda = 2\pi\xi/d(hkl) = \varphi$$

or
$$\lambda = 2d(hkl) \sin\theta \qquad \text{(Bragg's Law, Eq. 1.14)}$$

When this condition is satisfied,

$$F(\theta) = |A(hkl)| \int_0^{2\pi} (\cos\varphi \cos 2\pi\alpha + \sin\varphi \sin 2\pi\alpha) \exp(i\varphi)\, d\varphi$$

which is proportional to

$$|A(hkl)|\, (\cos 2\pi\alpha + i \sin 2\pi\alpha)$$

The intensity $I(\theta)$ is then proportional to

$$F(\theta)F^*(\theta) = A^2(hkl) \tag{1.40}$$

which is what we set out to show.

By applying Eq. 1.40, the Fourier series expansion (Eq. 1.38) becomes

$$\rho(\mathbf{r}) = \frac{1}{V} \sum |F(\mathbf{H})| \cos 2\pi(\mathbf{H}\cdot\mathbf{r} - \alpha(\mathbf{H})) \tag{1.41}$$

$$= \frac{1}{V} \sum |F(\mathbf{H})|\, [\cos 2\pi\mathbf{H}\cdot\mathbf{r} \cos 2\pi\alpha(\mathbf{H}) + \sin 2\pi\mathbf{H}\cdot\mathbf{r} \sin 2\pi\alpha(\mathbf{H})]$$

The right side can be rewritten as

$$\frac{1}{V} \sum |F(\mathbf{H})|\, [\cos 2\pi\alpha(\mathbf{H}) + i \sin 2\pi\alpha(\mathbf{H})]\, [\cos 2\pi\mathbf{H}\cdot\mathbf{r} - i \sin 2\pi\mathbf{H}\cdot\mathbf{r}]$$

because the imaginary terms vanish on summing over positive and negative values of **H**, owing to the relationship (Fourier Transforms, theorem 4): $\alpha(\mathbf{H}) = -\alpha(-\mathbf{H})$, whence $\cos 2\pi\alpha(\mathbf{H}) = \cos 2\pi\alpha(-\mathbf{H})$ and $\sin 2\pi\alpha(\mathbf{H}) = -\sin 2\pi\alpha(-\mathbf{H})$. The phase angles $\alpha(\mathbf{H})$ can now be incorporated with the magnitudes $|F(\mathbf{H})|$ to rewrite Eq. 1.41 in terms of complex quantities

$$\rho(\mathbf{r}) = \frac{1}{V} \sum_{\mathbf{H}} F(\mathbf{H}) \exp(-2\pi i\mathbf{H}\cdot\mathbf{r})$$

with
$$F(\mathbf{H}) = |F(\mathbf{H})| \exp[2\pi i\alpha(\mathbf{H})] \tag{1.42}$$

Eq. 1.42 is equivalent to Eq. 1.4 when the integration over the continuous variable $F(\mathbf{R})$ in the latter is replaced by a sum over the discrete quantities $F(\mathbf{H})$. The periodic electron-density function $\rho(\mathbf{r})$ in a crystal is thus the inverse Fourier transform of $F(\mathbf{R})$ sampled at points in reciprocal space that have integral components on the reciprocal basis vectors.

The expansion of the electron-density distribution $\rho(xyz)$ as a sum of component density waves is fundamental for the X-ray analysis of complex crystals. Since the intensities can be measured, the relative magnitudes of the Fourier coefficients, the structure amplitudes $|F(hkl)|$, can be obtained from a suitable experiment. However, to reconstruct the electron-density distribution by adding the component waves we need to know not only their amplitudes but also their relative phases $\alpha(hkl)$, which are not experimentally available. In the absence of the phase-angle information, we do not know whether each component density wave starts at the origin with a maximum, a minimum, or some intermediate value. This is the fundamental problem of X-ray analysis—the phase problem. For centrosymmetric structures the origin can be taken at a center of symmetry so that the phase angles are restricted to the values 0 or π. The Fourier coefficients are then real quantities, and the phase problem is reduced to an ambiguity of sign for each coefficient. This is a great simplification, but it still leaves the problem unanswered. We return to it in Chapter 3.

With normal light or with electrons the phase problem can be circumvented by using a suitable lens system. The radiation scattered by an object in different directions is refocused by the lens and effectively recombined to give an appropriately magnified image of the object. For X-rays the recombination process cannot be carried out by lenses since (1) the refractive index of X-rays is virtually unity (to within one or two parts in 10^5) for all known materials, and (2) X-rays are not refracted by electric or magnetic fields. Instead, the intensity of scattering in different directions has to be measured and the lost phase information recovered somehow or other so that the recombination can be carried out mathematically.

In practice the electron-density function obtained by Fourier summation is always distorted because of random and systematic errors in the measured intensities (discussed in more detail in Chapter 6). Even when the phase problem can be solved (by methods discussed in Chapter 3), errors are unavoidably present in the values of the phase angles actually assigned to the Fourier coefficients. This is less troublesome for centrosymmetric structures because of the limitation to only two discrete possibilities (0 or π) for each coefficient. However, for non-centrosymmetric structures, where phase angles are not a priori restricted, there is always some uncertainty in the values adopted.

In some cases it may be possible to derive signs or approximate phase angles for some but not all of the Fourier coefficients. If the subset of component waves included in the partial synthesis is chosen systematically, the relationship of the resulting density distribution to the actual distribution can sometimes be inferred. For example, if we include only waves with even values of h in the Fourier summation, the resulting distribution $\rho'(xyz)$ has the same value at $x+a/2$, y, z as at x, y, z since each component wave repeats an even number of times along the **a** axis. Moreover,

$$\begin{aligned}\rho\left(x+\frac{a}{2}, y, z\right) &= \frac{1}{V}\sum_h\sum_k\sum_l |F(hkl)| \cos 2\pi\left[h\left(\frac{x}{a}+\frac{1}{2}\right)+\frac{ky}{b}+\frac{lz}{c}-\alpha(hkl)\right] \\ &= \frac{1}{V}\sum_h\sum_k\sum_l |F(hkl)| \cos 2\pi\left(\frac{hx}{a}+\frac{ky}{b}+\frac{lz}{c}-\alpha(hkl)\right)\cos \pi h \\ &= S_E - S_O\end{aligned}$$

where S_E and S_O are the partial sums over even and odd h values, respectively. However,

$$\rho(x, y, z) = S_E + S_O$$

$$S_E = [\rho(x, y, z) + \rho(x + a/2, y, z)]/2$$

$$S_O = [\rho(x, y, z) - \rho(x + a/2, y, z)]/2$$

The partial sum S_E over the even values of h thus corresponds to the superposition of the electron densities at x, y, z and $x + a/2$, y, z and the partial sum S_O corresponds to the difference between these densities. Similar relationships can be derived for partial sums over other systematically selected subsets of the component waves.

The ideal crystal does not exist, and the Fourier series representation applies only to the averaged repeating unit. It is only the Fourier components of this averaged cell that give rise to Bragg reflections. However, we can still describe deviations from perfect periodicity in terms of component waves with nonintegral values of h, k, l, waves that do not "fit" exactly into the unit cell. The transition from a discrete set of component waves with sharply defined wavelengths to a continuum of waves with all possible wavelengths corresponds to the transition from the Fourier transform of a periodic function to that of an arbitrary nonperiodic one. Just as we can think of each component wave of the periodic structure as reflecting the incident beam at a definite angle to produce a Bragg reflection, we can also think of each of the waves with nonintegral h, k, l values as having the same property; it is these waves that produce the diffuse, non-Bragg scattering.

If the disturbances from perfect periodicity are not too severe, they can be decomposed into a spectrum of modulations of the undisturbed structure, each with its own period. Consider, for example, a one-dimensional case where the undisturbed and modulated electron-density distributions are:

$$\rho_o(x) = \sum_h A_h \cos(2\pi hx/a)$$

$$\rho(x) = \rho_o(x)\,[1 + p\cos(2\pi x/Qa)]$$

with Qa the period of the modulation. We have

$$\rho(x) = \sum_h A_h \cos(2\pi hx/a) + p\sum_h A_h \cos(2\pi hx/a)\cos(2\pi x/Qa)$$

$$= \sum_h A_h \cos(2\pi hx/a) + \frac{p}{2}\sum_h A_h \cos[2\pi x(h+1/Q)/a]$$

$$+\frac{p}{2}\sum_h A_h \cos[2\pi x(h-1/Q)/a]$$

In addition to the component waves corresponding to the undisturbed structure, there are also waves with period $a/(h+1/Q)$ and $a/(h-1/Q)$ that produce a pair of satellite reflections, one on either side of the main Bragg reflection. The intensity of the satellites is proportional to p^2, the square of the amplitude of the modulation. The larger the period of the modulation, the closer the satellites occur to the main reflection. Clearly, the superposition of many such modulations with different periods, large compared with a, will produce a diffuse halo around each Bragg reflection (see also earlier section entitled "Disorder").

The Patterson Function

We have seen that the electron-density distribution in an averaged unit cell of a crystal can be reconstructed by summing the Fourier series (Eq. 1.42)

$$\rho(\mathbf{r}) = \frac{1}{V}\sum_{\mathbf{H}} F(\mathbf{H}) \exp(-2\pi i\mathbf{H}\cdot\mathbf{r})$$

However, because only the amplitudes $|F(\mathbf{H})|$ and not the phase angles

$\alpha(\mathbf{H})$ of the coefficients can be measured, the desired reconstruction cannot be carried out directly. But since the amplitudes $|F(\mathbf{H})|$ depend on the structure, that is, on the nature and positions of the atoms within the unit cell, a knowledge of the amplitudes alone should give some kind of information about the structure. What kind of information is this?

Some insight can be gained by writing the structure factor as a sum of contributions from all atoms contained in the unit cell

$$F(\mathbf{H}) = \sum_i f_i(\mathbf{H}) \exp(2\pi i\mathbf{H}\cdot\mathbf{r}_i) \tag{1.43}$$

The right side involves the scattering powers $f_i(\mathbf{H})$ (including thermal vibration) and the positions $\mathbf{r}_i$ of the atoms, referred to some origin. We recall that although a change of origin alters the phase of a structure factor, it leaves its magnitude unchanged. Hence $|F(\mathbf{H})|$ must depend on the *relative positions* of the atoms, that is, on the vectorial distances between them, rather than on the actual position vectors $\mathbf{r}_i$. This notion can be made a little more precise:

$$\begin{aligned} |F(\mathbf{H})|^2 = F(\mathbf{H})F^*(\mathbf{H}) &= \sum_{i,j} f_i f_j \exp[2\pi i\mathbf{H}\cdot(\mathbf{r}_i - \mathbf{r}_j)] \\ &= \sum_i f_i^2 + 2\sum_{i<j} f_i f_j \cos 2\pi\mathbf{H}\cdot(\mathbf{r}_i - \mathbf{r}_j) \end{aligned} \tag{1.44}$$

The quantity $|F(\mathbf{H})|^2$ depends on the scattering powers $f_i(\mathbf{H})$ and on the interatomic vectors. Hence a knowledge of the amplitudes gives information about the distribution of such vectors. Indeed, as Patterson[16] was the first to show, this information can be obtained by carrying out a Fourier synthesis with amplitudes $|F(\mathbf{H})|^2$ and all phase angles zero.

$$P(\mathbf{u}) = \frac{1}{V} \sum_{\mathbf{H}} |F(\mathbf{H})|^2 \exp(2\pi i\mathbf{H}\cdot\mathbf{u}) \tag{1.45}$$

Consider the function $P(\mathbf{u})$ defined as

$$P(\mathbf{u}) = V \int \rho(\mathbf{r})\, \rho(\mathbf{r} + \mathbf{u})\, d\mathbf{r} \tag{1.46}$$

and substitute the corresponding Fourier syntheses for the electron-density distributions

$$P(\mathbf{u}) = \frac{1}{V} \int \sum_{\mathbf{H}} \sum_{\mathbf{H}'} F(\mathbf{H}) \exp(-2\pi i\mathbf{H}\cdot\mathbf{r})\, F(\mathbf{H}') \exp[-2\pi i\mathbf{H}'\cdot(\mathbf{r}+\mathbf{u})]\, d\mathbf{r}$$

[16] A. L. Patterson, *Z. Kristallogr.* **90**, 517 (1935).

$$= \frac{1}{V} \sum_{\mathbf{H}} \sum_{\mathbf{H}'} \int F(\mathbf{H})F(\mathbf{H}') \exp\left[-2\pi i(\mathbf{H}+\mathbf{H}')\cdot\mathbf{r}\right] \exp\left(-2\pi i\mathbf{H}'\cdot\mathbf{u}\right) d\mathbf{r}$$

Since the integration runs over a complete unit cell and since $\mathbf{H}$ and $\mathbf{H}'$ have integral components on the reciprocal basis vectors, all terms are zero except those with $\mathbf{H}' = -\mathbf{H}$, leading to

$$P(\mathbf{u}) = \frac{1}{V} \sum_{\mathbf{H}} F(\mathbf{H})\, F(-\mathbf{H}) \exp\left(2\pi i\mathbf{H}\cdot\mathbf{u}\right)$$

$$= \frac{1}{V} \sum_{\mathbf{H}} |F(\mathbf{H})|^2 \exp\left(2\pi i\mathbf{H}\cdot\mathbf{u}\right)$$

since $F(\mathbf{H})$ and $F(-\mathbf{H})$ are complex conjugates, $\rho(\mathbf{r})$ being real.

The same result could have been derived by using the convolution theorem (Eq. 1.17). The set of coefficients $F(\mathbf{H})$ is the Fourier transform, sampled at the reciprocal lattice points, of $\rho(\mathbf{r})$. Similarly, the set of coefficients $F^*(\mathbf{H})$ is the Fourier transform of $\rho(-\mathbf{r})$. The set of products $F(\mathbf{H})F^*(\mathbf{H}) = |F(\mathbf{H})|^2$ is then the Fourier transform of the convolution (Eq. 1.16) of $\rho(\mathbf{r})$ and $\rho(-\mathbf{r})$, which is the same function

$$P(\mathbf{u}) = V \int \rho(\mathbf{r})\, \rho(\mathbf{r}+\mathbf{u})\, d\mathbf{r}$$

as defined in Eq. 1.46. The factor V (volume of the unit cell) is necessary if $P(\mathbf{u})$ is to be expressed in units of (electrons)^2 per unit volume.

We can obtain some insight into the relationship between $\rho(\mathbf{r})$ and $P(\mathbf{u})$ by thinking of the integral as a sum. To evaluate $P(\mathbf{u})$, we have to multiply the electron density at $\mathbf{r}$ and at $\mathbf{r} + \mathbf{u}$ and add the resulting products for all possible positions of $\mathbf{r}$. If $\rho(\mathbf{r})$ is built from discrete peaks $\rho_i(\mathbf{r}-\mathbf{r}_i)$, centered at $\mathbf{r}_i$, then $P(\mathbf{u})$ can also be regarded as being built from peaks, one for every interatomic vector.

We can decompose $\rho(\mathbf{r})$ into "atomic" peaks by substituting

$$F(\mathbf{H}) = \sum_i f_i(\mathbf{H}) \exp\left(2\pi i\mathbf{H}\cdot\mathbf{r}_i\right) \tag{1.43}$$

into

$$\rho(\mathbf{r}) = \frac{1}{V} \sum_{\mathbf{H}} F(\mathbf{H}) \exp\left(-2\pi i\mathbf{H}\cdot\mathbf{r}\right) \tag{1.42}$$

to obtain

$$\rho(\mathbf{r}) = \frac{1}{V} \sum_{\mathbf{H}} \sum_i f_i(\mathbf{H}) \exp\left[-2\pi i\mathbf{H}\cdot(\mathbf{r}-\mathbf{r}_i)\right]$$

$$= \sum \rho_i(\mathbf{r}-\mathbf{r}_i)$$

with
$$\rho_i(\mathbf{r}-\mathbf{r}_i) = \frac{1}{V}\sum_{\mathbf{H}} f_i(\mathbf{H}) \exp\,[-2\pi i\mathbf{H}\cdot(\mathbf{r}-\mathbf{r}_i)] \tag{1.47}$$

If each peak is now referred to a new origin at its center, this becomes

$$\rho_i(\mathbf{r}') = \frac{1}{V}\sum_{\mathbf{H}} f_i(\mathbf{H}) \exp\,(-2\pi i\mathbf{H}\cdot\mathbf{r}') \tag{1.48}$$

with
$$\int \rho_i(\mathbf{r}')\, d\mathbf{r}' = Z_i$$

the total number of electrons in the ith atom.

In a similar way, we can decompose the Patterson function into separate peaks. We substitute

$$|F(\mathbf{H})|^2 = F(\mathbf{H})F^*(\mathbf{H}) = \sum_{i,\,j} f_i(\mathbf{H})f_j(\mathbf{H}) \exp\,[2\pi i\mathbf{H}\cdot(\mathbf{r}_i-\mathbf{r}_j)]$$

into Eq. 1.45 to obtain

$$\begin{aligned} P(\mathbf{u}) &= \frac{1}{V}\sum_{\mathbf{H}}\sum_{i,\,j} f_i(\mathbf{H})f_j(\mathbf{H}) \exp\,(2\pi i\mathbf{H}\cdot[\mathbf{u}-(\mathbf{r}_j-\mathbf{r}_i)]) \\ &= \sum_{i,\,j} P_{ij}[\mathbf{u}-(\mathbf{r}_j-\mathbf{r}_i)] \end{aligned} \tag{1.49}$$

where each peak can be expressed in the alternative forms

$$P_{ij}(\mathbf{v}) = \frac{1}{V}\sum f_i(\mathbf{H})f_j(\mathbf{H}) \exp\,(2\pi i\mathbf{H}\cdot\mathbf{v}) \tag{1.50}$$

or
$$P_{ij}(\mathbf{v}) = V\int \rho_i(\mathbf{r})\,\rho_j(\mathbf{r}+\mathbf{v})\, d\mathbf{r} \tag{1.51}$$

with $\mathbf{v} = \mathbf{u}-(\mathbf{r}_j-\mathbf{r}_i)$, that is, with a new origin at $\mathbf{u} = \mathbf{r}_j-\mathbf{r}_i$, the vector distance between atom i and atom j, for each peak. We see that there are just as many Patterson peaks as there are i, j pairs, i.e., N^2. The N peaks corresponding to i, i pairs occur at the origin of the Patterson function, the remaining $N(N-1)$ being centrosymmetrically distributed around the origin since for every i, j vector at $\mathbf{u} = \mathbf{r}_j-\mathbf{r}_i$ there is a j, i vector at $\mathbf{u} = \mathbf{r}_i-\mathbf{r}_j$.

By Fourier transforming Eq. 1.50 we obtain

$$f_i(\mathbf{H})f_j(\mathbf{H}) = \int P_{ij}(\mathbf{v}) \exp\,(-2\pi i\mathbf{H}\cdot\mathbf{v})\, d\mathbf{v}$$

or
$$f_i(0)f_j(0) = Z_iZ_j = \int P_{ij}(\mathbf{v})\, d\mathbf{v} \tag{1.52}$$

Hence the *weight* of each Patterson peak is proportional to the product of the atomic numbers of atoms i and j. This is approximately true also for the height of the peak, which is, from Eqs. 1.50 and 1.51:

$$P_{ij}(0) = \frac{1}{V} \sum_{\mathbf{H}} f_i(\mathbf{H}) f_j(\mathbf{H}) = \int \rho_i(\mathbf{r}) \rho_j(\mathbf{r})\, d\mathbf{r} \qquad (1.53)$$

and it would be exactly true if $\rho_i(\mathbf{r})$ and $\rho_j(\mathbf{r})$ had exactly the same shape. The height of the peak at the origin of the Patterson function is:

$$P(0) = \sum_{i} P_{ii}(0) + \sum_{i,\,j} P_{ij}(\mathbf{r}_i - \mathbf{r}_j)$$

$$= \frac{1}{V} \sum_{i} \sum_{\mathbf{H}} f_i^2(\mathbf{H}) + \frac{1}{V} \sum_{i,\,j} \sum_{\mathbf{H}} f_i(\mathbf{H}) f_j(\mathbf{H}) \exp\,[2\pi i \mathbf{H} \cdot (\mathbf{r}_i - \mathbf{r}_j)]$$

If we assume that the atomic peaks all have the same shape, scaled up to give the appropriate number of electrons (not a bad approximation), then the individual form factors can be expressed in terms of a common form factor, $f_i(\mathbf{H}) = Z\hat{f}(\mathbf{H})$. Suppose also that the atomic peaks do not overlap significantly. Then

$$P(0) = k \sum Z_i^2 \qquad \text{where} \qquad k = \frac{1}{V} \sum_{\mathbf{H}} \hat{f}^2(\mathbf{H}) \qquad (1.54)$$

Similarly, under these conditions, the height of an i, j peak (Eq. 1.53) becomes

$$P_{ij}(0) = kZ_iZ_j$$

In principle, the atomic positions $\mathbf{r}_i$ can be derived from the vector map if the individual peaks P_{ij} are clearly resolved from one another. If Patterson peaks occur at

$$\mathbf{u}_{ij} = \mathbf{r}_j - \mathbf{r}_i$$

$$\mathbf{u}_{ik} = \mathbf{r}_k - \mathbf{r}_i$$

then

$$\mathbf{u}_{ik} - \mathbf{u}_{ij} = \mathbf{r}_k - \mathbf{r}_j = \mathbf{u}_{jk}$$

In other words, if two Patterson peaks arise from interactions involving a common atom, then the vector between them corresponds to a third Patterson peak. By successively displacing a copy of the Patterson function so that its origin coincides in turn with each peak of the original function, the atomic positions can be found in principle, as illustrated in Fig. 1.20.

This is more easily said than done. For N atoms in the unit cell, the number of Patterson peaks, excluding the origin peak, is $N(N-1)$. As N

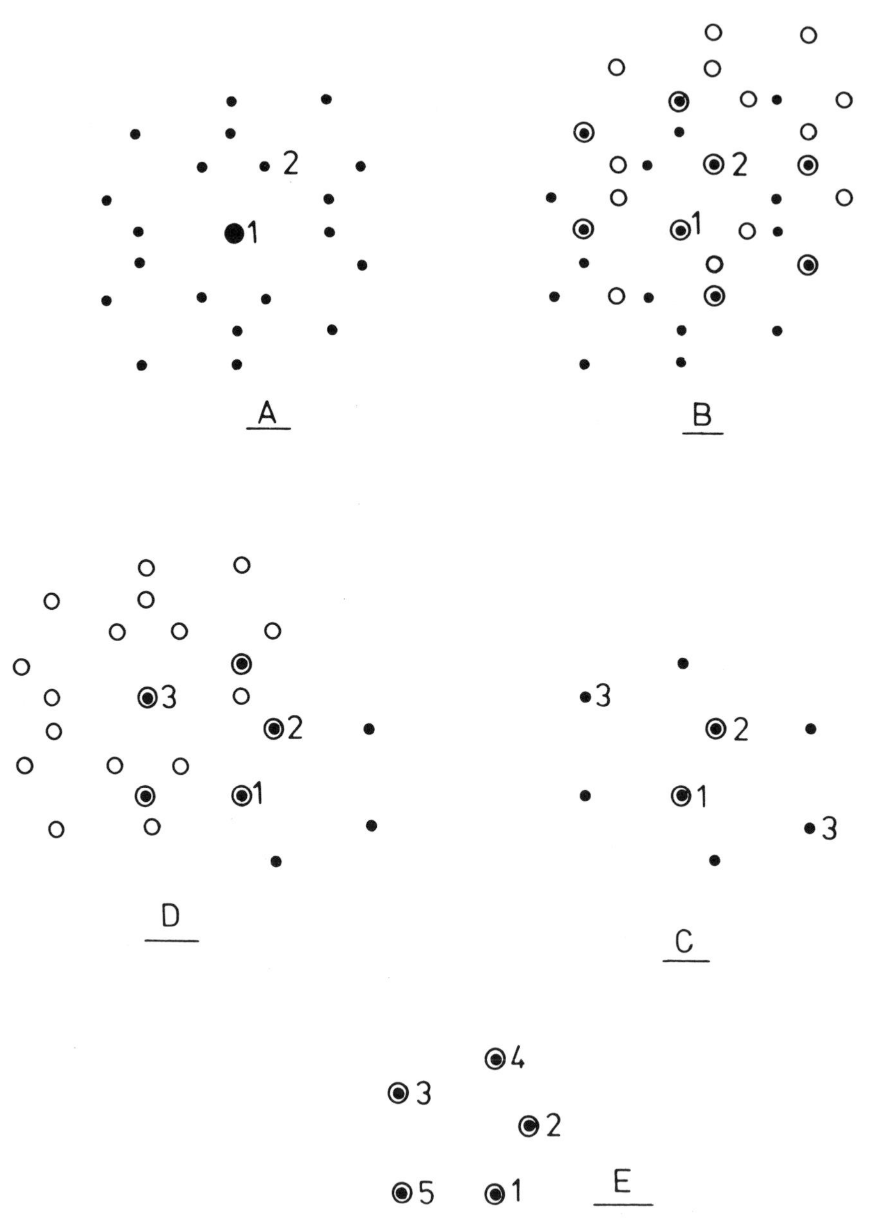

Figure 1.20. (A) Patterson function (vector map) containing 5.4 = 20 peaks centrosymmetrically arranged about the origin. Locate one atom (1) at origin, a second (2) at any peak. (B) Displace origin of vector map to peak 12 (open circles) and superimpose on original map (black circles). Coincidences indicate possible positions of additional atoms. (C) Coincidences marked by black circles. Displace origin of vector map to one of these positions (3) and search for further coincidences (D). (E) Result: an irregular pentagon of points that produces the vector map A. If, at stage C, a different coincidence 3′ had been chosen as new origin, the enantiomorphic pentagon would have resulted.

increases, the volume V of the unit cell increases linearly so that the density of atomic peaks N/V remains roughly constant. However, the density of *Patterson* peaks N^2/V increases linearly with N. As a result, N does not have to be very large before the Patterson peaks begin to overlap and are no longer resolved from one another. For all but the simplest structures (containing up to say, 15–20 atoms in the unit cell), the derivation of the atomic positions by analysis of the Patterson function is usually not feasible without prior information about the relative positions of at least some of the atoms in the molecule.

The resolution of Patterson peaks can be artificially increased by suitable measures—i.e., by computing the Patterson function with suitably modified coefficients. Normal Patterson peaks are about twice as broad as the atomic peaks in the corresponding electron density synthesis. If the actual atoms could be replaced by point atoms occupying the same positions, the effect on the structure factors $F(\mathbf{H})$ would be approximately to multiply them by $1/f(\mathbf{H})$, where $f(\mathbf{H})$ is an averaged, normalized form factor (including the mean temperature factor; we assume again that the atoms have roughly the same shape). We can then calculate the Patterson function with modified coefficients F^2/f^2. Much the same result is obtained by using $E^2(\mathbf{H})$ values (defined in Chapter 2 under "Intensity statistics") as coefficients. However, with a finite set of coefficients we cannot produce infinitely sharp Patterson peaks. The actual F^2 values are only measurable to some limiting radius of the scattering vector $\mathbf{R}$ in reciprocal space. The Fourier transform of a point peak is a constant whereas the Fourier transform of our sharpened Patterson peak can be described as:

$$f(\mathbf{H}) = c \qquad |\mathbf{H}| \leq |\mathbf{H}_{\text{max}}|$$

$$f(\mathbf{H}) = 0 \qquad |\mathbf{H}| > |\mathbf{H}_{\text{max}}|$$

This is the inverse of the problem that was discussed earlier in connection with the Fourier transform of a crystal of finite size, and the result is obvious. By analogy with Fig. 1.9, the Patterson peak drops sharply to zero but is surrounded by a series of negative and positive ripples of gradually decreasing amplitude. The ripples associated with some peaks will be superimposed on the central maxima of others. Indeed, small peaks may be completely obliterated by a chance reinforcement of negative ripples from larger ones. Thus the additional resolution is always accompanied by some distortion of the information inherent in the original diffraction pattern.

Even when the complete interpretation of the Patterson function is not feasible, a partial interpretation of its more prominent features is sometimes possible, and this may yield sufficient structural information to overcome the phase problem. Such circumstances arise, for example, where the unit cell contains a few atoms of high atomic number because peaks arising from vectors between these atoms will stand out above the general background of the Patterson function. Thus the relative arrangement of the heavy atoms may be derived, and this information may be enough to assign approximate phase angles to the observed structure amplitudes $|F(\mathbf{H})|$, as discussed in more detail in Chapter 3.

A partial interpretation is also possible if the unit cell contains regularly repeating fragments of known or approximately known structure—e.g., long zigzag chains of atoms, planar hexagons of atoms in aromatic molecules, and so forth. In such situations the superposition of Patterson peaks due to systems of parallel interatomic vectors can lead to a resultant peak that is prominent and easy to identify. A particularly favorable situation arises when a pair of centrosymmetric molecules or molecular fragments—e.g., benzene rings, are related by a center of symmetry. If the $2N$ atoms of one such fragment lie at $\mathbf{x}_M \pm \mathbf{x}_i (i = 1 \text{ to } N)$, and those of the other lie at $-\mathbf{x}_M \pm x_i$, we must have a pair of Patterson peaks at $\pm 2\mathbf{x}_M$, each of weight proportional to $2N$, the number of atoms in the centrosymmetric fragment. If N is large, this pair of peaks will be prominent and easy to recognize so that the midpoint of the molecular fragment can be located. With any luck, the orientation of the fragment can also be recognized by comparing the distributions of Patterson peaks around $\pm 2\mathbf{x}_M$ and around the origin. Further aspects of the interpretation of Patterson functions are discussed in Chapter 3.

2. Internal Symmetry of Crystals

Alice was very nearly getting up and saying, "Thank you, sir, for your interesting story," but she could not help thinking there *must* be more to come, so she sat still and said nothing.

Historical Background

So far we have hardly mentioned symmetry, in spite of the fact that it is one of the most characteristic properties of crystals and has been recognized since antiquity. The first suggestion that crystal symmetry might be the expression of an underlying order in the geometric arrangement of invisible, constituent particles seems to have come from Kepler,[1] in his speculations about the origin of the beautifully regular, hexagonal forms of snowflakes. Similar ideas were advanced by other scientists during the latter half of the 17th century, notably by Hooke and Huygens, but in the absence of an atomic theory of matter, the nature of the constituent particles remained obscure. By the middle of the 18th century, it had been recognized that although natural crystals of a given substance could grow in different forms, they show the same interfacial angles, and in 1774 Haüy introduced the concept of space lattices to explain this constancy in the form of the Law of Rational Indices. Haüy and others used the term *molecule* to describe the content of the repeating units, but the nature of these units remained obscure until much later when chemists recognized that the structural units in crystals could be identified with particular arrangements of atoms. Meanwhile, it was realized that only certain point symmetries of crystals are compatible with the lattice concept. These were enumerated in 1830 by Hessel.

[1]J. Kepler, *De Nive Sexangula,* Frankfurt am Main, 1611 (translated as *The Six-Cornered Snowflake,* by Colin Hardie, with essays by L. L. Whyte and B. F. J. Mason, Oxford University Press, 1966). Apparently Kepler did not take his own ideas too seriously—they would have led him to atomism, for which the world was not yet ready. His light-hearted attitude toward his theme is indicated by his play with the similarity between the words nix (snow) and nichts (nothing). "Nam si a Germano quaeras nix quid sit, respondebit Nihil, siquidem Latine possit;" translated by Hardie as "Ask a German what Nix means, and he will answer 'nothing' (if he knows Latin)." Kepler was not above making puns.

The geometrical theory of space lattices was completed by the end of the 19th century, culminating in the recognition that there are only a finite number of ways of combining elements of point symmetry and translational symmetry to form space groups. The 230 possibilities were enumerated independently and almost simultaneously by Fedorow, Schoenflies, and Barlow. By 1912, when von Laue discovered X-ray diffraction, the mathematical theory of crystals was essentially complete, even though nothing was known about the actual structure of the repeating units.[2]

Space Lattices

As mentioned in Chapter 1, an idealized crystal consists of some structural unit that is repeated in a perfectly regular pattern in three dimensions. A wallpaper pattern is an obvious two-dimensional analogy.[3] The one illustrated in Fig. 2.1 can serve as an example. We choose some arbitrary point in one unit and consider the arrangement of all equivalent points. This arrangement constitutes a net that has the same periodicity as the pattern. Alternatively, we can think of the net (two-dimensional lattice) as the set of translation operations that convert one unit of pattern into another. The distinction between thinking of the lattice as a set of points or as a set of translation operations may seem fine. However, a point is usually defined in geometry as having position, and since the choice of the representative point in the unit of pattern is arbitrary, the position occupied by the lattice has no special significance. It is only the relative arrangement of lattice points that matters. The lattice is best pictured as a mental device for selecting equivalent regions in a periodic pattern; when it is superimposed on the pattern, every lattice point has exactly the same environment. For the infinitely extending pattern, the set of translation operations forms a group known as the translation group or lattice.

In three dimensions a group of translations is described by

$$\mathbf{T} = u\mathbf{a}_1 + v\mathbf{a}_2 + w\mathbf{a}_3$$

where $\mathbf{a}_1$, $\mathbf{a}_2$, $\mathbf{a}_3$ are three noncoplanar translations (basis vectors), and

[2]Many references to the early literature on crystals are given in J. R. Partington, *An Advanced Treatise on Physical Chemistry,* Vol. 3, *The Properties of Solids,* Longmans, London, 1952.

[3]See C. H. Macgillavry, *Symmetry Aspects of M. C. Escher's Periodic Drawings,* Oosthoek, Utrecht, 1965, for many examples of two-dimensional repeating patterns of aesthetic and intellectual interest.

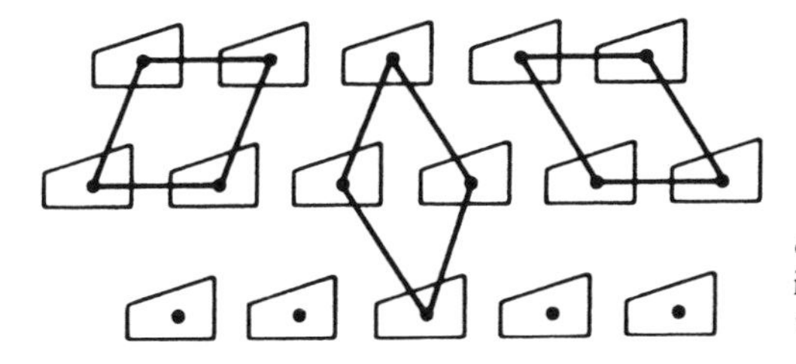

Figure 2.1. Two-dimensional net of equivalent points describing the periodicity of the pattern. Some alternative ways of choosing a unit cell are indicated.

u, v, w take on all positive and negative integral values. If the parallelepiped (unit cell) defined by the basis vectors contains one lattice point, then all the other lattice points are generated by the group of translations **T**. In this case the lattice (and the corresponding unit cell) is said to be primitive. Nonprimitive unit cells contain other lattice points in addition to those at the corners of the unit cell. For reasons of symmetry, basis vectors defining nonprimitive lattices sometimes give a more convenient coordinate system than those defining primitive lattices.

A lattice may have other kinds of symmetry besides translational symmetry. For example, every lattice is centrosymmetric since $\mathbf{T}(u, v, w)$ and $\mathbf{T}(-u, -v, -w)$ correspond to lattice translations. A lattice may also be left invariant by reflection or rotation through an angle of $2\pi/n$ about certain directions. However, only certain values of n are possible.

In Fig. 2.2, $\mathbf{a}$ is a unit translation of a two-dimensional net perpendicular to a rotation axis. Additional lattice points must occur at the termini of vectors $-\mathbf{a}$ (inversion), $\mathbf{a}'$ (rotation of $\mathbf{a}$ through $2\pi/n$), and $\mathbf{a}''$ (rotation of $-\mathbf{a}$ through $-2\pi/n$). The vector $\mathbf{a}'-\mathbf{a}''$ must also be a lattice translation, and since it is parallel to $\mathbf{a}$

$$\mathbf{a}' - \mathbf{a}'' = m\mathbf{a} \qquad (m = \text{integer})$$

or, dividing by the length of $\mathbf{a}$

$$2 \cos \alpha = m$$

The only solutions are:

m	$\cos \alpha$	α	n
-2	-1	180°	2
-1	$-1/2$	120	3
0	0	90	4
1	1/2	60	6
2	1	0	1

These are the only possible values of n for a crystal based on a three-

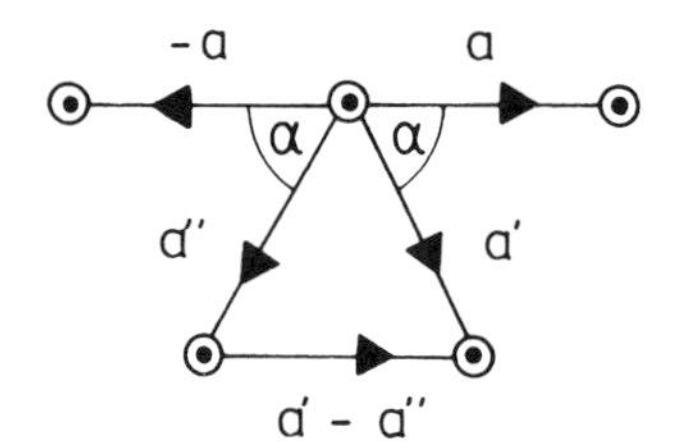

Figure 2.2. Part of a net of lattice points perpendicular to an n-fold rotation axis.

dimensional lattice. However, lattices with $n = 5, 8, 12$ are possible in four-dimensional space.[4]

In two dimensions, a symmetry classification gives four kinds of net—oblique, rectangular, square, and hexagonal. For the rectangular case, the conventional cell may be primitive or centered, giving five types of net (Fig. 2.3).

In three dimensions, seven kinds of lattice symmetry are possible. These provide the basis for classifying crystals in terms of seven "systems"—triclinic, monoclinic, orthorhombic, tetragonal, trigonal, hexagonal, and cubic or isometric—each with its own rules for the choice of a conventional coordinate system (although for reasons that will become apparent later, the modern practice is to combine the trigonal and hexagonal systems into a single trigonal-hexagonal system). When the different possibilities for nonprimitive unit cells are taken into account, we have a total of 14 kinds of lattice, usually known as the Bravais lattices. These are shown in Fig. 2.4 and listed in Table 2.1.

The last column of Table 2.1 describes the point-group symmetry associated with the points of the various lattice types. Two systems for describing point-group symmetries are in use—the Hermann-Mauguin or International system, standard in crystallography,[5] and the Schoenflies system, standard almost everywhere else and probably more familiar to chemists (e.g., C_n, C_{nh}, C_{nv}, D_n, D_{nd}, D_{nh}, S_n, T, O, and so forth).[6] In the Hermann-Mauguin system an n-fold rotation axis is represented simply by the number n, and n-fold rotatory inversion axis by $\bar{n}$, except that reflection planes are written as m (mirror plane) rather than $\bar{2}$. The symbol for a point group is obtained by combining the symbols for the individual classes of operations in a prescribed sequence, but it is not

[4]C. Hermann, *Acta Crystallogr.* **2,** 139 (1949).

[5]For a full description see *International Tables for X-ray Crystallography,* 3rd ed., Vol. I, Kynoch Press, Birmingham, 1969.

[6]For details see any book on group theory for chemists, e.g., F. A. Cotton, *Chemical Applications of Group Theory,* Wiley-Interscience, New York; 1st ed., 1963; 2nd ed. 1971. D. S. Schonland, *Molecular Symmetry,* Van Nostrand-Reinhold, London, 1965. D. M. Bishop, *Group Theory and Chemistry,* Oxford University Press, 1973.

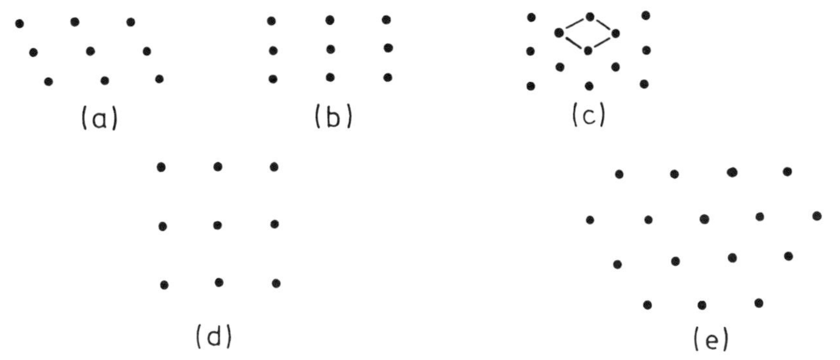

Figure 2.3. The five types of net: (a) oblique, (b) rectangular primitive, (c) rectangular centered (the primitive cell outlined is a rhombus), (d) square, (e) hexagonal.

necessary to specify every class, only the combination that uniquely characterizes the group. The combination of a rotation axis and a perpendicular mirror plane is indicated by a slanted line after the number n—e.g., $2/m$ (C_{2h}). Cubic groups, characterized by the presence of four 3-fold rotation axes, are symbolized by a 3 in the second place—e.g., $23(T)$ and $m3$ (T_d) are symbols for cubic groups, whereas 32 (D_3) and $3m(C_{3v})$ are symbols for noncubic ones.

The main difference between the two systems is in the description of alternating axes; they are regarded as rotatory-reflection axes (S_n) in the Schoenflies system but as rotatory-inversion axes ($\bar{n}$) in the Hermann-Mauguin system. Thus, $\bar{3} \equiv S_6$ and $\bar{6} \equiv C_{3h}$. (It is clearly arbitrary to assign the former point group to the trigonal system and the latter to the hexagonal system.) To describe point groups there is little to choose between the Schoenflies and Hermann-Mauguin symbols, but, as shown later, the latter are more suitable for describing space groups.

Some of the conventions used in the standard description of the Bravais lattices (Table 2.1) are not always followed. For example, although monoclinic lattices are usually defined in the "b-axis unique" coordinate system, the "c-axis unique" orientation is often used instead, in which case the conventional centered lattice becomes B, not C. In both orientations, moreover, the centered face could also be chosen as A. Similarly, the centered orthorhombic lattice is often taken as A or B instead of the conventional C.

The 14 Bravais lattices listed in Table 2.1 exhaust all the possibilities. Other imaginable possibilities can either be reduced to one of the 14 standard lattices by changing the coordinate system or else they are not lattices at all because the environments of all points are not equivalent. For example, an I-centered or F-centered monoclinic lattice is converted

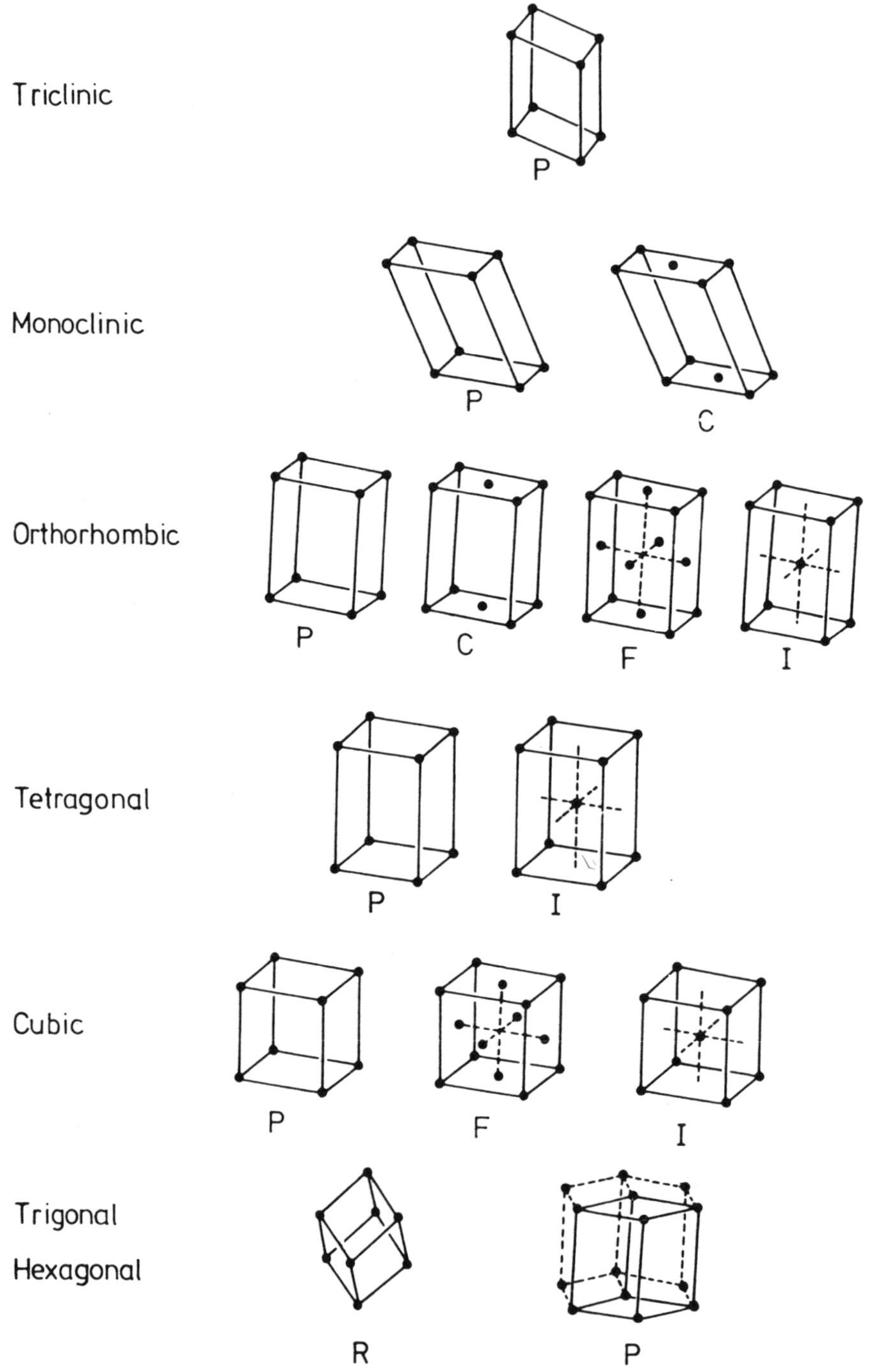

Figure 2.4. The 14 Bravais lattices. For the hexagonal lattice, two additional unit cells (dashed lines) are shown to bring out the hexagonal symmetry.

Table 2.1. The 14 Bravais lattices

System	Lattice symbol[a]	Nature of unit-cell axes and angles[b]	Lattice symmetry[c]
Triclinic	P	$a \neq b \neq c$ $\alpha \neq \beta \neq \gamma$	$\bar{1}(C_i)$
Monoclinic[d]	P C	$a \neq b \neq c$ $\alpha = \gamma = 90° \neq \beta$	$2/m(C_{2h})$
Orthorhombic	P C F I	$a \neq b \neq c$ $\alpha = \beta = \gamma = 90°$	$mmm(D_{2h})$
Tetragonal	P I	$a = b \neq c$ $\alpha = \beta = \gamma = 90°$	$4/mmm(D_{4h})$
Cubic	P F I	$a = b = c$ $\alpha = \beta = \gamma = 90°$	$m3m(O_h)$
Trigonal	R	$a = b = c$ $\alpha = \beta = \gamma \neq 90°$	$\bar{3}m(D_{3d})$
Hexagonal	P[e]	$a = b \neq c$ $\alpha = \beta = 90°$ $\gamma = 120°$	$6/mmm(D_{6h})$

[a]P (primitive, lattice points at 0, 0, 0). C (centered on one pair of opposite faces, lattice points at 0, 0, 0; $\frac{1}{2}$, $\frac{1}{2}$, 0). F (all faces centered, lattice points at 0, 0, 0; $\frac{1}{2}$, $\frac{1}{2}$, 0; $\frac{1}{2}$, 0, $\frac{1}{2}$; 0, $\frac{1}{2}$, $\frac{1}{2}$). I (body centered, lattice points at 0, 0, 0; $\frac{1}{2}$, $\frac{1}{2}$, $\frac{1}{2}$). R (primitive rhombohedral lattice).
[b]Symbols = and ≠ refer to symmetry equivalence. Accidental equalities of lengths and angles may occur.
[c]Hermann-Mauguin (and Schoenflies) symbols.
[d]With **b** as unique axis (usually but not always adopted).
[e]The primitive hexagonal *lattice* is common to both the trigonal and hexagonal *systems*. It seems preferable to combine these in a common trigonal-hexagonal system. (see text).

into the C-centered lattice by changing the coordinate system, while a "tetragonal lattice centered on the A and B faces" is not a lattice but a superposition of two interpenetrating primitive tetragonal lattices (Fig. 2.5).

Before 1912 the point-group symmetries of thousands of crystals had been determined.[7] This could be done by examining the symmetry of their external faces and physical properties, such as electrical and thermal conductivity as well as dielectric, optical, and elastic properties. A classification into crystal systems could therefore be made, but no method was available for establishing lattice type—primitive or non-

[7]See P. Groth, *Chemische Kristallographie,* Engelmann; Leipzig, five volumes, 1906-1919, for a monumental tabulation.

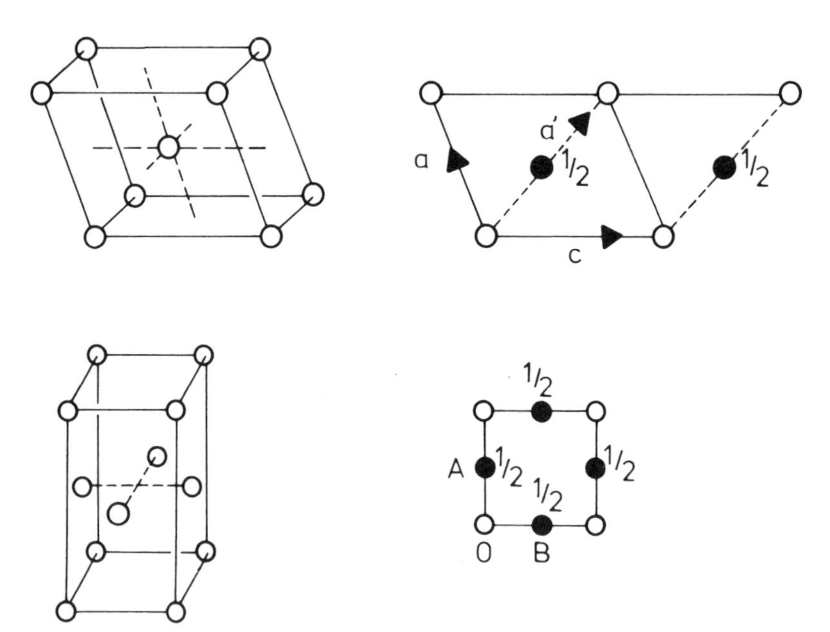

Figure 2.5. Top: the body-centered monoclinic cell (I) is converted into a side-centered cell (C) by a change of axes ($\mathbf{a}' = \mathbf{a} + \mathbf{c}$). Bottom: the tetragonal cell centered on the A and B faces does not correspond to a lattice since the translation that converts point O into point A does not convert point B into a lattice point.

primitive. Only with the discovery of X-ray diffraction could this problem be solved. Although certain crystals classified as belonging to the trigonal system are indeed based on a rhombohedral lattice, others are based on a primitive hexagonal lattice. Thus, if the distinction between the trigonal and hexagonal systems is preserved, the hexagonal lattice can serve as basis for both. This seems to be a source of serious confusion.[8]

The hexagonal and rhombohedral lattices contain two-dimensional hexagonal nets; the difference is in their stacking. In the hexagonal lattice, the nets lie directly over one another, and the unit cell is conventionally defined by the two shortest translation vectors of the net plus the translation perpendicular to the net. In the rhombohedral lattice, the nets are displaced so that points of one net lie above and below the centers of triangles in the two neighboring nets (Fig. 2.6). As a result, the lattice symmetry is reduced from $6/mmm$ (D_{6h}) to $\bar{3}m$ (D_{3d}). The con-

[8]Indeed, some authors like to make a distinction between the primitive hexagonal lattice and the "primitive trigonal lattice," thus obtaining 15 Bravais lattices, "two of which are identical." See, for example, A. Nussbaum, *Applied Group Theory for Chemists, Physicists, and Engineers,* Prentice-Hall, Englewood Cliffs, N.J., 1971.

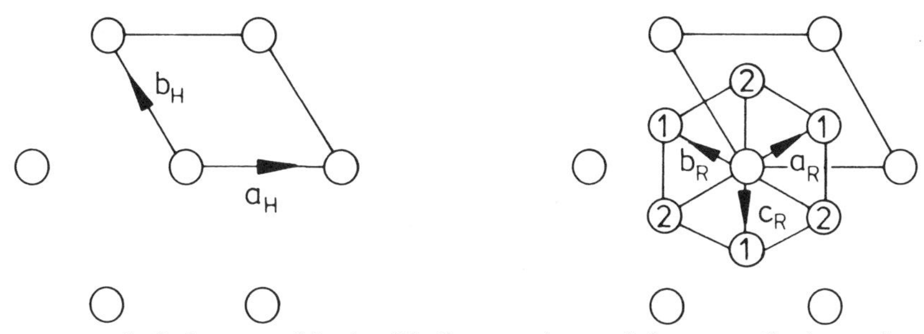

Figure 2.6. Left: hexagonal lattice (*P*). Because the **c**-axis is perpendicular to the plane of the paper, all hexagonal layers are exactly superimposed. Right: rhombohedral lattice; the hexagonal layers are displaced so that only every third layer is in register with the original one. The hexagonal cell marked extends three layers in the **c** direction and contains three lattice points.

ventional unit cell is defined by the three vectors from a point in one hexagonal net to the three nearest points in an adjacent parallel net (these vectors are equal and make equal angles with one another). The rhombohedral cell defined in this way is primitive. Alternatively, the same *R* lattice can be referred to a hexagonal coordinate system, leading to a nonprimitive cell. The transformation is:

$$\begin{aligned}
\mathbf{a}_H &= \mathbf{a}_R - \mathbf{b}_R \\
\mathbf{b}_H &= \qquad\ \mathbf{b}_R - \mathbf{c}_R \\
\mathbf{c}_H &= \mathbf{a}_R + \mathbf{b}_R + \mathbf{c}_R
\end{aligned}$$

and the volume of the hexagonal cell is

$$V_H = [\mathbf{a}_H\mathbf{b}_H\mathbf{c}_H] = \begin{vmatrix} 1 & -1 & 0 \\ 0 & 1 & -1 \\ 1 & 1 & 1 \end{vmatrix} [\mathbf{a}_R\mathbf{b}_R\mathbf{c}_R] = 3V_R$$

so it contains three lattice points (at 0, 0, 0; $\frac{2}{3}, \frac{1}{3}, \frac{1}{3}$; $\frac{1}{3}, \frac{2}{3}, \frac{2}{3}$, referred to the above hexagonal axes).

The confusion arises because in the early classifications it was sometimes assumed that all crystals with a 3 or $\bar{3}$ axis could be based on a rhombohedral lattice. However, with the development of space-group theory, it was realized that these symmetries are just as compatible with a primitive hexagonal lattice as with a primitive rhombohedral lattice, which, as we have seen, corresponds to a special kind of nonprimitive hexagonal lattice. As far as X-ray diffraction is concerned, the two lattice types are easily distinguished, and it seems best to refer both to a hexagonal coordinate system, the unit cell being primitive in one case, nonprimitive in the other, with the possibility of using rhombohedral axes when this seems desirable.

Space Groups

A lattice is a group of translation operations that converts a given unit of a periodic pattern into all translationally equivalent units. However, the unit of pattern itself may have certain elements of symmetry. A space group is the group of all operations that convert an asymmetric unit of a periodic pattern into all equivalent units. Whereas point groups apply to finite figures (where at least one *point* remains invariant to the symmetry operations), space groups apply to infinite (or quasi-infinite) periodic patterns. Thus, the symmetry of a macroscopic crystal is described by a point group, the symmetry of its internal structure by a space group. The symmetry elements of space groups include translation operations, elements of point-group symmetry, and certain combinations of these that occur only in periodic patterns, such as glide planes and screw axes.

As mentioned, the symmetry elements that occur in periodic patterns are limited, the only permissible rotation axes being of the order 2, 3, 4, or 6. Combination of these with elements of inversion and reflection in all possible ways leads to the 32 crystallographic point groups or crystal classes listed in Table 2.2, where both Hermann-Mauguin and Schoenflies symbols are given. Table 2.2 also shows the assignment of each point group to one of the six crystal systems.

Each point group can be associated with a set of equivalent positions, the positions into which an arbitrary point at x, y, z is transformed by the symmetry operations of the group. The number of equivalent positions clearly equals the order of the group.

Space groups are obtained by combining elements of point-group

Table 2.2. The 32 crystallographic point groups and their assignment to the six crystal systems

System	Essential symmetry			
Triclinic	none	$1(C_1)$	$\bar{1}(C_i)$	
Monoclinic	2 or m	$2(C_2)$	$m(C_s)$	$2/m(C_{2h})$
Orthorhombic	222 or $mm2$	$222(D_2)$	$mm2(C_{2v})$	$mmm(D_{2h})$
Trigonal-hexagonal				
P or R	3 or $\bar{3}$	$3(C_3)$ $32(D_3)$ $\bar{3}(S_6)$ $3m(C_{3v})$ $\bar{3}m(D_{3d})$		
P	6 or $\bar{6}$	$6(C_6)$ $622(D_6)$ $\bar{6}(C_{3h})$ $6/m(C_{6h})$ $6mm(C_{6v})$ $\bar{6}2m(D_{3h})$ $6/mmm(D_{6h})$		
Tetragonal	4 or $\bar{4}$	$4(C_4)$ $422(D_4)$ $\bar{4}(S_4)$ $4/m(C_{4h})$ $4mm(C_{4v})$ $\bar{4}2m(D_{2d})$ $4/mmm(D_{4h})$		
Cubic	23	$23(T)$ $432(O)$ $\bar{4}3m(T_d)$ $m3(T_h)$ $m3m(O_h)$		

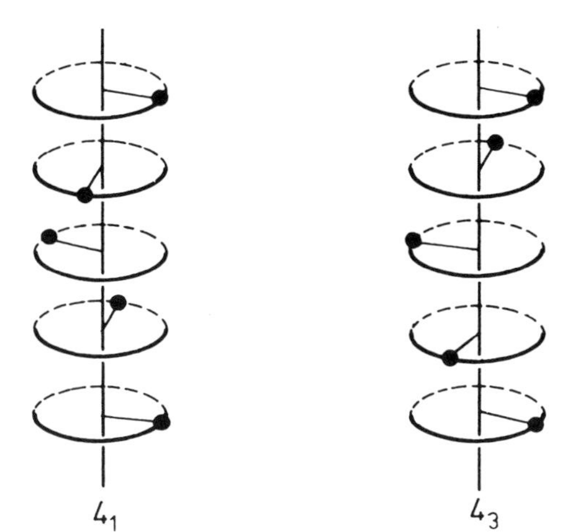

Figure 2.7. Operation of 4_1 and 4_3 screw axes.

symmetry with those of translational symmetry. The simplest combinations are those in which a given point-group symmetry is associated with the points of a lattice, e.g., $P\bar{1}$, $P2/m$, $C2/m$, $Pmmm$, and so forth. The equivalent positions of the space group are then simply the equivalent positions of the corresponding point group with the origin taken at each lattice point in turn. In periodic patterns, however, additional symmetry elements, particular combinations of rotations or reflections with translational operations, known as screw axes and glide planes, are possible.

A screw axis involves rotation about an axis through an angle, $2\pi/n$, followed by a translation parallel to the axis. If the translation in question equals m/n of the identity period along the axis, where m is an integer less than n, then repetition of this operation n times is equivalent to a translation along m periods. In the Hermann-Mauguin notation, a screw axis is symbolized by n_m, e.g., 4_1 corresponds to rotation through $2\pi/4 = 90°$, followed by a translation of one-fourth of the identity period along the axis. The operations 4_1 and 4_3 are enantiomorphous, the former corresponding to a right-handed screw, the latter to a left-handed one. The set of equivalent positions for the space group $P4_1$ is then: $x, y, z; \bar{y}, x, \frac{1}{4} + z; \bar{x}, \bar{y}, \frac{1}{2} + z; y, \bar{x}, \frac{3}{4} + z$ (Fig. 2.7).[9]

A glide plane involves reflection across a plane followed by translation parallel to the plane by a distance equal to half of a lattice translation. Repetition of the operation is then equivalent to a translation. In the Hermann-Mauguin notation, a glide plane is symbolized by a, b, or c (if the translation element is parallel to a crystal axis) or n (if it is parallel

[9]In crystallographic texts a negative value of a variable is often indicated by a bar placed over the symbol, e.g., $\bar{x} = -x$.

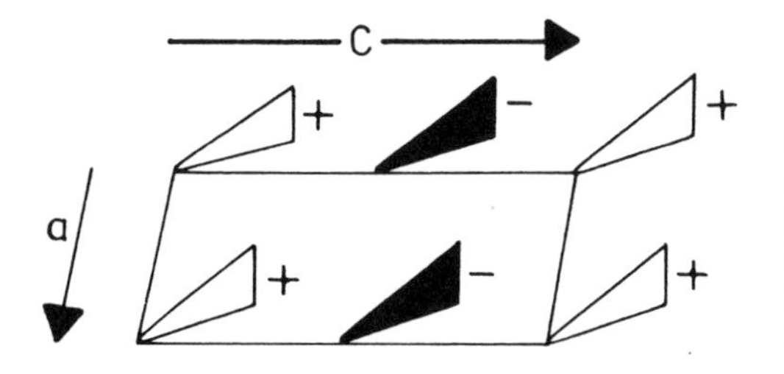

Figure 2.8. Glide plane c; the open triangles (at $+y$) are related to the filled ones (at $-y$) by reflection in the plane of the paper (normal to the **b**-axis) followed by translation by **c**/2. This operation forms a group with the lattice translations.

to a face diagonal) or d (if the lattice translation involved corresponds to a centering of the unit cell). The set of equivalent positions for the space group Pc (reflection across the plane perpendicular to **b**) is then: $x, y, z; x, \bar{y}, \frac{1}{2} + z$ (Fig. 2.8).

Crystals containing screw axes or glide planes have the same macroscopic symmetry as those having rotation axes and reflection planes; they belong to the same crystal class. However, the presence of these additional symmetry elements is easily recognized by the characteristic absence of certain classes of reflection in the X-ray diffraction diagram.

The standard compilation of the 17 plane groups and the 230 three-dimensional space groups is contained in Volume I of *International Tables for X-ray Crystallography.*[10] The tables include standard space-group symbols, diagrams of equivalent positions and of symmetry elements, lists of equivalent and special positions, and additional information. The standard symbols used for symmetry elements in the diagrams are shown in Table 2.3. Short listings are given in Tables 2.4 and 2.5.

Unless otherwise indicated, the diagrams are drawn as projections *down* the z-axis, which is supposed to point up out of the plane of the paper. The x- and y-axes are supposed to be in the plane of the paper with the origin at the top left corner of the diagram, x pointing downward, y to the right. The coordinate system is thus conventionally taken as right handed.[11]

To focus attention on some points that arise, we now discuss a few space groups in some detail. We select three that occur frequently among organic compounds and note that the distribution of crystalline

[10] *International Tables for X-ray Crystallography,* Kynoch Press, Birmingham. This publication supersedes the earlier *Internationale Tabellen zur Bestimmung von Kristallstrukturen,* Berlin, 1935.

[11] This convention must be regarded as binding when the absolute structure of a chiral crystal is to be described. The planes **H** and −**H** are then unambiguously distinguishable. The question left to be answered in defining the enantiomorph is whether the atomic coordinates are given by $\mathbf{r}_i$ or by $-\mathbf{r}_i$ (see chapter 3 for a discussion of how this problem may be solved).

Table 2.3. Standard symbols for symmetry elements of space groups

Symmetry element	Symbol	Graphical symbol: Normal to projection plane	Graphical symbol: Parallel to projection plane
Reflection (mirror) plane	m		
Axial glide plane	a, b, c		
Diagonal glide plane	n		
"Diamond" glide plane	d		
Center of inversion	$\bar{1}$		
Rotation axis	2, 3, 4, 6		
Rotatory-inversion axis	$\bar{3}, \bar{4}, \bar{6}$		
Screw axis	2_1 $3_1, 3_2$ $4_1, 4_2, 4_3$ $6_1, 6_2, 6_3, 6_4, 6_5$		

Table 2.4. The 17 plane groups (two-dimensional space groups)

Oblique	*p*1, *p*2
Rectangular	*pm, pg, cm, pmm, pmg, pgg, cmm*
Square	*p*4, *p*4*m*, *p*4*g*
Hexagonal	*p*3, *p*3*m*1, *p*31*m*, *p*6, *p*6*m*

substances among the 230 space groups is by no means uniform. A survey[12] has shown that as far as organic compounds are concerned, only about six space groups account for more than 60% of the crystals whose space groups could be regarded as reasonably well established (about 3200 in 1963, according to the second edition of *Crystal Data, Determinative Tables*).[13] Most space groups occur rarely, and about 20 have never been observed. According to Mackay,[14] the number of possi-

[12]W. Nowacki, T. Matsumoto, and A. Edenharter, *Acta Crystallogr.* **22,** 935 (1967).

[13]J. D. H. Donnay, G. Donnay, E. G. Cox, O. Kennard, and M. V. King. *Crystal Data, Determinative Tables,* 2nd ed., American Crystallographic Association Monograph No. 5, 1963.

[14]A. L. Mackay, *Acta Crystallogr.* **22,** 329 (1967).

Table 2.5. The 230 three-dimensional space groups arranged by crystal classes and systems[a]

No.	Triclinic	
1	$1(C_1)$	$P1$
2	$\bar{1}(C_i)$	$P\bar{1}$
No.	**Monoclinic**	
3–5	$2(C_2)$	$P2$, $P2_1$, $C2$
6–9	$m(C_s)$	Pm, Pc, Cm, Cc
10–15	$2/m(C_{2h})$	$P2/m$, $P2_1/m$, $C2/m$, $P2/c$, $\mathbf{P2_1/c}$, $C2/c$
No.	**Orthorhombic**	
16–24	$222(D_2)$	$P222$, $\mathbf{P222_1}$, $\mathbf{P2_12_12}$, $\mathbf{P2_12_12_1}$, $\mathbf{C222_1}$, $C222$, $F222$, $I222$, $I2_12_12_1$
25–46	$mm2(C_{2v})$	$Pmm2$, $Pmc2_1$, $Pcc2$, $Pma2_1$, $Pca2_1$, $Pnc2_1$, $Pmn2_1$, $Pba2$, $Pna2_1$, $Pnn2$, $Cmm2$, $Cmc2_1$, $Ccc2$, $Amm2$, $Abm2$, $Ama2$, $Aba2$, $Fmm2$, $\mathbf{Fdd2}$, $Imm2$, $Iba2$, $Ima2$
47–74	$mmm(D_{2h})$	$Pmmm$, $\mathbf{Pnnn}$, $Pccm$, $\mathbf{Pban}$, $Pmma$, $\mathbf{Pnna}$, $Pmna$, $\mathbf{Pcca}$, $Pbam$, $\mathbf{Pccn}$, $Pbcm$, $Pnnm$, $Pmmn$, $\mathbf{Pbcn}$, $\mathbf{Pbca}$, $Pnma$, $Cmcm$, $Cmca$, $Cmmm$, $Cccm$, $Cmma$, $\mathbf{Ccca}$, $Fmmm$, $\mathbf{Fddd}$, $Immm$, $Ibam$, $\mathbf{Ibca}$, $Imma$
No.	**Tetragonal**	
75–80	$4(C_4)$	$P4$, $\mathbf{P4_1}$, $P4_2$, $\mathbf{P4_3}$, $I4$, $\mathbf{I4_1}$
81–82	$\bar{4}(S_4)$	$P\bar{4}$, $I\bar{4}$
83–88	$4/m(C_{4h})$	$P4/m$, $P4_2/m$, $\mathbf{P4/n}$, $\mathbf{P4_2/n}$, $I4/m$, $\mathbf{I4_1/a}$
89–98	$422(D_4)$	$P422$, $\mathbf{P42_12}$, $\mathbf{P4_122}$, $\mathbf{P4_12_12}$, $\mathbf{P4_222}$, $\mathbf{P4_22_12}$, $\mathbf{P4_322}$, $\mathbf{P4_32_12}$, $I422$, $\mathbf{I4_122}$
99–110	$4mm(C_{4v})$	$P4mm$, $P4bm$, $P4_2cm$, $P4_2nm$, $P4cc$, $P4nc$, $P4_2mc$, $P4_2bc$, $I4mm$, $I4cm$, $I4_1md$, $\mathbf{I4_1cd}$
111–122	$\bar{4}m(D_{2d})$	$P\bar{4}2m$, $P\bar{4}2c$, $P\bar{4}2_1m$, $\mathbf{P\bar{4}2_1c}$, $P\bar{4}m2$, $P\bar{4}c2$, $P\bar{4}b2$, $P\bar{4}n2$, $I\bar{4}m2$, $I\bar{4}c2$, $I\bar{4}2m$, $I\bar{4}2d$
123–142	$4/mmm(D_{4h})$	$P4/mmm$, $P4/mcc$, $\mathbf{P4/nbm}$, $\mathbf{P4/nnc}$, $P4/mbm$, $P4/mnc$, $\mathbf{P4/nmm}$, $\mathbf{P4/ncc}$, $P4_2/mmc$, $P4_2/mcm$, $\mathbf{P4_2/nbc}$, $\mathbf{P4_2/nnm}$, $P4_2/mbc$, $P4_2/mnm$, $\mathbf{P4_2/nmc}$, $\mathbf{P4_2/ncm}$, $I4/mmm$, $I4/mcm$, $\mathbf{I4_1/amd}$, $\mathbf{I4_1/acd}$,
No.	**Trigonal-hexagonal**	
143–146	$3(C_3)$	$P3$, $P3_1$, $P3_2$, $R3$
147–148	$\bar{3}(C_{3i})$	$P\bar{3}$, $R\bar{3}$
149–155	$32(D_3)$	$P312$, $P321$, $\mathbf{P3_112}$, $\mathbf{P3_121}$, $\mathbf{P3_212}$, $\mathbf{P3_221}$, $R32$
156–161	$3m(C_{3v})$	$P3m1$, $P31m$, $P3c1$, $P31c$, $R3m$, $R3c$
162–167	$\bar{3}m(D_{3d})$	$P\bar{3}1m$, $P\bar{3}1c$, $P\bar{3}m1$, $P\bar{3}c1$, $R\bar{3}m$, $R\bar{3}c$
168–173	$6(C_6)$	$P6$, $\mathbf{P6_1}$, $\mathbf{P6_5}$, $\mathbf{P6_2}$, $\mathbf{P6_4}$, $P6_3$
174	$\bar{6}(C_{3h})$	$P\bar{6}$
175–176	$6/m(C_{6h})$	$P6/m$, $P6_3/m$
177–182	$62(D_6)$	$P622$, $\mathbf{P6_122}$, $\mathbf{P6_522}$, $\mathbf{P6_222}$, $\mathbf{P6_422}$, $\mathbf{P6_322}$
183–186	$6m(C_{6v})$	$P6mm$, $P6cc$, $P6_3cm$, $P6_3mc$
187–190	$\bar{6}m(D_{3h})$	$P\bar{6}m2$, $P\bar{6}c2$, $P\bar{6}2m$, $P\bar{6}2c$
191–194	$6/mmm(D_{6h})$	$P6/mmm$, $P6/mcc$, $P6_3/mcm$, $P6_3/mmc$

Table 2.5. (cont.)

No.	Cubic	
195–199	23(*T*)	*P*23, *F*23, *I*23, $\mathbf{P2_13}$, $I2_13$
200–206	$m3(T_h)$	*Pm*3, **Pn3**, *Fm*3, **Fd3**, *Im*3, **Pa3**, **Ia3**
207–214	43(*O*)	*P*432, $\mathbf{P4_232}$, *F*432, $\mathbf{F4_132}$, *I*432, $\mathbf{P4_332}$, $\mathbf{P4_132}$, $I4_132$
215–220	$\bar{4}3m(T_d)$	$P\bar{4}3m$, $F\bar{4}3m$, $I\bar{4}3m$, $P\bar{4}3n$, $F\bar{4}3c$, $\mathbf{I\bar{4}3d}$
221–230	$m3m(O_h)$	*Pm3m*, **Pn3n**, *Pm3n*, **Pn3m**, *Fm3m*, *Fm3c*, **Fd3m**, **Fd3c**, *Im3m*, **Ia3d**

[a]Space groups (or enantiomorphous pairs) that are uniquely determinable from the symmetry of the diffraction pattern and systematic absences are shown in boldface.

ble space groups can be estimated as 220 from the statistics of the observed distribution; the actual number for this comparison is 219.[15] Mackay concludes that since the number of unobserved space groups agrees with statistical estimates, their absence can be attributed to chance and not to any intrinsic impossibility of occurrence for physical reasons. Impossible—no. Improbable—yes. As shown later, the few space groups that occur frequently in molecular crystals are just the ones for which close-packed arrangements of molecules are possible.

Space Group $P\bar{1}$ (C_i^1)

Fig. 2.9 shows the simple space group $P\bar{1}$ as given in the *International Tables for X-ray Crystallography.*[10] The origin is conventionally taken at a center of inversion so that symmetry-equivalent positions have coordinates x, y, z; $\bar{x}, \bar{y}, \bar{z}$. An atom at x, y, z is represented by an open circle associated with a plus sign (indicating that the atom is above the xy plane, the plane of the paper); the symmetry-equivalent atom at $\bar{x}, \bar{y}, \bar{z}$ is represented by a circle enclosing a comma and associated with a minus sign. The comma means that the symmetry operation involved—in this case, inversion through a point—is one that converts a chiral object into its mirror image—e.g., a right-handed screw into a left-handed one. Thus,

[15]The 230 space groups include 11 enantiomorphous pairs—$P3_1(P3_2)$, $P3_112(P3_212)$, $P3_121(P3_221)$, $P4_1(P4_3)$, $P4_122(P4_322)$, $P4_12_12(P4_32_12)$, $P6_1(P6_5)$, $P6_2(P6_4)$, $P6_122(P6_522)$, $P6_222(P6_422)$, $P4_132(P4_332)$. If the (+) isomer of an optically active molecule crystallizes in one of these pairs, the (−) isomer will crystallize in the other. Although these pairs can be distinguished by X-ray diffraction (using anomalous dispersion), they can be counted together for classification purposes if the absolute chirality sense is unknown or irrelevant, leading to 219 "distinct" space groups. However, once the chirality sense of the asymmetric unit is specified, repetition of that unit in one enantiomorphous space group is quite distinct from repetition in the other. It is therefore most improbable that a given enantiomer could ever crystallize in two enantiomorphous space groups because the resulting packing arrangements and hence the corresponding packing energies would be completely different.

optically active molecules cannot crystallize in the space group $P\bar{1}$, or in any other space group that contains rotatory-inversion axes, reflection planes, or glide planes.

Such operations—those that transform a chiral object into its mirror image—are referred to as improper operations (or operations of the second kind), in contrast to proper rotations and translations (operations of the first kind). These two types of operation are distinguished by the sign of the determinant of the corresponding transformation matrix; the sign is positive for a proper operation and negative for an improper one. For the inversion center, for example,

$$\begin{bmatrix} -x \\ -y \\ -z \end{bmatrix} = \begin{bmatrix} -1 & 0 & 0 \\ 0 & -1 & 0 \\ 0 & 0 & -1 \end{bmatrix} \begin{bmatrix} x \\ y \\ z \end{bmatrix}$$

so the value of the determinant equals -1. On the other hand, for a proper rotation through an angle α about the z-axis

$$\begin{bmatrix} x' \\ y' \\ z' \end{bmatrix} = \begin{bmatrix} \cos\alpha & \sin\alpha & 0 \\ -\sin\alpha & \cos\alpha & 0 \\ 0 & 0 & 1 \end{bmatrix} \begin{bmatrix} x \\ y \\ z \end{bmatrix}$$

so the value of the determinant equals $+1$.

Fig. 2.9 also shows the arrangement of symmetry elements in the space group $P\bar{1}$. The presence of inversion centers at the lattice points implies that additional inversion centers must be present at the midpoint of every lattice translation—i.e., at the center of the cell and of every face and edge, as well as at the corners. Of course, these additional inversion centers do not have the same environment as those at the lattice points.

The eight nonequivalent sets of inversion centers are called special positions of the space group, in contrast to the two general positions, x, y, z and $\bar{x}, \bar{y}, \bar{z}$. If the unit cell were known to contain one centrosymmetric molecule, the molecular center would have to coincide with one of the special positions, the other atoms occupying related pairs of general positions. In such a case, where a molecular symmetry element coincides with a symmetry element of the space group, we would say that the molecule had crystallographic symmetry—here a crystallographic center of symmetry. If the cell were known to contain two centrosymmetric molecules, the molecular centers could be placed either at a related pair of general positions or at two of the nonequivalent inversion centers. In the former case, the environment of the molecules—their site symmetry—would not be centrosymmetric, and the molecules would not

Triclinic $\bar{1}$ $\quad$ $P\bar{1}$ $\quad$ No. 2 $\quad$ $P\bar{1}$ $\quad$ C_i^1

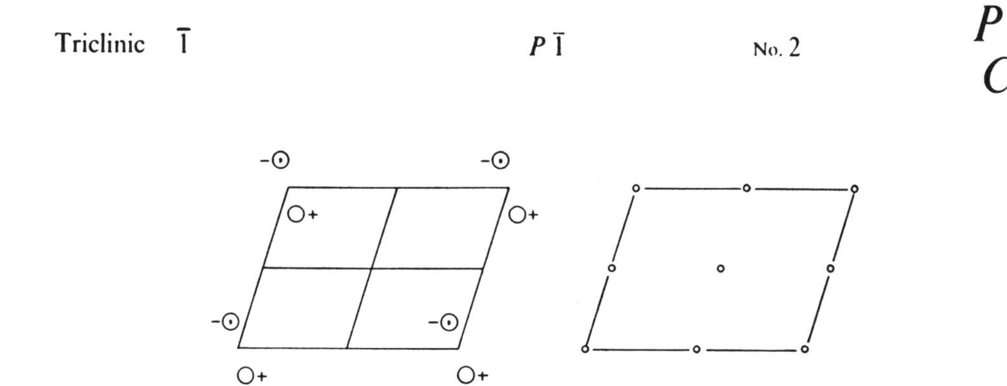

Origin at $\bar{1}$

Number of positions, Wyckoff notation, and point symmetry			Co-ordinates of equivalent positions	Conditions limiting possible reflections
				General:
2	*i*	1	$x,y,z;\ \bar{x},\bar{y},\bar{z}.$	No conditions
				Special:
1	*h*	$\bar{1}$	$\frac{1}{2},\frac{1}{2},\frac{1}{2}.$	No conditions
1	*g*	$\bar{1}$	$0,\frac{1}{2},\frac{1}{2}.$	
1	*f*	$\bar{1}$	$\frac{1}{2},0,\frac{1}{2}.$	
1	*e*	$\bar{1}$	$\frac{1}{2},\frac{1}{2},0.$	
1	*d*	$\bar{1}$	$\frac{1}{2},0,0.$	
1	*c*	$\bar{1}$	$0,\frac{1}{2},0.$	
1	*b*	$\bar{1}$	$0,0,\frac{1}{2}.$	
1	*a*	$\bar{1}$	$0,0,0.$	

Symmetry of special projections

All *p*2

Figure 2.9. Space group $P\bar{1}$. From *International Tables for X-ray Crystallography,* Vol. I, Kynoch Press, Birmingham, 1969.

be expected to retain an exact center of symmetry; small deviations between chemically equivalent pairs of bond lengths or bond angles could occur. However, the two molecules would be symmetry-equivalent. In the latter case, the two molecules would not be equivalent by symmetry, but each would retain an exact center of symmetry.

The geometric structure factor, the reduced structure factor expression (Eq. 1.35) for a set of unit point atoms placed at the general equivalent positions of a space group, is also given in the *International Tables for X-ray Crystallography*. For $P\bar{1}$ this is simply

$$A = 2\cos 2\pi(hx + ky + lz) \qquad B = 0 \tag{2.1}$$

obtained by summing real and imaginary parts over equivalent scattering points at x, y, z and $\bar{x}, \bar{y}, \bar{z}$. Note that Eqs. (2.1) hold only when the standard origin, at a center of inversion, is adopted.

Space Group $P2_1/c(C_{2h}^5)$

This is by far the most common space group for optically inactive organic compounds. In *Structure Reports* for 1970 the crystal structures of about 700 organic compounds are described;[16] about 250 of these fall into the space group $C_{2h}^5(P2_1/c$ and alternative orientations).[17] The only other space groups that score 20 or more in this count are: $C_i^1(P\bar{1})$, 88; $D_2^4(P2_12_12_1)$, 88; $C_2^2(P2_1)$, 49; $C_{2h}^6(C2/c$, etc.), 43; $D_{2h}^{15}(Pbca$, etc.), 33.

As an exercise it is worthwhile to derive the equivalent positions from first principles. We first choose an origin at the intersection of the 2_1 axis (parallel to **b**) with the glide plane (perpendicular to **b**). The operation of these symmetry elements gives equivalent positions (Fig. 2.10, left) at, say:

A	x, y, z	identity
B	$-x, \frac{1}{2}+y, -z$	screw axis
A'	$x, -y, \frac{1}{2}+z$	glide plane
B'	$-x, \frac{1}{2}-y, \frac{1}{2}-z$	screw axis plus glide plane

[16]*Structure Reports,* a continuation of the former *Strukturbericht,* is an annual series produced for the International Union of Crystallography by a team of editors and published by Oosthoek, Utrecht. It was originally intended to provide critical abstracts of all published crystal structures but now records cell dimensions, space groups, and atomic positions with brief comments on the structures and how they were determined.

[17]The Schoenflies symbol is useful for describing space groups without reference to any coordinate system. The Hermann-Mauguin symbols $P2_1/c$, $P2_1/b$, $P2_1/a$, and $P2_1/n$ all refer to the same space group (C_{2h}^5) but differ in the orientation of the coordinate system with respect to the symmetry elements of the group. In describing actual crystal structures the coordinate system must, of course, be specified, which accounts for the popularity of the Hermann-Mauguin system among X-ray crystallographers. The other disadvantage of the Schoenflies nomenclature is the difficulty of remembering which superscript refers to which particular space group in the crystal class defined by the remainder of the symbol. For example, how should one know that D_{2h}^{15} means *Pbca*? Only from a tabulation (Table 2.5).

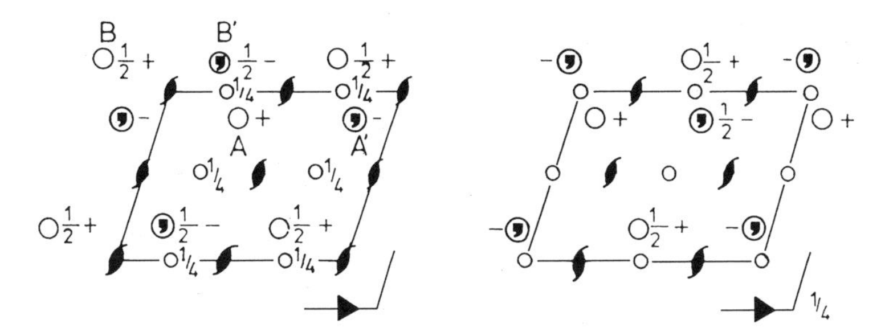

Figure 2.10. Left: space group $P2_1/c$ with origin at intersection of glide plane with screw axis. The operation 2_1 converts A into B, the glide plane (at $y = 0$) converts A into A', B into B'. The inversion center relating A to B' and A' to B is at 0, $\frac{1}{4}$, $\frac{1}{4}$. Right: same space group referred to standard origin at inversion center. The glide plane is now at $y = \frac{1}{4}$. The projections are drawn with the **a**- and **c**-axes in the plane of the paper, **c** being horizontal. In this projection, pairs of equivalent points related in space by the glide plane appear to be related by a translation of **c**/2.

The pairs A, B' and A', B are related by a center of inversion at 0, $\frac{1}{4}$, $\frac{1}{4}$. This symmetry element is not specified in the space group symbol, but it is a necessary consequence of combining the two other symmetry elements. Transforming to a new origin at this inversion center, we obtain

A	$x, y-\frac{1}{4}, z-\frac{1}{4}$	B'	$-x, \frac{1}{4}-y, \frac{1}{4}-z$
A'	$x, -y-\frac{1}{4}, \frac{1}{4}+z$	B	$-x, \frac{1}{4}+y, -z-\frac{1}{4}$

or writing $y-\frac{1}{4} = y'$, $z-\frac{1}{4} = z'$

A	x, y', z'	B'	$-x, -y', -z'$
A'	$x, -\frac{1}{2}-y', \frac{1}{2}+z'$	B	$-x, \frac{1}{2}+y', -\frac{1}{2}-z'$.

The addition of a unit translation [0, 1, 0] or [0, 0, 1] to the coordinates of A' and B merely displaces these points to equivalent points in adjacent unit cells. When the primes are dropped, we obtain the standard set of equivalent positions referred to the conventional origin at the inversion center: $x, y, z; \bar{x}, \bar{y}, \bar{z}; x, \frac{1}{2}-y, \frac{1}{2}+z; \bar{x}, \frac{1}{2}+y, \frac{1}{2}-z$, as shown in Fig. 2.11 where the **a**- and **b**-axes are now supposed to lie in the plane of the diagram.

The screw axes and glide planes cannot give rise to special positions since the corresponding symmetry operations are not point-group operations. Because they have translational components, they do not leave any point in space unaltered. Thus, the only special positions are the inversion centers (as was the case for $P\bar{1}$), but they are now equivalent in pairs (Fig. 2.11, positions a, b, c, d). If only two molecules are present in the unit cell, they must be centrosymmetric, with their centers at one of these pairs of special positions. Four molecules can occupy the general

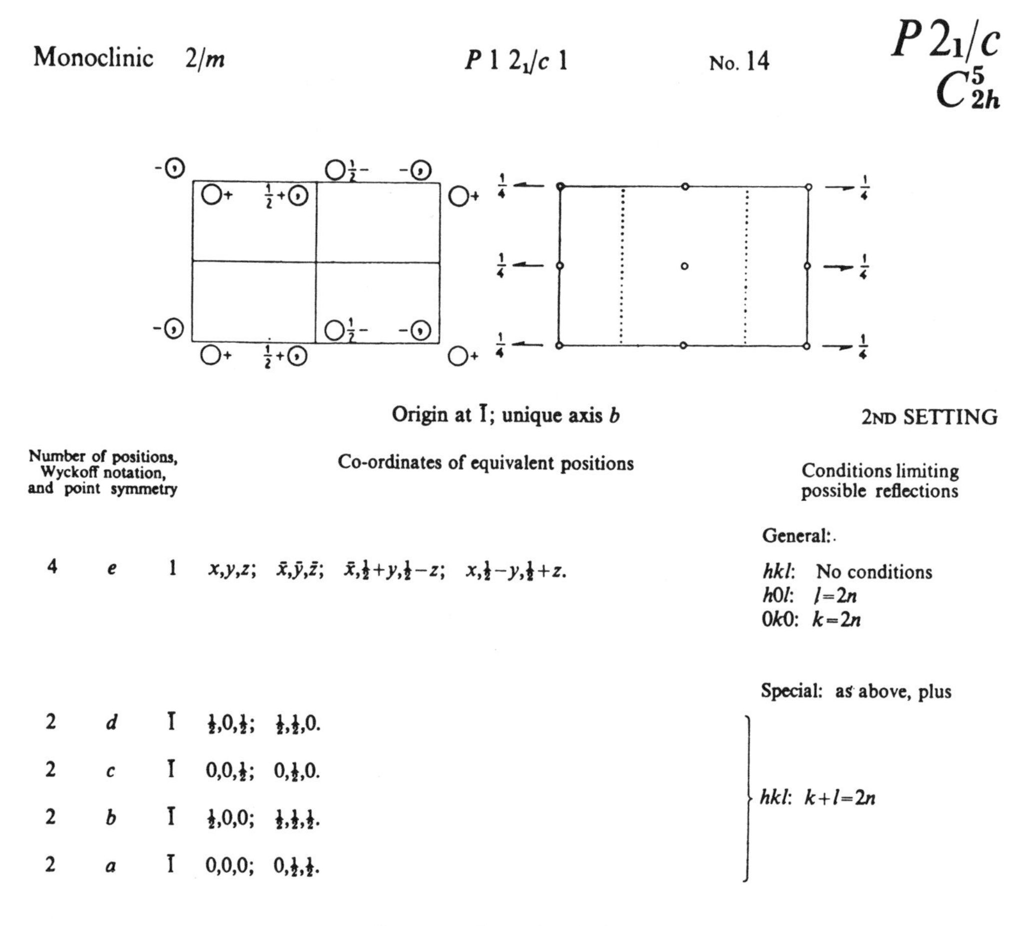

Figure 2.11. Space group $P2_1/c$. From *International Tables for X-ray Crystallography,* Vol. I, Kynoch Press, Birmingham, 1969.

positions. A few crystals are known with six molecules in a $P2_1/c$ cell—four in general positions, two at special positions; diphenylene is an example.[18]

The geometric structure factor for $P2_1/c$ is

$$A = 2[\cos 2\pi(hx + ky + lz) + \cos 2\pi(hx + k(\tfrac{1}{2}-y) + l(\tfrac{1}{2}+z))]$$
$$B = 0$$

[18] J. Waser and C. S. Lu, *J. Am. Chem. Soc.* **66,** 2035 (1944); T. C. W. Mak and J. Trotter, *J. Chem. Soc.* **1962,** 1; J. K. Fawcett and J. Trotter, *Acta Crystallogr.* **20,** 87 (1966).

with the origin at one of the inversion centers. The two cosine terms can be combined to give

$$A = 4\cos 2\pi\left(hx+lz+\frac{k+l}{4}\right)\cos 2\pi\left(ky-\frac{k+l}{4}\right)$$

or

$$\begin{aligned} A &= 4\cos 2\pi(hx+lz)\cos 2\pi\, ky & k+l &= 2n \\ &= -4\sin 2\pi(hx+lz)\sin 2\pi\, ky & k+l &= 2n+1 \end{aligned} \tag{2.2}$$

The $(h0l)$ reflections with l odd and $(0k0)$ reflections with k odd belong to the second class with one of the sine terms equal to zero. For these reflections then, $A = B = 0$. Hence, this space group automatically extinguishes certain classes of reflections. The systematic absence of $(h0l)$ reflections with l odd is due to the presence of the glide plane perpendicular to **b** with translation component **c**/2. When the crystal structure is viewed in projection down the **b** axis, equivalent positions at x, y, z and $x, \frac{1}{2}-y, \frac{1}{2}+z$ degenerate into x, z and $x, \frac{1}{2}+z$ and thus appear to be related by a translation of half the real periodicity along **c** (Fig. 2.10). In a similar way, if the structure is projected on the **b**-axis, equivalent positions related by the screw axis operation degenerate into y and $\frac{1}{2}+y$, so that the systematic absence of $(0k0)$ reflections with k odd is an expression of the presence of the screw axis. In general, any symmetry element that has a translation component will lead to a systematic absence. Conversely, the occurrence of systematic absences in the diffraction pattern makes it possible to recognize any centerings, screw axes, or glide planes that may be present in the space group.

Space Group $P2_12_12_1(D_2^4)$

This is the most common space group for crystals composed of homochiral molecules. Let us derive the equivalent positions and the arrangement of symmetry elements of this orthorhombic space group from first principles.

Suppose that two of the screw axes intersect at a point and hence lie in a common plane. It is clear from Fig. 2.12, left, that the successive operation of these two symmetry elements on a representative point unavoidably produces a set of pure rotation axes normal to the plane in which the screw axes lie. The space group corresponding to this arrangement is then $P2_12_12$ not $P2_12_12_1$.

Since no two sets of screw axes may intersect, they must be interleaved. Choose the set parallel to **a** to lie at $z = 0, \frac{1}{2}$, the set parallel to **b** to lie at $z = \frac{1}{4}, \frac{3}{4}$. This arrangement has the desired property of producing a third set of nonintersecting screw axes running in the **c** direction, as

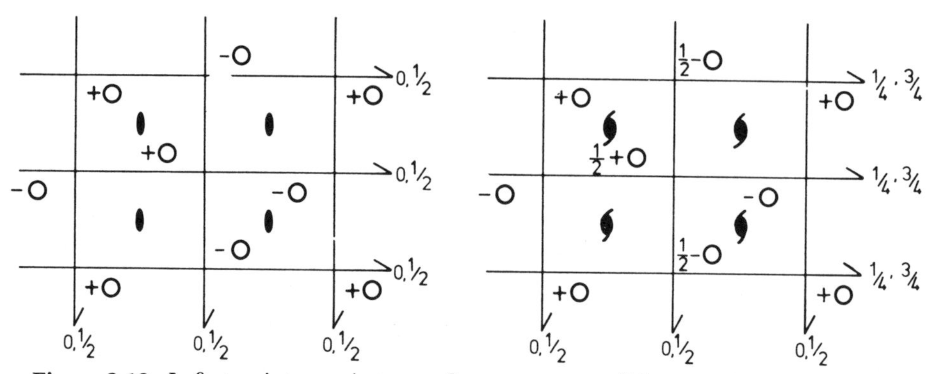

Figure 2.12. Left: two intersecting sets of screw axes parallel to **a** and **b** produce a set of rotation axes parallel to **c**. The space group is $P2_12_12$. Right: two nonintersecting sets of screw axes parallel to **a** and **b** produce a third set of screw axes parallel to **c**. The space group is $P2_12_12_1$.

seen from Fig. 2.12, right, and leads to the four equivalent positions:

$$x, y, z; \quad \tfrac{1}{2}+x, \bar{y}, \bar{z}; \quad \bar{x}, \tfrac{1}{2}+y, \tfrac{1}{2}-z; \quad \tfrac{1}{2}-x, \tfrac{1}{2}-y, \tfrac{1}{2}+z$$

There are no special positions, and clearly the choice of origin is somewhat arbitrary.

The standard origin, as adopted in the *International Tables for X-ray Crystallography,* is taken midway between three pairs of nonintersecting screw axes at $x, \tfrac{1}{4}, 0$, etc.; $0, y, \tfrac{1}{4}$, etc.; $\tfrac{1}{4}, 0, z$, etc., thus preserving the cyclic order of the symbols. The resulting arrangement of symmetry elements is shown in Fig. 2.13, where it is seen that the cyclic order of symbols is also retained in the equivalent positions:

$$x, y, z; \quad \tfrac{1}{2}-x, \bar{y}, \tfrac{1}{2}+z; \quad \tfrac{1}{2}+x, \tfrac{1}{2}-y, \bar{z}; \quad \bar{x}, \tfrac{1}{2}+y, \tfrac{1}{2}-z$$

Since the space group $P2_12_12_1$ is noncentrosymmetric, the geometric structure factor is a complex quantity. Use of the trigonometric identities

$$\cos\alpha + \cos\beta + \cos\gamma + \cos(\alpha+\beta+\gamma) = 4\cos\frac{\alpha+\beta}{2}\cos\frac{\beta+\gamma}{2}\cos\frac{\gamma+\alpha}{2}$$

$$\sin\alpha + \sin\beta + \sin\gamma + \sin(-\alpha-\beta-\gamma) = 4\sin\frac{\alpha+\beta}{2}\sin\frac{\beta+\gamma}{2}\sin\frac{\gamma+\alpha}{2}$$

with

$$\alpha = h(\tfrac{1}{2}-x)-ky + l(-\tfrac{1}{2}+z)$$

$$\beta = h(-\tfrac{1}{2}+x) + k(\tfrac{1}{2}-y) - lz$$

$$\gamma = -hx + k(-\tfrac{1}{2}+y) + l(\tfrac{1}{2}-z)$$

$$\alpha+\beta+\gamma = -(hx+ky+lz)$$

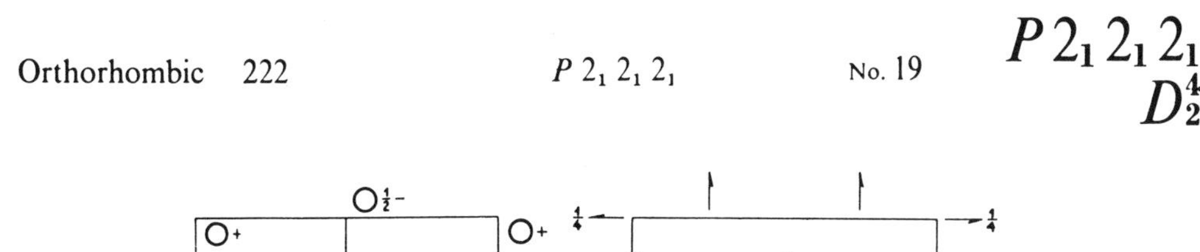

Orthorhombic 222 $P\,2_1\,2_1\,2_1$ No. 19 $P\,2_1\,2_1\,2_1$ D_2^4

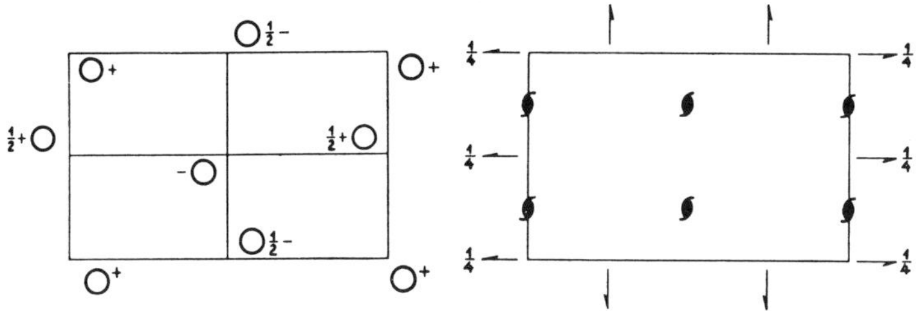

Origin halfway between three pairs of non-intersecting screw axes

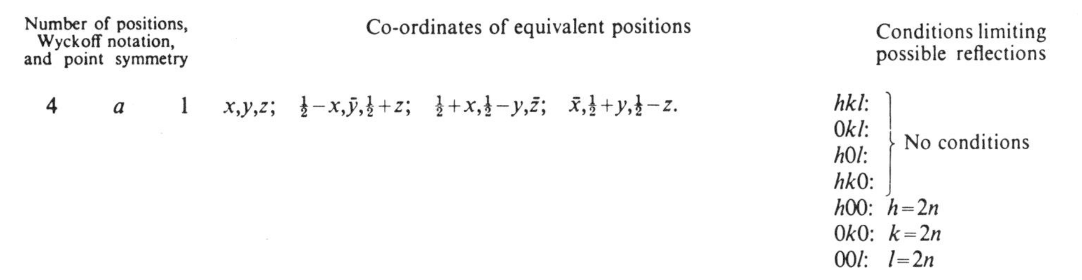

Number of positions, Wyckoff notation, and point symmetry			Co-ordinates of equivalent positions	Conditions limiting possible reflections
4	a	1	$x,y,z;\ \tfrac{1}{2}-x,\bar{y},\tfrac{1}{2}+z;\ \tfrac{1}{2}+x,\tfrac{1}{2}-y,\bar{z};\ \bar{x},\tfrac{1}{2}+y,\tfrac{1}{2}-z.$	hkl: $0kl$: $h0l$: $hk0$: No conditions
				$h00$: $h=2n$
				$0k0$: $k=2n$
				$00l$: $l=2n$

Symmetry of special projections

(001) pgg; $a'=a$, $b'=b$ (100) pgg; $b'=b$, $c'=c$ (010) pgg; $c'=c$, $a'=a$

Figure 2.13. Space group $P2_12_12_1$. From *International Tables for X-ray Crystallography*, Vol. I, Kynoch Press, Birmingham, 1969.

leads immediately to the expressions

$$A = 4 \cos 2\pi \left(hx-\frac{h-k}{4}\right) \cos 2\pi \left(ky-\frac{k-l}{4}\right) \cos 2\pi \left(lz-\frac{l-h}{4}\right)$$

$$B = -4 \sin 2\pi \left(hx-\frac{h-k}{4}\right) \sin 2\pi \left(ky-\frac{k-l}{4}\right) \sin 2\pi \left(lz-\frac{l-h}{4}\right)$$

which can be further simplified for particular combinations of the indices:

$$\begin{array}{llll} h+k=2n & A = & 4 \cos 2\pi hx \cos 2\pi ky \cos 2\pi lz & \\ k+l=2n & B = & -4 \sin 2\pi hx \sin 2\pi ky \sin 2\pi lz & \\ h+k=2n & A = & -4 \cos 2\pi hx \sin 2\pi ky \sin 2\pi lz & \\ k+l=2n+1 & B = & 4 \sin 2\pi hx \cos 2\pi ky \cos 2\pi lz & (2.3) \end{array}$$

$$\begin{array}{ll} h + k = 2n + 1 & A = -4 \sin 2\pi hx \cos 2\pi ky \sin 2\pi lz \\ k + l = 2n & B = \ \ 4 \cos 2\pi hx \sin 2\pi ky \cos 2\pi lz \end{array}$$

$$\begin{array}{ll} h + k = 2n + 1 & A = -4 \sin 2\pi hx \sin 2\pi ky \cos 2\pi lz \\ k + l = 2n + 1 & B = \ \ 4 \cos 2\pi hx \cos 2\pi ky \sin 2\pi lz \end{array}$$

The only systematic absences are:

($h00$) when h is odd
($0k0$) when k is odd
($00l$) when l is odd

produced by the three screw axes. For ($0kl$) reflections, the structure factor is real or imaginary, depending on whether the index k is even or odd, and similarly for the ($h0l$) and ($hk0$) reflections, depending on whether the index after the zero (cyclic order) is even or odd. The three two-dimensional projections, viewed along each of the coordinate axes, are centrosymmetric since the screw axis perpendicular to each projection plane acts like a center of symmetry in projection. However, the standard origin, midway between the screw axes, does not project on these "centers." A different origin would be required for each projection to make the corresponding structure factors real.

Determination of Space Group

The determination of the unit cell dimensions and space group is an important first step in any crystal structure analysis. Certain space groups (those in bold face in Table 2.5) can be uniquely assigned from the symmetry of the diffraction pattern and systematic absences. As mentioned earlier, in normal scattering the diffraction pattern is always centrosymmetric, whether the structure itself has a center of inversion or not. Other symmetry elements of the space group are associated with corresponding symmetries of the diffraction pattern, screw axes and glide planes acting in this respect as rotation axes and mirror planes, respectively. As a result, only 11 possible diffraction symmetries (Laue classes) can be distinguished:

triclinic	$\bar{1}(C_i)$
monoclinic	$2/m(C_{2h})$
orthorhombic	$mmm(D_{2h})$
tetragonal	$4/m(C_{4h})$, $4/mmm(D_{4h})$
trigonal-hexagonal	$\bar{3}(S_6)$, $\bar{3}m(D_{3d})$, $6/m(C_{6h})$, $6/mmm(D_{6h})$
cubic	$m3(T_h)$, $m3m(O_h)$

The remaining 21 crystal classes (Table 2.2) are assigned to the appropriate Laue class by adding an inversion center. For example, the space groups *P*222, *Pmm*2, and *Pmmm* belong to the Laue class *mmm* and cannot be differentiated from the symmetry of their diffraction patterns (or from systematic absences).

The systematic absences allow lattice centerings, glide planes, and screw axes to be detected. Some examples have been noted. The effect of a lattice centering is readily seen by considering its effect on the structure factor. If a unit of pattern with transform $\mathcal{F}(hkl)$ is repeated at lattice points 0, 0, 0 and $\frac{1}{2}, \frac{1}{2}, \frac{1}{2}$, we obtain

$$\begin{aligned} F(hkl) &= \mathcal{F}(hkl)(1 + \exp[\pi i(h + k + l)]) \\ &= 2\mathcal{F}(hkl) \qquad & h+k+l = 2n \\ &\text{or } 0 \qquad & h+k+l = 2n + 1 \end{aligned}$$

A body-centered lattice (I) thus causes all reflections with $h+k+l$ odd to be extinguished. Other types of centering also give rise to characteristic absences, reflections occurring only when the following conditions are satisfied:

C	$h + k = 2n$
A	$k + l = 2n$
B	$l + h = 2n$
F	$h + k = 2n,\ k + l = 2n,\ l + h = 2n$
R	$-h + k + l$ (referred to hexagonal axes) $= 3n$

The translational component of a glide plane acts like a lattice-centering in projection, so the corresponding systematic absence affects only a particular plane of reciprocal lattice points—the plane that is parallel to the glide plane and passes through the origin of reciprocal space. Similarly, the systematic absence due to the translational component of a screw axis affects only a particular central row of reflections in reciprocal space. Examples were seen in the detailed discussion of space groups $P2_1/c$ and $P2_12_12_1$. In detecting possible screw axes, it may not be easy to decide whether an observed pattern of absences is really "systematic" or only accidental—the reflections in question are few and may be too weak to be observed. This difficulty is much less serious for lattice centerings and glide planes since the systematic absences involved affect a much larger number of reflections.

If the crystal class (point-group symmetry) and not just the Laue class of a crystal can be established, the space group follows from the systematic absences. In some cases, nondiffraction methods can be useful.

The crystal class can be assigned, in principle, from the morphological properties of the macroscopic crystal (thousands of crystals were assigned to their respective classes in just this way before the advent of X-ray diffraction methods). However, for this purpose, well developed crystals are required, and these are often not available for organic compounds. Crystals that lack a center of symmetry should have pyro- or piezoelectrical properties;[19] they also show the phenomenon of second harmonic generation (SHG)[20]—i.e., when monochromatic light passes through a noncentrosymmetric crystal, a fraction of the light emerges at twice its initial frequency. Very large light intensities—e.g., from a pulsed laser—are needed for the second harmonic to be easily observed. Crystals that have only proper symmetry elements should have chiro-optical properties—i.e., they should rotate the plane of linearly polarized light. However, these physical diagnostic tests are seldom used for organic crystals, perhaps because they require specialized skills, knowledge, and equipment or perhaps because a negative result is inconclusive—the effect in question may be present but too weak to be detected. To resolve space-group ambiguities that depend on the presence or absence of a center of symmetry, X-ray crystallographers usually fall back on the diffraction pattern itself. Although the diffraction pattern is centrosymmetric under normal scattering conditions, the statistical distribution of intensities is different for centrosymmetric and noncentrosymmetric crystals. The difference depends on the fact that the structure amplitudes are real in the former case but complex in the latter, as discussed in the next section. Centrosymmetric and noncentrosymmetric space groups that belong to the same Laue class can thus be distinguished, in principle, by statistical analysis of the intensity distribution, as first pointed out by Wilson.[21] They can also be distinguished from the distribution of interatomic vector peaks in the Patterson function.[22] For example, space groups $P2$, Pm, and $P2/m$ should yield quite different concentrations of vectors between symmetry-related atoms. For $P2$ such a concentration must occur in a plane ($u0w$) normal to the rotation axis, for Pm it should be along a line ($0v0$) normal to the mirror plane, while for $P2/m$ both concentrations must occur. In principle, all 219 distinct space groups (i.e., disregarding the distinction between the 11 enantio-

[19] J. F. Nye, *Physical Properties of Crystals,* Oxford, University Press, 1957.

[20] S. C. Abrahams, *J. Appl. Crystallogr.* **5,** 143 (1972); J. P. Dougherty and S. K. Kurtz, *ibid.* **9,** 145 (1976).

[21] A. J. C. Wilson, *Acta Crystallogr.* **2,** 318 (1949).

[22] M. J. Buerger, *Acta Crystallogr.* **3,** 465 (1950).

morphous pairs) are distinguishable on the basis of the vector concentrations they produce.[22] Finally, we should not forget that an optically active compound consisting of homochiral molecules cannot crystallize in a space group that contains improper symmetry operations.

Intensity Statistics

Suppose we have a large number N of independent random variables $x_i (i = 1 \text{ to } N)$ with mean values $\langle x_i \rangle$ and variances σ_i^2. The Central Limit Theorem of statistics states that regardless of the probability distribution functions of the individual variables x_i (within certain general restrictions), when $N \rightarrow \infty$, the sum of the variables $X = \Sigma x_i$ tends to the normal probability distribution with mean value $\langle X \rangle = \Sigma \langle x_i \rangle$ and variance $\sigma^2 = \Sigma \sigma_i^2$. For large N the distribution of X is then

$$P(X) = (2\pi\sigma^2)^{-\frac{1}{2}} \exp\left(-[(X-\langle X \rangle)^2/2\sigma^2]\right)$$

If the atoms in the unit cell are regarded as being arranged randomly, we can use this theorem to derive the probability distribution of structure amplitudes.

For a noncentrosymmetric structure,

$$F_{\mathbf{H}} = A_{\mathbf{H}} + iB_{\mathbf{H}}$$

$$|F_{\mathbf{H}}|^2 = F_{\mathbf{H}} F_{\mathbf{H}}^* = A_{\mathbf{H}}^2 + B_{\mathbf{H}}^2$$

where

$$A_{\mathbf{H}} = \sum_{i=1}^{N} f_i \cos 2\pi \mathbf{H}\cdot\mathbf{r}_i$$

$$B_{\mathbf{H}} = \sum_{i=1}^{N} f_i \sin 2\pi \mathbf{H}\cdot\mathbf{r}_i$$

If all atomic positions are equally probable, the average value of each trigonometric term is zero; hence,

$$\langle A \rangle = \langle B \rangle = 0$$

The variance of A is $\langle A^2 \rangle - \langle A \rangle^2 = \langle A^2 \rangle$:

$$A_{\mathbf{H}}^2 = \sum_{i}^{N} f_i^2 \cos^2 2\pi \mathbf{H}\cdot\mathbf{r}_i + 2 \sum_{i<j}^{N} f_i f_j \cos 2\pi \mathbf{H}\cdot\mathbf{r}_i \cos 2\pi \mathbf{H}\cdot\mathbf{r}_j$$

The average value of the second term is zero, and the average value of $\cos^2 x$ is $\frac{1}{2}$, so that

$$2\langle A^2\rangle = \sum_i^N f_i^2 = \Sigma$$

and similarly,
$$2\langle B^2\rangle = \sum_i^N f_i^2 = \Sigma$$

whence
$$\langle F^2\rangle = \langle A^2\rangle + \langle B^2\rangle = \Sigma \tag{2.4}$$

We can now write the probability distributions of A and B as

$$P(A) = (\pi\Sigma)^{-\frac{1}{2}} \exp(-A^2/\Sigma)$$
$$P(B) = (\pi\Sigma)^{-\frac{1}{2}} \exp(-B^2/\Sigma)$$

The joint probability that A lies between A and dA, and B lies between B and dB is

$$\begin{aligned} P(A, B)dAdB &= (\pi\Sigma)^{-1} \exp[-(A^2+B^2)/\Sigma]\, dAdB \\ &= (\pi\Sigma)^{-1} \exp(-|F|^2/\Sigma)\, dAdB \end{aligned}$$

To obtain the probability that the structure amplitude lies between $|F|$ and $|F|+d|F|$, we transform to polar coordinates (area element $=$ $|F|\, d|F|\, d\alpha$) and integrate over the phase angle α. Since this integration merely introduces a factor 2π, we obtain

$$P(|F|)d|F| = 2\Sigma^{-1}|F| \exp(-|F|^2/\Sigma)\, d|F| \tag{2.5}$$

This probability is a function of Σ—i.e., it depends on the atomic form factors (including thermal motion effects), which, in turn, depend on the scattering angle. It is convenient at this stage to remove this dependence by introducing the normalized structure factor, $E_{\mathbf{H}}$, defined by

$$E_{\mathbf{H}} = \frac{F_{\mathbf{H}}}{\langle F^2\rangle^{\frac{1}{2}}} = \frac{F_{\mathbf{H}}}{\Sigma^{\frac{1}{2}}} \tag{2.6}$$

$E_{\mathbf{H}}$ is thus the ratio of $F_{\mathbf{H}}$ to its root-mean-square value in the appropriate region of reciprocal space. Normalized structure factors are much used in so-called direct methods of X-ray analysis (Chapter 3). The point of introducing them here is that the probability distribution of $|E|$ takes on a simple form since it has a variance of unity.

$$P(|E|) = 2|E| \exp(-|E|^2) \tag{2.7}$$

The average value of $|E|$ is, by definition, the integral:

$$\langle |E|\rangle = \int_0^\infty |E|\, P(|E|)d|E|$$

$$= 2\int_0^\infty |E|^2 \exp(-|E|^2)\, d|E|$$

$$= 2\left(\frac{\sqrt{\pi}}{4}\right) = 0.886 \qquad (2.8)$$

For a centrosymmetric structure, the calculation follows the same lines but is somewhat simpler:

$$F_{\mathbf{H}} = 2\sum_i^{N/2} f_i \cos 2\pi\mathbf{H}\cdot\mathbf{x}_i$$

$$\langle F\rangle = 0$$

$$\langle F^2\rangle = \Sigma$$

$$P(F) = (2\pi\Sigma)^{-\frac{1}{2}} \exp(-F^2/2\Sigma) \qquad (2.9)$$

$$P(E) = (2\pi)^{-\frac{1}{2}} \exp(-E^2/2) \qquad (2.10)$$

$$\langle |E|\rangle = 2(2\pi)^{-\frac{1}{2}} \int_0^\infty E \exp(-E^2/2)\, dE$$

$$= (2/\pi)^{\frac{1}{2}} = 0.798 \qquad (2.11)$$

The expectation value of $\langle |E|\rangle = \langle |F|\rangle/\langle |F|^2\rangle^{\frac{1}{2}}$ is thus 0.798 for a centrosymmetric structure and 0.886 for a noncentrosymmetric structure, and the difference is sufficient to allow the distinction to be made in many cases.

However, it is always a little dangerous to base a test on the value of a single number. It is safer to look at the actual distributions and compare them with the theoretical ones (Fig. 2.14). The centrosymmetric distribution yields a higher proportion of very large and very small $|E|$ values than the noncentrosymmetric one. For example, the proportion

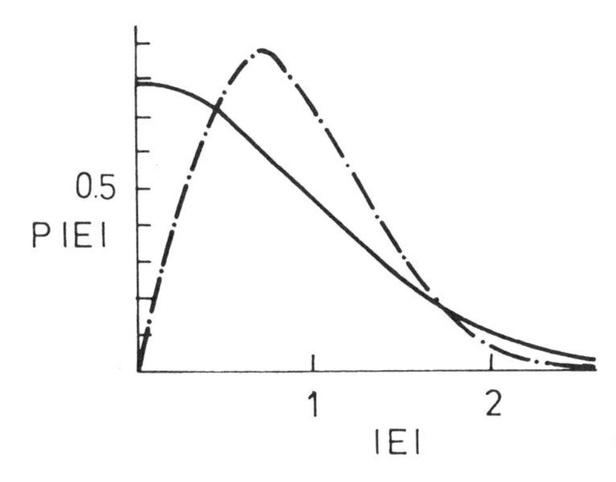

Figure 2.14. Probability distribution of $|E|$ for centrosymmetric (solid line) and noncentrosymmetric structures (broken line).

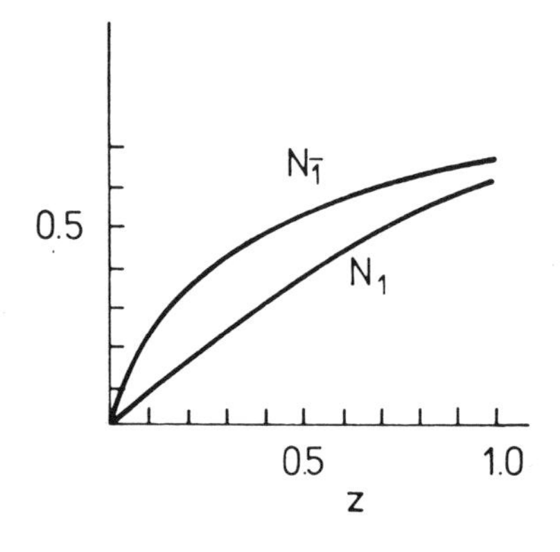

Figure 2.15. Cumulative $N(z)$ distributions for centrosymmetric and noncentrosymmetric structures.

of reflections with $|E| \geq 2$ is 0.046 for $\bar{1}$ but only 0.018 for 1. The distributions can also be used to estimate the proportion of reflections with $|E|^2$ less than or equal to a given number z, the $N(z)$ test:[23]

$$N_{\bar{1}}(z) = (2\pi)^{-\frac{1}{2}} \int_{-\sqrt{z}}^{\sqrt{z}} \exp(-E^2/2)\, dE$$

$$N_1(z) = 2 \int_0^{\sqrt{z}} E \exp(-E^2)\, dE$$

$$= 1 - \exp(-z)$$

These functions are shown in Fig. 2.15 where the higher accumulation of reflections with low values of $|E|^2$ in $N_{\bar{1}}(z)$ is obvious. The so-called hypercentric distribution[24] has an even higher proportion of reflections with low $|E|^2$ values; it applies to structures containing centrosymmetric molecules in general positions of centrosymmetric space groups.

All the tests described can also be expected to apply to two-dimensional zones of reflections, e.g., $(h0l)$, but only if the atoms do not overlap too seriously in the corresponding two-dimensional projection of the structure. When atoms overlap, sums of the type

$$\sum_{i<j} f_i f_j \cos 2\pi \mathbf{H}\cdot\mathbf{r}_i \cos 2\pi \mathbf{H}\cdot\mathbf{r}_j$$

do not necessarily average to zero, so that it would no longer be true that $\langle F^2 \rangle = \Sigma$. In general, the overlapping would be most serious in projections down long unit-cell axes. In principle, the application to individual projections allows the detection of other symmetry elements besides inversion centers. For example, in space group $P2$, the $(h0l)$ reflections

[23] E. R. Howells, D. C. Phillips, and D. Rogers, *Acta Crystallogr.* **3**, 210 (1950).
[24] H. Lipson and M. M. Woolfson, *Acta Crystallogr.* **6**, 439 (1952).

should have a centrosymmetric distribution, all other reflections a noncentrosymmetric one. In space group *Pm*, indistinguishable from *P*2 on the basis of Laue symmetry and systematic absences, none of the projections is centrosymmetric. Thus, in principle, *P*2, *Pm*, and *P*2/*m* can all be distinguished by appropriate statistical tests. Statistical tests are useful guides, but they are not infallible. Structures in noncentrosymmetric space groups may be nearly but not quite centrosymmetric.

Space Groups and Molecular Symmetry

Given the unit cell volume of a crystal and the molecular weight, the number of molecules (Z) per unit cell, which must be an integer, can be established from a knowledge of the density of the crystals. The mass (in grams) of the unit cell contents is MZ/N, where M is the molecular weight (in daltons) and N is Avogadro's number (6.0226×10^{23}). If the unit cell volume is V (in $Å^3$), the density (g/cm^3) is

$$d = \frac{MZ}{V \times 10^{-24} \times 6.0226 \times 10^{23}}$$

whence

$$Z = dV/1.660\ M$$

The density is conveniently measured by flotation of the crystals in suitable liquid mixtures.

For molecules that have no nontrivial symmetry, Z is usually equal to the number of general positions of the space group, although it may be an integral multiple of this number. Conversely, if Z equals the number of general positions or an integral multiple thereof, no molecular symmetry is required, but neither is it prohibited. On the other hand, if Z is a submultiple of the number of general positions, the molecules must have some symmetry element or elements in common with the space group—i.e., the molecules must occur at special positions of the space group with the appropriate multiplicity. The molecular site symmetry in a crystal may correspond to the full intrinsic molecular symmetry or to any subgroup of this symmetry down to mere identity, but it cannot be higher than the molecular symmetry, at least not in an ordered crystal.

In general, space-group determination provides a fairly quick and easy method for recognizing elements of molecular symmetry; it can be especially useful for distinguishing between alternative stereoisomers with different symmetries and avoids the trouble and expense of a complete structure analysis. For example, one of the four possible stereoisomers of 1,2,3,4-tetraphenylcyclobutane crystallizes in space group

$P2_1/c$ with $Z = 2$. The only sites of multiplicity less than four are the inversion centers of multiplicity two; hence the molecular centers must occupy these sites, and the isomer in question must be the centrosymmetric isomer.[25]

Unfortunately, the method is not infallible, mainly because the molecular arrangement in the crystal may be disordered. Thus, *p*-chlorobromobenzene crystallizes in $P2_1/a$ (alternative orientation of C_{2h}^5) with $Z = 2$, and the structure was correctly recognized many years ago[26] as disordered although the crystals do not show pronounced diffuse reflections. *p*-Chlorobenzylidene-*p*-chloroaniline crystallizes in space group *Pccn* with $Z = 4$.[27] This space group has eight general positions, and according to the rules the molecules should display either a center of inversion or a twofold axis; both possibilities are obviously incompatible with the molecular constitution. The crystals are disordered about the twofold axes.[28] Another crystal modification has a triclinic unit cell with $Z = 1$. In this case, an ordered structure could occur in space group $P1$, but Bernstein and Schmidt showed that the molecules are almost certainly disordered in space group $P\bar{1}$.[29]

Perhaps the most famous case, certainly the most confusing one, is that of azulene, cited earlier as an example of a disordered structure. The first X-ray crystallographic study of azulene appears to be that of Misch and van der Wyk;[30] they found the crystals to be monoclinic but overlooked the systematic absences that show the existence of a glide plane. The space group was given as $P2_1/a$ ($Z = 2$) by Günthard, Plattner, and Brandenberger;[31] they noted that the apparent presence of a molecular center of symmetry was incompatible with the molecular formula. Günthard[32] considered the possibility that the true space-group symmetry might be lower to allow the two molecules to occupy general positions, but he later showed that the anomaly is probably the result of orientational disorder in the crystals.[33] The entropy of fusion of azulene is about 4.5 cal/mole-deg lower than that of naphthalene, suggesting that azulene has a higher entropy in the crystalline state.

[25] J. D. Fulton and J. D. Dunitz, *Nature (London)* **160,** 161 (1947); J. D. Dunitz, *Acta Crystallogr.* **2,** 1 (1949).
[26] S. B. Hendricks, *Z. Kristallogr.* **84,** 85 (1933); A. Klug, *Nature (London)* **160,** 570 (1947).
[27] H. B. Bürgi, J. D. Dunitz, and C. Züst, *Acta Crystallogr.* **B24,** 463 (1968).
[28] J. Bernstein, *Acta Crystallogr.* **A31,** 118 (1975).
[29] J. Bernstein and G. M. J. Schmidt, *J. Chem. Soc., Perkin II* **1972,** 951.
[30] L. Misch and A. J. A. van der Wyk, *C. R. Séances Soc. Phys. Hist. Natur. Genève* **54,** 106 (1937) (cited in Ref. 31).
[31] H. Günthard, P. A. Plattner, and E. Brandenberger, *Experientia* **4,** 425 (1948).
[32] H. H. Günthard, Dissertation, E. T. H. Zürich, 1949.
[33] E. Kováts, H. H. Günthard, and P. A. Plattner, *Helv. Chim. Acta* **38,** 1912 (1955).

On the other hand, two independent crystal structure analyses[34,35] using two-dimensional data were based on the supposition of an ordered crystal with molecules in the general positions of $Pa(C_s^2)$, as suggested in one case by intensity statistics of the $(h0l)$ reflections, in the other by analysis of the corresponding Patterson projection. The assignment of the crystal class as C_s, rather than C_{2h}, was also supported by observations of the crystal form, which did not show the presence of a C_2 axis.[36] However, when the three-dimensional analysis was carried out a few years later,[37] it became obvious that something was fundamentally wrong with the postulated ordered structure, which failed to refine below an R factor (Chapter 4, first section) of about 0.22. Much better agreement ($R = 0.065$) was obtained for the disordered structure in $P2_1/a$, with the azulene molecules at general positions but assigned half-weight to give the required $Z = 2$.

In the examples considered so far, the straightforward interpretation of the space-group symmetry is obviously incompatible with the established formula of a more-or-less rigid molecule. The azulene saga shows, moreover, that it can be very difficult to decide between the alternatives: admit a lower symmetry space group than that indicated by the systematic absences, or assume the structure to be disordered. These and other similar examples warn us against drawing firm conclusions about molecular symmetry from space-group evidence alone. The following account of some experiences with medium-ring compounds is a cautionary tale.

In 1952 Prelog and Schenker tentatively assigned the configuration of the two cyclodecane-1,6-diols melting at 152 ° and 144 °C as *trans* and *cis*, respectively.[38] This assignment was subsequently supported by Sicher, Zavadá, and Svoboda,[39] who were able to relate the configurations of the diols to those of the 1,6-dibromides by the reaction scheme:

Br → SC_6H_5 ← OTs ← OH

Br SC_6H_5 OTs OH

[34]J. M. Robertson and H. M. M. Shearer, *Nature (London)* **177,** 885 (1956).

[35]Y. Takeuchi and R. Pepinsky, *Science* **124,** 126 (1956).

[36]J. D. Bernal, *Nature (London)* **178,** 40 (1956).

[37]J. M. Robertson, H. M. M. Shearer, G. A. Sim, and D. G. Watson, *Nature (London)* **182,** 177 (1958); *Acta Crystallogr.* **15,** 1 (1962).

[38]V. Prelog and K. Schenker, *Helv. Chim. Acta* **35,** 2044 (1952).

[39]J. Sicher, J. Zavadá, and M. Svoboda, *Collect. Czech. Chem. Commun.* **27,** 1927 (1962).

The configurations of the dibromides were assigned on the basis of their dipole moments (later confirmed by X-ray analysis of the 1,6-*trans*-dibromide).[40] Several years later we became interested in these diols as possible candidates for a neutron diffraction study. Preliminary X-ray examination showed that both crystals were monoclinic, space group $P2_1/c$, with $Z = 4$ (*trans* isomer, m.p. 152 °C) and $Z = 6$ (*cis* isomer, m.p. 144 °C). Now $Z = 4$ neither demands nor precludes any molecular symmetry, but $Z = 6$ implies four molecules in general positions plus two in special positions, the inversion centers; only the *trans* isomer can have the molecular symmetry thereby imposed. The obvious conclusion is that the previously assigned configurations should be reversed. However, when the detailed structure of the higher melting isomer (once deemed *trans,* now *cis*) was worked out, the four molecules in the cell were found to occupy two independent pairs of inversion centers and to be indubitably *trans* after all.[41] What about the other isomer—once deemed *cis,* then *trans,* now by elimination *cis*? Analysis of the lower melting crystals showed that the four molecules in general positions have the *cis* configuration, the two in special positions the *trans.*[42] In other words, this crystal form is actually a molecular compound of the *cis* and *trans* isomers in the ratio 2:1. This molecular compound seems to be always formed by crystallization from a solution containing both isomers.[43] When the pure *cis* isomer was finally prepared (from the corresponding ditosyl ester) it proved to be isomorphous with the 2:1 compound, the two formally centrosymmetric molecules at the special positions of the space group being disordered.[43]

Any assignment of molecular symmetry based on space-group considerations should be regarded as tentative until verified by a complete structure analysis.

Molecular Packing Arrangements

The reasons for the frequent occurrence of some space groups and the virtual nonoccurrence of others have been discussed by Kitaigorodsky[44] in terms of a purely geometrical model. Kitaigorodsky starts from the observation that the packing coefficient in molecular crystals—the

[40] J. D. Dunitz and H. P. Weber, *Helv. Chim. Acta* **47,** 951 (1964).

[41] O. Ermer and J. D. Dunitz, *Chem. Commun.* **1971,** 178; O. Ermer, J. D. Dunitz, and I. Bernal, *Acta Crystallogr.* **B29,** 2278 (1973).

[42] O. Ermer and J. D. Dunitz, unpublished results.

[43] H. H. Westen and O. Ermer, unpublished results.

[44] A. I. Kitaigorodsky, *Molecular Crystals and Molecules,* Academic Press, New York, 1973.

ratio of the volume occupied by the molecules to the unit cell volume—is generally quite high (0.65–0.75), similar to that obtainable by close packing of spheres and ellipsoids. Such a high packing efficiency is only possible if the "bumps" of one molecule fit into the "hollows" of others. In Kitaigorodsky's model, which is quite independent of any detailed theory of intermolecular forces, the molecules are regarded simply as irregularly shaped, geometric figures. In a close packed layer each figure would have a coordination number (CN) of 6—i.e., it would be in contact with six of its neighbors. The optimal packing of such layers in three dimensions would then give a CN of 12, with only slightly less efficient packing for 10 or 14. These are, in fact, the CNs observed in actual crystals. An analysis of the symmetry conditions for packing irregularly shaped figures, first into close packed layers and then into three-dimensional arrangements leads to the following conclusions.

Close packed layers of figures of arbitrary shape are not possible in plane groups that contain mirror planes or rotation axes of order greater than 2. They are possible for the four remaining plane groups, *p*1, *p*2, *pg*, and *pgg* (examples are shown in Figs. 2.16, 2.17, and 2.18). In addition, close packed layers of symmetrical figures such that the figure symmetry is retained in its environment can be formed in plane groups: *pmg* and *cm* for figures with *m* symmetry (Fig. 2.19); *cmm* for figures with *mm* symmetry; *p*2 and *pgg* for centrosymmetric figures.

If layers with CN 6 are to be stacked to achieve close packing in three dimensions, they may be related by a translation at an oblique angle to the layer or by inversion centers, glide planes, or screw axes—certainly not by mirror planes. Kitaigorodsky finds that close packing is possible only in the space groups $P1$, $P2_1$, $P2_1/c$, $Pca2_1$, $Pna2_1$, $P2_12_12_1$ for objects of arbitrary shape; $P\bar{1}$, $P2_1/c$, $C2/c$, $Pbca$, for centrosymmetric objects.

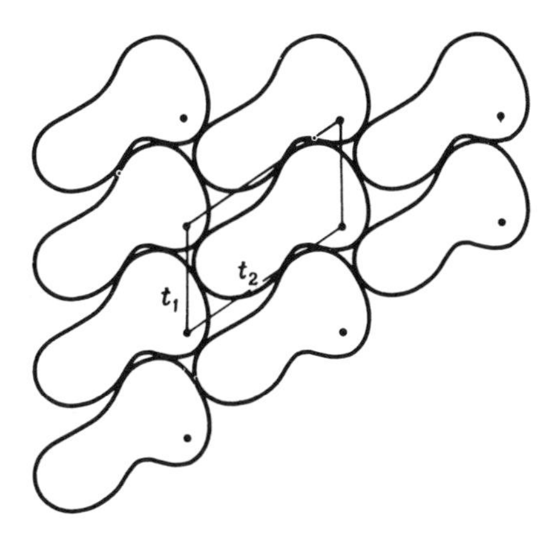

Figure 2.16. Close packed layer in plane group *p*1. From Kitaigorodsky, *Molecular Crystals and Molecules,* Academic Press, New York, 1973.

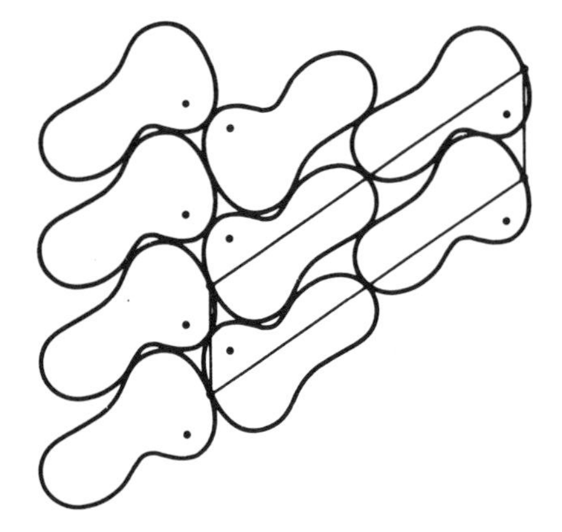

Figure 2.17. Close packed layer in plane group *p*2. From Kitaigorodsky, *Molecular Crystals and Molecules,* Academic Press, New York, 1973.

The retention of other molecular symmetries leads unavoidably to less efficient packing, equivalent to a decrease of about 0.02–0.03 in the packing coefficient. Kitaigorodsky has derived those space groups that allow optimal packing of more symmetrical figures, under the supplementary condition that the symmetry in question is preserved. Objects that retain a twofold rotation axis pack best in $C2/c$, $P2_12_12$ and $Pbcn$; those that retain a mirror plane pack best in $Pnma$. The results are summarized in Table 2.6.

The only molecular symmetry element that can be preserved in close

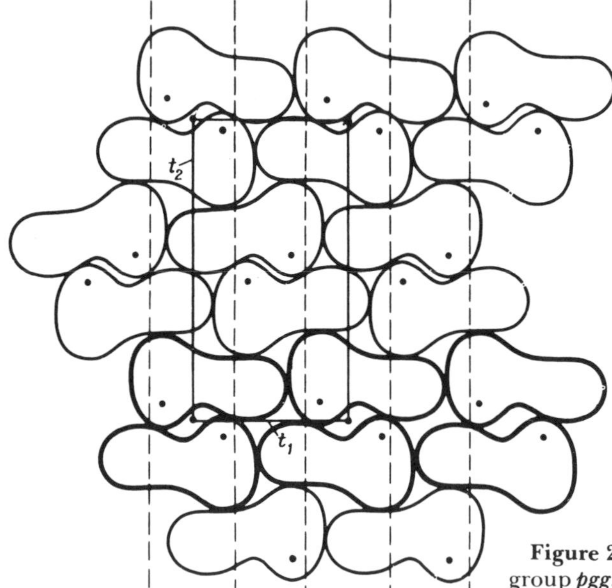

Figure 2.18. Close packed layer in plane group *pgg*. From Kitaigorodsky, *Molecular Crystals and Molecules,* Academic Press, New York, 1973.

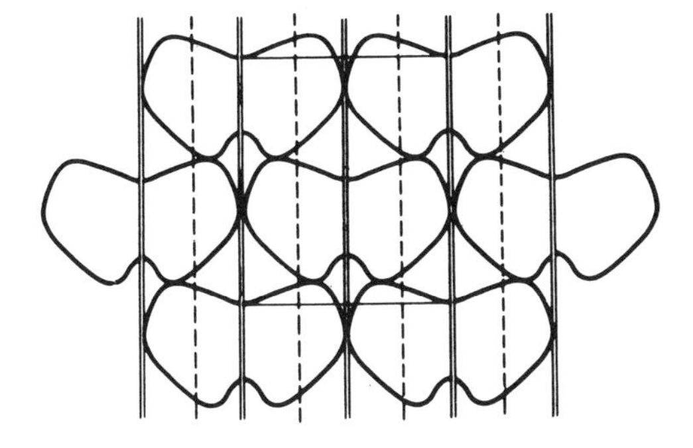

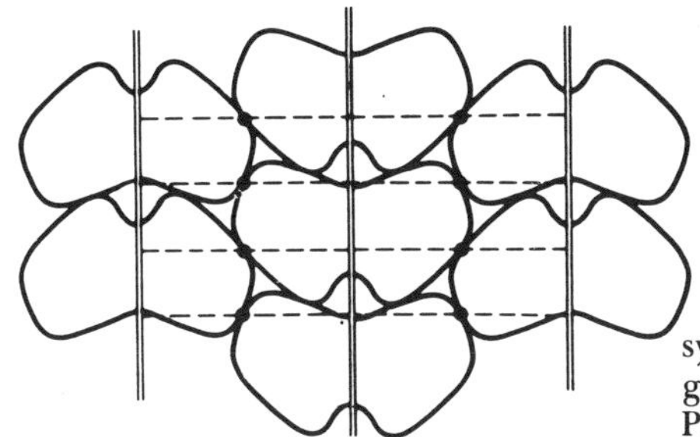

Figure 2.19. Close packed layers of figures with *m* symmetry in plane groups *cm* and *pmg*. From Kitaigorodsky, *Molecular Crystals and Molecules,* Academic Press, New York, 1973.

packed arrangements is thus a center of inversion. Efficient packing of molecules with other kinds of symmetry usually entails a lowering of the intrinsic molecular symmetry. Aromatic hydrocarbons such as benzene and naphthalene are good examples. For benzene, with intrinsic $6/mmm$ (D_{6h}) molecular symmetry, the crystallographic site symmetry is only $\bar{1}$ (C_i), and similarly for naphthalene and anthracene, with intrinsic mmm (D_{2h}) molecular symmetry. In most cases, the effect of the reduction in site symmetry on chemically equivalent molecular structure parameters is hardly detectable—i.e., the virtual molecular symmetry (as opposed to the crystallographic symmetry) is equal to the intrinsic molecular symmetry within observational error—but not always. As the accuracy attainable in crystal structure analysis improves, the experimental significance of small deviations from expected molecular symmetries evoked by packing effects becomes clearer. The interpretation of such deviations as expressions of intermolecular interactions may teach us something about the nature and strength of these interactions and of the restoring forces within molecules.[45]

[45]This opinion was not shared by Kitaigorodsky, at least not in 1973. In his book he argued that differences between chemically equivalent molecular parameters observed in crystals are *in all cases* due to experimental errors or to artifacts (Ref. 44, pp. 187–190).

Table 2.6. Space groups that allow closest packing and those compatible with optimal packing arrangements of symmetrical objects[a]

Molecular site symmetry	Closest packing space groups	Optimal packing space groups
1	$P\bar{1}$, $P2_1/c$, $Pca2_1$, $Pna2_1$	
1 [b]	$P2_1$, $P2_12_12_1$	
2	none	$C2/c$, $Pbcn$
2 [b]	none	$P2_12_12$
m	none	$Pmc2_1$, $Cmc2_1$, $Pnma$
$\bar{1}$	$P\bar{1}$, $P2_1/c$, $C2/c$, $Pbca$	
$mm2$	none	$Fmm2$, $Pnma$, $Pmnn$
$2/m$	none	$C2/m$, $Pbaa$, $Cmca$
222	none	$Ccca$
222 [b]	none	$C222$, $F222$, $I222$
mmm	none	$Cmmm$, $Fmmm$, $Immm$

[a]After Kitaigorodsky, *Molecular Crystals and Molecules,* Academic Press, New York, 1973.
[b]Homochiral objects or molecules.

The geometrical considerations outlined above are, of course, very general and cannot be expected to apply without qualification to every crystal structure. Nevertheless, the possibilities and restrictions that emerge summarize a large body of experimental information about molecular packing. The relative constancy of the packing coefficient corresponds to the observation that the shortest intermolecular contact distances between specified kinds of atoms do not vary much from one crystal to another. These distances can be regarded, to a first approximation at least, as sums of characteristic packing radii (the radii used by

Table 2.7. Volume increments for common atomic groups in organic molecules[a]

Group		ΔV(Å^3)	Group		ΔV(Å^3)
Methyl	$-CH_3$	23.5	nitro	$-NO_2$	23.0
Methylene	$>CH_2$	17.1	amino	$-NH_2$	19.7
Methylene	$=CH_2$	13.1	fluoro	$-F$	9.6
Methine	$\geq CH$	11.1	chloro	$-Cl$	19.9
Aromatic CH	$\geq C-H$	14.7	bromo	$-Br$	26.0
Quaternary C	$>C<$	5.0	cyano	$-CN$	15.9
Trigonal C	$>C-$	8.4	carboxyl	$-COOH$	23.1

[a]Adapted from Kitaigorodsky, *Molecular Crystals and Molecules,* Academic Press, New York, 1973.

Kitaigorodsky[44] are $R_H = 1.17$ Å, $R_C = 1.80$ Å, $R_N = 1.58$ Å, $R_O = 1.52$ Å, somewhat different from the van der Waals radii adopted by Pauling.[46]

An effective molecular volume can then be estimated from these radii, used together with standard bond distances and angles. For this purpose it is convenient to express the volume as a sum of volume increments for characteristic atomic groupings (the values used by Kitaigorodsky are given in Table 2.7). From a knowledge of the unit-cell volume and the numbers of molecules in the cell, the packing coefficient can be calculated. It is usually 0.65–0.75 for crystals and 0.5–0.6 for liquids.

In accounting for these regularities and for the regularities in molecular packing arrangements, the geometrical model provides a first step towards an organic chemical crystallography. Large discrepancies between ideal and actual packing arrangements as well as unusually short intermolecular contact distances merely point to the limitations inherent in the purely geometrical approach. Such discrepancies can be regarded as indicators for specific types of intermolecular interactions that cannot so easily be incorporated into the purely geometrical model.

[46] L. Pauling, *The Nature of the Chemical Bond,* 3rd ed., Cornell University Press, Ithaca, N.Y., 1960.

3. Methods of Crystal Structure Analysis

"And when I found the door was locked,
I pulled and pushed and kicked and knocked."

Statement of the Problem

Problem: given an experimental set of structure amplitudes $|F_{\mathbf{H}}|$ or their squares, to find the distribution of scattering matter. If we knew the structure amplitudes *and* their relative phase angles, there would be no problem; the electron-density distribution could be calculated immediately by summation of the Fourier synthesis (Eq. 1.41)

$$\rho(\mathbf{r}) = \frac{1}{V} \sum_{\mathbf{H}} |F_{\mathbf{H}}| \cos 2\pi[\mathbf{H}\cdot\mathbf{r} - \alpha(\mathbf{H})]$$

the kind of task that requires mere arithmetical accuracy, provided nowadays by electronic computers. Unfortunately, we do not know the phase angles, so in order to use Eq. 1.41, we must infer the missing information. Of course, the phase angles can be calculated by Fourier inversion of $\rho(\mathbf{r})$, but this is not very useful since we need the phase angles to calculate $\rho(\mathbf{r})$ in the first place.

Even in the absence of phase-angle information the problem must be soluble *in principle,* but it is much more complicated. From Eq. 1.44 a system of simultaneous equations

$$|F_{\mathbf{H}}|^2 = \sum_i f_i^2 + 2\sum_{i<j} f_i f_j \cos 2\pi\mathbf{H}\cdot(\mathbf{r}_i - \mathbf{r}_j)$$

can be constructed. The phase angles do not appear here. The atomic scattering factors f_i are known, at least to within a few percent, so the only unknowns are the atomic position vectors $\mathbf{r}_i$, each involving three coordinates. Since we can measure a large number of $|F_{\mathbf{H}}|$ values for different $\mathbf{H}$ vectors (about 100 or so per atom), the system of equations is heavily overdetermined, even allowing for errors in the experimental $|F|$ values and in the fs. However, because the unknowns appear in the arguments of trigonometric functions, the equations are severely nonlinear. Thus, the solution, if it is to be found at all, must be found in two stages: first, a rough solution, yielding an approximately correct set of atomic positions, and second, the adjustment of the values of the parameters describing this rough model to produce optimal agreement with the experimental data.

In this chapter we discuss ways of grappling with the first part of the problem—the derivation of a model structure; refinement methods are discussed in Chapter 4. We assume a knowledge of the cell dimensions, crystal density, and empirical formula, so that the number of formula units per unit cell can be estimated. In most cases, we can also assume that the space group is known or at least restricted to a few possibilities on the basis of the Laue symmetry and systematic absences (see Chapter 2). In addition, fairly reliable information may be available about the approximate arrangement of atoms in the molecules or in some particular molecular fragments. It is no accident that so many of the successful, early crystal-structure determinations of organic compounds were concerned with aromatic compounds.

Trial-and-Error Analysis

These early structures were mostly solved by a trial-and-error procedure; this method consisted of postulating a model, an atomic arrangement consistent with the cell dimensions, space group, and any prior knowledge about the molecular structure, and checking whether the structure amplitudes calculated for the model were in qualitative agreement with the observed magnitudes for a few selected reflections. A patently incorrect model would soon lead to some violent discrepancies, in which case the model would be rejected and a new one proposed. If only minor discrepancies were uncovered, they could be corrected by slight modifications to the model. Once a respectable level of agreement between calculated and observed $|F_{\mathrm{H}}|$ values was achieved, the model structure could be regarded as correct in essence if not in all details. The next step consisted of calculating an incomplete Fourier summation, usually in two dimensions,[1] using those observed $|F_{\mathrm{H}}|$ terms whose phase angles or signs could be reasonably well determined from the

[1]Before computers, calculations were usually carried out for two-dimensional projections of the crystal structure, rather than for the three-dimensional structure itself. Far fewer measurements were required, and the calculations could be done in a fraction of the time needed for three-dimensional work. As long as there was not too much atomic overlapping in the projections, the three-dimensional structure could be constructed by combining the results of two or more such projections.

model. The resulting electron-density projection, based partially on experimental data, could be expected to reveal further modifications necessary for the trial model. Incorporation of these adjustments then led to an improved model, to new and presumably better phase angles, and hence to an improved Fourier synthesis. For centrosymmetric structures, the process could be terminated when the cycle of operations did not lead to any changes in the signs associated with the $|F_{\mathbf{H}}|$ values. For noncentrosymmetric structures, the iterations could be continued until the patience of the investigator was exhausted.

As an example we can look at the analysis of hexamethylbenzene, the first aromatic compound whose crystal structure was determined.[2] The crystals are triclinic and can be referred to a primitive unit cell, with $a = 9.01$ Å, $b = 8.93$ Å, $c = 5.34$ Å, $\alpha = 44.5°$, $\beta = 116.7°$, $\gamma = 119.6°$, containing one molecule, that was assumed to be centrosymmetric. The crystals cleave easily along the (001) plane, and the (00*l*) reflections show a gradual, uniform decrease in intensity, similar to that observed for graphite, suggesting that the carbon atoms lie in a common plane, parallel to (001). The intensities of the (*hk*0) reflections show marked pseudohexagonal symmetry, suggesting that the carbon atoms lie at the vertices of two concentric hexagons. It is then possible to describe the (001) projection of the structure in terms of three parameters: the common orientation of the two hexagons and their radii, which could be defined within fairly narrow limits by comparing observed $|F_{\mathbf{H}}|$ values with those calculated for various trial structures. In 1928, when this analysis was carried out, the planarity of the benzene ring was not generally accepted; indeed, Lonsdale's work furnished one of the first experimental proofs for this characteristic feature of the structure of aromatic compounds, which was abundantly confirmed by the subsequent studies of the Robertson school. A later application of iterative Fourier syntheses to hexamethylbenzene[3] confirmed the essential correctness of the Lonsdale structure and established the dimensions of the molecule within still narrower limits.

For a second example of the trial-and-error approach, let us refer to the analysis of the structure of the centrosymmetric isomer of 1,2,3,4-tetraphenylcyclobutane.[4] The crystals are monoclinic, $a = 17.02$ Å, $b = 5.775$ Å, $c = 12.35$ Å, $\beta = 127°$, space group $P2_1/a$ (variant of $P2_1/c$), $Z = 2$. The molecules must be centrosymmetric, so the asymmetric unit

[2] K. Lonsdale, *Nature (London)* **122**, 810 (1928).
[3] L. O. Brockway and J. M. Robertson, *J. Chem. Soc.* **1939**, 1324.
[4] J. D. Dunitz, *Acta Crystallogr.* **2**, 1, (1949).

consists of two phenyl groups plus two adjacent atoms of the four-membered ring. The shortness of the **b**-axis ensures that atoms belonging to different phenyl groups cannot overlap in the (010) projection; furthermore, if the perpendicular distance between translationally equivalent phenyl groups related by the **b**-axis is to be about 3.4 Å, as in graphite, the rings must be tilted steeply (by about 50°) out of the (010) plane. With these preliminary assumptions, a trial structure could be based on the presence of a few outstandingly strong $(h0l)$ reflections, whose geometric structure factor is of the form $\cos 2\pi(hx+lz)$. For outstandingly strong reflections of this type, most of the atoms must lie on the traces of the planes $hx+lz = n$ (integer) or $hx+lz = n/2$, where their contributions to the structure factors have a maximum positive or negative value. These traces can be drawn on a sheet of paper and a model of the molecule adjusted until the above conditions are satisfied as closely as possible.

Inspection of the $(h0l)$ intensities showed that the low-order reflections $(20\bar{3})$ and $(40\bar{1})$ and also the high-order reflections $(20, 0, \bar{13})$, $(18, 0, \bar{14})$, (605) and (10, 0, 4) were very strong. With the help of a rough molecular model, various positions and orientations of the phenyl groups were tested to explain these observations. Here, the molecular center of symmetry can be placed at the origin of the cell, which restricts the possibilities considerably. The strength of the $(20\bar{3})$ and $(40\bar{1})$ reflections can be explained if the traces of these planes pass approximately through the midpoints of the projected phenyl groups, and the orientations of these groups can then be adjusted to account for the strength of the high-order reflections. The final adjustment—to make $(20, 0, \bar{13})$ strongly positive and $(18, 0, \bar{14})$, (605), and (10, 0, 4) strongly negative—is shown in Fig. 3.1. For comparison, the corresponding final electron-density projection is also shown in Fig. 3.1, and in expanded form to show the entire unit cell in Fig. 3.2. In this case the trial model can be regarded as a very good approximation to the actual structure.

These examples indicate the strengths and weaknesses of the trial-and-error method. Obviously, to solve a crystal structure by trial and error, one had to know a lot about it in advance. Also, especially for centrosymmetric arrangements, even a rough structural model is enough to fix most of the phase angles within quite narrow limits. Once this could be done, the details of the molecular structure (bond lengths and angles, and so forth) could be determined from the available experimental data.

The prerequisite that fairly safe assumptions had to be made about the relative positions of major groups of atoms was, of course, a factor

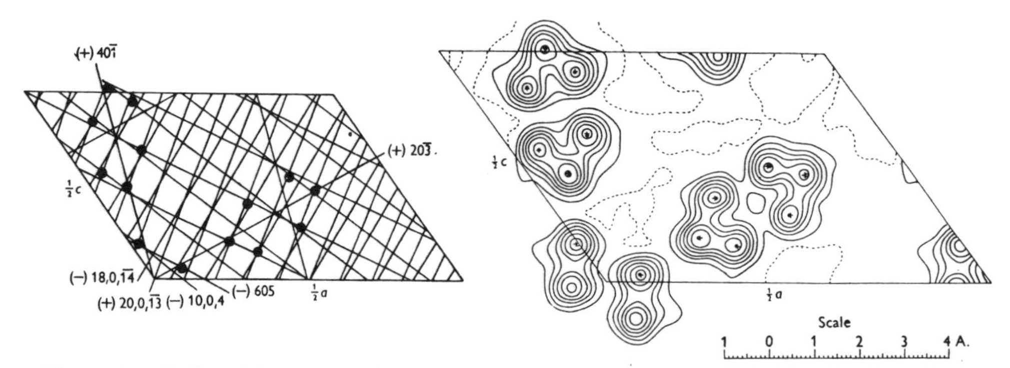

Figure 3.1. Left: trial structure based on outstandingly strong ($h0l$) reflections. Right: final electron-density projection on (010) showing the asymmetric crystal unit. Contours are drawn at electron-density increments of approximately one electron per Å^3; the one-electron line is dotted. From Dunitz, *Acta Crystallogr.* **2,** 1 (1949).

that severely restricted the range of the trial-and-error approach. Applications to natural products of unknown or only partially known chemical constitution, or even to molecules of known constitution but containing many torsional degrees of freedom, were more or less impossible and were generally avoided.

Trial-and-error analysis required not only persistence and care but also the same kind of intuitive discernment that seems to be involved in recognizing a barely distinguishable pattern, in drawing inferences from shreds and scraps of disconnected clues, some of which are false or misleading. The experimental structure amplitudes contained a message that could be decoded in stages once the general meaning could be guessed. However, there was no systematic method for improving or amending one's guesses, so that it was not always easy to decide how to proceed. A seriously incorrect model could be recognized easily enough from the poor agreement it produced, but a model that gave good agreement for some reflections and bad agreement for others might be

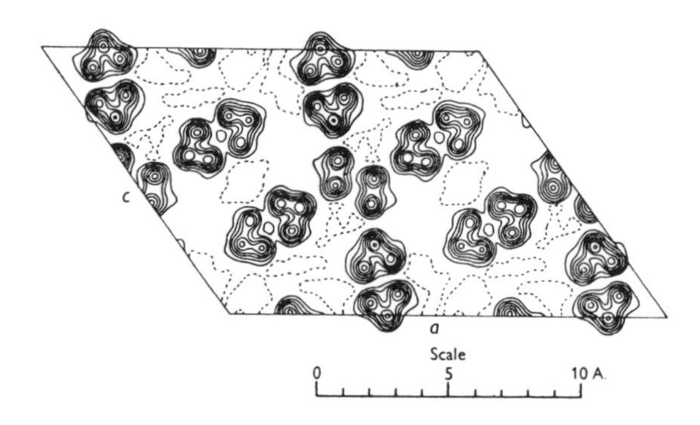

Figure 3.2. Same as Fig. 3.1, right, but expanded to show a complete unit cell containing two molecules of 1,2,3,4-tetraphenylcyclobutane, one complete. From Dunitz, *Acta Crystallogr.* **2,** 1 (1949).

manipulated into an essentially correct structure, or it might contain some unredeemable feature. One had to be persistent in adjusting a model that seemed promising, but one also had to be adaptable enough to switch to a radically new model at the first hint that the old model was incorrect. One also had to know when to abandon the problem as hopeless.

The use of the optical transform technique[5] would have been enormously helpful for rapid testing of structural models in trial-and-error analysis. This technique uses the close analogy between the diffraction of X-rays and of visible light. A parallel, monochromatic light beam is passed through a two-dimensional arrangement of holes punched in an opaque screen, and the resulting diffraction pattern is recorded photographically. Thus, the effect of changes in the model (arrangement of holes) on the diffraction pattern as a whole can readily be assessed. In this way, a variety of possible models can be tested fairly rapidly. Although a number of structures were solved by this method, the technical difficulties (optical design, preparation of accurate, small-scaled masks) were really overcome only when trial-and-error analysis was already being replaced by other, more powerful methods.

Heavy-Atom Methods

As shown in Chapter 1, the complete interpretation of the Patterson function is generally extremely difficult for all but the simplest structures. As long as the $N(N-1)$ interatomic vector peaks in the Patterson function are resolved from one another, the vector set can be unscrambled into the point set in a fairly straightforward manner. However, as the density of vectors increases, the unscrambling becomes more and more difficult and ultimately impossible in the absence of extraneous information about the structure of the atomic aggregrates. This limit is reached soon enough in three dimensions and even earlier in the two-dimensional Patterson projections to which crystallographers were restricted before computing aids became available.

Even in these unfavorable circumstances, a partial interpretation of the Patterson function becomes possible as soon as a few heavy atoms are present. This is because the height of the Patterson peak associated with an interatomic vector $\mathbf{r}_j - \mathbf{r}_i$ is approximately proportional to the product of the atomic numbers Z_iZ_j. Peaks corresponding to vectors between

[5] C. A. Taylor and H. Lipson, *Optical Transforms,* Bell, London, 1964.

heavy atoms therefore stand out prominently above an undulating background due to the numerous interactions involving lighter atoms.

As an example, consider an organic molecule containing N carbon atoms ($Z = 6$) and one iodine atom ($Z = 53$; we neglect the hydrogens) and suppose there are two such molecules in the unit cell of a crystal with space group $P2_1$. The carbon atom positions can be taken as:

$$\begin{array}{ll} x_1,\ y_1,\ z_1 & -x_1,\ \tfrac{1}{2}+y_1,\ -z_1 \\ x_2,\ y_2,\ z_2 & -x_2,\ \tfrac{1}{2}+y_2,\ -z_2 \\ \cdot\ \ \cdot\ \ \cdot & \cdot\ \ \cdot\ \ \cdot \\ \cdot\ \ \cdot\ \ \cdot & \cdot\ \ \cdot\ \ \cdot \\ \cdot\ \ \cdot\ \ \cdot & \cdot\ \ \cdot\ \ \cdot \\ x_N,\ y_N,\ z_N & -x_N,\ \tfrac{1}{2}+y_N,\ -z_N \end{array}$$

with the iodine atom at

$$x_I,\ y_I,\ z_I \qquad -x_I,\ \tfrac{1}{2}+y_I,\ -z_I$$

We expect N pairs of peaks corresponding to vectors between symmetry-related carbon atoms at $\pm(2x_i,\ \tfrac{1}{2},\ 2z_i)$ (weight $6 \times 6 = 36$), and another $2N(N-1)$ pairs of peaks corresponding to vectors between atoms not related by symmetry—i.e., $N(N-1)/2$ pairs at each of:

$$\begin{array}{l} \pm[(x_i-x_j),\ (y_i-y_j),\ (z_i-z_j)] \\ \pm[(x_i-x_j),\ (-y_i+y_j),\ (z_i-z_j)] \\ \pm[(x_i+x_j),\ (\tfrac{1}{2}+y_i-y_j),\ (z_i+z_j)] \\ \pm[(x_i+x_j),\ (\tfrac{1}{2}-y_i+y_j),\ (z_i+z_j)] \end{array}$$

In addition, we expect $4N$ pairs of stronger peaks due to I···C vectors (weight $6 \times 53 = 318$) at:

$$\begin{array}{l} \pm[(x_I-x_i),\ (y_I-y_i),\ (z_I-z_i)] \\ \pm[(x_I-x_i),\ (-y_I+y_i),\ (z_I-z_i)] \\ \pm[(x_I+x_i),\ (\tfrac{1}{2}+y_I-y_i),\ (z_I+z_i)] \\ \pm[(x_I+x_i),\ (\tfrac{1}{2}-y_I+y_i),\ (z_I+z_i)] \end{array}$$

and finally, a single pair of very strong peaks due to the symmetry-related I···I interaction (weight $53 \times 53 = 2809$) at $\pm(2x_I,\ \tfrac{1}{2},\ 2z_I)$. Check that

$$\begin{aligned} 2\{N + 2N(N-1) + 4N + 1\} &= 2(2N^2 + 3N + 1) \\ &= (2N + 2)(2N + 1) \end{aligned}$$

that is, we have correctly counted all the vectors from each of the $2N + 2$ atoms in the unit cell to all the other atoms.

Even when N is quite large, say 50, the I···I peaks should be clearly recognizable. Once they are identified, the positions of the iodine atoms are known. (We could take the y coordinate of one of the I atoms as zero or any other value we choose since the origin in this direction is arbitrary.)

At this stage we could proceed in one of two ways:

1. The $4N$ pairs of I···C peaks, which are relatively strong, form images of the molecule (and of its enantiomer) as seen from each of the two I atoms. These four images are, of course, superimposed in the Patterson function, and the individual images will not, in general, be immediately apparent by inspection of the peak pattern. We can resort to the device of displacing a copy of the Patterson function so that its origin coincides in turn with the I positions (see Fig. 1.20); every coincidence of peaks in the displaced and undisplaced Patterson functions will then indicate a possible site for a carbon atom. Whereas the space-group symmetry of the original Patterson function is $P2/m$, the peak pattern thus obtained (Fig. 3.3) has space group $P2_1/m$; it is still centrosymmetric with spurious mirror planes passing through the I positions. This is unavoidable because, although the actual structure in space group $P2_1$ is noncentrosymmetric, the partial structure consisting of the two I atoms alone is centrosymmetric. To destroy the spurious center (or mirror plane), we have to make an arbitrary choice. We can select one of a pair of matched peaks related by either of the spurious symmetry operations as genuine and disregard the other. With the help of stereochemical criteria, it may then be possible to select other individual peaks from

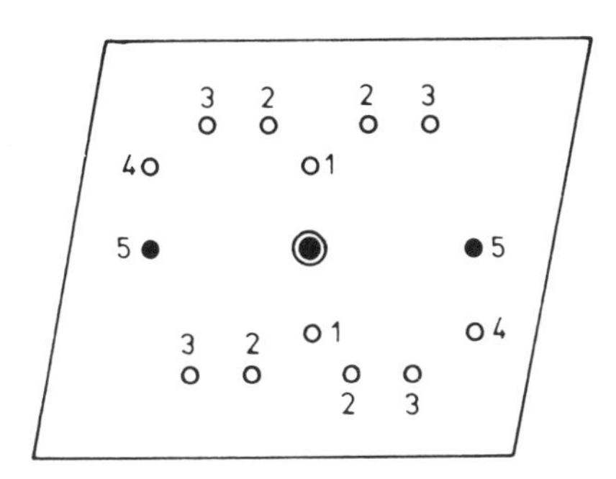

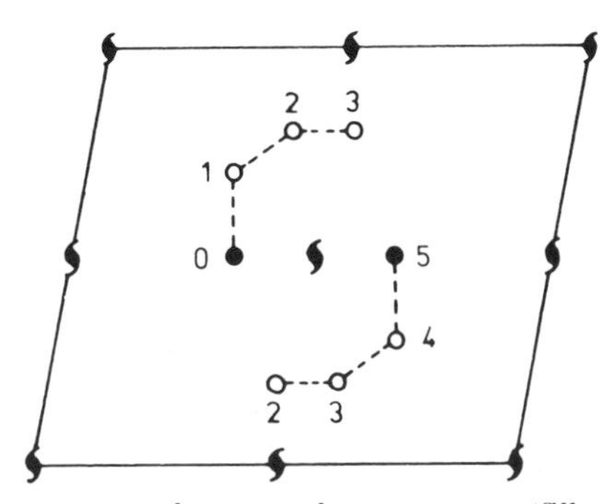

Figure 3.3. Left: Patterson function showing vectors between heavy atoms (filled circles) and between heavy atoms and light atoms (open circles). The origin peak is indicated by a double circle, and the numbers give the heights of the peaks above *and* below the projection plane in tenths of the unit period. Right: pattern of peaks obtained by displacing the origin of the Patterson function to a filled circle and noting coincidences. The numbers again give heights (in tenths) above *and* below the projection plane.

the remaining matched pairs, but some uncertainty about these choices might remain. However, if the actual structure had belonged to a centrosymmetric space group, the unscrambling of the Patterson peaks due to heavy atom–light atom interactions could be carried out in the fairly straightforward way indicated above and would present far less serious problems.

2. We could regard the partial structure consisting of the I atoms alone as a trial structure and proceed to calculate a Fourier synthesis using the observed $|F_H|$ values with phase angles derived from this trial structure. These phase angles are certainly incorrect and some may be seriously in error. However, the heavy-atom contribution to the structure factor corresponds to a vector of length f_h in a known direction, while the contributions of the light atoms correspond to many vectors of length f_l in a variety of still unknown directions[6] (cf., Fig. 1.13). If f_h/f_l is sufficiently large, the phase of the resultant will tend to be approximately that of the heavy-atom contribution. At small scattering angles this ratio is the ratio of the atomic numbers Z_h/Z_l, but at high angles the relative decrease in the form factors is greater for light atoms than for heavy ones, thus making heavy atoms even "heavier" than they would appear to be from their atomic numbers. The form factors for C and I are, for example:[7]

$\sin\theta/\lambda$ (Å^{-1})	0	0.10	0.20	0.30	0.40	0.50	0.60	0.70
f_C	6.00	5.126	3.581	2.502	1.950	1.685	1.536	1.426
f_I	53.0	49.36	42.79	37.14	32.38	28.51	25.29	22.63
f_I/f_C	8.83	9.63	11.92	14.84	16.61	16.92	16.46	15.87

The reason is, of course, that the electrons in the heavy atom are relatively more concentrated than in the light atom and that the Fourier transforms of the electron-density distributions (the f curves) show the inverse behavior. The heavy atom has to be heavy enough so that its contribution will dominate the phase angles of the structure factors, but it must not be too heavy, or the net light-atom contribution to F_H will tend to be swamped by experimental error in the $|F_H|$ measurements. As a crude guide, Σf_h^2 should be roughly equal to Σf_l^2—i.e., a single I atom should function reasonably well as phase determiner with about 100 light atoms in the molecule.

Since the phase angles based on the I positions are better than a

[6]In this context, subscripts h and l are not indices but abbreviations for heavy and light, respectively.

[7]*International Tables for X-ray Crystallography,* Vol. III, Kynoch Press, Birmingham, 1962, pp. 202, 211.

random selection, the distribution function represented by the resulting Fourier synthesis should be an improvement on the trial structure consisting of the iodine atoms alone, an improvement that will manifest itself in a tendency for small accumulations of density to occur at or near the positions of the light atoms. If some of these can be recognized, the phase angles can be recalculated using this extra structural information and the process repeated, just as in the trial-and-error method. For centrosymmetric structures, where the phase angles are restricted to the values 0 or π, the process is fairly efficient, but for noncentrosymmetric structures some difficulties may arise.

The first difficulty is the same as the one encountered in trying to unscramble the Patterson function—spurious symmetry. In our example in space group $P2_1$, the partial structure consisting of the two I atoms alone is centrosymmetric. If we take the origin at the midpoint of the vector between these atoms, the I positions become $\pm(x_1, \frac{1}{4}, z_1)$, and the calculated phase angles are restricted to 0 or π. The resulting Fourier synthesis is therefore centrosymmetric, with spurious mirror planes through the I positions at $y = \pm\frac{1}{4}$. For every density peak at x, y, z there would be an exactly equivalent one at $x, \frac{1}{2}-y, z$. The density distribution is thus the superposition of the distorted density associated with the two possible enantiomers reflected across the spurious mirror planes; again an arbitrary selection has to be made of one of a pair of symmetry-related peaks, or better (if possible) a group of peaks that makes stereochemical sense. This difficulty cannot be circumvented by choosing a different origin, one for which the phase angles are not restricted to 0 or π, because this would merely displace the spurious mirror planes along the y-axis but not remove them. Phases derived from a centrosymmetric trial structure must lead to a centrosymmetric density distribution, whether the origin is taken at the center of symmetry or not. The spurious symmetry has to be broken by some arbitrary choice of one peak or group of peaks as genuine, the symmetry-related one as false. The relationships between peak heights in the actual electron density distribution and in that calculated with phase angles corresponding to a partial structure have been treated by Luzzati,[8] who also considers how heavy a heavy atom should be to be useful in phasing calculations.

For centrosymmetric structures, the process of identifying certain density peaks as those corresponding to true atomic positions and including them in the next set of phasing calculations is largely self-correcting

[8]V. Luzzati, *Acta Crystallogr.* **6**, 142 (1953).

in the sense that an incorrectly chosen atom tends to disappear in the subsequent calculated density distributions. For noncentrosymmetric structures, however, incorrectly chosen atoms tend to persist in the subsequent distributions, and genuine atoms not included in the phasing calculations often produce only weak, diffuse peaks that are not always easy to recognize.

One of the best accounts of the analysis of a complex molecule of largely unknown structure by the heavy-atom method is that of the hexacarboxylic acid obtained by degradation of vitamin B_{12}.[9] The cobalt of the corrin nucleus served as the heavy atom. After its position in the unit cell (space group $P2_12_12_1$) was determined from the Patterson function, the positions of the remaining atoms were gradually derived from a series of 10 successive three-dimensional Fourier syntheses; the first was phased only on the Co positions, the last on 73 nonhydrogen atoms, thus accounting for the molecule itself ($C_{46}H_{58}O_{13}N_6CoCl$) plus two water molecules and one acetone molecule. Needless to say, the calculations were carried out with a digital computer—the analysis would hardly have been possible a decade earlier. Even with this aid, however, considerable skill, experience, and imagination were required to recognize correct groupings of atoms and to disregard the spurious peaks that arose in the early and intermediate stages of the analysis.

The development and increased availability of digital computers during the late 1950s and early 1960s transformed the heavy-atom method into a more-or-less routine tool by which the molecular structures of complex natural products could be determined by X-ray analysis, in many cases faster and more efficiently than by the classical methods of chemical degradation. Without computers, the heavy-atom method was essentially limited to structures that could be solved by analysis of two-dimensional data. Three-dimensional analyses were possible in principle, but they required so much time and effort that they could be carried out only in drastically simplified form—e.g., calculation of the electron density along specified lines in three-dimensional space—and even then only for a few key structures of special importance. This situation was changed by computers and a little later by automated diffractometers for efficient measurement of the intensity data. The determination of the structure of a complex natural product by X-ray analysis was no longer a major enterprise to be undertaken only when all other methods failed. Once the intensities could be measured and

[9]D. C. Hodgkin, J. Pickworth, J. H. Robertson, R. J. Prosen, R. A. Sparks, and K. N. Trueblood, *Proc. Roy. Soc., London* **A251,** 306 (1959).

the computations carried out in a reasonable time, the main bottleneck was often that of preparing a suitable heavy-atom derivative and obtaining it in a suitable crystal form. The difficulties of interpreting imperfectly phased Fourier syntheses based on heavy-atom contributions can also be overcome to a large extent by the intervention of computers. The three-dimensional density distribution is stored in the computer, which locates the peaks and selects from them a set of possible atomic sites consistent with reasonable bond distances and angles. The postulated structure can be refined, wrong atoms deleted, and new ones inserted by an iterative sequence of successive least-squares calculations and Fourier syntheses with a minimum of human intervention at any stage. Several quite complex heavy-atom structures have been solved in this way.[10]

Method of Isomorphous Replacement

In the heavy-atom method the phase assigned to a given structure amplitude $|F_H|$ is the phase of the heavy-atom contribution. The isomorphous replacement method is a variant of this, where the phase angle is estimated from the change in $|F_H|$ that occurs when one or more heavy atoms are introduced into the unit cell without appreciable changes in the overall structure of the crystal. The heavy atoms may simply replace lighter atoms (e.g., Pt for Ni, Br for Cl, Rb for K), or they may be introduced as additional atoms; the latter possibility is especially important in protein crystallography, where heavy atoms can often be attached to specific sites on the protein molecule without causing appreciable changes in the structure of the protein or its packing arrangement. Crystals in which the atomic positions are essentially the same but differ only in the nature of the atoms that occupy these positions are called isomorphous;[11] they are recognized by their similar cell dimensions, same space group, and similar diffraction patterns.

The simplest case involves a centrosymmetric structure. Suppose we have two centrosymmetric crystals A and B that differ only in the scattering powers f_A and f_B ($f_B > f_A$) of a group of atoms. We can express the structure factors of crystals A and B as

[10] H. Koyama, K. Okada, and C. Itoh, *Acta Crystallogr.* **B26,** 444 (1970); H. Koyama and K. Okada, *ibid.* **A31,** 518 (1975).

[11] In English two adjectives are essentially synonymous: isomorphic and isomorphous. It is customary to use the former in mathematical contexts and the latter in crystallography. Thus, we speak of isomorphic groups but of isomorphous crystals.

$$\begin{aligned} F_{\mathbf{H}}^{\mathrm{A}} &= F_{\mathbf{H}}^{\mathrm{R}} + f_{\mathrm{A}}\, T_{\mathbf{H}} \\ F_{\mathbf{H}}^{\mathrm{B}} &= F_{\mathbf{H}}^{\mathrm{R}} + f_{\mathrm{B}}\, T_{\mathbf{H}} \end{aligned} \tag{3.1}$$

where $F_{\mathbf{H}}^{\mathrm{R}}$ is the contribution of the atoms common to both structures and $T_{\mathbf{H}}$ is a trigonometric factor that depends only on the positions of the replaceable atoms. All the quantities are real numbers, $|F_{\mathbf{H}}^{\mathrm{A}}|$ and $|F_{\mathbf{H}}^{\mathrm{B}}|$ can be measured, and f_{A}, f_{B} and $T_{\mathbf{H}}$ can be assumed to be known. The few atomic positions on which $T_{\mathbf{H}}$ depends can usually be determined from the Patterson function of the crystal containing the heavier atoms (B) or, even better, from the difference Patterson function calculated with coefficients $(F^{\mathrm{B}})^2 - (F^{\mathrm{A}})^2$. The only peaks present in this function correspond to vectors among the replaceable atoms and to vectors between these atoms and the common ones; peaks corresponding to vectors among the common atoms are subtracted out, which makes the interpretation much simpler.

From the Eqs. 3.1 we obtain

$$F_{\mathbf{H}}^{\mathrm{B}} - F_{\mathbf{H}}^{\mathrm{A}} = (f_{\mathrm{B}} - f_{\mathrm{A}})\, T_{\mathbf{H}} = \delta f_{\mathrm{AB}}\, T_{\mathbf{H}}$$

where δf_{AB} is positive. If we know the magnitude and sign of $T_{\mathbf{H}}$, it is easy to deduce the signs of $F_{\mathbf{H}}^{\mathrm{B}}$ and $F_{\mathbf{H}}^{\mathrm{A}}$, even allowing for some experimental uncertainty in their magnitudes. For example, if

$$|F_{\mathbf{H}}^{\mathrm{B}}| = 82, \quad |F_{\mathbf{H}}^{\mathrm{A}}| = 50, \quad \delta f_{\mathrm{AB}} T_{\mathbf{H}} = -28.6$$

it is clear that $F_{\mathbf{H}}^{\mathrm{A}}$ and $F_{\mathbf{H}}^{\mathrm{B}}$ must both be negative; however, if

$$|F_{\mathbf{H}}^{\mathrm{B}}| = 10, \quad |F_{\mathbf{H}}^{\mathrm{A}}| = 40, \quad \delta f_{\mathrm{AB}} T_{\mathbf{H}} = -28.6$$

both must be positive, i.e., they have the sign opposite to $T_{\mathbf{H}}$. (In the heavy-atom method we would assume that $F_{\mathbf{H}}$ *always* has the same sign as $T_{\mathbf{H}}$).

For noncentrosymmetric crystals, however, $F_{\mathbf{H}}^{\mathrm{A}}$, $F_{\mathbf{H}}^{\mathrm{B}}$, $F_{\mathbf{H}}^{\mathrm{R}}$ and $T_{\mathbf{H}}$ are complex numbers, where $|F_{\mathbf{H}}^{\mathrm{A}}|$ and $|F_{\mathbf{H}}^{\mathrm{B}}|$ are experimentally available and $T_{\mathbf{H}}$ can be calculated, in magnitude and phase, from the positions of the replaceable atoms. The unknown quantities can be obtained from a simple graphical construction (Fig. 3.4) based on the vector equation

$$\mathbf{F}^{\mathrm{R}} = -f_{\mathrm{A}}\mathbf{T} + \mathbf{F}^{\mathrm{A}} = -f_{\mathrm{B}}\mathbf{T} + \mathbf{F}^{\mathrm{B}} \tag{3.2}$$

(where we have dropped the subscript **H** on the understanding that the equation refers to a particular reflection). We draw two circles of radii $|F^{\mathrm{A}}|$ and $|F^{\mathrm{B}}|$ with their centers at the termini of the known vectors $-f_{\mathrm{A}}\mathbf{T}$ and $-f_{\mathrm{B}}\mathbf{T}$, respectively. There are two possible solutions, corresponding

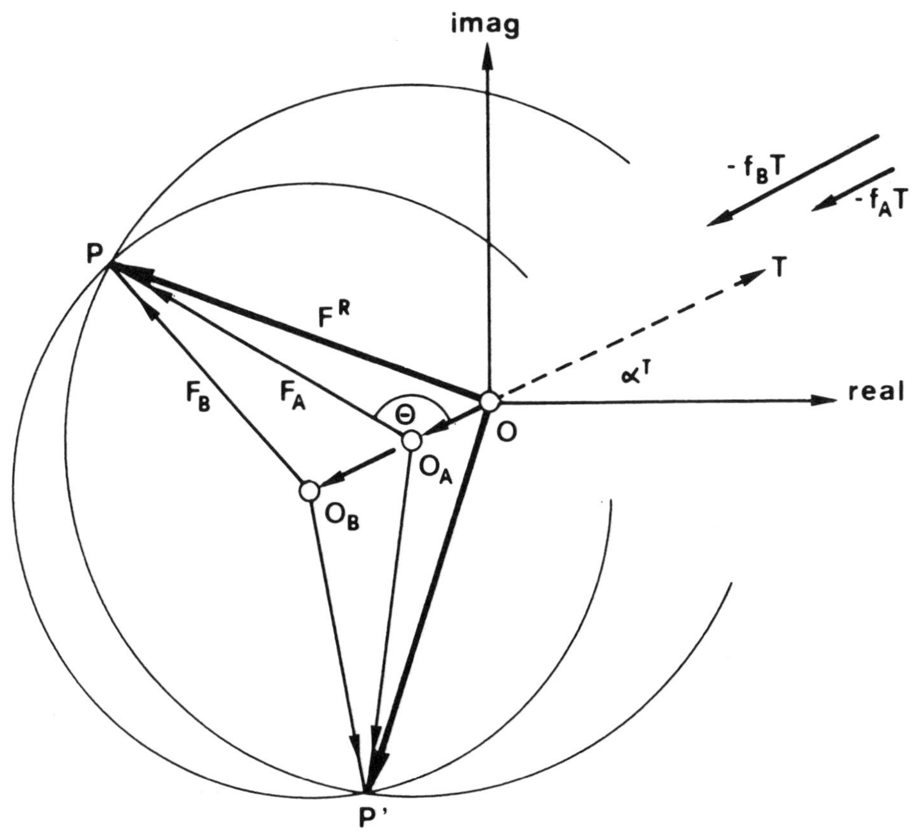

Figure 3.4. Graphical solution of Eq. 3.2. From origin O draw the collinear vectors $-f_A\mathbf{T}$ and $-f_B\mathbf{T}$. With the termini O_A and O_B of these vectors as centers, construct circles of radius $|F^A|$ and $|F^B|$, respectively, which intersect at P and P′. The two possible solutions for $\mathbf{F}^R$ are then the vectors OP and OP′. The phase angle of $\mathbf{F}^A$ is $\alpha^A = \alpha^T \pm \theta$.

to the two intersections of the circles—the figure is obviously symmetrical across the direction of **T**.

The analytical solution can be obtained by considering the triangle formed by $\mathbf{F}_A$, $\mathbf{F}_B$, and $(f_B - f_A)\mathbf{T}$ in Fig. 3.4:

$$\cos\theta = \frac{|F^B|^2 - |F^A|^2 - (f_B - f_A)^2|T|^2}{2(f_B - f_A)|T||F^A|} \tag{3.3}$$

where $\theta = \alpha^A - \alpha^T$. We can thus derive the phase angle of $\mathbf{F}^A$, $\alpha^A = \alpha^T \pm \theta$, apart from the ambiguity that Eq. 3.3 determines $\cos\theta$ rather than θ itself, leaving the sign of the latter open.

There are thus two possible phase angles, say α and β, for each reflection, and the best one can do under these circumstances is to calculate the Fourier synthesis with both possibilities—i.e.,

$$\rho'_A(\mathbf{r}) = \frac{1}{V}\sum |F^A_{\mathbf{H}}|\{\cos(2\pi\mathbf{H}\cdot\mathbf{r}-\alpha^A_{\mathbf{H}}) + \cos(2\pi\mathbf{H}\cdot\mathbf{r}-\beta^A_{\mathbf{H}})\}$$

$$= \frac{1}{V}\sum |F^A_{\mathbf{H}}|\{\cos(2\pi\mathbf{H}\cdot\mathbf{r}-\alpha^T_{\mathbf{H}}-\theta_{\mathbf{H}}) + \cos(2\pi\mathbf{H}\cdot\mathbf{r}-\alpha^T_{\mathbf{H}}+\theta_{\mathbf{H}})\}$$

$$= \frac{2}{V}\sum |F^A_{\mathbf{H}}|\cos\theta_{\mathbf{H}}\cos(2\pi\mathbf{H}\cdot\mathbf{r}-\alpha^T_{\mathbf{H}}) \qquad (3.4)$$

as suggested by Kartha.[12] The terms with the correct phases will build up the correct electron density $\rho_A(\mathbf{r})$ with peaks at the atomic positions, whereas those with the wrong phases produce a large number of spurious peaks at positions related to the true structure but in a complicated way. If the set of replaceable atoms is noncentrosymmetric—i.e., if the phase angles α^T are not restricted to 0 and π, the spurious peaks contribute mainly a general background to the true peaks and should not interfere seriously with the identification of the latter.

However, if the set of replaceable atoms is centrosymmetric but the complete structure is not, one of the two possible phase angles corresponds to the genuine structure, and the other corresponds to the enantiomorph obtained by inversion across the local center of symmetry of the replaceable atoms. In this case, $\rho'_A(\mathbf{r})$ is centrosymmetric, and the two sets of peaks—the genuine ones and the spurious ones—occur in matched pairs of equal height and cannot be distinguished without introducing stereochemical criteria, just as in a Fourier synthesis phased from the contributions of a centrosymmetric arrangement of heavy atoms. Nevertheless, the function $\rho'_A(\mathbf{r})$, however difficult to interpret, will still be an improvement on the heavy-atom Fourier synthesis with coefficients $|F^A_{\mathbf{H}}|$ (or $|F^B_{\mathbf{H}}|$) and phase angles $\alpha^T_{\mathbf{H}}$. As seen from Eq. 3.4, $\rho'_A(\mathbf{r})$ can be regarded as the synthesis with coefficients $|F^A_{\mathbf{H}}|\cos\theta_{\mathbf{H}}$ and phase angles $\alpha^T_{\mathbf{H}}$. The factor $\cos\theta$ ensures that coefficients whose actual phase angles differ markedly from the heavy-atom phase angles will be appropriately down-weighted or even reversed in sign.

The phase-angle ambiguity can also be resolved by making a second isomorphous replacement, as suggested by Bokhoven, Schoone, and Bijvoet[13] and developed by Harker.[14] If the same set of atoms is replaced by atoms of still different scattering power, say, f_C, then Fig. 3.4 is simply expanded into three circles whose centers lie on a common line and which have two common intersection points related by the mirror sym-

[12]G. Kartha, *Acta Crystallogr.* **14,** 680 (1961).
[13]C. Bokhoven, J. C. Schoone, and J. M. Bijvoet, *Acta Crystallogr.* **4,** 275 (1951).
[14]D. Harker, *Acta Crystallogr.* **9,** 1 (1956).

metry of the figure. If, however, the second replacement involves a *different* set of atomic positions, the mirror symmetry will be destroyed and the three circles will have only a single intersection point corresponding to a unique solution for the phase angle. The situation for three isomorphous crystals A, B, C, where B and C contain additional heavy atoms (of scattering power f_B and f_C) not present in A, is shown in Fig. 3.5. This kind of multiple isomorphous replacement is very important in protein crystallography.

In multiple isomorphous replacement problems can arise concerning the choice of origin, when this is not fixed by the space-group symmetry, and concerning the choice of a common reference frame, when the heavy-atom arrangements are themselves chiral. The relative positions of the additional heavy atoms in crystal B, say, can be determined from a suitable difference Patterson function, and the same can be done for crystal C. However, if the origin of B is different from C, or if the reference frames are enantiomorphic, the three circles will not generally intersect at a common point. Returning to Fig. 3.5 in this connection, we note that displacing the origin of C by $\mathbf{r}'$ is equivalent to rotating the vector $-f_C\mathbf{T}_C$ through an angle $2\pi\mathbf{H}\cdot\mathbf{r}'$ radians for the reflection $\mathbf{H}$, while inversion of the reference frame reflects the vector across the real axis.

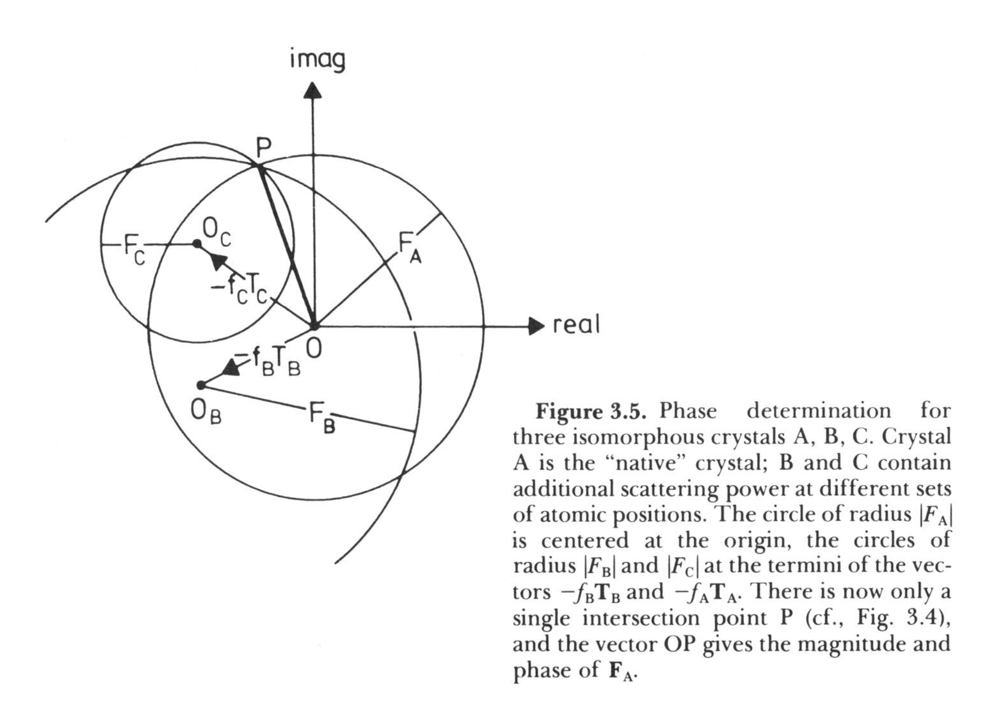

Figure 3.5. Phase determination for three isomorphous crystals A, B, C. Crystal A is the "native" crystal; B and C contain additional scattering power at different sets of atomic positions. The circle of radius $|F_A|$ is centered at the origin, the circles of radius $|F_B|$ and $|F_C|$ at the termini of the vectors $-f_B\mathbf{T}_B$ and $-f_A\mathbf{T}_A$. There is now only a single intersection point P (cf., Fig. 3.4), and the vector OP gives the magnitude and phase of $\mathbf{F}_A$.

The most useful methods for ensuring that relative positions of replaceable atoms are referred to the same origin in double isomorphous replacement involve so-called correlation functions—convolutions of various kinds of Patterson function. Here we mention only one, due to Rossmann,[15] which consists of calculating the "Patterson function" with coefficients

$$(|F^{B}| - |F^{C}|)^2 = |F^{B}|^2 + |F^{C}|^2 - 2|F^{C}||F^{B}|$$

The first two terms simply yield a superposition of the normal Patterson functions of B and C, respectively, with the strongest peaks corresponding to vectors between the heavy atoms in these crystals. The third term yields approximately the cross-Patterson of B and C with negative sign; strong negative peaks will occur at $\mathbf{r}_i^{B} - \mathbf{r}_j^{C}$, due to interactions between heavy atoms at $\mathbf{r}_i$ in B and at $\mathbf{r}_j$ in C. Consider an example for space group $P2_1$, where the origin along the y-axis is arbitrary, and suppose that the heavy atoms in B and C are situated at:

$$x_B, y_B, z_B \qquad -x_B, \tfrac{1}{2}+y_B, -z_B$$

$$x_C, y_C, z_C \qquad -x_C, \tfrac{1}{2}+y_C, -z_C$$

The Patterson functions of B and C contain peaks at

$$\pm 2x_B, \tfrac{1}{2}, \pm 2z_B \qquad \pm 2x_C, \tfrac{1}{2}, \pm 2z_C$$

which establish the relative heavy-atom arrangements in the two crystals separately but not their relative displacement along the y-axis. This information is given by the third term, the cross-Patterson, which should have strong negative peaks at

$$\pm[(x_B - x_C), (y_B - y_C), (z_B - z_C)]$$

$$\pm[(x_B + x_C), (\tfrac{1}{2} + y_B - y_C), (z_B + z_C)]$$

More detailed discussion of the properties of this and other kinds of correlation function can be found elsewhere.[16]

In practice, there is always some uncertainty in the phase angles determined by isomorphous replacement. This arises not only from experimental error in the measurements but also from the lack of perfect isomorphism from one derivative to another. The assumption that the crystal coordinates of the common atoms remain unchanged on substitu-

[15] M. G. Rossmann, *Acta Crystallogr.* **13,** 221 (1960).

[16] For example, G. N. Ramachandran and R. Srinivasan, *Fourier Methods in Crystallography,* Wiley-Interscience, New York, 1970. See particularly Chapter 10 and references therein.

tion or addition of other atoms is an idealization that holds only approximately at best. Despite these drawbacks the isomorphous replacement method is clearly far more powerful than the heavy-atom method, at the cost of a correspondingly greater outlay of experimental work. Even when the phase angles cannot be determined uniquely, as in single replacement (Fig. 3.4), those reflections whose phases differ greatly from those of the heavy-atom phases are easily identifiable and can be treated accordingly.

Anomalous Scattering: Determination of Absolute Configuration

So far we have assumed the validity of Friedel's Law, which states that the X-ray diffraction pattern of a crystal is centrosymmetric, whether the crystal structure itself is centrosymmetric or not. Friedel's Law depends on the assumption that phase differences between waves scattered at different points depend only on path differences, which would imply that any intrinsic phase change connected with the actual scattering must be the same for all scattering centers. This assumption is nearly but not quite correct. Depending on the energy of the incident X-rays, some of the atoms may scatter slightly out of phase with the others. For non-centrosymmetric crystal structures, this leads to a slight difference between the intensities of (hkl) and $(\bar{h}\bar{k}\bar{l})$ reflections; these would be equal according to Friedel's Law. Such intensity differences can be utilized in structure analysis since they help to determine the phases of the relevant pair of reflections. More important, they provide an experimental basis for defining the absolute reference frame used to describe the structure; for chiral structures, this is equivalent to determining the chirality sense (or absolute configuration).

The possibility of utilizing the breakdown of Friedel's Law to provide a bridge between macroscopic and molecular chirality was first proposed by J. M. Bijvoet in 1949.[17] At that time it was not known which of the two possible enantiomorphic structures for any optically active molecule corresponded to the dextrorotatory isomer and which to the levorotatory. All that the then available chemical methods could do was to relate the configurations of many optically active compounds among one another—i.e., to establish their configurations relative to some reference compound, usually taken as (+)-glyceraldehyde, arbitrarily assigned the configuration I and conventionally represented by

[17] J. M. Bijvoet, *Proc. Koninkl. Ned. Akad. Wetenschap* **B52**, 313 (1949); *Comptes-Rendus*, 5th Congress of Pure and Applied Chemistry, Amsterdam, Sept. 5–10, 1949, p. 73.

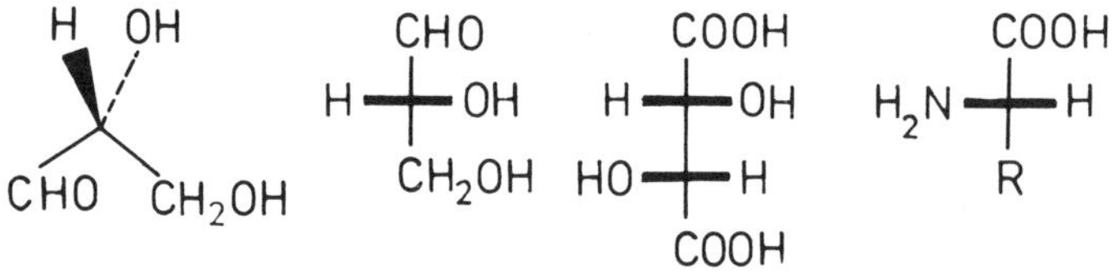

H OH CHO COOH COOH

H—OH H—OH H_2N—H

CHO CH_2OH CH_2OH HO—H R

COOH

I II III IV

the projection formula II (Fischer projection). Within this system, for example, the configuration of (+)-tartaric acid was known to be III and that of the naturally occurring amino acids to be IV, but there was no way to establish the actual configuration of the reference molecule. Within a couple of years, Bijvoet's proposal had been put into practice.[18] X-ray photographs of NaRb-(+)-tartrate crystals were made using Zr radiation (to excite the anomalous scattering of the Rb atoms), and analysis of the intensity differences between Friedel pairs showed clearly that the absolute configuration of the (+)-tartrate ion was indeed III. Hence (+)-glyceraldehyde was indeed I. A few years later, a similar analysis of isoleucine HBr confirmed the correctness of IV for the configuration of the natural amino acids.[19]

Since 1950 the absolute configurations of about 500 optically active compounds have been determined using Bijvoet's method or minor modifications of it. In almost all cases where comparisons are possible, the results agree with those obtained by internal chemical correlation. The Bijvoet method is still probably the only reliable one for deciding on which side of the mirror plane the pictures of the chiral molecules of life on this planet stand. Which of a pair of enantiomorphic structures corresponds to milk and which to looking-glass milk?[20] This problem may have no utilitarian significance, but it must have perplexed many chemists and natural philosophers during the last 100 years.[21]

The principle behind the use of anomalous scattering for phase determination and enantiomorph discrimination is perhaps best explained by a simple example. Imagine a crystal referred to a right-handed coordinate system **a, b, c** (Fig. 3.6). The law of vector multiplication ensures

[18] J. M. Bijvoet, A. F. Peerdeman, and A. J. van Bommel, *Nature (London)* **168,** 271 (1951); A. F. Peerdeman, A. J. van Bommel, and J. M. Bijvoet, *Proc. Koninkl. Ned. Akad. Wetenschap* **B54,** 16 (1951).

[19] J. Trommel and J. M. Bijvoet, *Acta Crystallogr.* **7,** 703 (1954).

[20] "Perhaps Looking-glass milk isn't good to drink," wondered Alice.

[21] It is curious that Bijvoet's transformation of stereochemistry from a relative to an absolute basis is not even mentioned in Martin Gardner's otherwise excellent book, *The Ambidextrous Universe* (Basic Books, New York, London 1964), which deals in a semipopular fashion with problems of asymmetry in chemistry, biology, and physics.

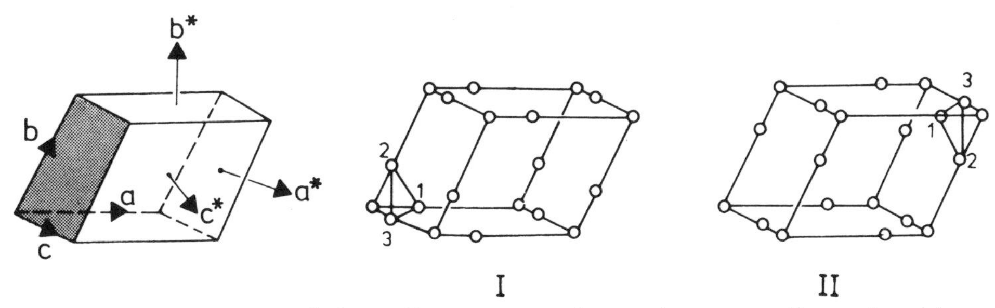

Figure 3.6. Left: right-handed coordinate system **a**, **b**, **c** and corresponding reciprocal lattice vectors **a***, **b***, **c***. The shaded face is $(\bar{1}00)$; the opposite one is (100). Middle and right: enantiomorphic structures I and II based on the same unit cell, related by inversion through the origin. In the absence of anomalous scattering $F_{\mathbf{H}}$ and $F_{\bar{\mathbf{H}}}$ are complex conjugate and $F^{\mathrm{I}}_{\mathbf{H}} = F^{\mathrm{II}}_{\bar{\mathbf{H}}}$.

that the reciprocal axes, **a***, **b***, **c***, defined by

$$\mathbf{a} = \frac{\mathbf{b}^* \times \mathbf{c}^*}{V} \qquad \mathbf{b} = \frac{\mathbf{c}^* \times \mathbf{a}^*}{V} \qquad \mathbf{c} = \frac{\mathbf{a}^* \times \mathbf{b}^*}{V}$$

also form a right-handed system (see footnote 11, Chapter 2), so that the directions of the vectors

$$\mathbf{H}(hkl) = h\mathbf{a}^* + k\mathbf{b}^* + l\mathbf{c}^*$$

with respect to the crystal are uniquely defined for any set of positive or negative integers (hkl). Assume that the unit cell contains a heavy atom (scattering power f_h) at the origin, plus a chiral arrangement of other atoms at positions $\mathbf{r}_j$ (fractional coordinates x_j, y_j, z_j with respect to the **a**, **b**, **c** axes defined above). In Fig. 3.7, left, we have the phase diagram for a Friedel pair of reflections. Since we have fixed the **a**, **b**, **c** coordinate system used to describe the crystal, there is no difficulty in deciding which reflection corresponds to **H** and which to $\bar{\mathbf{H}}$. However, in the absence of anomalous scattering we expect to find, within experimental error, $|F_{\mathbf{H}}| = |F_{\bar{\mathbf{H}}}|$ since

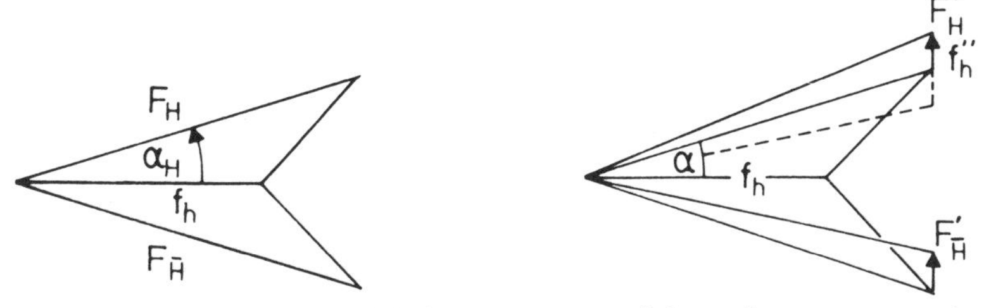

Figure 3.7. Phase diagram for chiral structure containing a heavy atom at the origin. Left: normal scattering conditions. Right: heavy atom as anomalous scatterer.

$$F_{\mathbf{H}} = f_h + \sum f_j \exp(2\pi i \mathbf{H}\cdot\mathbf{r}_j)$$
$$F_{\bar{\mathbf{H}}} = f_h + \sum f_j \exp[2\pi i(-\mathbf{H})\cdot\mathbf{r}_j]$$

are complex conjugates.

In principle we could determine the magnitude of the phase angle $\alpha_{\mathbf{H}}$ from isomorphous replacement measurements (as explained in the previous section), but its sign would still be uncertain. In other words, we would still not know which resultant vector corresponded to reflection **H** and which to $\bar{\mathbf{H}}$; this is another way of saying that the chirality sense of the structure cannot be determined. Reversing the chirality is equivalent to reversing the signs of the x, y, z coordinates of all the atoms while preserving the directions of the **a**, **b**, **c** axes. Since

$$\exp[2\pi i\mathbf{H}\cdot(-\mathbf{r})] = \exp[2\pi i(-\mathbf{H})\cdot\mathbf{r}]$$

the structure factors $F_{\mathbf{H}}$ for a chiral arrangement equal the structure factors $F_{\bar{\mathbf{H}}}$ of the enantiomorph in magnitude and phase.

In Chapter 1 we noted that if the incident wave is written as:

$$\exp[2\pi i(\nu t - R/\lambda)]$$

then an increase in path by ΔR corresponds to a phase difference of $-2\pi\Delta R/\lambda$. Thus far our discussion has been based on the assumption that the only phase differences involved in scattering are those arising from path differences. Suppose that in our example the heavy atom at the origin is an anomalous scatterer—i.e., that there is an additional phase change on scattering caused by a special interaction between this atom and the incident wave, apart from any path differences involved. This intrinsic phase change can be taken into account by expressing the scattering factor of the heavy atom as a complex number

$$f_h = |f_h| \exp(i\phi) = f'_h + if''_h \tag{3.5}$$

Here f'_h is real and positive (like the scattering factor of a "normal" atom). The theory of the scattering process shows that f''_h must also be positive; in other words, the phase angle ϕ must lie between 0° and +90°, corresponding to an apparent retardation of the wave scattered by the anomalous scatterer in relation to the waves scattered by the other atoms.

Fig. 3.7 shows the effect of adding the imaginary contribution if''_h to the normal phase diagram. The vector resultants $F'_{\mathbf{H}}$ and $F'_{\bar{\mathbf{H}}}$ are no longer equal in magnitude, so the intensities of the Friedel pair of reflections **H** and $\bar{\mathbf{H}}$ will be different. Assume that we have already determined the structure, say by the heavy-atom method, apart from its sense of chirality. That is, we have a set of atomic coordinates x_j, y_j, z_j referred to a right-handed system, but we do not know whether or not to invert through the origin (see structures I and II of Fig. 3.6). We now compute for structure I:

$$F'_{\mathbf{H}} = f'_h + if''_h + \sum f_j \exp(2\pi i \mathbf{H}\cdot\mathbf{r}_j)$$

$$F'_{\bar{\mathbf{H}}} = f'_h + if''_h + \sum f_j \exp[2\pi i(-\mathbf{H})\cdot\mathbf{r}_j]$$

and obtain, say, $|F'_{\mathbf{H}}| > |F'_{\bar{\mathbf{H}}}|$, as shown in Fig. 3.7. The corresponding intensities are experimentally available quantities. If $I_{\mathbf{H}} > I_{\bar{\mathbf{H}}}$, the set of atomic coordinates describing structure I is indicated as being in the correct reference frame. If $I_{\mathbf{H}} < I_{\bar{\mathbf{H}}}$, contrary to the calculated result, then the structure has to be inverted ($x_j, y_j, z_j \rightarrow -x_j, -y_j, -z_j$).

In practice, of course, the calculation of $|F'_{\mathbf{H}}|$ and $|F'_{\bar{\mathbf{H}}}|$ is made for several pairs of reflections, and a conclusion regarding the chirality of the structure can be drawn only if the results are self-consistent. Table 3.1, based on the original analysis of NaRb-(+)-tartrate,[18] shows that the discrimination in this case was quite clear-cut. The configuration derived from chemical correlation, based on the conventional assignment of configuration I to represent (+)-glyceraldehyde, is confirmed, and hence the element of arbitrariness previously present in writing structures of chiral molecules is eliminated.

Table 3.1. Sodium rubidium (+)-tartrate: comparison of $I_{\mathbf{H}}$ and $I_{\bar{\mathbf{H}}}$

h k l	Observed[a]	$I_{\mathbf{H}}/I_{\bar{\mathbf{H}}}$[b]
1 5 1	?	1.08
1 6 1	>	1.30
1 7 1	<	0.83
1 8 1	>	1.25
1 9 1	>	1.41
1 10 1	>	1.19
1 11 1	<	0.66
2 6 1	>	1.01
2 7 1	>	3.00
2 8 1	>	1.07
2 9 1	?	1.02
2 10 1	<	0.84
2 11 1	?	1.07
2 12 1	<	0.92
2 13 1	<	0.51

[a]Zr radiation; > means that the reflection **H** was observed to be more intense than $\bar{\mathbf{H}}$, < less intense.

[b]Calculated for the tartrate molecule according to chemical convention with f''(Rb) = +3.0.

Returning to our example, if the structure in question were unknown, the phases of the reflections could be determined (or at least restricted to certain values) from the experimentally available values of $|F'_{\mathbf{H}}|$ and $|F'_{\bar{\mathbf{H}}}|$ plus a knowledge of f''_h, which can be estimated theoretically.[22] From Fig. 3.7 we have:

$$|F'_{\mathbf{H}}|^2 = |F_{\mathbf{H}}|^2 + (f''_h)^2 - 2|F_{\mathbf{H}}|(f''_{\mathbf{H}}) \cos (90° + \alpha)$$

$$|F'_{\bar{\mathbf{H}}}|^2 = |F_{\mathbf{H}}|^2 + (f''_h)^2 - 2|F_{\mathbf{H}}|(f''_h) \cos (90° - \alpha)$$

$$|F'_{\mathbf{H}}|^2 - |F'_{\bar{\mathbf{H}}}|^2 = 4|F_{\mathbf{H}}|(f''_h) \sin \alpha$$

$$\sin \alpha = \frac{|F'_{\mathbf{H}}|^2 - |F'_{\bar{\mathbf{H}}}|^2}{4|F_{\mathbf{H}}|(f''_h)} \tag{3.6}$$

$$|F_{\mathbf{H}}|^2 = \frac{|F'_{\mathbf{H}}|^2 + |F'_{\bar{\mathbf{H}}}|^2}{2} - (f''_h)^2 \tag{3.7}$$

Given sin α, there are two possible values for α itself, one being the supplement of the other. In our case, with a heavy atom at the origin, we would choose the angle closer to zero; in general, the angle closer to the phase of the heavy-atom contribution. Moreover, the magnitude of α (but not its sign) can be determined from isomorphous replacement measurements. The combination—isomorphous replacement to determine cos α and anomalous scattering to determine sin α—suffices to determine the phase angles $\alpha_{\mathbf{H}}$ unambiguously, apart from uncertainty introduced by errors in the experimental intensity measurements. With $|F_{\mathbf{H}}|$ and $\alpha_{\mathbf{H}}$ determined for a sufficient number of reflections, the resulting Fourier synthesis $\rho(\mathbf{r})$ should reveal the atomic positions already in the correct absolute reference frame.

With more than one anomalous scatterer in the cell, the phase diagram corresponding to Fig. 3.7 may be somewhat more complicated, but the same principles apply. As an example, the phase diagram for a noncentrosymmetric arrangement of anomalous scatterers is shown in Fig. 3.8. We want to derive the phase angle $\alpha_{\mathbf{H}}$ of $F(\mathbf{H})$ from the experimentally available values of $|F_a(\mathbf{H})|$ and $|F_a(\bar{\mathbf{H}})|$, measured under conditions where a group of heavy atoms scatter anomalously. If the heavy atom positions are known, their contribution to the structure factors can be calculated. Let their normal contribution be $|F_h(\mathbf{H})|$ with phase $\alpha'_{\mathbf{H}}$, the anomalous part $|F''_h(\mathbf{H})|$ with phase $\alpha'_{\mathbf{H}} + 90°$. For clarity the relevant triangles are shown in Fig. 3.8 (right), where the vector $F_a(\bar{\mathbf{H}})$ has been

[22]For a compilation, see *International Tables for X-Ray Crystallography,* Vol. III, Kynoch Press, Birmingham, 1962, pp. 213–216.

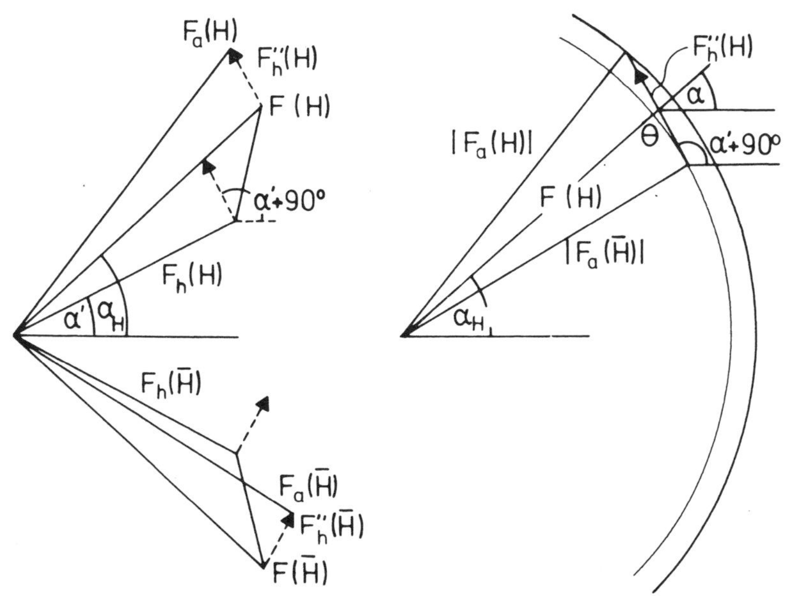

Figure 3.8. Left: phase diagram for a noncentrosymmetric group of heavy atoms with anomalous scattering contributions $F_h''(\mathbf{H})$ and $F_h''(\overline{\mathbf{H}})$. The normal contributions are $F_h(\mathbf{H})$ (with phase angle α') and $F_h(\overline{\mathbf{H}})$. We seek the phase angle $\alpha_{\mathbf{H}}$ of the total structure factor $F(\mathbf{H})$. Right: derivation of phase angle $\alpha_{\mathbf{H}}$.

reflected across the real axis of the Argand diagram. We have

$$|F_a(\mathbf{H})|^2 = |F(\mathbf{H})|^2 + |F_h''(\mathbf{H})|^2 + 2|F(\mathbf{H})||F_h''(\mathbf{H})| \cos\theta$$

$$|F_a(\overline{\mathbf{H}})|^2 = |F(\mathbf{H})|^2 + |F_h''(\mathbf{H})|^2 - 2|F(\mathbf{H})||F_h''(\mathbf{H})| \cos\theta$$

so that
$$\cos\theta = \frac{|F_a(\mathbf{H})|^2 - |F_a(\overline{\mathbf{H}})|^2}{4|F(\mathbf{H})||F_h''(\mathbf{H})|} \tag{3.8}$$

where
$$|F(\mathbf{H})|^2 = \frac{|F_a(\mathbf{H})|^2 + |F_a(\overline{\mathbf{H}})|^2}{2} - |F_h''(\mathbf{H})|^2 \tag{3.9}$$

From Fig. 3.8 it is clear that $\theta = \alpha'_{\mathbf{H}} + 90° - \alpha_{\mathbf{H}}$. However, as in the previous example, we can determine only $\cos\theta$, leaving the sign of θ open. We therefore obtain two solutions for the desired phase angle

$$\alpha_{\mathbf{H}} = \alpha'_{\mathbf{H}} + 90° \pm \theta \tag{3.10}$$

The ambiguity can be resolved by isomorphous replacement measurements, but if these are not available, the phase angle closer to the heavy-atom phase α' can be taken as the more likely. A more serious problem is that the chirality sense of the heavy-atom arrangement is not usually known at this stage of the analysis. Thus, the heavy-atom phase angle could be taken either as $\alpha'_{\mathbf{H}}$ or as $-\alpha'_{\mathbf{H}}$ (for the opposite enantiomorph). The probable phase angles $\alpha_{\mathbf{H}}$ for these two possibilities differ and lead to distinct (not enantiomorphic) Fourier syntheses. One of these syntheses would be an improvement on the normal heavy-atom phased Fourier; the other would be worse.

This problem was encountered in the analysis of the Cs salt of a degradation product of boromycin (an unusual antibiotic compound containing boron).[23] The space group is $P2_12_12_1$, and the set of four Cs atoms in general positions of this space group is itself chiral. Of the two electron-density maps calculated with phase angles derived from Bijvoet differences for the two enantiomorphic Cs arrangements, one was completely uninterpretable, and the other showed peaks corresponding to several chemically plausible fragments that could be recognized as defining the rough overall shape of the molecule. Thus, the absolute configuration of the molecule was established before its structural formula.

The arrangement *and* chirality of the heavy-atom anomalous scatterers—indeed the positions of all the atoms in an absolute reference frame—can, in principle, be deduced from a special kind of Patterson function introduced by Okaya, Saito, and Pepinsky.[24] If the form factors are complex quantities we may write

$$F_{\mathbf{H}} = \sum^{N} (f'_j + if''_j) \exp(2\pi i\mathbf{H}\cdot\mathbf{r}_j)$$

$$F^*_{\mathbf{H}} = \sum^{N} (f'_j - if''_j) \exp(-2\pi i\mathbf{H}\cdot\mathbf{r}_j)$$

$$|F_{\mathbf{H}}|^2 = F_{\mathbf{H}}F^*_{\mathbf{H}} = \sum_{j,\,k} (f'_jf'_k + f''_jf''_k - i(f'_jf''_k - f''_jf'_k)) \exp[2\pi i\mathbf{H}\cdot(\mathbf{r}_j - \mathbf{r}_k)]$$

to obtain

$$|F_{\mathbf{H}}|^2 + |F_{\overline{\mathbf{H}}}|^2 = 2\sum_{j,\,k} (f'_jf'_k + f''_jf''_k) \cos 2\pi\mathbf{H}\cdot(\mathbf{r}_j - \mathbf{r}_k)$$

$$|F_{\mathbf{H}}|^2 - |F_{\overline{\mathbf{H}}}|^2 = 2\sum_{j,\,k} (f'_jf''_k - f''_jf'_k) \sin 2\pi\mathbf{H}\cdot(\mathbf{r}_j - \mathbf{r}_k)$$

The Patterson function with $|F_{\mathbf{H}}|^2 + |F_{\overline{\mathbf{H}}}|^2$ as coefficients is essentially the normal Patterson function, the only difference being that the peaks corresponding to $\mathbf{r}_j - \mathbf{r}_k$ vectors have slightly modified weights. The differences $|F_{\mathbf{H}}|^2 - |F_{\overline{\mathbf{H}}}|^2$ are Fourier coefficients of an odd function

[23] J. D. Dunitz, D. M. Hawley, D. Mikloš, D. N. J. White, Y. Berlin, R. Marušić, and V. Prelog, *Helv. Chim. Acta* **54,** 1709 (1971).

[24] Y. Okaya, Y. Saito, and R. Pepinsky, *Phys. Rev.* **98,** 1857 (1955).

$$P_s(\mathbf{r}) = \frac{1}{V} \sum (|F_\mathbf{H}|^2 - |F_{\overline{\mathbf{H}}}|^2) \sin 2\pi \mathbf{H}\cdot\mathbf{r} \tag{3.11}$$

with peaks of weight $f_j'f_k'' - f_j''f_k'$ at the termini of vectors from atom k to atom j. For simplicity, suppose that there are only two kinds of atoms present, one kind, N, being normal, and the other kind, A, being anomalous scatterers. Peaks between pairs of N atoms and between pairs of A atoms have zero weight since $f_j'f_k'' - f_j''f_k' = 0$ for both cases. The only peaks present in $P_s(\mathbf{r})$ then correspond to vectors between different kinds of atoms. Peaks corresponding to vectors $\mathbf{r}_N - \mathbf{r}_A$ ($\overrightarrow{AN}$, *from* an A atom *to* an N atom) are positive, of weight $+f_N'f_A''$; those in the reverse direction are negative. If only a single A atom is present in the unit cell, then the whole structure including its chirality sense could, in principle, be read off directly from the difference function, apart from the fact that some of the atoms might not be recognized because of chance overlapping of positive and negative peaks.

Fig. 3.9 illustrates how the correct chirality of the arrangement of heavy-atom anomalous scatterers A can be recognized from the P_s function. From the normal Patterson function the arrangement itself can be derived, provided the AA peaks are strong enough to show above the background. However, because the function is centrosymmetric the choice of enantiomorph is arbitrary. The P_s function, however, is antisymmetric across the origin. Its negative peaks, corresponding to $\overrightarrow{NA}$ vectors, contain images of the A atoms as seen *from* the normal atoms, which ensures that these images occur with the same chirality sense as

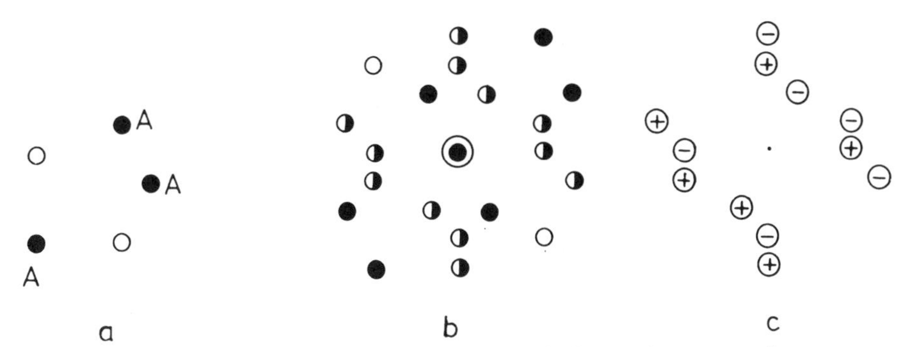

Figure 3.9. (a) Arrangement of five atoms, of which three (A) are anomalous scatterers, and two (N) are normal. (b) Normal Patterson function; AA vectors are filled circles, AN and NA vectors are half-filled circles, and NN vectors are open circles. The pattern is centrosymmetric and contains images of the actual AAA arrangement and of its enantiomorph. (c) Odd Patterson function P_s. The $\overrightarrow{AN}$ vectors give positive peaks, and $\overrightarrow{NA}$ vectors give negative ones. The pattern of negative peaks contain only images of the correct enantiomorph.

that of the actual arrangement. Images of the N atoms, as seen from the A atoms, are to be found in the arrangement of positive peaks $(\overrightarrow{AN})$

Although several structures including absolute configuration have been determined by using the P_s function,[25] the main interest of the method would seem to be in illustrating the principles involved rather than in its utility. In practice, serious difficulties may arise in unravelling the pattern of positive and negative peaks and in allowing for the effect of possible overlaps. The procedure most often used to determine absolute configuration is still the original one of first determining the structure by the heavy-atom method and then establishing its correct chirality by comparison of Bijvoet pairs, as indicated in Table 3.1.

It should be stressed that the determination of chirality by anomalous scattering of X-rays stands or falls on the correctness of the positive sign of f''_h in Eq. 3.5. If f''_h, the imaginary part of the atomic scattering factor, were negative, the Fischer convention would be incorrect and all absolute configurations of molecules determined so far would have to be reversed. At least the method would be self-consistent. Since there does not seem to be any other method for determining absolute configuration that gives entirely reliable results, it seems worthwhile to discuss the reasons why the sign of f''_h has to be taken as positive. For this we need to fill in some gaps that were left in the brief introductory paragraph of Chapter 1.

The main qualitative aspects of the theory of scattering of electromagnetic waves by atoms can be derived from a completely classical model which leads to well-known differential equations. In this model each electron is regarded as a particle of mass m, charge e, subject to the alternating electric field $E = E_0 \exp(i\omega t)$ of an electromagnetic wave with frequency ω and bound to its nucleus by a restoring force. The external force on the electron is eE, so the equation of motion is

$$mx'' + kx = E_0 \, e \exp(i\omega t) \tag{3.12}$$

and the steady-state solution is of the form

$$\begin{aligned} x &= A \exp(i\omega t) \\ x'' &= -\omega^2 A \exp(i\omega t) \end{aligned} \tag{3.13}$$

Substituting Eq. 3.13 into Eq. 3.12

$$A = \frac{E_0 \, e}{k - m\,\omega^2} \tag{3.14}$$

[25]See, for example, R. Pepinsky, *Rec. Chem. Progr.* **17**, 145 (1956); Y. Okaya and R. Pepinsky, *Computing Methods and the Phase Problem in X-ray Analysis*, R. Pepinsky, J. M. Robertson, and J. C. Speakman, Eds., Pergamon Press, Oxford, 1961, p. 273ff.

For a free electron, $k = 0$ and the displacement is given by

$$x = \frac{-E_o\, e}{m\, \omega^2} \exp (i\omega t) = \frac{E_o\, e}{m\, \omega^2} \exp [i(\omega t+\pi)] \tag{3.15}$$

That is, the electron oscillates with the same frequency as the electromagnetic wave, but its displacement is π out of phase with the external force. For a bound electron

$$x = \frac{E_o\, e}{k-m\omega^2} \exp (i\omega t) \tag{3.16}$$

The amplitude of the scattered wave is proportional to the displacement x. We recall that to eliminate proportionality constants, the structure factor was defined earlier as the ratio of the amplitude of the radiation scattered by a distribution $\rho(\mathbf{r})$ to that scattered by a free electron at the origin. For the structure factor of the bound electron we then have

$$f = \frac{m\omega^2}{m\omega^2-k} = \frac{\omega^2}{\omega^2-\omega_o^2} \tag{3.17}$$

where $\omega_o = (k/m)^{\frac{1}{2}}$ is called the characteristic frequency of the oscillator. For X-rays, the frequency ω corresponds to an energy of several thousand eV. The frequency ω_o can be regarded as roughly equivalent to the ionization potential of the electron, which is typically much smaller than ω, except for K and L electrons of heavy atoms. For first-row elements, $\omega^2 \gg \omega_o^2$ holds for all electrons, so that each electron contributes +1 to the form factors of these atoms. For heavy atoms, $\omega^2 \gg \omega_o^2$ will hold for all but the most tightly bound electrons, which may then scatter with opposite phase to the others; in this case $f_i(0) < Z_i$.

The case $\omega^2 = \omega_o^2$ leads to an infinite displacement, velocity, and acceleration of the electron in question. To avoid this catastrophe in the classical model, we introduce a damping coefficient g and rewrite the equation of motion as

$$mx'' + gx' + kx = E_o e \exp (i\omega t) \tag{3.18}$$

leading to

$$A = \frac{E_o\, e}{k-m\omega^2+ig\omega}$$

and

$$f = \frac{m\omega^2}{m(\omega^2-\omega_o^2)-ig\omega}.$$

On separating f into real and imaginary parts, we get

$$f = m\omega^2 \left(\frac{m(\omega^2-\omega_o^2)+ig\omega}{m^2(\omega^2-\omega_o^2)^2+g^2\omega^2} \right) = f' + if'' \tag{3.19}$$

Since m, g, and ω are all positive, the coefficient f'' of the imaginary term must be positive, which is what we wanted to show. The scattering power can be represented by a vector $|f| \exp(i\phi)$ where

$$|f| = \frac{m\omega^2}{[m^2(\omega^2-\omega_o^2)^2+g^2\omega^2]^{\frac{1}{2}}}$$
$$\tan \phi = \frac{g\omega}{m(\omega^2-\omega_o^2)} \ . \qquad (3.20)$$

The angle ϕ through which the vector is rotated from the real axis is positive and proportional to the damping coefficient g, which, within the framework of this mechanical model, might be thought of as a kind of energy dissipation factor analogous to friction. A large value of g should therefore imply a loss of energy or, what is the same, absorption of the incident wave. This is about all that can be said about g, which of course was introduced somewhat artificially to avoid the infinity catastrophe when $\omega = \omega_o$. In particular, there is no way to estimate the magnitude of g for particular electrons from known or calculated atomic properties.

The classical model is therefore unsuitable for calculating the magnitudes of f' and f''. However, one would expect that the signs of these quantities are correct and should be reproduced in a quantum-mechanical treatment. Although classical mechanics is completely inadequate for describing certain kinds of phenomena, and its description of other phenomena is valid only in the "classical limit," it would be remarkable if processes that can be described in classical terms turned out to run backwards according to quantum mechanics. Indeed, the only important difference between the two models is that according to quantum mechanics, f'' is zero when $\omega < \omega_o$, whereas the classical model gives a finite value. The scattering factor has an imaginary component only when the incident wavelength is close to but less than that of an absorption edge, a sudden discontinuity in an atomic absorption coefficient as function of wavelength. For example, Fe atoms strongly absorb radiation of wavelength less than 1.74 Å, at which there is a large, sudden drop in the absorption coefficient (Fig. 3.10). They strongly absorb Cu$K\alpha$ radiation (λ = 1.542 Å), whereas Ni and the following elements have a much smaller absorption for this radiation (Fig. 3.11). However, Ni absorbs Cu$K\beta$ radiation (λ = 1.392 Å) very strongly (which is why thin sheets of Ni are used to filter out this wavelength from Cu radiation).

Phenomena of this kind are mysterious from a classical viewpoint, but they can be interpreted readily in terms of a quantum description—in fact, they are the kind of phenomena that showed the need for quantum

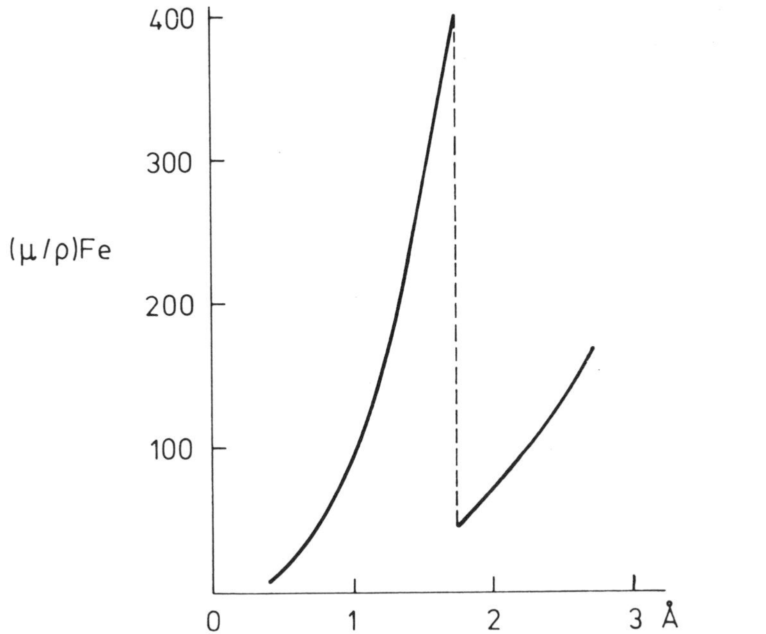

Figure 3.10. Mass absorption coefficient μ/ρ of Fe as function of wavelength (in Å).

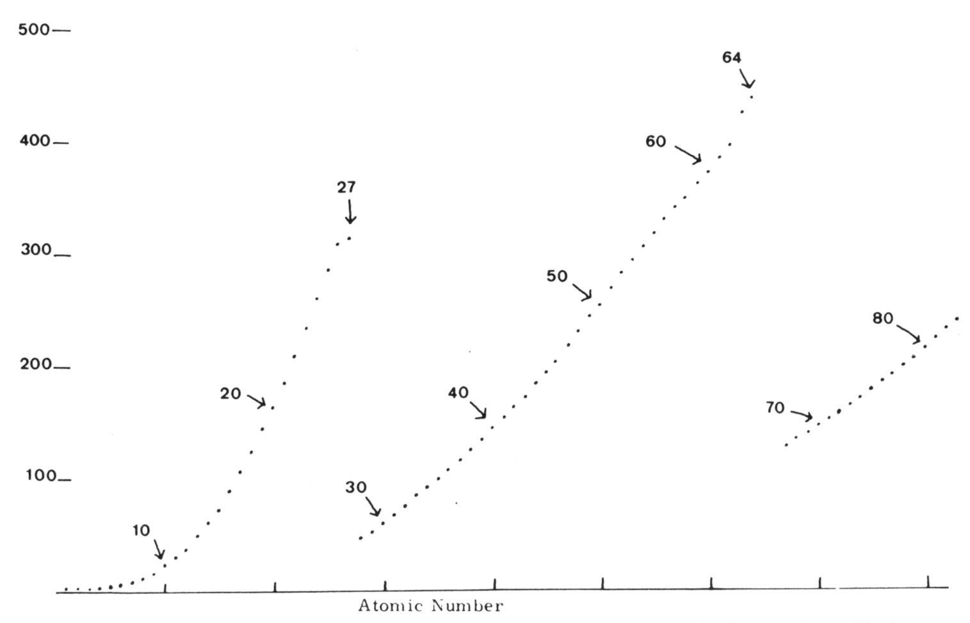

Figure 3.11. Mass absorption coefficients (μ/ρ) of elements (1–83) for CuKα radiation (λ = 1.542 Å).

theory. During absorption, when the X-ray photon disappears, its energy is used to excite a low-energy electron to a higher quantum state or to eject the electron from the atom completely; excess energy is carried away by the photoelectron as kinetic energy. The probability of ejecting a K electron is maximal when the energy of the incident photons is just great enough for the process to occur (analogous to the infinity catastrophe for $\omega = \omega_o$ in the classical model). However, the probability drops to zero if the energy of the incident photon is even slightly less than that required to promote the electron to the lowest unoccupied level. For Fe this energy corresponds to a wavelength of 1.74 Å ($\approx 7.1 \times 10^3$ eV). The atoms excited by absorption of a photon eliminate most of their excess energy by emitting a photon of secondary X-radiation (fluorescence), but since this is of lower frequency than the incident radiation it plays no part in interference phenomena. The secondary, fluorescent radiation from a crystal contributes to the background scattering only.

This is not the place to discuss the quantum-mechanical description of anomalous scattering in detail, nor is the writer capable of fathoming the physical arguments, approximations, and mathematical complications involved. The original approach, due to Hönl,[26] is dealt with in James's monograph,[27] and more recent developments are summarized by Wagenfeld[28] and Groenewege.[29] In the quantum-mechanical model, the atomic form factor for scattering in the positive direction can be expressed in the form

$$f = f_o - \frac{1}{2} \sum g_{nm}\, \omega_{nm} \left(\frac{1}{\omega_{nm} - \omega} + \frac{1}{\omega_{nm} + \omega} \right)$$

where f_o is the normal X-ray scattering factor, g_{nm} is the oscillator strength for the transition from state n to state m and $\omega_{nm} = 2\pi(E_m - E_n)/h$. For $\omega \gg \omega_{nm}$ (transitions from K shell to continuum in light atoms) the term in parentheses becomes

$$-\frac{1}{\omega} + \frac{1}{\omega} = 0$$

so $f = f_o$. For $\omega \ll \omega_{nm}$ (transitions from K shell to continuum in heavy atoms) the correction terms become

[26]H. Hönl, *Z. Phys.* **84,** 11 (1933); *Ann. Phys.* **18,** 625 (1933).

[27]R. W. James, *Optical Principles of the Diffraction of X-rays; The Crystalline State,* Vol. 2, Bell, London, 1958.

[28]H. Wagenfeld, *Anomalous Scattering,* S. Ramaseshan and S. C. Abrahams, Eds., Munksgaard, Copenhagen, 1974, p. 13 (published for IUCr.).

[29]M. P. Groenewege, *Anomalous Scattering,* S. Ramaseshan and S. C. Abrahams, Eds., Munksgaard, Copenhagen, 1974, p. 25.

$$-\frac{1}{2}\sum g_{nm}\,\omega_{nm}\left(\frac{1}{\omega_{nm}}+\frac{1}{\omega_{nm}}\right)=-\sum g_{nm}$$

so

$$f=f_o-\sum g_{nm}$$

Radiation damping for $\omega \approx \omega_{nm}$ can be taken into account by writing the relevant correction terms as

$$-\frac{1}{2}\,g_{nm}\,\omega_{nm}\left(\frac{1}{\hat{\omega}_{nm}-\omega}+\frac{1}{\hat{\omega}^*_{nm}+\omega}\right)$$

with $\hat{\omega}_{nm} = \omega_{nm} + \frac{1}{2}i\Gamma$ where Γ is a positive constant. This expression can be rewritten as

$$-\frac{1}{2}\,g_{nm}\,\omega_{nm}\,\frac{(\hat{\omega}^*_{nm}+\hat{\omega}_{nm})}{\hat{\omega}_{nm}\hat{\omega}^*_{nm}-\omega^2+\omega(\hat{\omega}_{nm}-\hat{\omega}^*_{nm})}$$

$$=\frac{g_{nm}\,\varphi^2_{nm}}{\omega^2-\omega^2_{nm}-\Gamma^2/4-i\omega\Gamma}$$

The real and imaginary parts of each correction term are then:

$$\delta'_{nm}=\frac{g_{nm}\omega^2_{nm}(\omega^2-\omega^2_{nm}-\Gamma^2/4)}{(\omega^2-\omega^2_{nm}-\Gamma^2/4)^2+\omega^2\Gamma^2}$$

$$\delta''_{nm}=\frac{g_{nm}\omega^2_{nm}\omega\Gamma}{(\omega^2-\omega^2_{nm}-\Gamma^2/4)^2+\omega^2\Gamma^2}$$

The real part of each correction term now changes sign when $\omega^2 = \omega^2_{nm} + \Gamma^2/4$, that is, it becomes negative when ω is still slightly greater than ω_{nm}. The imaginary part is positive as before and is proportional to the oscillator strength of the transition or to the photoelectric absorption cross-section of the electron involved. It is zero if $\omega < \omega_k$ where ω_k is the ionization potential of the electron. Detailed calculations for estimating the magnitudes and angular dependence of f' and f'' in Eq. 3.5 involve expansion of transition probability matrix elements for suitable eigenfunctions into electromagnetic multipole transitions. Semi-empirical calculations based on the observed positions of absorption edges provide a useful alternative method.

Tanaka caused quite a flurry when he claimed in 1972 that the sign of f'' should be reversed and hence that all absolute configurations determined by X-ray analysis should be changed into their antipodes. His claim was based on an apparent contradiction between the absolute configurations of certain triptycene and 9,10-dihydroanthracene derivatives as derived from interpretation of circular-dichroism measurements and as determined by X-ray analysis,[30] and it was backed up by a re-examination of the theory of anomalous scattering in terms of a quantum-field model. This conclusion planted seeds of doubt in many minds. If the theoreticians could not agree about the sign of f'', then the absolute basis for stereochemistry, which had seemed to emerge from Bijvoet's work and had been unquestioned for 20 years, might prove illusory. What was clearly needed was an independent, *experimental* proof that the Bijvoet method gave correct results, but it was not clear from what direction this proof might come.[31] It came from an unexpected quarter—from noble gas ion reflection mass-spectrometry[32] from opposite faces of ZnS and CdS crystals.

Cubic crystals of ZnS (sphalerite) can be regarded as being built from an alternating sequence of layers consisting of Zn and S atoms. These layers, which are perpendicular to the [111] axis, are not equally spaced but are arranged as in Fig. 3.12, say S layers at 0, $\frac{4}{12}$, $\frac{8}{12}$, . . . of the periodicity along [111], Zn layers at $\frac{1}{12}$, $\frac{5}{12}$, $\frac{9}{12}$, . . . to form a polar double-layer structure. On one of the {111} faces the top layer will consist of Zn atoms; on the opposite face, it will consist of S atoms. In fact, opposite {111} faces differ in appearance and in chemical properties—e.g., etching behavior and crystal growth.

The polarity sense of ZnS crystals was determined more than 40 years ago by Coster, Knol, and Prins;[33] they were the first to demonstrate the breakdown of Friedel's Law under conditions of anomalous scattering. They used Au$L\alpha$ radiation, which consists of a doublet, Au$L\alpha_1$ (λ = 1.276 Å) and Au$L\alpha_2$ (λ = 1.288 Å). The K absorption edge of Zn lies just between these wavelengths at 1.283 Å, so that the α_1 component should excite anomalous scattering of the Zn atoms, whereas the α_2 component has just insufficient energy to do this. Other experiments were carried out with W$L\beta$ radiation, which also consists of several components, some on either side of the Zn absorption edge.

The two opposite {111} faces of a ZnS crystal are often distinguishable by eye. One face tends to be well developed with a brilliant, shiny

[30]J. Tanaka, *Acta Crystallogr.* **A28,** 229 (1972); J. Tanaka, F. Ogura, M. Kuritami, and M. Nakagawa, *Chimia* **26,** 471 (1972); J. Tanaka, C. Katayama, F. Ogura, H. Tatemitsu, and M. Nakagawa, *Chem. Commun.* **1973,** 21; J. Tanaka, K. Ozeki-Minakata, F. Ogura, and M. Nakagawa, *Nature, Phys.* **241,** 22 (1973).

[31]My colleague, Professor D. Arigoni, suggested the possibility of observing the helicity sense of DNA by electron microscopy. A given helicity sense would be compatible with only one chirality sense of D-2-deoxyribose. As far as I am aware, this experiment has not yet been carried out.

[32]H. H. Brongersma and P. M. Mul, *Chem. Phys. Lett.* **19,** 217 (1973).

[33]D. Coster, K. S. Knol, and J. A. Prins, *Z. Phys.* **63,** 345 (1930).

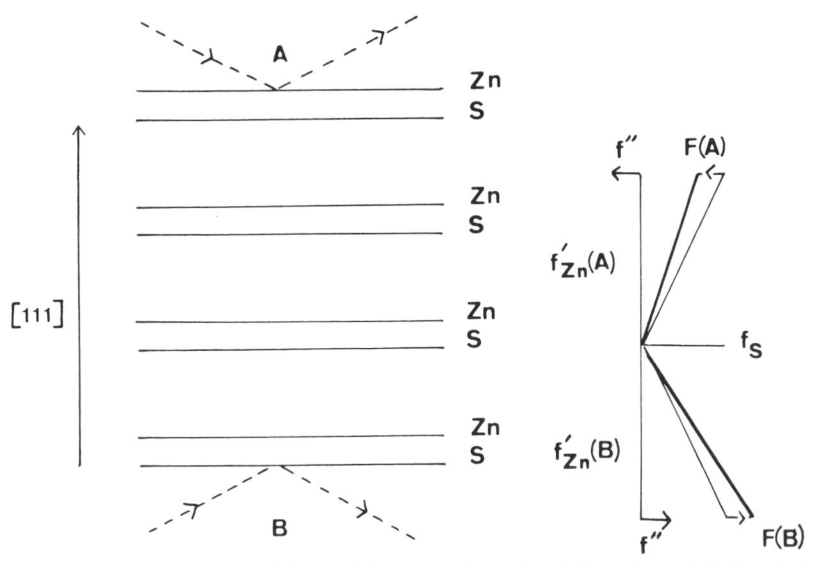

Figure 3.12. Double layers of Zn and S atoms along [111] direction of ZnS (sphalerite). Phase diagram for first-order reflection from opposite A and B faces, assuming $f_{Zn} = f'_{Zn} + if''_{Zn}$ with f''_{Zn} positive.

appearance; the other is poorly developed and dull. For the pair of first-order reflections from these opposite faces, the intensity reflected from the dull face was less than the intensity reflected from the shiny face under anomalous scattering conditions. For normal scattering the intensities were roughly equal. For the third-order reflections (anomalous scattering) the intensity relationship was reversed. If, for the Zn atoms, $f_{Zn} = f'_{Zn} + if''_{Zn}$ (Eq. 3.5), then for the layer arrangement shown in Fig. 3.12,

$$F(A) = f_S + i(f'_{Zn} + if''_{Zn}) = f_S - f''_{Zn} + if'_{Zn}$$

$$F(B) = f_S - i(f'_{Zn} + if''_{Zn}) = f_S + f''_{Zn} - if'_{Zn}$$

should hold for the first-order reflection, leading to $|F^2(A)| < |F^2(B)|$ if f''_{Zn} is positive. For the third-order reflection, the relation is reversed. Hence A (reflection from the side that terminates in a layer of Zn atoms) must be identified with reflection from the dull face, B with reflection from the shiny face, and the polarity sense of the crystals is established. It was Bijvoet who realized nearly 20 years later that the principles introduced here to connect macroscopic with molecular polarity were all that are required to determine the absolute configuration of chiral crystals. Polarity is just one-dimensional chirality.

Tanaka's claim that f'' might be negative can therefore be tested by an independent determination of the polarity sense of a polar crystal. Brongersma and Mul[32] found a way to do this. They reflected beams of positive ions with known incident energy from opposite faces of freshly cleaned and etched ZnS and CdS crystals and analyzed the energy of the reflected ions in a mass spectrometer. During reflection the ions lose an amount of energy that depends on the mass of the atoms with which they collide. For reflection at a 45° angle the calculation is straightforward:

	Before collision		After collision	
	Ions	Atoms	Ions	Atoms
Energy	$E_0 = \frac{1}{2} m v_0^2$	0	$E_1 = \frac{1}{2} m v_1^2$	$\frac{1}{2} MV^2$
Momentum ‖ plane	$\frac{mv_0}{\sqrt{2}}$	0	$\frac{mv_1}{\sqrt{2}}$	MV_x
Momentum ⊥ plane	$\frac{mv_0}{\sqrt{2}}$	0	$-\frac{mv_1}{\sqrt{2}}$	MV_y

where the meaning of the symbols is obvious. From conservation of energy and momentum:

$$M^2V^2 = 2M(E_0 - E_1)$$
$$MV_x = m(v_0 - v_1)/\sqrt{2}$$
$$MV_y = m(v_0 + v_1)/\sqrt{2}$$

Combining the two momentum equations,

$$M^2V^2 = M^2(V_x^2 + V_y^2) = m^2(v_0^2 + v_1^2)$$
$$= 2m(E_0 + E_1)$$

Hence
$$\frac{M}{m} = \frac{E_0 + E_1}{E_0 - E_1}$$

or
$$\frac{M - m}{M + m} = \frac{E_1}{E_0}$$

In one experiment He^+ ions ($m = 4$) with $E_0 = 2000$ eV were reflected from opposite faces of a CdS crystal at incidence angle of 45°. Collision with Cd atoms ($M = 112$) gives $E_1 = 2000 \times 108/116 = 1862$ eV, while collision with S atoms ($M = 32$) gives $E_1 = 2000 \times 28/36 = 1556$ eV. The experimental result (Fig. 3.13) leaves no doubt which face is which. On

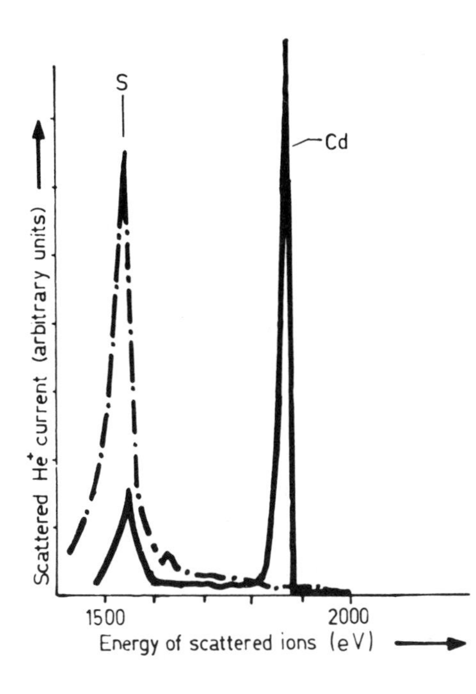

Figure 3.13. Solid and broken curves show the energy distributions obtained by reflecting 2000 eV He^+ ions at 45° angle of incidence from opposite faces of a CdS crystal. From Brongesma and Mul, *Chem. Phys. Lett.* **19,** 217 (1973).

one side only sulfur atoms are detected, while the Cd peak is the prominent feature on the opposite face. A similar experiment with ZnS, using Ne^+ ions of energy 1000 eV at $22\frac{1}{2}°$ angle of incidence gave an equally clear-cut identification of the opposite polar faces. In both cases the sense of polarity found was *in agreement* with that derived by anomalous scattering, assuming *a positive sign for* f''. Brongersma and Mul conclude that "the commonly accepted absolute configuration assignments for all asymmetric molecules and noncentrosymmetric crystals remain unchanged."

The disagreement found by Tanaka between conclusions based on circular-dichroism measurements and those based on crystal structure analysis did not pass unchallenged by chiropticians.[34] More correct interpretations of the circular-dichroism measurements restore the agreement. It thus appears on all sides that Tanaka's proposal to reverse all absolute configurations determined by anomalous scattering because of an error in the sign adopted for f'' was ill founded and can be rejected. The flurry it caused was short lived but useful in that it stimulated a more critical re-examination of what had been taken for granted and led

[34]S. F. Mason, *Chem. Commun.* **1973,** 239; A. F. Beecham, A. C. Hurley, A. Mc L. Mathieson, and J. A. Lamberton, *Nature, Phys.* **224,** 30 (1973); A. M. F. Hezemans and M. P. Groenewege, *Tetrahedron* **29,** 1223 (1973).

to new, conclusive experimental tests of the Bijvoet method for determining absolute configurations.

Direct Methods

In certain cases the phase problem can be solved by interpretation of the Patterson function—i.e., by translating information about the distribution of interatomic vectors into information about atomic positions. The method is particularly successful for structures containing a few heavy atoms per molecule, but even here the interpretation of the subsequent Fourier syntheses phased by the heavy-atom contributions may not be quite straightforward without a modicum of chemical structural information. Besides, if heavy atoms are present, the X-ray intensities are less sensitive to the nature and positions of the lighter atoms that usually make up the more interesting part of the molecule, with a consequent loss of accuracy in determining the latter. In principle, the phase problem can always be solved by multiple isomorphous replacement methods, but in practice enough derivatives showing the required degree of isomorphism are seldom available.

It would be convenient to have methods of deriving phases for the structure factors or of placing limitations on the possible phases that do not depend on the presence of heavy atoms and do not involve any structural assumptions. This is obviously asking too much because *any* assignment of phases to the Fourier coefficients corresponds to *some* electron-density distribution, and we must have criteria to distinguish correct or at least reasonable distributions from unreasonable ones. Two general criteria that any physically reasonable electron-density distribution must satisfy are: (1) The electron density is everywhere nonnegative. (2) The electron density consists of discrete, approximately spherical peaks, whose number and relative weights are usually known in advance. (This condition strictly applies only if the density function is sufficiently well resolved—i.e., only if sufficiently high-order Fourier coefficients are available.)

If the structure amplitudes are known, these criteria are sufficient to restrict the phase angles of many of the larger Fourier coefficients. The infinity of possible electron-density maps is thereby reduced to a small number, and these can be calculated individually and examined for stereochemically reasonable patterns of peaks. In favorable cases, one or more of the calculated distributions will be close enough to the true electron density to allow the approximate atomic positions, or at least a large fraction of them, to be recognized without much difficulty. Methods based on such phase-angle restrictions are known as direct methods.

A Simple Example.

Consider a simple example—a two-dimensional structure consisting of a single pair of point atoms related by a center of symmetry. The structure factors (Fourier coefficients) are

$$F(h,k) = 2 \cos 2\pi(hx_1 + ky_1)$$

where x_1, y_1 are the fractional coordinates along **a** and **b**—i.e., the structure factors are just a set of cosine fringes sampled at the reciprocal lattice points (h, k integral). Given the magnitudes of $F(h, k)$, their signs can be inferred by interpolating the zeroes of the cosine function, which occur at $hx_1 + ky_1 = (4n \pm 1)/4$. For the relative $|F(h, k)|$ values shown in Fig. 3.14, the first zero-trace cuts the **a***-axis at $h \approx -2.2$, the **b***-axis at $k \approx 0.75$, leading to

$$-2.2x_1 + 0(y_1) = \tfrac{1}{4} \qquad 0(x_1) + 0.75y_1 = \tfrac{1}{4}$$

or

$$x_1 = \frac{0.25}{-2.2} = -0.114 \qquad y_1 = \frac{0.25}{0.75} = 0.33$$

close to the coordinates that were actually used to calculate the $|F|$ values ($x_1 = -0.110$, $y_1 = 0.330$). The signs can thus be derived and the structure solved by inspection.

The example is trivial, but the method can be used in more complicated cases where, for example, the structure contains a pair of heavy atoms plus a number of lighter ones. Fig. 3.15 shows the intensity distribution of the $(h0l)$ reflections from a crystal of 1,6-*trans*-dibromocyclodecane with traces of the zeroes of an assumed set of cosine fringes superimposed.[35] The Fourier synthesis calculated with about 80 coefficients whose signs could thus be inferred showed not only the heavy bromine atoms but also clear indications of the carbon skeleton. For comparison the final electron-density projection is also shown in Fig. 3.15.

In Fig. 3.14, bottom, the signs associated with the large Fourier coefficients are marked. It is seen that if $F(\mathbf{H})$, $F(\mathbf{H}')$, and $F(\mathbf{H}+\mathbf{H}')$ are large, the product of their signs is positive (the signs associated with the

[35]H. P. Weber, Dissertation No. 3494, E.T.H., Zurich, 1964.

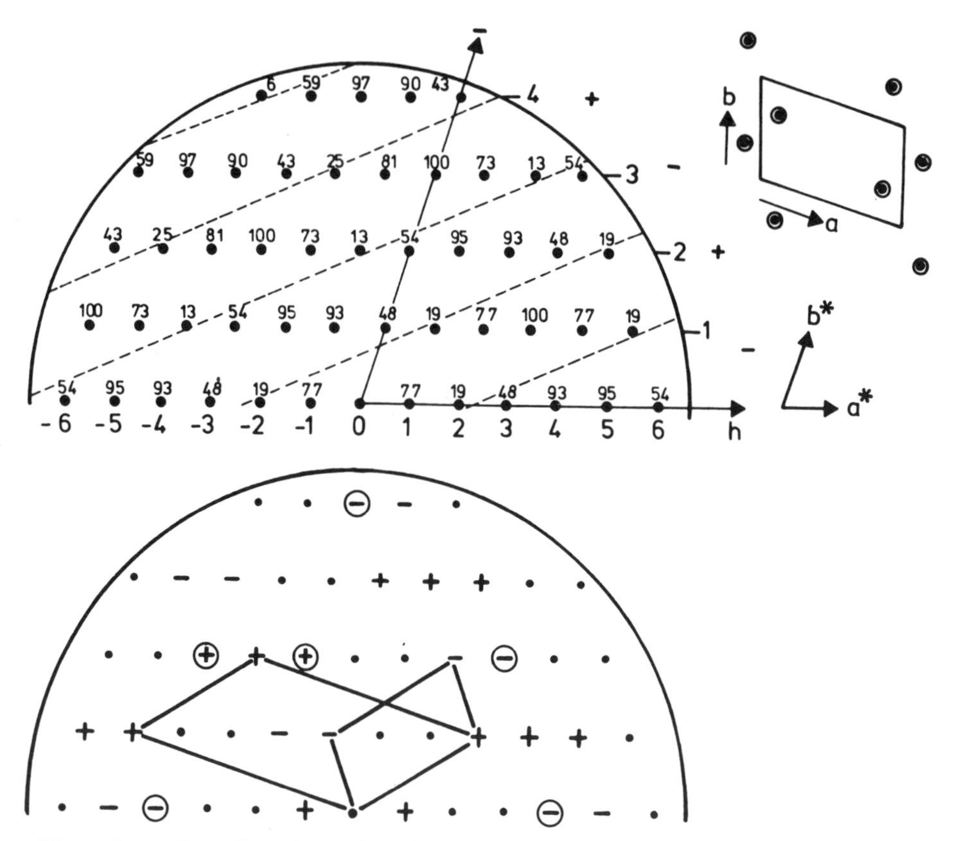

Figure 3.14. Top: the value of $|F(\mathbf{H})|$ for the point-atom structure on the right is shown for each R. L. point, and the zero lines of a set of possible cosine fringes are indicated. Bottom: signs of very large $F(\mathbf{H})$ values. Note that the product of signs of any three such values related as $\mathbf{H}$, $\mathbf{H}'$, $\mathbf{H}+\mathbf{H}'$ (vector parallelogram) is always positive. R. L. points with h and k both even are marked by circles.

vertices of the vector parallelogram are either $+++$ or $+--$). This restriction follows since

$$(h+h')x_1 + (k+k')y_1 = (hx_1+ky_1) + (h'x_1+k'y_1)$$

If $F(\mathbf{H})$ is large, then hx_1+ky_1 must be nearly integral (+) or half-integral (−) and similarly for $F(\mathbf{H}')$ and $F(\mathbf{H}+\mathbf{H}')$. Thus, either none or two such sums can be half-integral, and correspondingly, either none or two signs can be negative.

The array of fringes shown in Fig. 3.14 is not unique. For example, we could have chosen an alternative set of fringes such that the first zero trace cuts the **a***-axis at $h \approx 2.2$, the **b***-axis at $k \approx 1.5$, yielding $x_1 = +0.114$, $y_1 = 0.167$. These coordinates are related to the former ones by

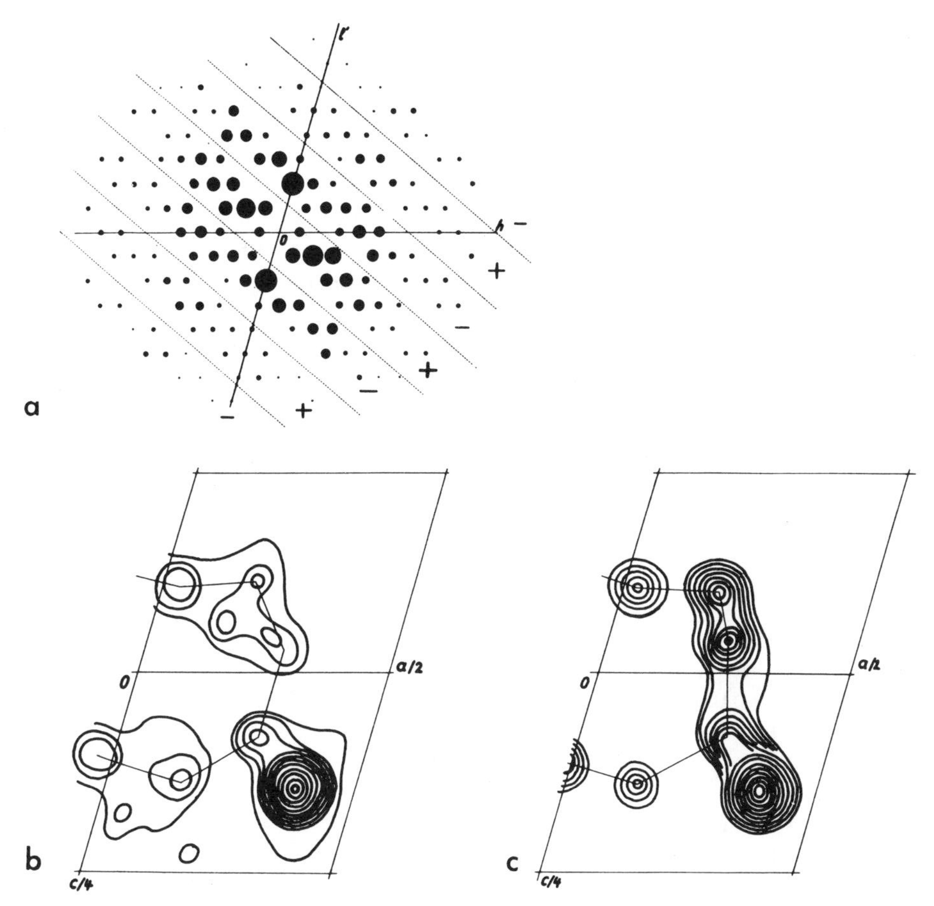

Figure 3.15. (a) Weighted reciprocal lattice for $(h0l)$ reflections of 1,6-*trans*-dibromocyclodecane. The radii of the circles are proportional to $|F|_{\text{obs}}$. (b) Fourier synthesis calculated with signs of large coefficients derived from (a). (c) Final Fourier synthesis.

a displacement of the origin ($x' = -x$, $y' = \frac{1}{2}-y$). The effect of the origin displacement on $F(h, k)$ is easily seen by expanding

$$F(h, k) = 2 \cos 2\pi[hx + k(y - \tfrac{1}{2})]$$

$$= 2 \cos 2\pi(hx + ky) \cos \pi k$$

The magnitude of $F(h, k)$ is unaltered, but its sign is reversed for k odd. This means that we may choose the sign of one $F(h, k)$ with k odd as we please; the choice will merely reflect the arbitrary choice of placing the origin at one set of symmetry centers or at another set displaced by $y = \frac{1}{2}$. Similarly the sign of one $F(h, k)$ with h odd can be chosen as we please. The choice of two such signs fixes the origin uniquely in two dimensions. However, the sign of an $F(h, k)$ with h and k both even is independent of the choice of origin and cannot be assigned arbitrarily. These reflections are marked with a circle in Fig. 3.14, bottom.

Let us now denote the sign of a Fourier coefficient by $S(\mathbf{H})$ and see how far we get by applying the $S(\mathbf{H})\,S(\mathbf{H}')\,S(\mathbf{H}+\mathbf{H}')$ rule to derive signs for the large coefficients listed in Table 3.2, assuming that we had not recognized the cosine fringe system. We are allowed to choose $S(1, 0) = +$ and $S(2, 1) = +$, whence $S(3,1) = S(4, 1) = +$ also. To make any further progress, we have to introduce a symbolic sign, say $S(\bar{1}, 1) = a$, whence $S(\bar{2}, 2) = a^2 = +$, and so on, leading eventually to the results of Table 3.2 where all signs have been derived as either + or a. That is as far as we can go.

We could now calculate two Fourier syntheses with the large $F(\mathbf{H})$ coefficients, taking a as positive in one case and negative in the other. For the first possibility, all the signs would be +, and the resulting Fourier synthesis would therefore have a large peak at the origin plus

Table 3.2. Derivation of signs of large $F(h, k)$ coefficients by $S(\mathbf{H})\,S(\mathbf{H}')\,S(\mathbf{H}+\mathbf{H}') = +$ rule

h k	|*F*|	*h k*	|*F*|	*h k*	|*F*|	*h k*	|*F*|	*h k*	|*F*|	
1 0	77	$\bar{6}$ 1	100	$\bar{4}$ 2	81	$\bar{5}$ 3	97	$\bar{2}$ 4	97	
4 0	93	$\bar{5}$ 1	73	$\bar{3}$ 2	100	$\bar{4}$ 3	90	$\bar{1}$ 4	90	R. L. points associated with large Fourier coefficients
5 0	95	$\bar{2}$ 1	95	$\bar{2}$ 2	73	$\bar{1}$ 3	81			
		$\bar{1}$ 1	93	1 2	95	0 3	100			
		2 1	77	2 2	93	1 3	73			
		3 1	100							
		4 1	77							

1 0			+
2 1			+
3 1	1 0(+)	2 1(+)	+
4 1	1 0(+)	3 1(+)	+
$\bar{1}$ 1			a
$\bar{2}$ 1	$\bar{1}$ 0(+)	$\bar{1}$ 1(a)	a
2 2	3 1(+)	$\bar{1}$ 1(a)	a
1 2	2 1(+)	$\bar{1}$ 1(a)	a
$\bar{2}$ 2	$\bar{1}$ 1(a)	$\bar{1}$ 1(a)	+
$\bar{3}$ 2	$\bar{2}$ 2(+)	$\bar{1}$ 0(+)	+
$\bar{4}$ 2	$\bar{3}$ 2(+)	$\bar{1}$ 0(+)	+
$\bar{5}$ 1	$\bar{3}$ 2(+)	$\bar{2}$ $\bar{1}$(+)	+
$\bar{6}$ 1	$\bar{5}$ 1(+)	$\bar{1}$ 0(+)	+
4 0	5 $\bar{1}$(+)	$\bar{1}$ 1(a)	a
5 0	4 0(a)	1 0(+)	a
$\bar{5}$ 3	$\bar{4}$ 2(+)	$\bar{1}$ 1(a)	a
$\bar{4}$ 3	$\bar{5}$ 3(a)	1 0(+)	a
0 3	1 2(a)	$\bar{1}$ 1(a)	+
$\bar{1}$ 3	0 3(+)	$\bar{1}$ 0(+)	+
1 3	0 3(+)	1 0(+)	+
$\bar{1}$ 4	0 3(+)	$\bar{1}$ 1(a)	a
$\bar{2}$ 4	$\bar{1}$ 3(+)	$\bar{1}$ 1(a)	a

two smaller peaks at $\pm(2x_1, 2y_1)$—it would closely resemble the Patterson function, in fact. The second possibility, as seen from comparing Table 3.2 with Fig. 3.14, bottom, is the correct one, and the structure could be recognized easily from the corresponding Fourier synthesis. Until we decide which Fourier synthesis is the more reasonable, our structure analysis could be described as "direct"—it is based on routine application of a mathematical relationship and not on any preconceived structural model. Even if we were unable to decide between the two possible structures—the correct one and the one consisting of a heavy atom and a pair of lighter ones—both could be accepted provisionally and subjected to mathematical refinement procedures (as described in Chapter 4). This should enable us to make a final decision.

Unitary and Normalized Structure Factors

In general, information about the main features of the atomic arrangement in a crystal is contained in the very strong and very weak reflections. Here, strong and weak are relative to the average strength of a reflection at a given scattering angle, and allowance has to be made for the decrease of the atomic form factors with increasing scattering angle. This can be done by modifying the observed $|F_H|$ values so that they correspond to a hypothetical structure in which the actual atoms with more or less diffuse electron densities are replaced by point atoms. Two main kinds of modifications are in use. In the first, we convert the structure factors to unitary structure factors, defined as

$$U_H = \frac{F_H}{\sum_j f_j} \tag{3.21}$$

where the f_j symbols include an overall temperature factor—i.e.,

$$f_j = f_j^0 \exp(-B \sin^2\theta/\lambda^2) \tag{3.22}$$

U_H is then the ratio of F_H to its maximum possible value, the value it would have if all atoms were to scatter exactly in phase. In the second kind of modification we convert to normalized structure factors, E_H, which were introduced in the discussion of intensity statistics (Chapter 2):

$$E_H = \frac{F_H}{(\epsilon \sum f_j^2)^{\frac{1}{2}}} \tag{3.23}$$

where the f_j symbols are defined as above. E_H is thus the ratio of F_H to its root-mean-square expectation value, rather than to its maximum possible value, and the quantity ϵ is introduced to allow for the fact that for certain classes of reflections this expectation value may be different from $(\Sigma f_j^2)^{\frac{1}{2}}$ by some integral multiple.

The expressions for $U_{\mathbf{H}}$ and $E_{\mathbf{H}}$ imply that the $F_{\mathbf{H}}$ values are given on an absolute scale. If, as is often the case, the experimentally available values are on some arbitrary scale, the proportionality factor (and the overall temperature factor) can be derived simply.[36] Let k be the unknown scale factor that would convert the relative F values to absolute values:

$$k(F_{\mathbf{H}})_{\mathrm{rel}} = F_{\mathbf{H}}$$

From Eqs. 2.4 and 3.22

$$k^2\langle F_{\mathbf{H}}^2\rangle_{\mathrm{rel}} = \sum_j f_j^2 = \sum_j (f_j^0)^2 \exp(-2B \sin^2\theta/\lambda^2)$$

$$\ln \left\{\frac{\langle F_{\mathbf{H}}^2\rangle_{\mathrm{rel}}}{\sum (f_j^0)^2}\right\} = -2(\ln k + B \sin^2\theta/\lambda^2) \tag{3.24}$$

By calculating $\langle F_{\mathbf{H}}^2\rangle_{\mathrm{rel}}$ and $\Sigma(f_j^0)^2$ in various $\sin\theta/\lambda$ ranges and plotting the quantity on the left side of Eq. 3.24 against $\sin^2\theta/\lambda^2$, an approximately straight line should be obtained. From its slope the overall temperature factor coefficient B can be obtained and from its intercept at $\sin\theta/\lambda = 0$, the scale factor k can be obtained. In the following we assume that $F_{\mathbf{H}}$ values are on an absolute scale.

When all N atoms in the unit cell can be assumed to have the same form factor (equal-atom case),

$$\sum f_j = Nf$$

$$\sum f_j^2 = Nf^2 = \langle F_{\mathbf{H}}^2\rangle$$

Under this condition, taking $\epsilon = 1$

$$E_{\mathbf{H}} = \frac{F_{\mathbf{H}}}{N^{\frac{1}{2}}f} \qquad U_{\mathbf{H}} = \frac{F_{\mathbf{H}}}{Nf} \tag{3.25}$$

so that $E_{\mathbf{H}} = N^{\frac{1}{2}}U_{\mathbf{H}}$. Most modern work with direct methods is done with normalized structure factors or E values, introduced by Karle and Hauptman, but many references to unitary structure factors or U values can be found in the early literature. One advantage of using E values is that if the atoms are randomly arranged, the probability distribution of E is normal with mean value zero and variance of unity. According to Eqs. 2.10 and 2.7

[36] A. J. C. Wilson, *Nature (London)* **150,** 152 (1942).

$$P(|E|) = 2(2\pi)^{-\frac{1}{2}} \exp(-E^2/2)$$

for centrosymmetric arrangements (E real), while for noncentrosymmetric ones (E complex)

$$P(|E|) = 2|E| \exp(-|E|^2)$$

since the distribution of E itself is bivariate. The use of E statistics for distinguishing between centrosymmetric and noncentrosymmetric space groups was mentioned in Chapter 2 under Intensity Statistics. The expectation values are

	Centrosymmetric	Noncentrosymmetric
$\langle \lvert E^2 \rvert \rangle$	1.000	1.000
$\langle \lvert E \rvert \rangle$	0.798	0.886
$\langle \lvert E^2-1 \rvert \rangle$	0.968	0.736

Atomic arrangements in crystals are not random—characteristic interatomic distances and angles are repeated over and over again, larger or smaller atomic groupings may be coplanar, and so forth, apart from regularities connected with space-group symmetry. Any pronounced regularity in the atomic arrangement will be expressed by more-or-less marked deviations from the above expectation values. The effect of space-group symmetry elements is taken into account by the factor ϵ that was introduced in Eq. 3.23.

For instance, any symmetry element with a translational component will cause certain classes of reflections to be systematically absent and lead to an increase in the average intensity of other classes of reflections. As an example, consider a structure with space group $P2_1/c$ containing N atoms in the unit cell. For the general (hkl) reflections we have the usual condition that

$$\langle F_{\mathbf{H}}^2 \rangle = \sum_N f_j^2$$

However, because of the glide plane, the ($h0l$) reflections have to be considered separately. For these

$$F(h0l) = 2 \sum_{N/4} f_j(\cos 2\pi(hx+lz) + \cos 2\pi(hx+l(\tfrac{1}{2}+z)))$$

$$= 0, \qquad l \text{ odd}$$

or

$$= 4 \sum_{N/4} f_j \cos 2\pi(hx+lz), \qquad l \text{ even}$$

Since $\langle \cos^2 x \rangle = \frac{1}{2}$, we obtain for the latter case

$$\langle F^2(h0l)\rangle = 16 \sum_{N/4} f_j^2/2 = 2 \sum_{N} f_j^2$$

The average intensity of $(h0l)$ reflections with l even is thus twice as large as that of (hkl) reflections, and this is allowed for in Eq. 3.23 by setting $\epsilon = 2$ for this special class of reflections. Note that the presence of a center of symmetry has no effect on $\langle |E|^2 \rangle$.

The appropriate ϵ value for any special class of reflections for any space group can be obtained from the form of the relevant geometric structure factor expression or, alternatively, from the general positions of the space group. For example:

$P2$, general positions $x, y, z; \bar{x}, y, \bar{z}$

Each pair of symmetry-related atoms is projected on the **b** axis as a single atom with double weight. For the $(0k0)$ reflections, $\epsilon = 2$ since

$$\langle F^2 \rangle = \sum_{N/2} (2f_j)^2 = 2 \sum_{N} f_j^2$$

Alternatively, from the geometric structure factor expression,

$$F(0k0) = 2 \sum_{N/2} f_j(\cos 2\pi ky + i \sin 2\pi ky)$$

$$FF^* = 4 \sum_{N/2} f_j^2(\cos^2 2\pi ky + \sin^2 2\pi ky)$$

$$\langle |F|^2 \rangle = (4)(\tfrac{1}{2})(\tfrac{1}{2}+\tfrac{1}{2}) \sum_{N} f_j^2 = 2 \sum_{N} f_j^2$$

The effects of other symmetry elements can be summarized:

Element	ϵ	Class of reflections affected
$2(z)$, $\bar{4}(z)$	2	$00l$
$3(z)$, $\bar{3}(z)$, $\bar{6}(z)$	3	$00l$
$4(z)$	4	$00l$
$6(z)$	6	$00l$
$2_1(z)$	2	$00l$ ($l = 2n$)
$3_1(z)$	3	$00l$ ($l = 3n$)
4_1, 4_2	4	$00l$ ($l = 4n$, $l = 2n$)
6_1, 6_2, 6_3	6	$00l$ ($l = 6n$, $l = 3n$, $l = 2n$)
glide $a \perp c$	2	$hk0$ ($h = 2n$)

Equalities and Inequalities

Consider a 3×3 determinant of a certain kind formed by the Fourier coefficients of the simple model structure of Fig. 3.14 (the coefficients shown are actually $U_{\mathbf{H}}$ values times 100). The determinant in question is

$$D_3 = \begin{vmatrix} U_0 & U_{-\mathbf{H}} & U_{-\mathbf{K}} \\ U_{\mathbf{H}} & U_0 & U_{\mathbf{H-K}} \\ U_{\mathbf{K}} & U_{\mathbf{K-H}} & U_0 \end{vmatrix}$$

The value of U_0 is necessarily unity. As an example, with $\mathbf{H} = (2, 1)$, $\mathbf{K} = (1, 2)$, $\mathbf{H-K} = (1, \bar{1})$, we obtain from Fig. 3.14

$$D_3 = \begin{vmatrix} 1 & 0.77 & -0.95 \\ 0.77 & 1 & -0.93 \\ -0.95 & -0.93 & 1 \end{vmatrix}$$

$$= 1(1-0.865)-0.77(0.77-0.884)-0.95(-0.716+0.95)$$

$$= 0.001$$

where we have used the fact that for a centrosymmetric structure $U_{-\mathbf{H}} = U_{\mathbf{H}}$. The value of the determinant is zero, allowing for rounding-off errors, and it is also zero for any other choice of **H** and **K**. To see this, recall that the determinant

$$\begin{vmatrix} 1 & \cos\alpha & \cos\beta \\ \cos\alpha & 1 & \cos\gamma \\ \cos\beta & \cos\gamma & 1 \end{vmatrix} = V^2$$

is the square of the volume of the parallelepiped formed by three unit vectors making angles of α, β, γ with each other. In our case $U_{\mathbf{H}} = \cos 2\pi\mathbf{H}\cdot\mathbf{r}$, so the general determinant D_3 can be written

$$D_3 = \begin{vmatrix} 1 & \cos 2\pi\mathbf{H}\cdot\mathbf{r} & \cos 2\pi\mathbf{K}\cdot\mathbf{r} \\ \cos 2\pi\mathbf{H}\cdot\mathbf{r} & 1 & \cos 2\pi(\mathbf{H}-\mathbf{K})\cdot\mathbf{r} \\ \cos 2\pi\mathbf{K}\cdot\mathbf{r} & \cos 2\pi(\mathbf{H}-\mathbf{K})\cdot\mathbf{r} & 1 \end{vmatrix}$$

but now the three angles are linearly dependent so the three unit vectors of our parallelepiped are coplanar and the determinant must be zero.

For more than a single pair of centrosymmetric atoms in the unit cell, the angles $\alpha = \cos^{-1}(U_{\mathbf{H}})$, $\beta = \cos^{-1}(U_{\mathbf{K}})$, and $\gamma = \cos^{-1}(U_{\mathbf{H-K}})$ are not linearly dependent because each U is a sum over the positions of several atoms. In this case the value of the determinant must be positive since it corresponds to the square of a volume. Expansion of D_3 then gives:

$$1 - U_{\mathbf{H}}^2 - U_{\mathbf{K}}^2 - U_{\mathbf{H-K}}^2 + 2U_{\mathbf{H}}U_{\mathbf{K}}U_{\mathbf{H-K}} \geq 0 \qquad (3.26)$$

or

$$U_{\mathbf{H}}U_{\mathbf{K}}U_{\mathbf{H-K}} \geq \tfrac{1}{2}(U_{\mathbf{H}}^2 + U_{\mathbf{K}}^2 + U_{\mathbf{H-K}}^2 - 1) \qquad (3.27)$$

Now if $|U_{\mathbf{H}}|$, $|U_{\mathbf{K}}|$ and $|U_{\mathbf{H-K}}|$ are sufficiently large, the right side of the inequality must be positive, and the product $U_\mathbf{H}U_\mathbf{K}U_{\mathbf{H}-\mathbf{K}}$ must also be positive—i.e., $S_\mathbf{H}S_\mathbf{K}S_{\mathbf{H}-\mathbf{K}} = +$.

In 1950 Karle and Hauptman[37] showed that the condition for a nonnegative electron density is that the Fourier coefficients satisfy the determinant inequality

$$D_m = \begin{vmatrix} F_0 & F_{-\mathbf{H}_1} & F_{-\mathbf{H}_2} & \cdots\cdots & F_{-\mathbf{H}_{m-1}} \\ F_{\mathbf{H}_1} & F_0 & F_{\mathbf{H}_1-\mathbf{H}_2} & \cdots\cdots & F_{\mathbf{H}_1-\mathbf{H}_{m-1}} \\ F_{\mathbf{H}_2} & F_{\mathbf{H}_2-\mathbf{H}_1} & F_0 & \cdots\cdots & F_{\mathbf{H}_2-\mathbf{H}_{m-1}} \\ \cdots & \cdots\cdots & \cdots\cdots & \cdots\cdots & \cdots\cdots \\ \cdots & \cdots\cdots & \cdots\cdots & \cdots\cdots & \cdots\cdots \\ F_{\mathbf{H}_{m-1}} & F_{\mathbf{H}_{m-1}-\mathbf{H}_1} & F_{\mathbf{H}_{m-1}-\mathbf{H}_2} & \cdots\cdots & F_0 \end{vmatrix} \geq 0 \tag{3.28}$$

The subscripts $\mathbf{H}_i$ must be different but are not restricted in any other way, and the determinant may be of any order. Eq. 3.28 is the generalization of the very special case considered above. Goedkoop[38] showed that if the unit cell contains N point atoms (replace the F terms in Eq. 3.28 by U or E terms), then the inequality reduces to an equality when $m > N$. For example, $D_3 = 0$ for $N = 2$ as we have already seen. We thus have the conditions that the Fourier coefficients must satisfy if the electron density is to be nonnegative everywhere and built from discrete point atoms. Unfortunately, the equations that would apply for $m > N$ are insoluble except for trivial cases, so we are left with the lower-order inequality conditions, which are either too complicated or too weak to provide additional knowledge of signs or phases except when the U values are very large.

Inequality relationships were actually introduced by Harker and Kasper,[39] who outlined the first practicable approach to the direct solution of simple crystal structures in 1948. Their method of deriving inequalities was based on Cauchy's inequality

$$\left|\sum_N a_j b_j\right|^2 \leq \left(\sum_N |a_j|^2\right)\left(\sum_N |b_j|^2\right) \tag{3.29}$$

and can be illustrated by a simple example. For a centrosymmetric structure

$$U_\mathbf{H} = \sum_N n_j \cos 2\pi\mathbf{H}\cdot\mathbf{r}_j$$

where $n_j = f_j/\Sigma f_j$ may be called the unitary scattering factor, the fraction

[37] J. Karle and H. Hauptman, *Acta Crystallogr.* **3,** 181 (1950).
[38] J. A. Goedkoop, *Acta Crystallogr.* **3,** 374 (1950).
[39] D. Harker and J. S. Kasper, *Acta Crystallogr.* **1,** 70 (1948).

of the total scattering power associated with atom j. Now write

$$a_j = (n_j)^{\frac{1}{2}} \qquad b_j = (n_j)^{\frac{1}{2}} \cos 2\pi \mathbf{H}\cdot\mathbf{r}_j$$

so that

$$|U_{\mathbf{H}}|^2 = |\sum_N a_j b_j|^2$$

Since

$$\sum_N |a_j|^2 = 1$$

and

$$\sum_N |b_j|^2 = \sum_N n_j \cos^2 2\pi \mathbf{H}\cdot\mathbf{r}_j$$

$$= \tfrac{1}{2} \sum_N n_j (1+\cos 4\pi \mathbf{H}\cdot\mathbf{r}_j)$$

$$= \tfrac{1}{2}(1+U_{2\mathbf{H}})$$

It follows from Cauchy's inequality that

$$|U_{\mathbf{H}}|^2 \leq \tfrac{1}{2}(1+U_{2\mathbf{H}}) \tag{3.30}$$

The same result could have been derived from Eq. 3.26 by writing $\mathbf{K} = 2\mathbf{H}$ to obtain

$$1 - 2U_{\mathbf{H}}^2 - U_{2\mathbf{H}}^2 + 2U_{\mathbf{H}}^2 U_{2\mathbf{H}} \geq 0$$

$$(1 + U_{2\mathbf{H}})(1 - U_{2\mathbf{H}}) - 2U_{\mathbf{H}}^2(1 - U_{2\mathbf{H}}) \geq 0$$

identical to Eq. 3.30 since $(1-U_{2\mathbf{H}})$ is necessarily positive for any real structure. Many other inequalities of this kind can be derived in a similar way, including special inequalities valid for particular space groups. The elucidation of the decaborane ($B_{10}H_{14}$) structure by the use of inequalities soon provided a striking success for the new method.[40] However, it was also becoming clear that the usefulness of the method was limited to structures of modest complexity, as first noted by Hughes.[41] This is because unless the $|U|$ values contained in the inequalities are very large, the inequalities are satisfied by any sign combination. However, large $|U|$ values are unlikely to occur for complex structures. For example, Eq. 3.27 can yield a sign restriction only if the three $|U|$ values involved are about 0.6 or more, but for N atoms in the unit cell the root-mean-square value of $|U|$ is $N^{-\frac{1}{2}}$. For $N = 20$, say, this is 0.22 and, assuming a Gaussian distribution, only about one reflection in 300 could be considered a possible candidate for inclusion in the inequality.

Thus far we have considered only centrosymmetric structures, but the

[40] J. S. Kasper, C. M. Lucht, and D. Harker, *Acta Crystallogr.* **3**, 436 (1950).
[41] E. W. Hughes, *Acta Crystallogr.* **2**, 34 (1949).

Karle-Hauptman inequality (Eq. 3.28) holds also for noncentrosymmetric structures where the F terms are complex quantities. For the noncentrosymmetric case, D_3 can be written in terms of U values as

$$D_3 = \begin{vmatrix} 1 & |U_{\mathbf{H}}| \exp(-i\phi_{\mathbf{H}}) & |U_{\mathbf{K}}| \exp(-i\phi_{\mathbf{K}}) \\ |U_{\mathbf{H}}| \exp(i\phi_{\mathbf{H}}) & 1 & |U_{\mathbf{H-K}}| \exp(i\phi_{\mathbf{H-K}}) \\ |U_{\mathbf{K}}| \exp(i\phi_{\mathbf{K}}) & |U_{\mathbf{H-K}}| \exp(-i\phi_{\mathbf{H-K}}) & 1 \end{vmatrix} \geq 0$$

which can be expanded to

$$1-|U_{\mathbf{H}}|^2-|U_{\mathbf{K}}|^2-|U_{\mathbf{H-K}}|^2+2|U_{\mathbf{H}}U_{\mathbf{K}}U_{\mathbf{H-K}}| \cos \varphi \geq 0 \qquad (3.31)$$

where $\varphi = \varphi_{-\mathbf{H}} + \varphi_{\mathbf{K}} + \varphi_{\mathbf{H-K}}$. When $|U_{\mathbf{H}}|$, $|U_{\mathbf{K}}|$, $|U_{\mathbf{H-K}}|$ are small, there is no restriction on the phase angles, but if they are large enough, the cosine term must approach its maximum positive value (+1) for the inequality to be satisfied, implying that $\varphi \approx 0$. Although the individual phase angles depend on the choice of origin, the angle $\varphi = \varphi_{-\mathbf{H}} + \varphi_{\mathbf{K}} + \varphi_{\mathbf{H-K}}$ does not, as is easily confirmed. A displacement of the origin by an amount $\Delta\mathbf{r}$ simply adds an amount $-2\pi\mathbf{H}\cdot\Delta\mathbf{r}$ to the phase of $U_{\mathbf{H}}$. Hence the change in $\varphi = \varphi_{-\mathbf{H}} + \varphi_{\mathbf{K}} + \varphi_{\mathbf{H-K}}$ is

$$\Delta\varphi = -2\pi\Delta\mathbf{r}(-\mathbf{H}+\mathbf{K}+\mathbf{H}-\mathbf{K}) = 0$$

As Kitaigorodsky has shown,[42] there is an elegant geometrical interpretation of the Karle-Hauptman determinant. Each reciprocal lattice point $\mathbf{H}_p (p = 0, 1, 2 \ldots m-1)$ is associated with an N-dimensional vector

$$\mathbf{G}_p = \sum_N \mathbf{e}_j n_j^{\frac{1}{2}} \exp(2\pi i \mathbf{H}_p \cdot \mathbf{r}_j)$$

where the $\mathbf{e}_j$ axes are a set of orthonormal unit vectors of an N-dimensional space, and the n_j values are unitary scattering factors (all positive). The vectors $\mathbf{G}_p$ are unit vectors since

$$\mathbf{G}_p \cdot \mathbf{G}_p^* = \sum_N n_j = 1$$

and each scalar product $\mathbf{G}_p \cdot \mathbf{G}_q^*$ can be identified with a unitary structure factor U_{p-q}:

$$\mathbf{G}_p \cdot \mathbf{G}_q^* = \sum_N n_j \exp[2\pi i(\mathbf{H}_p - \mathbf{H}_q)\cdot\mathbf{r}_j] = U_{p-q}$$

The Karle-Hauptman determinant of order m (in terms of U) is then

[42] A. I. Kitaigorodsky, *The Theory of Crystal Structure Analysis*, Consultants Bureau, New York, 1961 (translated from Russian by D. and K. Harker).

equivalent to

$$D_m = \begin{vmatrix} \mathbf{G}_0 \cdot \mathbf{G}_0^* & \mathbf{G}_0 \cdot \mathbf{G}_1^* & \mathbf{G}_0 \cdot \mathbf{G}_2^* & \cdots\cdots & \mathbf{G}_0 \cdot \mathbf{G}_{m-1}^* \\ \mathbf{G}_1 \cdot \mathbf{G}_0^* & \mathbf{G}_1 \cdot \mathbf{G}_1^* & \mathbf{G}_1 \cdot \mathbf{G}_2^* & \cdots\cdots & \mathbf{G}_1 \cdot \mathbf{G}_{m-1}^* \\ & & & & \\ \mathbf{G}_{m-1} \cdot \mathbf{G}_0^* & \mathbf{G}_{m-1} \cdot \mathbf{G}_1^* & \mathbf{G}_{m-1} \cdot \mathbf{G}_2^* & \cdots\cdots & \mathbf{G}_{m-1} \cdot \mathbf{G}_{m-1}^* \end{vmatrix}$$

This is known to mathematicians as the Gram determinant of the vectors $\mathbf{G}_p$. The value of this determinant is the square of the N-dimensional hypervolume of the N-dimensional figure spanned by the vectors $\mathbf{G}_p$ and cannot be negative. It is zero if the m vectors $\mathbf{G}_p$ are linearly dependent, i.e., when $N < m$. In this interpretation $U_{\mathbf{H}-\mathbf{K}}$ is the cosine of the angle between vectors $\mathbf{G}_\mathbf{H}$ and $\mathbf{G}_\mathbf{K}^*$, and the Cauchy inequality (Eq. 3.29) is seen to be no more than the statement that the scalar product of two unit vectors is less than or equal to unity. Of course, there may be certain problems about visualizing these vectors—not only are they N-dimensional vectors but their components on the basis vectors $\mathbf{e}_j$ are complex numbers in general.

Sayre's Equation

Another kind of equality between structure factors exists if the electron-density distribution $\rho(\mathbf{r})$ can be regarded as being built from a number of equal, resolved peaks.[43] In this case, the electron density squared $\rho^2(\mathbf{r})$ also consists of equal, resolved peaks, located at the same positions as the peaks of $\rho(\mathbf{r})$. The only difference is that the peaks of $\rho^2(\mathbf{r})$ are sharper, i.e., the "atoms" have different scattering factors. If the Fourier coefficients of $\rho(\mathbf{r})$ and $\rho^2(\mathbf{r})$ are $F_\mathbf{H}/V$ and $G_\mathbf{H}/V$ respectively, we have

$$F_\mathbf{H} = f \sum \exp\,(2\pi i \mathbf{H} \cdot \mathbf{r}_j)$$

$$G_\mathbf{H} = g \sum \exp\,(2\pi i \mathbf{H} \cdot \mathbf{r}_j)$$

where the summations involve the same position vectors $\mathbf{r}_j$ so that $G_\mathbf{H} = (g/f)F_\mathbf{H}$. We now use the convolution theorem (Chapter 1) to derive another relationship between $F_\mathbf{H}$ and $G_\mathbf{H}$. Since $\rho^2(\mathbf{r})$ is the product of $\rho(\mathbf{r})$ with itself, each Fourier coefficient $G_\mathbf{H}/V$ of $\rho^2(\mathbf{r})$ is the convolution of $F_\mathbf{H}/V$ with itself—i.e.,

[43] D. Sayre, *Acta Crystallogr.* **5**, 60 (1952).

$$\frac{G_{\mathbf{H}}}{V} = \left(\frac{g}{f}\right)\frac{F_{\mathbf{H}}}{V} = \frac{1}{V^2}\sum_{\mathbf{K}} F_{\mathbf{K}}F_{\mathbf{H-K}}$$

or

$$F_{\mathbf{H}} = \left(\frac{f}{gV}\right)\sum_{\mathbf{K}} F_{\mathbf{K}}F_{\mathbf{H-K}} \tag{3.32}$$

This is Sayre's equation. It is not immediately useful as it stands because in order to calculate the sign or phase of $F_{\mathbf{H}}$, we must know the signs or phases associated with all the products on the right. However, if one of these products is large, its value will tend to dominate the value of the sum. For a centrosymmetric structure, if $F_{\mathbf{H}}$, $F_{\mathbf{K}}$, $F_{\mathbf{H-K}}$ are all large, it follows that

$$S_{\mathbf{H}} \sim S_{\mathbf{K}}S_{\mathbf{H-K}}$$

leading to

$$S_{\mathbf{H}}S_{\mathbf{K}}S_{\mathbf{H+K}} \sim +1 \tag{3.33}$$

since $S_{\mathbf{H}} = S_{-\mathbf{H}}$, etc. The symbol $\sim$ here means "probably equals." This suggests that even if the U values are not quite large enough to establish the positive sign of the triple product, the positive sign may be regarded as probable. Similar conclusions, reached by different lines of argument, were drawn by Cochran,[44] Zachariasen,[45] and Hughes[46] at about the same time. For the noncentrosymmetric case, if $F_{\mathbf{H}}$, $F_{\mathbf{K}}$, $F_{\mathbf{H-K}}$ are all large, we might expect that

$$\phi_{\mathbf{H}} \sim \phi_{\mathbf{K}} + \phi_{\mathbf{H-K}} \tag{3.34}$$

or, since $\phi_{\mathbf{H}} = -\phi_{-\mathbf{H}}$,

$$\phi_{-\mathbf{H}} + \phi_{\mathbf{K}} + \phi_{\mathbf{H-K}} \sim 0 \text{ (modulo } 2\pi) \tag{3.34a}$$

suggesting that the analogous condition, $\cos(\phi_{-\mathbf{H}} + \phi_{\mathbf{K}} + \phi_{\mathbf{H-K}}) = +1$, might also be regarded as probably true even when it is not strictly imposed by the inequality relationship (Eq. 3.31). The question is: how probable?

Probability Relationships

When Sayre's Eq. 3.32 is written in terms of U or E, the sum of products on the right contains an infinite number of terms, whereas only a few of the relevant magnitudes are experimentally available. By using appropriate averaging assumptions[47] (similar to those introduced in

[44] W. Cochran, *Acta Crystallogr.* **5**, 65 (1952).
[45] W. Zachariasen, *Acta Crystallogr.* **5**, 68 (1952).
[46] E. W. Hughes, *Acta Crystallogr.* **6**, 871 (1953).
[47] M. M. Woolfson, *Acta Crystallogr.* **7**, 61 (1954).

Chapter 2 under Intensity statistics), the Sayre equation can be written in the modified forms

$$\begin{aligned} U_{\mathbf{H}} &= N\langle U_{\mathbf{K}} U_{\mathbf{H-K}} \rangle_{\mathbf{K}} \\ E_{\mathbf{H}} &= N^{\frac{1}{2}}\langle E_{\mathbf{K}} E_{\mathbf{H-K}} \rangle_{\mathbf{K}} \end{aligned} \tag{3.35}$$

where the bracket signifies that the average value of the products on the right has to be taken. Woolfson showed that, in the centrosymmetric equal-atom case, if **H** and **K** are varied but $U_{\mathbf{K}}U_{\mathbf{H-K}}$ is regarded as fixed, $U_{\mathbf{H}}$ is normally distributed about this fixed value with variance $1/N$. For given values of $U_{\mathbf{K}}$ and $U_{\mathbf{H-K}}$ the probability distribution of $U_{\mathbf{H}}$ is then

$$P(U_{\mathbf{H}}) = (N/2\pi)^{\frac{1}{2}} \exp\left[-\tfrac{1}{2}N(U_{\mathbf{K}}U_{\mathbf{H-K}} - U_{\mathbf{H}})^2\right]$$

Now, either $U_{\mathbf{H}}$ has the same sign as $U_{\mathbf{K}}U_{\mathbf{H-K}}$, or it has the opposite sign. The ratio of the two probabilities is

$$\begin{aligned} \frac{P_+}{P_-} &= \frac{\exp\left[-\tfrac{1}{2}N(|U_{\mathbf{K}}U_{\mathbf{H-K}}| - |U_{\mathbf{H}}|)^2\right]}{\exp\left[-\tfrac{1}{2}N(|U_{\mathbf{K}}U_{\mathbf{H-K}}| + |U_{\mathbf{H}}|)^2\right]} \\ &= \exp\left(2N|U_{\mathbf{H}}U_{\mathbf{K}}U_{\mathbf{H-K}}|\right) = e^{2x} \end{aligned}$$

However, the sum of the two probabilities is unity, so

$$\begin{aligned} P_+ &= \frac{e^{2x}}{1+e^{2x}} = \frac{e^{x}}{e^{x}+e^{-x}} = \frac{1}{2}\left(1 + \frac{e^{x}-e^{-x}}{e^{x}+e^{-x}}\right) \\ &= \frac{1}{2} + \frac{1}{2}\tanh\left(N|U_{\mathbf{H}}U_{\mathbf{K}}U_{\mathbf{H-K}}|\right) \end{aligned} \tag{3.36}$$

In terms of E values the corresponding expression is

$$P_+ = \frac{1}{2} + \frac{1}{2}\tanh\left(N^{-\frac{1}{2}}|E_{\mathbf{H}}E_{\mathbf{K}}E_{\mathbf{H-K}}|\right) \tag{3.37}$$

The form of the function

$$f(x) = \frac{1}{2} + \frac{1}{2}\tanh x$$

is shown in Fig. 3.16. For $N = 50$ and $|E_{\mathbf{H}}E_{\mathbf{K}}E_{\mathbf{H-K}}| = 8$, $x = 8/\sqrt{50} = 1.13$, giving a better than 90% probability that the sign product is positive. For a given $E_{\mathbf{H}}$, if we can find several triples involving known signs of the other members, the probability that $E_{\mathbf{H}}$ will be positive becomes

$$P_+ = \frac{1}{2} + \frac{1}{2}\tanh\left(N^{-\frac{1}{2}}|E_{\mathbf{H}}|\sum_{K} E_{\mathbf{K}}E_{\mathbf{H-K}}\right) \tag{3.38}$$

These formulas apply strictly only for the equal-atom case; for nonequal

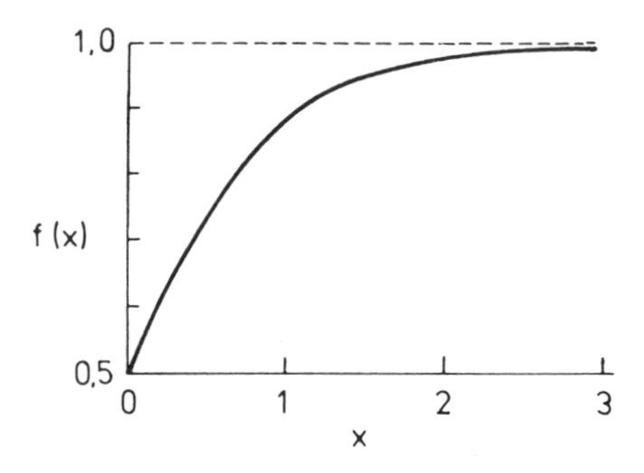

Figure 3.16. The function $f(x) = \frac{1}{2} + \frac{1}{2} \tanh x$.

atoms, the factor $N^{-\frac{1}{2}}$ must be amended slightly. However, when using Eqs. 3.37 and 3.38, a rough estimate of the probabilities involved is usually all that is needed.

For the noncentrosymmetric case, we have seen that

$$\phi_{\mathbf{H}} \sim \phi_{\mathbf{K}} + \phi_{\mathbf{H-K}}$$

if the corresponding structure amplitudes are large enough. For given values of $|E_{\mathbf{H}}|$, $|E_{\mathbf{K}}|$, $|E_{\mathbf{H-K}}|$ and known values of $\phi_{\mathbf{K}}$ and $\phi_{\mathbf{H-K}}$, the probability distribution of $\phi_{\mathbf{H}}$ for the equal-atom case, analogous to Eq. 3.36, is:[48]

$$P(\phi_{\mathbf{H}}) = \frac{\exp\left[4x \cos\left(\phi_{\mathbf{H}} - \phi_{\mathbf{K}} - \phi_{\mathbf{H-K}}\right)\right]}{\int_0^{2\pi} \exp\left(4x \cos \gamma\right) d\gamma}$$

where x is as defined above. The value of the definite integral is $2\pi I_0(4x)$, where I_0 is a modified Bessel function of the second kind. This distribution (shown for a particular case in Fig. 3.17), is nearly normal about zero, and its variance, as a function of x, has been estimated by Karle and Karle,[49] leading to the results shown in Fig. 3.18. For $x = 1$ (e.g., $N = 30$, $|E_{\mathbf{H}}| = |E_{\mathbf{K}}| = |E_{\mathbf{H-K}}| = \sqrt[3]{(30)^{\frac{1}{2}}} = 1.76$), the standard deviation in the estimate of $\phi_{\mathbf{H}}$ (assuming $\phi_{\mathbf{K}}$ and $\phi_{\mathbf{H-K}}$ to be known) is about 50°.

The numerical probabilities obtained in this way should not be taken too seriously. They are based on the assumption that the atoms are arranged randomly—an assumption that is often seriously violated in actual crystal structures. Nevertheless, they provide rough probability estimates that can be used to decide whether a sign or phase angle derived from such triples or combinations of triples is reasonably trustworthy.

[48] W. Cochran, *Acta Crystallogr.* **8**, 473 (1955).
[49] J. Karle and I. L. Karle, *Acta Crystallogr.* **21**, 849 (1966).

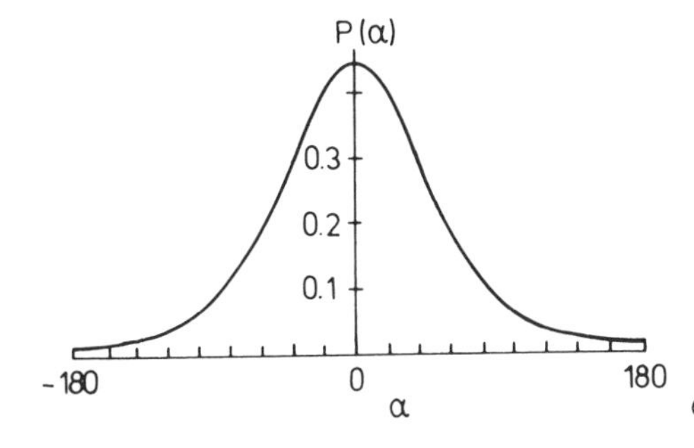

Figure 3.17. Distribution of $\alpha = \phi_{\mathbf{H}} - \phi_{\mathbf{K}} - \phi_{\mathbf{H-K}}$ for $x = 0.81$ ($N = 30$, $|E_{\mathbf{H}}| = |E_{\mathbf{K}}| = |E_{\mathbf{H-K}}| = 1.64$). After Cochran, *Acta Crystallogr.* **8**, 473 (1955).

Practical Application of Direct Methods

For centrosymmetric structures, the process of developing the probable signs of the strong reflections is usually fairly straightforward. The first step is to prepare a list of strong triples and their associated triple products $|E_{\mathbf{H}}E_{\mathbf{K}}E_{\mathbf{H-K}}|$. This process is laborious without a computer, but the subsequent sign-development process can be carried out by hand although, of course, it too is normally done by computer these days. Up to three reflections (depending on the space group) are given arbitrary signs to fix the origin, and a few other reflections are associated with a symbolic sign (as in Table 3.2). The strong triples lead to highly probable sign assignments for other reflections that can be used in turn to develop still further signs. Certain reflections will be involved in several strong triples, leading to multiple sign indications that can be used to eliminate symbolic signs (e.g., from different chains of triples the sign of a certain reflection may be given as − and as *b,* enabling *b* to be eliminated). As weaker and weaker triples are included in the sign-developing chains, various contradictions are likely to be encountered. The symbolic signs must be chosen so that contradictions are kept to a minimum. Proceeding in this way, probable signs of most of the strong reflections, say those with $|E| > 1.5$, can usually be derived without too much difficulty, and the atomic arrangement can often be recognized immediately from the

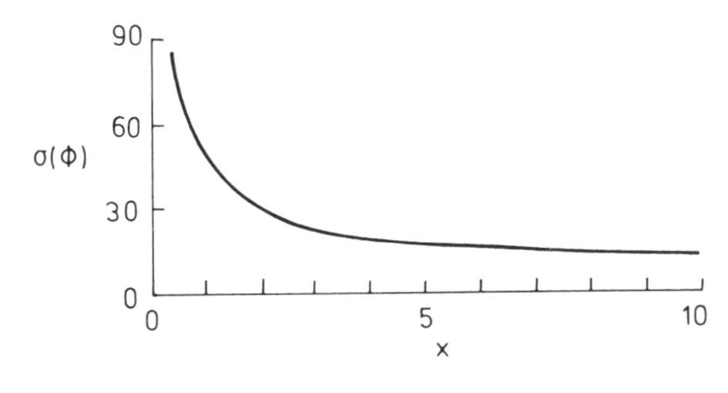

Figure 3.18. Standard deviation of $\alpha = \phi_{\mathbf{H}} - \phi_{\mathbf{K}} - \phi_{\mathbf{H-K}}$ from mean value zero as function of x. After Karle and Karle, *Acta Crystallogr.* **21**, 849 (1966).

arrangement of peaks in the corresponding E map (a Fourier synthesis where the coefficients are experimental $|E|$ values associated with derived signs). If one or two of the symbolic signs cannot be eliminated without causing many contradictions, it may be advisable to compute several E maps, one for each residual symbol combination; with any luck, one of these E maps will be more easily interpretable than the others.

A word is necessary here concerning what might be called the "triclinic catastrophe," although it is by no means confined to triclinic structures. For the monoclinic space group $P2_1/c$, the geometric structure factor has the form (Eq. 2.2):

$$\begin{aligned} A &= 4\cos 2\pi(hx+lz)\cos 2\pi ky \qquad & k+l = 2n \\ &= -4\sin 2\pi(hx+lz)\sin 2\pi ky \qquad & k+l = 2n+1 \\ B &= 0 \end{aligned}$$

so that, for reflections with $(k+l)$ odd, there are relations of the kind

$$F(hkl) = -F(h\bar{k}l) = -F(\bar{h}k\bar{l})$$

This ensures that negative signs as well as positive ones are involved in the sign-determining process. However, for the triclinic space group $P\bar{1}$, we have, (Eq. 2.1):

$$\begin{aligned} A &= 2\cos 2\pi(hx+ky+lz) \\ B &= 0 \end{aligned}$$

and there are no relations of the above kind. If the origin-determining signs are taken as positive, then exhaustive elimination of all symbolic signs must necessarily lead for this space group to the solution with all signs positive, producing an E map that resembles the Patterson function rather than the actual structure, with a large peak at the origin (as in the example discussed under Direct Methods, A Simple Example). Choosing one or more of the origin-determining signs as negative is no way to circumvent this problem; the E map is simply displaced so that its major peak occurs at one of the alternative origins. To avoid the "triclinic catastrophe," at least one of the symbolic signs must be given the sign opposite to that derived by straightforward, stubborn application of the rules. Indeed, if it is known on general structural grounds that no atom can actually be present on a center of symmetry, the positive and negative E values must be distributed so that the following conditions are roughly fulfilled:

$$\Sigma\, E(hkl) \approx 0$$

$$\Sigma\, (-1)^p\, E(hkl) \approx 0$$

where $p = h, k, l, h+k, k+l, h+l, h+k+l$ in turn, to eliminate each of the special positions of the space group. This test may help one to decide which symbols have to be replaced by a negative sign, but if it is equivocal, several E maps should be calculated for the various remaining symbolic sign combinations. In general, great caution must be exercised in eliminating the symbols in the triclinic case and in other space groups that lack symmetry elements involving translations (glide planes or screw axes).

Although the symbolic sign method was introduced by Zachariasen in 1952 in his successful analysis of the structure of metaboric acid ($P2_1/a$, 12 molecules HBO_2 per unit cell),[45] it did not come into general use until more than a decade later. During this period, several reasonably complex centrosymmetric crystal structures were solved by direct methods, most notably that of *p,p'*-dimethoxybenzophenone ($P2_1/c$, eight molecules $C_{15}H_{14}O_3$ per unit cell),[50] but much effort was wasted in developing elaborate algorithms for solving structures in projection using two-dimensional data. In fact, three-dimensional data are almost a prerequisite for the successful application of direct methods. The main reason is that the chain process of developing phase angles has to be initiated from a sufficient number of triples involving very large $|E|$ values, ca. 2.5 or greater. For a centrosymmetric structure, only about 1% of the reflections are as strong as this; for noncentrosymmetric structures only about 0.2%. With a set of two-dimensional data, comprising ca. 200 reflections, it is therefore very difficult to develop the sign chain in terms of only a few symbols. In general, many symbols must be introduced, and once introduced they are not easy to eliminate again. Moreover, if several E maps have to be computed, it is much more difficult to recognize the right one because of possible overlapping of the atomic peaks in projection. The main hindrance in using three-dimensional data was probably the labor involved in sorting out the strong triples from the enormous number possible—the so-called Σ_2 list—without the aid of high-speed computers. For two-dimensional data, this sorting-out step could be surmounted with modest computing facilities or even graphically. It was only when high-speed computers became generally available that the symbolic sign method came into routine use for solving centrosymmetric crystal structures.

[50] I. L. Karle, H. Hauptman, J. Karle, and A. B. Wing, *Acta Crystallogr.* **11**, 257 (1958).

For noncentrosymmetric structures the problem is much more difficult. The *sign* allotted to a reflection on the basis of a strong triple is either right or wrong but usually right; a phase angle allotted on the same basis is at best approximately correct, so that in the process of deriving further phases from a chain of triples, large errors unavoidably accumulate. Thus, many crystallographers felt that the solution of noncentrosymmetric structures by direct methods was more or less impossible, even when their application to centrosymmetric structures was well accepted. For example, in a monograph[51] on direct methods published in 1961, noncentrosymmetric structures are not mentioned, even in a final paragraph entitled, "What does the future hold?" Any doubts about the possibility of applying direct methods to noncentrosymmetric structures were dispelled in 1964 with the successful analysis of L-arginine dihydrate ($P2_12_12_1$, four molecules $C_6H_{14}N_4O_2 \cdot 2H_2O$ per unit cell).[52]

All that was necessary, it turned out, was to carry through the phase-determining procedure not once but many times iteratively, using the now-famous tangent formula proposed several years earlier.[53] This formula can be derived simply from the Sayre equation (Eq. 3.32), expressed in the form

$$|F_\mathrm{H}| \exp (i\phi_\mathrm{H}) = c \sum_K |F_\mathrm{K} F_\mathrm{H-K}| \exp [i(\phi_\mathrm{K}+\phi_\mathrm{H-K})]$$

or

$$|F_\mathrm{H}| \cos \phi_\mathrm{H} = c \sum_K |F_\mathrm{K} F_\mathrm{H-K}| \cos (\phi_\mathrm{K}+\phi_\mathrm{H-K})$$

$$|F_\mathrm{H}| \sin \phi_\mathrm{H} = c \sum_K |F_\mathrm{K} F_\mathrm{H-K}| \sin (\phi_\mathrm{K}+\phi_\mathrm{H-K})$$

In terms of $|E|$ values, this leads to

$$\tan \phi_\mathrm{H} = \frac{\sum_K |E_\mathrm{K} E_\mathrm{H-K}| \sin (\phi_\mathrm{K}+\phi_\mathrm{H-K})}{\sum_K |E_\mathrm{K} E_\mathrm{H-K}| \cos (\phi_\mathrm{K}+\phi_\mathrm{H-K})} \qquad (3.39)$$

where the sums are taken over all available terms. Starting with a set of assumed or derived initial phases for a small number of reflections, the formula is used iteratively to improve the initial phases and to obtain additional ones.

As in the centrosymmetric case, up to three phases can be set arbitrarily

[51] M. M. Woolfson, *Introduction to Direct Methods,* Oxford University Press, Oxford, 1961.
[52] I. L. Karle and J. Karle, *Acta Crystallogr.* **17,** 835 (1964).
[53] J. Karle and H. Hauptman, *Acta Crystallogr.* **9,** 635 (1956).

to define the origin, and a further phase can be restricted to lie between 0 and π or between π and 2π to define the chirality (or polarity) of the structure, as discussed in the next section. To expand the initial set of phases, two main approaches have been developed. One method—the symbolic addition procedure[54]—uses a set of symbolic phase angles in a way similar to that outlined above for centrosymmetric structures. The numerical and symbolic phases are combined using

$$\phi_{\mathbf{H}} \sim \phi_{\mathbf{K}} + \phi_{\mathbf{H}-\mathbf{K}} \tag{3.34}$$

or if several triples involving a given phase angle are available

$$\phi_{\mathbf{H}} \sim \langle \phi_{\mathbf{K}} + \phi_{\mathbf{H}-\mathbf{K}} \rangle_{\mathbf{K}}$$

The combinations obtained are developed in a chain process. Besides the intrinsic uncertainty in the derived sums, the cyclic property of the phase angles (modulo 2π) introduces large additional uncertainties in evaluating the averages and in eliminating symbols. Nevertheless, the process can be used, with due caution, to expand a small initial set of assumed or symbolic phases into a somewhat larger set to allow the tangent formula machinery to take over.

The other possibility is to begin right away with a multi-solution approach.[55] This approach is based on the premise (confirmed by experience) that if an assumed or derived phase angle is within about 45° of its true value, it will in turn lead to useful new information. To supplement the origin-defining phases, various permutations of phases are tested for a few reflections chosen among those that are linked most frequently with others by strong triples. For three such reflections, these permutations might be based on the trial phase angles:

reflection 1:	45°	135°	(set as positive to define the enantiomorph)	
2:	−135°	−45°	45°	135°
3:	−135°	−45°	45°	135°

giving a total of $2.4^2 = 32$ permutations to be tested for this starting set. For each such permutation, additional phase angles are derived from strong triples and weighted according to the corresponding value of $|E_{\mathbf{H}}E_{\mathbf{K}}E_{\mathbf{H}-\mathbf{K}}|$, and each is further expanded and refined by using the tangent formula with appropriate weights:

[54] J. Karle and I. L. Karle, *Acta Crystallogr.* **21,** 849 (1966).
[55] G. Germain and M. M. Woolfson, *Acta Crystallogr.* **B24,** 91 (1968); G. Germain, P. Main, and M. M. Woolfson, *ibid.,* **B26,** 274 (1970).

$$\tan \phi_{\mathbf{H}} = \frac{\sum_K \omega_{\mathbf{K}}\omega_{\mathbf{H-K}}|E_{\mathbf{K}}E_{\mathbf{H-K}}|\sin(\phi_{\mathbf{K}}+\phi_{\mathbf{H-K}})}{\sum_K \omega_{\mathbf{K}}\omega_{\mathbf{H-K}}|E_{\mathbf{K}}E_{\mathbf{H-K}}|\cos(\phi_{\mathbf{K}}+\phi_{\mathbf{H-K}})} = \frac{T_{\mathbf{H}}}{B_{\mathbf{H}}}$$

$$\omega_{\mathbf{H}} = \tanh\{N^{-\frac{1}{2}}|E_{\mathbf{H}}|(T_{\mathbf{H}}^2+B_{\mathbf{H}}^2)^{\frac{1}{2}}\}$$

to produce 32 sets of possible phases.

Various criteria have been proposed to recognize the correct or nearly correct sets among these. One possible criterion is

$$R = \frac{\Sigma(|E_{\mathbf{H}}^{\mathrm{obs}}|-|E_{\mathbf{H}}^{\mathrm{calc}}|)}{\Sigma|E_{\mathbf{H}}^{\mathrm{obs}}|}$$

where $E_{\mathbf{H}}^{\mathrm{calc}}$ is calculated from an equation of the Sayre type; combinations leading to low R values are regarded as more likely.

Another criterion is the so-called figure of merit[56]

$$Z = \sum \alpha_{\mathbf{H}} = N^{-\frac{1}{2}}\sum|E_{\mathbf{H}}|(T_{\mathbf{H}}^2+B_{\mathbf{H}}^2)^{\frac{1}{2}}$$

which should be high (the factor $N^{-\frac{1}{2}}$ applies only for the equal-atom case and is slightly different otherwise), and still another is the absolute figure of merit

$$M_{\mathrm{abs}} = \frac{Z - \sum\langle\alpha^2\rangle_r^{\frac{1}{2}}}{\sum\langle\alpha^2\rangle_e^{\frac{1}{2}} - \sum\langle\alpha^2\rangle_r^{\frac{1}{2}}}$$

where $\langle\alpha^2\rangle_e^{\frac{1}{2}}$ is the expectation value of α for a correct set of phases, and $\langle\alpha^2\rangle_r^{\frac{1}{2}}$ the expectation value for a random set of phases. These criteria are included in MULTAN,[56,57] an automatic direct-method computer program based on the multi-solution approach.

In the end, one or many E maps may have to be computed and scrutinized for the occurrence of stereochemically plausible arrangements of peaks. Once the molecule can be recognized in rough outline, the phase problem is solved and the structure has only to be refined by conventional methods. If only some substantial fraction of the molecule can be recognized, it can be regarded as a "trial structure" for phasing the coefficients of a subsequent Fourier synthesis. Alternatively, the phase angles calculated for the fragment can be used as input to further cycles of tangent formula refinement.

[56]G. Germain, P. Main, and M. M. Woolfson, *Acta Crystallogr.* **A27,** 368 (1971).

[57]P. Main, L. Lessinger, M. M. Woolfson, G. Germain, and J. P. Declercq, *MULTAN 77, A System of Computer Programs for the Automatic Solution of Crystal Structures from X-ray Diffraction Data,* Universities of York, England, and Louvain, Belgium, 1977.

For complex structures, an increase in the size of the starting set of reflections is obviously desirable, but if four trial phases (±45°, ±135°) are tested for each additional reflection, the total number of phase permutations to be tested soon becomes prohibitively large. One promising new approach that allows tentative phases to be assigned to a rather large starting set of reflections involves the use of "magic integers."[58] The basic idea is that approximate values of several phase angles (described in cycles rather than in radians) can be expressed in terms of a single parameter x by writing

$$\varphi_r = n_r x \text{ (modulo 1)}$$

for a suitably chosen set of integers n_i. The integers 3, 4, 5 constitute such a set. For example, with $\varphi_1 = 0.8$, $\varphi_2 = 0.6$, $\varphi_3 = 0.0$, and $x = 0.622$, we have:

$$3x = 1.866 = 0.866 \text{ (modulo 1)}$$

$$4x = 2.488 = 0.488 \text{ (modulo 1)}$$

$$5x = 3.110 = 0.110 \text{ (modulo 1)}$$

The errors are 0.066, 0.112, and 0.110 cycles, or 24°, 40°, and 40°, respectively. Errors of this magnitude can be tolerated easily in the phases of the starting group of reflections. With a set of five magic integers (e.g., 3, 5, 7, 11, 19), approximate phase angles of 15 reflections can then be expressed in terms of only three parameters—say x, y, z—and approximate phase angles of further reflections can be derived in the magic-integer representation by using Eq. 3.34.

We now come to the problem of finding best values of the parameters x, y, z, and two methods have been proposed. One method is based on the location of the highest positive peaks in a special kind of Fourier series.[57] Any relationship of the type in Eq. 3.34 involving phases in magic-integer representation takes the general form

$$Hx + Ky + Lz + b = 0$$

or

$$\cos 2\pi(Hx + Ky + Lz + b) = 1$$

where H, K, L are integers (sums of magic integers). For several such relationships, it is not possible, in general, to find x, y, z values that simultaneously satisfy all, but we can ask that the relationships be satis-

[58]P. S. White and M. M. Woolfson, *Acta Crystallogr.* **A31**, 53 (1975); J. P. Declercq, G. Germain, and M. M. Woolfson, *ibid.*, p. 367.

fied as well as possible. The best solutions can be identified with the x, y, z coordinates of the highest positive peaks of the function

$$\Psi(x,y,z) = \sum |E_{1r}E_{2r}E_{3r}| \cos 2\pi(\boldsymbol{H}_r x + \boldsymbol{K}_r y + \boldsymbol{L}_r z + b_r)$$

The second method involves the use of Karle-Hauptman determinants (Eq. 3.28).[59] As we have seen, this determinant can be written in terms of E values, and it must be nonnegative. Moreover, it has been shown that the most probable set of phases is that which maximizes the value of the determinant.[60] If the phases of all the elements can be expressed in magic-integer form, the value of the determinant is a function of x, y, z. Algorithms have been developed to find the x, y, z values that maximize such determinants of high order (ca. 20).[59] However, the magic-integer approach is still in an experimental stage, and neither its potentialities nor its limitations have been fully investigated.

Noncentrosymmetric crystal structures containing up to about 100 nonhydrogen atoms in the asymmetric unit have been solved more or less automatically by the methods described in this section using virtually no chemical information. For centrosymmetric structures the number of atoms may be even larger. On the other hand, structures that seem to be far less complex sometimes turn out to be very difficult; some of them have resisted all attempts. What factors lead to a successful outcome in one case and to failure in another? Luck? Some space groups present greater difficulty than others. Pseudosymmetry in the atomic arrangement can lead to problems. The presence of a few high-order reflections with large $|E|$ values is a great help. The unsuspected presence of a few strong triples that fail to fulfill the condition of Eq. 3.34 by a wide margin can produce misleading indications early in the phase-determining chain and lead to a set of completely wrong solutions (we return to this problem in the section on Abberant Structure Invariants). Favorable circumstances, one is tempted to say, have much to do with the matter, but it is still not clear what constitutes favorable circumstances, except in a *post hoc propter hoc* fashion—i.e., success means favorable circumstances, and failure can always be attributed to unfavorable ones.

Structure Invariants and Semi-Invariants: Origin and Enantiomorph Specification

So far we have passed over a number of problems connected with the specification of the origin of the unit cell. The magnitudes of the Fourier

[59] M. M. Woolfson, *Acta Crystallogr.* **A33,** 219 (1977).
[60] G. Tsoucaris, *Acta Crystallogr.* **A26,** 492 (1970).

coefficients $|F_{\mathrm{H}}|$, measurable quantities, depend only on the crystal structure; their phase angles, which are to be derived, depend not only on the structure but also on its sense of orientation with respect to the chosen coordinate system and on the choice of origin as well. In deriving explicit phase angles, the unknown structure must be tied to a definite origin by fixing the phases of certain Fourier coefficients or combinations thereof. In addition, for noncentrosymmetric structures, the orientation of the axial frame must be fixed by restricting the value of a suitably chosen phase angle as either positive ($0<\phi<\pi$) or negative ($0<-\phi<\pi$). In the absence of anomalous scattering measurements, the choice is arbitrary, and the actual chirality sense of the structure is left open.

As an example, consider the simple one-dimensional pattern of point atoms shown in Fig. 3.19. The origin can be chosen at any point of the pattern. As the origin is displaced through a complete period, the $|U_h|$ values are unchanged, but the phase angles ϕ_h vary through h complete cycles. The $|U_h|$ values are also unchanged when the positive direction of the coordinate axis is reversed, but the signs of the phase angles ϕ_h are reversed. Some representative values for different origins A, B, C, C′ (coordinate axis reversed, Fig. 3.19) are given in Table 3.3. Regardless of the polarity sense, there is just one set of translation-equivalent points where the phase angle ϕ_1 is zero, and fixing ϕ_1 at this value is equivalent to choosing the origin at one of these points (B in Fig. 3.19). Fixing the phase of any other coefficient $U_h (h \neq 1)$ at zero would not define a *unique* origin since there would be h equispaced points that all correspond to the same set value of ϕ_h. The problem of defining the positive sense of the coordinate system remains.

If ϕ_1 is fixed at zero, it is sufficient to restrict one of the other phase angles, say ϕ_2, to a positive value between 0 and π. In our example, this choice would actually give the pattern with reversed polarity (B′). For efficient discrimination between the two possible polarities, the phase angle chosen as positive should evidently not be too close to either 0 or

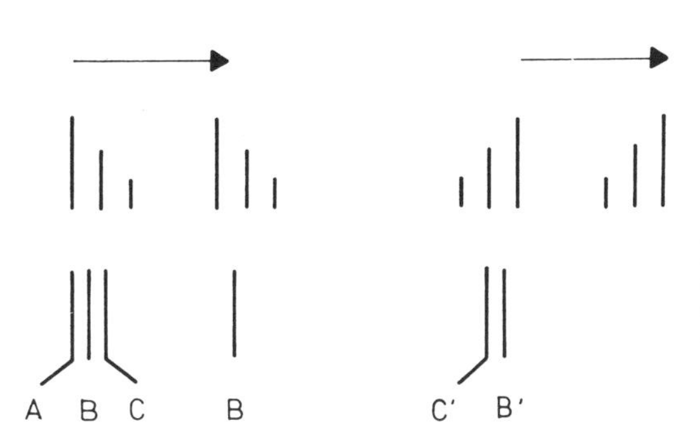

Figure 3.19. One-dimensional periodic pattern of point atoms with a selection of possible origins for each of the two alternative orientations of the coordinate system. Corresponding values of U_h and ϕ_h are given in Table 3.3.

Table 3.3. Values of $|U_h|$ and ϕ_h for different origins A, B, C of the simple point atom structure shown in Fig. 3.19[a]

h	$\lvert U_h\rvert$	ϕ_h(A)	ϕ_h(B)	ϕ_h(C)	ϕ_h(C′)
1	0.63	42°	0°	−42°	42°
2	0.28	8	−75	−159	159
3	0.28	−8	−132	103	−103
4	0.63	−42	152	−14	14
5	1.00	0	152	−56	56

[a]The origin C′ is at the same position as C, but the positive direction of the coordinate system is reversed.

π, but since the actual values of the phase angles are unknown, luck is involved in making a good choice.

If ϕ_1 is fixed at a value different from zero (or π), the two possible polarities correspond to different origins. For example, for the pattern on the left in Fig. 3.19, ϕ_1 is +42° for origin A and −42° for origin C (displaced by 84/360 along the positive direction of the coordinate system). Reversal of the polarity changes ϕ_1(C) from −42° back to +42° (pattern on right in Fig. 3.19). Hence, fixing ϕ_1 at +42° is equivalent to placing the origin either at A for one polarity sense, *or* at C for the reversed polarity sense (C′). Again, to discriminate between these two possibilities we must restrict another phase angle to lie within a certain range, and again there is an element of luck involved in making a good choice. Suppose the origin change A → C changes ϕ_1 into $-\phi_1$. It then changes ϕ_2 into $\phi_2 - 4\phi_1$, and reversal of polarity C → C′ changes this to $4\phi_1 - \phi_2$, which would be close to the original ϕ_2 if this happened to be close to $2\phi_1$.

The closer the pattern is to being centrosymmetric, the more difficult is the discrimination between the two possible polarities. If the pattern is so nearly centrosymmetric that no discrimination is possible, the derived phase angles will correspond to a centrosymmetric pattern that is a superposition of the original patterns with both possible polarity senses. Fixing ϕ_1 at zero (or π) ensures that the origin of this superposition pattern corresponds to the pseudocenter of the original pattern, whereas fixing it at some other value will produce an arbitrary superposition in which the main features of the original pattern may not be easily recognized.

In general, the procedure to follow for specifying the origin and orientation of the coordinate system depends on the nature and number of the symmetry elements of the space group. For each space group there is a standard functional form, a trigonometric expression for $F_\mathbf{H}$:

$$F(hkl) = A(h,k,l,x,y,z) + iB(h,k,l,x,y,z)$$

referred to a standard origin at a special position of the space group with a particular site symmetry (e.g., $\bar{1}$, $2/m$, 3_1, etc.). Once this site symmetry is specified, the functional forms of A and B are fixed and vice-versa. However, this does not usually fix the origin uniquely because the space group generally contains several translationally nonequivalent positions with the same site symmetry.

For example, in space group $C2/c$ there are two sets of inversion centers—one set on the glide planes, the second between them. The standard origin given in the *International Tables for X-ray Crystallography* belongs to the first set, and the standard functional form of A and B is:

$$A = 8\cos^2 2\pi\left(\frac{h+k}{4}\right)\cos 2\pi(hx+lz+l/4)\cos 2\pi(ky-l/4)$$

$$B = 0$$

but this does not fix the origin uniquely for we get exactly the same functional form for alternative origins at $(0, 0, \frac{1}{2})$, $(\frac{1}{2}, 0, \frac{1}{2})$, $(\frac{1}{2}, 0, 0)$. If the origin is taken at one of these positions, the only effect is to reverse the sign of A for half the reflections (those with l odd, $h + l$ odd, or h odd, respectively). Displacement of $(\frac{1}{2}, \frac{1}{2}, 0)$ does not change A in any way since origins related by this displacement are translationally equivalent.

A quantity, such as $\phi_{-\mathbf{H}} + \phi_\mathbf{K} + \phi_{\mathbf{H}-\mathbf{K}}$ or $F_{-\mathbf{H}}F_\mathbf{K}F_{\mathbf{H}-\mathbf{K}}$, whose value depends only on the structure, independent of the choice of origin (but not of the enantiomorph), is called a structure invariant. A quantity such as $F(2h, 2k, 2l)$ in $C2/c$, as in the last example, is not a structure invariant in the strict sense since its phase is altered by an arbitrary origin displacement. However, it is invariant with respect to all possible origins compatible with the standard functional form of the space group—the inversion centers situated on the glide planes. Such a quantity is called a structure semi-invariant. Some examples for a few specific space groups follow.

For $P1$ the functional form is:

$$A = \cos 2\pi(hx+ky+lz)$$

$$B = \sin 2\pi(hx+ky+lz)$$

There are no special positions, none of the individual phase angles are structure invariants, and the origin-definition procedure is closely analogous to that described for the simple one-dimensional example.

Here, the origin is uniquely specified by restricting the phase angles of any three reflections that define a primitive unit cell, e.g., (100), (010), (001) at zero or π. However, for a triclinic lattice there are an infinite number of ways of choosing a primitive cell. We could equally well pick any three other noncoplanar reciprocal lattice vectors:

$$\mathbf{H}_1 = h_1\mathbf{a}^* + k_1\mathbf{b}^* + l_1\mathbf{c}^*$$

$$\mathbf{H}_2 = h_2\mathbf{a}^* + k_2\mathbf{b}^* + l_2\mathbf{c}^*$$

$$\mathbf{H}_3 = h_3\mathbf{a}^* + k_3\mathbf{b}^* + l_3\mathbf{c}^*$$

such that the volume of the parallelepiped spanned by $\mathbf{H}_1$, $\mathbf{H}_2$, $\mathbf{H}_3$ is the same as that spanned by (100), (010), (001). The condition is

$$\frac{\mathbf{H}_1 \cdot \mathbf{H}_2 \mathbf{x} \mathbf{H}_3}{\mathbf{a}^* \cdot \mathbf{b}^* \mathbf{x} \mathbf{c}^*} = \begin{vmatrix} h_1 & k_1 & l_1 \\ h_2 & k_2 & l_2 \\ h_3 & k_3 & l_3 \end{vmatrix} = \pm 1$$

In addition, the phase angle of one other reflection can be restricted to the range $0<\phi<\pi$ to fix the chirality sense of the structure with respect to the coordinate system.

For $P\bar{1}$ the standard functional form is:

$$A = \cos 2\pi(hx+ky+lz)$$

$$B = 0$$

with the origin at a center of inversion. A shift of origin by

$$\begin{matrix} 0,\ 0,\ 0; & \tfrac{1}{2},\ 0,\ 0; & 0,\ \tfrac{1}{2},\ 0; & 0,\ 0,\ \tfrac{1}{2} \\ \tfrac{1}{2},\ \tfrac{1}{2},\ \tfrac{1}{2}; & 0,\ \tfrac{1}{2},\ \tfrac{1}{2}; & \tfrac{1}{2},\ 0,\ \tfrac{1}{2}; & \tfrac{1}{2},\ \tfrac{1}{2},\ 0 \end{matrix}$$

leaves the functional form unchanged and merely reverses the signs of certain classes of reflections, according to the parity of their indices

$$\begin{matrix} ggg & ugg & gug & ggu \\ uuu & guu & ugu & uug \end{matrix}$$

Reflections in the first parity class are structure semi-invariants since their signs are independent of which of the eight possible origins is chosen. The origin is specified by fixing the signs of any three reflections that belong to other parity classes provided these parity classes are linearly independent. This condition is necessary because once the signs of two reflections from different parity classes (say *uuu* and *uug*) are fixed, the sign of a third reflection from the parity class *ggu* is also determined from the fact that $S_\mathbf{H} S_\mathbf{K} S_{\mathbf{H}+K}$ is a structure invariant for centrosymmetric space groups. Hence, the third arbitrary sign would have to be chosen from one of the other parity classes (say *gug*).

Similar considerations apply to other space groups. In general, if the origin along any unit cell direction is restricted by the functional form, it may be fixed by *any* reflection in an appropriate parity class, whereas if the origin is completely arbitrary in any direction it must be fixed by a reflection whose Fourier wave has unit periodicity in that direction.

For $P222$ the standard functional form:

$$A = 4\cos 2\pi hx \cos 2\pi ky \cos 2\pi lz$$

$$B = -4\sin 2\pi hx \sin 2\pi ky \sin 2\pi lz$$

requires the origin to be taken at one of the eight special positions with 222 site symmetry:

$$0, 0, 0; \quad \tfrac{1}{2}, 0, 0; \quad 0, \tfrac{1}{2}, 0; \quad 0, 0, \tfrac{1}{2};$$
$$\tfrac{1}{2}, \tfrac{1}{2}, \tfrac{1}{2}; \quad 0, \tfrac{1}{2}, \tfrac{1}{2}; \quad \tfrac{1}{2}, 0, \tfrac{1}{2}; \quad \tfrac{1}{2}, \tfrac{1}{2}, 0$$

With the origin at one of these points, the phase angle of any particular (hkl) reflection is no longer arbitrary and cannot be set to zero. Suppose it has the value ϕ for one of the possible origins; for h odd, ϕ is changed to $\phi + \pi$ by a $\tfrac{1}{2}$, 0, 0 origin shift, whereas for h even, ϕ is left unaltered. The origin can thus be defined by restricting the phase angles of three reflections from linearly independent parity classes to be positive ($0 < \phi < \pi$), but it would be simpler to use zonal reflections, ($hk0$), ($h0l$), ($0kl$), whose phase angles are restricted by the functional form to be 0 or π since $B = 0$. (The projections down each of the three twofold axes are centrosymmetric.) The chirality is specified by restricting the phase angle of a (hkl) reflection to be positive (or negative).

For $P2_12_12_1$ the standard functional form [Chapter 2, section on Space group $P2_12_12_1(D_2^4)$] requires the origin to be taken at a point midway between the three nonintersecting screw axes. As in the previous example, it is convenient, although not obligatory, to use zonal reflections to define the origin; suitable choices might be:

$0\,g\,u$	0 or π		$0\,g\,u$	0 or π
$u\,0\,g$	0 or π	or	$u\,0\,u$	$\pm\,\pi/2$
$g\,u\,0$	0 or π		$u\,u\,0$	$\pm\,\pi/2$

For the standard functional form of this space group, the phases of zonal reflections are restricted to the values $\pm\,\pi/2$ when the index following the zero (in cyclic order) is odd. Thus, the phase of a suitably chosen fourth reflection can be set to one of these two possible values for efficient enantiomorph discrimination.

For $P3_1$ the standard origin is taken on one of the 3_1 axes situated at $0, 0, z; \frac{1}{3}, \frac{2}{3}, z; \frac{2}{3}, \frac{1}{3}, z$ (Fig. 3.20). To fix the origin in the z direction, the phase of a (hkl) reflection can be set to zero. A shift of origin from one 3_1 axis to another introduces a phase change of $\pm 2\pi(h-k)/3$ and hence leaves the phases of reflections with $(h-k) = 3n$ unaltered. The origin is therefore specified by restricting the phase angle of a $(hk0)$ reflection with $h-k \neq 3n$ to lie within the range $0 < \phi < 120°$. A special enantiomorph discrimination is superfluous here since the chirality is already defined by the general positions: $x,y,z;\ -y,x-y,z+\frac{1}{3};\ y-x,-x,z+\frac{2}{3}$, which fix the helicity sense of the screw axis (Fig. 3.20).

For centered space groups some or all of the possible origins compatible with the functional form are translationally equivalent and hence indistinguishable. The number of reflections whose phase angles can be fixed is therefore reduced, but so is the number of parity classes. For example, for a C-centered lattice, all reflections belonging to four of the parity classes, *ugg, ugu, gug, guu* are systematically absent; the origin must be specified by fixing the phase angles or signs of two reflections from the remaining *ggu, uug,* or *uuu* classes. For an F-centered lattice, the only reflections that are not systematically absent are those belonging to the *ggg* and *uuu* parity classes. The origin is specified by fixing the phase angle or sign of a single reflection from the latter class.

The complete theory of origin and enantiomorph specification in terms of structure invariants and semi-invariants is given in a series of papers by Karle and Hauptman.[61]

Aberrant Structure Invariants

As mentioned earlier the most probable value of the structure invariant $\phi_{-\mathbf{H}}+\phi_{\mathbf{K}}+\phi_{\mathbf{H}-\mathbf{K}}$ is zero. However, a given structure invariant of this type is not necessarily zero—it has some definite value that depends on the structure. For example, in the one-dimensional example discussed above (Fig. 3.19, Table 3.3), the value of the strongest structure invariant, $\phi_1+\phi_4+\phi_{-5}$, with $|U_1U_4U_5| = 0.40$, is indeed zero, but the next strongest, $\phi_1+\phi_3+\phi_{-4}$, with $|U_1U_3U_4| = 0.11$, is $+76°$.

Fig. 3.21 shows the Fourier transform of a centrosymmetric array of 20 point-atoms arranged as in the carbon skeletons of two naphthalene molecules related by a translation of $2\mathbf{t}$.[62] The three marked vectors

[61] H. Hauptman and J. Karle, *Acta Crystallogr.* **9,** 45 (1956); **12,** 93 (1959); J. Karle and H. Hauptman, *ibid.,* **14,** 217 (1961).

[62] W. E. Thiessen and W. R. Busing, *Acta Crystallogr.* **A30,** 814 (1974).

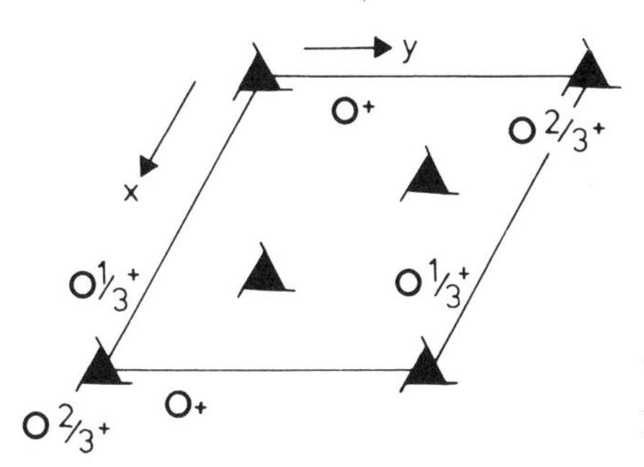

Figure 3.20. Arrangement of general positions and symmetry elements in space group $P3_1$. The 3_1 axes are right-handed screws (z axis upward).

emanating from the origin sum to zero and have the property that the product $E_{-\mathbf{H}}E_{\mathbf{K}}E_{\mathbf{H}-\mathbf{K}}$ is large and negative—i.e., the structure invariant $\phi_{-\mathbf{H}}+\phi_{\mathbf{K}}+\phi_{\mathbf{H}-\mathbf{K}}$ equals π. The transform shown in Fig. 3.21 is obtained from the naphthalene transform (Fig. 1.14) by multiplication by a set of cosine fringes, cos $2\pi\mathbf{R}\cdot\mathbf{t}$, and the reader can find other examples of aberrant triples in Fig. 1.14. If we knew in advance that the struc-

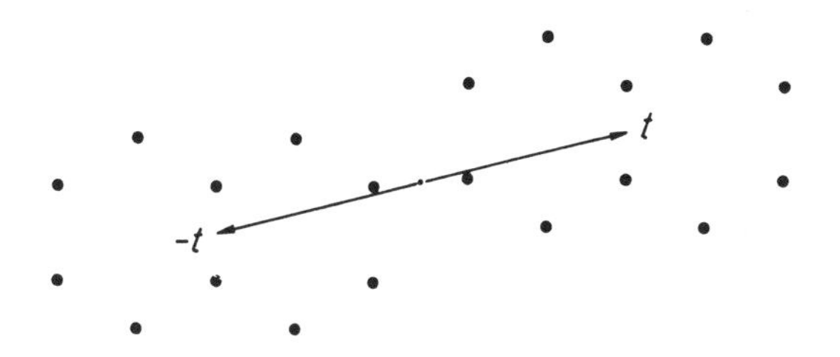

Figure 3.21. A planar array of 20 point atoms arranged as in the carbon skeletons of two naphthalene molecules related by a translation of 2**t** (above) and a parallel section through the corresponding Fourier transform. Negative contours are dashed, and the zero contour is dotted. A triple of vectors that sum to zero is superimposed on the transform. Two of the termini coincide with strong positive peaks, the third with a strong negative peak. The sign product is negative, so the corresponding structure invariant is aberrant. It would also be aberrant for the transform of a single benzene ring (cf., Fig. 1.14). From Thiessen and Busing, *Acta Crystallogr.* **A30,** 814 (1974).

ture contained regular patterns of six-membered rings or other similar molecular fragments related by noncrystallographic translations, we could expect aberrant triples to occur systematically in certain regions of reciprocal space and take appropriate steps. The safest procedure is probably simply to omit the suspect aberrant triples from the early stages of the phase-determining process where they can do most harm.

In general we do not know enough about the atomic arrangement to select possibly aberrant triples in this way. We only know that some of the stronger triples must be aberrant, but we do not know which ones. The difficulty of recognizing these aberrant triples is probably the main reason why direct methods do not always work. An error in sign or a large error in phase of one reflection, introduced during the early stages of the sign-determining process, is propagated through many other reflections. The same error introduced in later stages is much less serious. This means that success or failure may depend largely on the choice of the starting set of reflections. When direct methods seem to fail, it is often good to choose a quite different starting set and begin again.

Attempts have been made to use features of the observed pattern of $|E|$ values to identify the invariants that might be suspected to deviate markedly from zero. A complicated formula derived by Karle and Hauptman expresses the value of the cosine of the triple structure invariant in terms of statistical averages over various combinations of observed $|E|$ values.[63] More extensive studies are described in a recent monograph dealing with this problem.[64]

The following discussion shows that the presence of aberrant triple invariants can sometimes be diagnosed from certain features of the pattern of strong and weak reflections. Consider the four interconnected structure invariants

$$\phi_{-\mathbf{H}} + \phi_{\mathbf{K}} + \phi_{\mathbf{H}-\mathbf{K}} = \phi_1$$

$$\phi_{\mathbf{H}} + \phi_{-\mathbf{L}} + \phi_{\mathbf{L}-\mathbf{H}} = \phi_2$$

$$\phi_{-\mathbf{K}} + \phi_{\mathbf{L}} + \phi_{\mathbf{K}-\mathbf{L}} = \phi_3$$

$$\phi_{\mathbf{H}-\mathbf{L}} + \phi_{\mathbf{K}-\mathbf{H}} + \phi_{\mathbf{L}-\mathbf{K}} = \phi_4$$

Adding the four equations and using $\phi_{\mathbf{H}} = -\phi_{-\mathbf{H}}$, etc., gives

$$\phi_1 + \phi_2 + \phi_3 + \phi_4 = 0 \text{ (modulo } 2\pi) \tag{3.40}$$

[63]J. Karle and H. Hauptman, *Acta Crystallogr.* **11,** 264 (1958).

[64]H. Hauptman, *Crystal Structure Determination: The Role of the Cosine Semi-invariants,* Plenum Press, New York, 1972.

For simplicity, assume a centrosymmetric structure so that the individual phase angles and invariants are either 0 or π. Eq. 3.40 then says that the number of negative invariants ($\phi_i = \pi$) cannot be odd. Now suppose that one of the six reflections involved, say $|E_{\mathbf{H}}|$, is very weak and that the other five are strong. From the modified Sayre equation (Eq. 3.35)

$$E_{\mathbf{H}} \sim N^{\frac{1}{2}} \langle E_{\mathbf{K}} E_{\mathbf{H-K}} \rangle_{\mathbf{K}} \sim 0$$

the products $S_{\mathbf{K}}S_{\mathbf{H-K}}$ and $S_{\mathbf{L}}S_{\mathbf{H-L}}$ are likely to have opposite signs, which implies that one of the products $S_{\mathbf{H}}S_{\mathbf{K}}S_{\mathbf{H-K}}$ or $S_{\mathbf{H}}S_{\mathbf{L}}S_{\mathbf{H-L}}$ is negative—i.e., $\phi_1 + \phi_2 = \pi$. In this case it follows from Eq. 3.40 that $\phi_3 + \phi_4 = \pi$ also. In other words, of the two triple structure invariants involving exclusively strong reflections, one is probably negative. Of course, we do not know which one. However, if this pattern of weak and strong $|E|$ values were recognized at the start of a direct methods analysis, it would be advisable to exclude such triples from the sign derivation chain.

Summary of Direct-Method Procedures

The main steps to be followed in the usual process of deriving phase angles or signs by direct methods are:

1. Calculation of $|E|$ values and sorting of reflections in order of decreasing $|E|$.
2. Selection of all reflection triples, $|E_{\mathbf{H}}|$, $|E_{\mathbf{K}}|$, $|E_{\mathbf{H-K}}|$ such that each $|E|$ is greater than some minimum value (say 1.4).
3. Choice of reflections whose phase angles are to be locked or restricted to lie within certain ranges, to define the origin and possibly the orientation of the structure with respect to the coordinate system.
4. Assignment of symbolic phases to reflections with large $|E|$ values that are involved in many triples or permutation of actual phase angles (e.g., ±45°, ±135°) assigned to such reflections.
5. Phase-angle assignment to other reflections, first using triple structure invariants, then using tangent formula (for noncentrosymmetric structures) for each phase-angle permutation.
6. Use of tangent formula with derived phases as input to produce improved phases; repeated several times for each phase-angle permutation.
7. Calculation of E maps for solutions judged best according to chosen criteria. Selection of E maps (with peak distribution closest to that expected from stereochemical or other grounds) as provisional trial structures. If none of the E maps is acceptable as a trial structure or if the trial structures do not refine, return to step 3 with a different choice of reflections.

It may be advisable to include a search for possible negative triple structure invariants before embarking on step 5. The doubtful triples should be excluded from the subsequent analysis or downweighted, except that in certain cases they may suggest possible ways of averting the "triclinic catastrophe" alluded to earlier.

4. Methods of Crystal Structure Refinement

"It's too late to correct it," said the Red Queen, "when you've once said a thing, that fixes it, and you must take the consequences."

Structure Refinement: The R Factor

Chapter 3 dealt with methods of deriving a rough description of a crystal structure that is in qualitative agreement with the main features of the diffraction pattern. In this chapter we deal with methods of structure refinement, methods for adjusting the parameters that define the proposed structure to obtain optimal agreement with the observed X-ray data. Normally, the structural model is described in terms of a set of atomic positions, the atom types that occupy these positions, and a set of temperature factors that allow for atomic vibrations. From such a model, a set of structure factors,

$$F_{\mathbf{H}} = \sum_j f_j^0 T_j \exp(2\pi i \mathbf{H}\cdot\mathbf{r}_j)$$

where f_j^0 are the atomic scattering factors and T_j are assumed temperature factors, can be calculated, and their magnitudes $|F_{\mathbf{H}}|_c$ can be compared with the observed structure amplitudes $|F_{\mathbf{H}}|_o$, appropriately scaled.

The tradition has arisen of using the R factor, sometimes called the reliability index or residual,

$$R = \frac{\sum ||F_o| - |F_c||}{\sum |F_o|}$$

where F_o and F_c are the observed and calculated structure factors respectively, as a measure of agreement. The R factor may be a useful indicator in deciding whether any particular adjustment to the structural model is an improvement or not, but it is a poor criterion for the correctness of the model. A high R factor can be produced by an essentially correct structure whose parameters only need adjustment, while a low R factor can be produced by a structure that is incorrect in some essential feature (a few atoms missing; the molecule correctly oriented but displaced with respect to certain symmetry elements). A few large discrepancies between observed and calculated $|F|$ values for individual reflections can be hidden in a low R factor, but they may be enough to indicate that a structure is unacceptable.

Among other things, the R factor depends on the quality of the experimental data. For $|F_0|$ values derived from visual, photographic intensity estimates, final R factors in the range 0.1–0.15 could be deemed satisfactory; for diffractometer measurements, a value of about 0.05 or less could be expected, and a substantially higher value for a supposedly refined structure would be ground for suspicion. Heavy-atom structures tend to produce low R factors since the heavy-atom contributions alone are enough to produce a reasonable level of agreement. Noncentrosymmetric structures tend to produce lower R factors than centrosymmetric ones because they lead to a more uniform distribution of $|F|$ values—a lower proportion of very weak or very strong reflections. In the noncentrosymmetric equal-atom case, the expectation R factor for a completely random structure is 0.59, whereas it is 0.83 in the centrosymmetric case.[1]

R factors that are substantially lower than these values indicate at least that many of the interatomic vectors of the proposed structure coincide with those of the correct structure. This situation can occur, for example, if most of the atoms of the proposed structure are in the correct relative orientation but incorrectly placed with respect to the symmetry elements of the unit cell. In such cases, refinement can lead to fairly good agreement between high-order reflections but will leave a few marked discrepancies among the low-order ones. If these cannot be repaired by suitable adjustments to the model, it is almost certainly wrong in some respect.

The two main methods for refining a proposed structural model are (a) the $(F_0 - F_c)$ or difference synthesis; (b) the method of least squares. The former involves a step-by-step adjustment of the electron-density distribution expected for the model structure to the observed electron-density distribution; the adjustment is carred out in real space. The least-squares method involves a step-by-step adjustment of the calculated $|F|$ values to the observed ones; it is carried out in reciprocal space. In a sense the two methods are equivalent or can be made so by appropriate "weighting" of the observations in the least-squares method. In practice, however, each method has its advantages and limitations.

[1]A. J. C. Wilson, *Acta Crystallogr.* **3**, 397 (1950); D. C. Phillips, D. Rogers and A. J. C. Wilson, *ibid.*, p. 398.

The least-squares method is especially suited for automatic, iterative computer operation, but, working in reciprocal space, it is essentially blind. While it can efficiently optimize the values of the parameters used to describe the model structure, it cannot suggest the nature of any additional features whose inclusion might improve the level of agreement. The difference synthesis method is not as automatic; it demands intervention and interpretation by the investigator, but it can suggest the nature of possible inadequacies in the model (e.g., missing atoms, disorder, and so forth). We now discuss both methods in more detail.

The Difference Synthesis

Given a set of $(F_{\mathbf{H}})_c$ values for a proposed structure model, the electron-density distribution corresponding to the model is

$$\rho_c(\mathbf{r}) = \frac{1}{V} \sum_{\mathbf{H}} (F_{\mathbf{H}})_c \exp(-2\pi i \mathbf{H} \cdot \mathbf{r}) \tag{4.1}$$

which has to be compared with the experimental electron-density distribution

$$\rho_o(\mathbf{r}) = \frac{1}{V} \sum_{\mathbf{H}} (F_{\mathbf{H}})_o \exp(-2\pi i \mathbf{H} \cdot \mathbf{r}) \tag{4.2}$$

Both distributions differ from the true electron-density distribution. The function ρ_c includes the cumulative effect of all the deficiencies of the model—misplaced atomic positions, errors in assumed atomic scattering factors and temperature factors, possibly some atoms missing or superfluous atoms added—whereas ρ_o includes the effect of experimental inaccuracies in the magnitudes of the coefficients and of errors in the phase angles, which have to be assumed to be equal to those calculated for the model structure.

In principle both summations should include an infinite number of terms, but that for ρ_o is limited by the finite number of experimentally available observations. This situation introduces a termination-of-series error, whose main effect is to produce a set of approximately spherical ripples surrounding each atomic peak. Even when all phase angles are correct, the peak shapes in the series-terminated function ρ_o are convolutions of the actual peak shapes with the Fourier transform of a step function (Fig. 1.9), and the strength of the ripples is approximately proportional to the peak strength. Overlapping of weak peaks with ripples from strong ones can interfere seriously with the correct recognition or placing of light atoms in the presence of heavy ones. Whatever termination-of-series errors may be present in ρ_o, they should be present also in ρ_c if both summations are taken over exactly the same terms, and they should therefore cancel in the difference function

$$\rho_o(\mathbf{r})-\rho_c(\mathbf{r}) = \frac{1}{V}\sum_{\mathbf{H}} \{|F_{\mathbf{H}}|_o-|F_{\mathbf{H}}|_c\} \exp\left[-2\pi i\mathbf{H}\cdot\mathbf{r}+(\alpha_{\mathbf{H}})_c\right] \qquad (4.3)$$

which for the centrosymmetric case reduces to

$$\rho_o(\mathbf{r})-\rho_c(\mathbf{r}) = \frac{2}{V}\sum_{\mathbf{H}} \{(F_{\mathbf{H}})_o - (F_{\mathbf{H}})_c\} \cos 2\pi\mathbf{H}\cdot\mathbf{r} \qquad (4.4)$$

where $(F_{\mathbf{H}})_o$ is taken to have the same sign as $(F_{\mathbf{H}})_c$.

Obviously $|F_o|$ and $|F_c|$ values should be on the same scale. If the $|F_o|$ scale factor is too high, the difference map will emphasize ρ_o at the expense of ρ_c, with the result that the residual density close to the assumed atomic positions will tend to be positive. Conversely, a systematic negative residual density close to the assumed atomic positions would indicate that the $|F_o|$ scale factor is too low. However, as we shall see, errors in the assumed temperature factors can also cause the residual density at postulated atomic positions to be different from zero.

In the centrosymmetric case the signs attributed to $(F_{\mathbf{H}})_o$ are either right or wrong. If the model structure is a good approximation to the true structure, the overwhelming majority of signs will be right since small alterations in the model can hardly change $+F_{\mathbf{H}}$ into $-F_{\mathbf{H}}$ unless $F_{\mathbf{H}}$ is close to zero. Thus, if the model is good enough, $\rho_o(\mathbf{r})$ is more or less independent of the exact details of $\rho_c(\mathbf{r})$; apart from experimental inaccuracies and termination-of-series effects, it should represent the actual electron-density distribution in the crystal. On the other hand, for the noncentrosymmetric case, the phase angles attributed to $(F_{\mathbf{H}})_o$ in Eq. 4.2 are continuously variable quantities that differ by greater or smaller amounts from the true phase angles (Fig. 4.1). If $\rho_o(\mathbf{r})$ is calculated with the phase angles derived from the structure model, the resulting distribution is biased towards that of the model—it is approximately the average of the actual electron density and the electron density of the model structure. Thus, for example, if hydrogen atoms are present in the actual structure but are not included in the model, their electron-density peaks will occur in $\rho_o(\mathbf{r})$ and in $\rho_o(\mathbf{r})-\rho_c(\mathbf{r})$ with approximately correct strength in the centrosymmetric case but with approximately half strength in the noncentrosymmetric one. In what follows, the formulas derived refer to the centrosymmetric case, but we indicate briefly

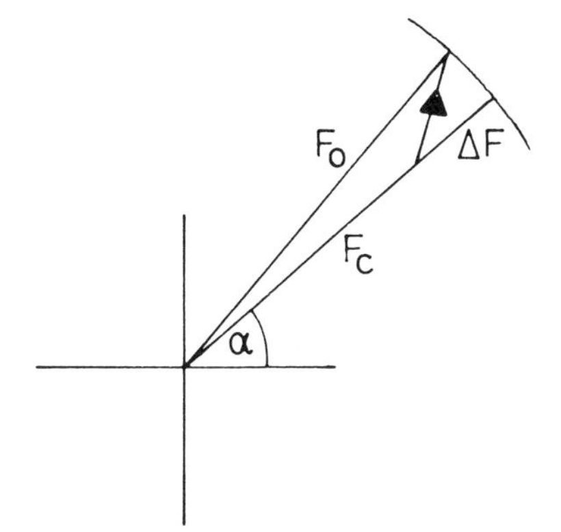

Figure 4.1. In the noncentrosymmetric case, ΔF has to be given the calculated phase angle α because the actual phase angle of F_0 is unknown.

how allowance can be made for the phase-angle bias in noncentrosymmetric structures.

From the prominent features of the difference synthesis, various kinds of errors in the model structure can be recognized and corrected. Perhaps the most important application is in finding the positions of hydrogen atoms. The mean contribution of an atom to the intensity of an X-ray reflection is roughly proportional to the square of its atomic number. Thus, the mean scattered intensity due to a hydrogen atom is only about 3% of that due to a carbon atom; this makes it very difficult to locate the hydrogen atoms during the early stages of an X-ray analysis before the positions of all the nonhydrogen atoms are known accurately. Once these are known, however, inspection of the residual electron density in a difference map usually reveals more-or-less well-defined positive peaks at or close to positions expected for hydrogen atoms on stereochemical grounds. Of course, the positions determined this way are not very accurate and they deviate slightly but significantly from the corresponding nuclear positions (which can be determined with high precision from neutron diffraction analysis). Nevertheless, the recognition of hydrogen atoms from difference maps is usually clear enough to allow a decision to be made between alternative possibilities—e.g., in distinguishing one tautomeric form of a molecule from another.

If most of the scattering power in the model structure is located correctly, one or two grossly misplaced atoms would be recognized by the presence of strong negative troughs at the assumed positions in the difference map, balanced by positive peaks at the correct positions. Less extreme positioning errors will be manifested by a nonzero gradient of the residual density at a postulated atomic position (Fig. 4.2). The misplaced atom should be moved along the direction of steepest ascent of

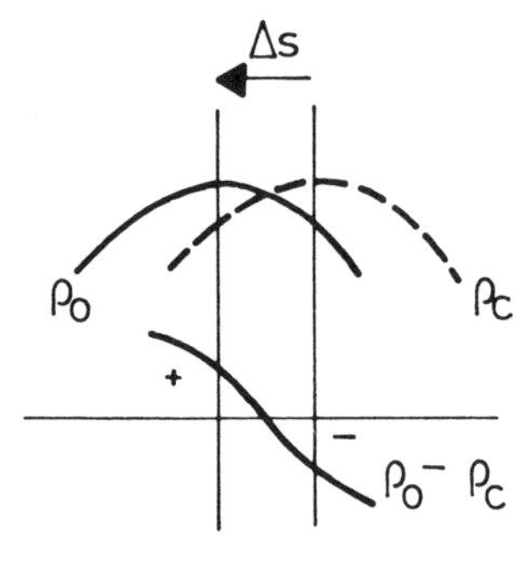

Figure 4.2. Profiles of $\rho_o(\mathbf{r})$ (full curve), $\rho_c(\mathbf{r})$ (dashed curve) and $D(\mathbf{r}) = \rho_o(\mathbf{r}) - \rho_c(\mathbf{r})$ along the direction of maximum slope of $D(\mathbf{r})$ near a postulated atomic position.

the residual density by an amount proportional to the gradient. To see this, recall that the peaks of $\rho_c(\mathbf{r})$ have their maxima at the postulated positions, whereas the peaks of $\rho_o(\mathbf{r})$ have their maxima at the "true" positions. Suppose that the true position is $\mathbf{r}' = \mathbf{r}_c + \Delta\mathbf{s}$ where $\Delta\mathbf{s}$ is the correction in the direction of maximum slope. At a postulated position:

$$\begin{aligned} D(\mathbf{r}_c) &= \rho_o(\mathbf{r}_c) - \rho_c(\mathbf{r}_c) \\ &= \rho_o(\mathbf{r}' - \Delta\mathbf{s}) - \rho_c(\mathbf{r}_c) \\ &= \rho_o(\mathbf{r}') - \left(\frac{\partial \rho_o}{\partial s}\right)_{\mathbf{r}_c} \Delta\mathbf{s} - \rho_c(\mathbf{r}_c) \end{aligned}$$

Taking the derivative along $\mathbf{s}$ at the postulated position,

$$\left(\frac{\partial D}{\partial s}\right)_{\mathbf{r}_c} = \left(\frac{\partial \rho_o}{\partial s}\right)_{\mathbf{r}'} - \left(\frac{\partial^2 \rho_o}{\partial s^2}\right)_{\mathbf{r}_c} \Delta\mathbf{s} - \left(\frac{\partial \rho_c}{\partial s}\right)_{\mathbf{r}_c}$$

and since the first derivatives of ρ_o and ρ_c at the respective maxima are both zero

$$\Delta\mathbf{s} = -\left(\frac{\partial D}{\partial s}\right)_{\mathbf{r}_c} \Big/ \left(\frac{\partial^2 \rho_o}{\partial s^2}\right)_{\mathbf{r}_c} \tag{4.5}$$

The denominator is negative—it is the second derivative of a peak close to its maximum—so the required displacement is along the positive direction of the gradient. The actual value of the proportionality factor can be estimated from measurements of the peak shape in a F_o Fourier synthesis. Peak shapes do not, in general, vary enormously. A typical profile through a peak maximum has the form[2]

[2]W. Cochran, *Acta Crystallogr.* **4,** 81 (1951).

$$\rho(\mathbf{r}) = Z\left(\frac{p}{\pi}\right)^{3/2} \exp(-pr^2) \tag{4.6}$$

with p about 4 to 6 Å^{-2}, depending on the temperature factor. The curvature at the peak center is then $-2Zp(p/\pi)^{3/2}$ or about $-10Z$ to $-30Z$ el. Å^{-5} or somewhat less (in el. Å^{-4}) for two-dimensional peaks with normalization factor p/π instead of $(p/\pi)^{3/2}$ in Eq. 4.6. Only the order of magnitude is important because the approach to the correct structure would normally be carried out as a step-by-step iterative process. In the noncentrosymmetric case, the shift indicated by Eq. 4.5 will be underestimated (by a factor of about 1.5–2) for reasons indicated earlier.

Eq. 4.5 also suggests how the standard deviation in an assumed atomic position can be estimated once the residual density gradient at such a position has been reduced to zero. The standard deviation in position is obviously proportional to the standard deviation in the gradient, and this can be estimated if one is prepared to make certain assumptions about the statistical distribution of errors among the F_0 values. It can also be estimated more directly from the difference map itself by taking it equal to the observed root-mean-square value of the residual gradient in regions that are far removed from any atoms—i.e., the intermolecular voids, where ρ_0 and ρ_c should both be approximately zero. The standard deviation in an atomic position is approximately inversely proportional to the atomic number, but it also depends on the temperature factor. Clearly, the sharper the electron-density peak (the more negative its curvature), the less the apparent position of its maximum will be perturbed by random errors in the measurement.

The assumed temperature factors can also be corrected from the shape of the residual density close to the postulated atomic positions. Assume, for simplicity, that the postulated position of an atom in the structure model has been adjusted so that the gradient of the residual density at the position in question is zero. If the assumed temperature factor is too high, the electron-density peak in ρ_c will be more diffuse than the corresponding peak in ρ_0, leading to a positive residual density at the peak center, surrounded by regions of negative residual density on either side (Fig. 4.3). Naturally, the opposite is true if the assumed temperature factor is too low. If spherically symmetrical temperature factors were assumed for the atoms of the postulated structure, the appearance of characteristic lobes in the residual density would indicate the presence of anisotropic thermal vibrations (Fig. 4.4).

Of course, the individual types of error (scale factor, missing or superfluous atoms, atoms incorrectly placed or assigned incorrect tempera-

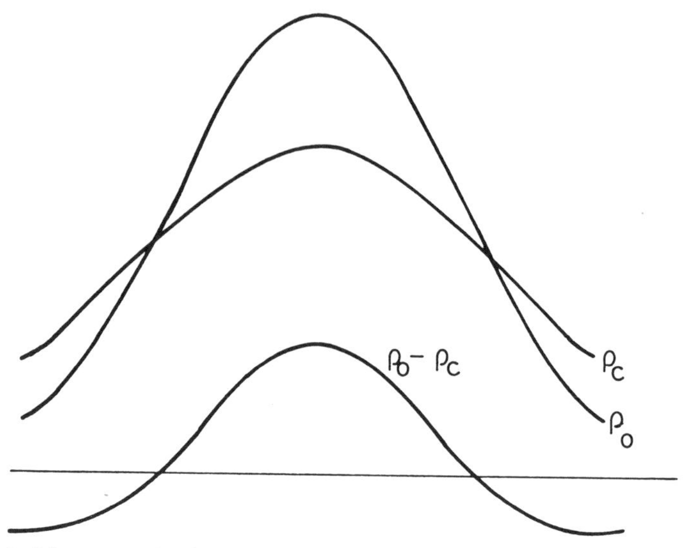

Figure 4.3. If an atom in the postulated structure is assigned too high a temperature factor, the corresponding peak in ρ_c will be too diffuse. The residual density $\rho_o - \rho_c$ will be positive at the peak center and negative on either side.

ture factors) do not occur in isolation but combine to produce complicated patterns in the residual density. In practice, we correct for the prominent features, include the corrections in a revised structure model, recalculate the difference synthesis, in which the features that have been corrected for will be less prominent, and proceed in this way until the residual density contains no significant systematic features. At this stage the structural model agrees as well as possible with the observed electron density, and the X-ray analysis can be regarded as concluded.

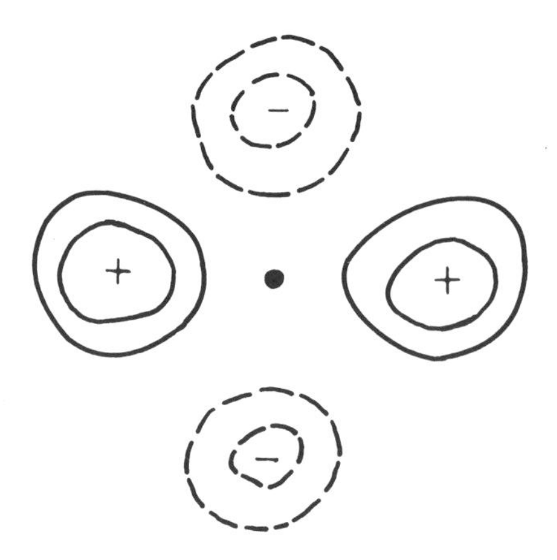

Figure 4.4. Characteristic residual density pattern showing that the assumed temperature factor is too low in one direction (positive lobes) and too high in the perpendicular direction (negative lobes).

Least-Squares Analysis

The method of least squares is often used in the physical and other sciences to fit a model to a given set of observed data. The method dates back to Legendre,[3] who proposed in 1806 that if we have a set of n linear equations in m unknowns

$$f_i - \sum_j a_{ij}x_j = 0, \qquad \begin{array}{l} i = 1,2,3 \ldots n \\ j = 1,2,3 \ldots m \end{array} \tag{4.7}$$

with $n > m$, that are not strictly compatible with each other, then, of all possible sets of values for the unknowns x_j, the most satisfactory is that for which

$$\sum_i (f_i - \sum_j a_{ij}x_j)^2 = \sum \Delta_i^2 \tag{4.8}$$

is a minimum. Legendre's principle was connected with the theory of probability by Gauss and Laplace. Each f_i in the above equations is regarded as an estimate of an observed quantity associated with a random error ϵ_i. The errors ϵ_i are, of course, unknown, but if they are normally distributed (more generally if they are drawn from populations with finite second moments[4]), then each f_i can, in principle, be associated with a variance $\sigma^2(f_i)$ and a corresponding weight w_i, inversely proportional to the variance. The best estimates of the x_j's are then those for which the weighted sum of squares of deviations

$$R = \sum_i w_i \Delta_i^2 \tag{4.9}$$

is minimized. This occurs if all the derivatives $\partial R/\partial x_j$ are set to zero, leading to m equations, each of the form

$$\sum_{i=1}^{n} w_i \Delta_i \frac{\partial \Delta_i}{\partial x_k} = 0 \qquad k = 1,2,3 \ldots m$$

From the definition of Δ_i we have $\partial\Delta_i/\partial x_k = -a_{ik}$ so

$$\sum_{i=1}^{n} w_i\left(f_i - \sum_{j=1}^{m} a_{ij}x_j\right)a_{ik} = 0$$

or

$$\sum_{j=1}^{m}\left(\sum_{i=1}^{n} w_i a_{ik} a_{ij} x_j\right) = \sum_{i=1}^{n} w_i a_{ik} f_i \tag{4.10}$$

[3]For an account of the early history of the least-squares method see E. T. Whittaker and G. Robinson, *The Calculus of Observations,* 2nd ed., Blackie, London and Glasgow, 1926.

[4]For discussion of this and the following points see W. C. Hamilton, *Statistics in Physical Science,* Ronald Press, New York, 1964.

There are m equations like this, one for each unknown, the so-called normal equations, and they are expressed most concisely in matrix form:

$$\mathbf{NX} = \mathbf{E} \tag{4.11}$$

where $\mathbf{X}$ and $\mathbf{E}$ are column matrices (vectors) of order m, and $\mathbf{N}$ is a symmetric square matrix of order $m \times m$. The elements of $\mathbf{N}$ and $\mathbf{E}$ are, respectively

$$N_{kj} = \sum w_i a_{ik} a_{ij} \qquad E_k = \sum w_i a_{ik} f_i \tag{4.12}$$

The initial weighted observational equations (Eq. 4.7) can also be expressed in matrix notation[5]

$$\mathbf{AX} = \mathbf{F} \tag{4.13}$$

with $A_{ij} = w_i^{\frac{1}{2}} a_{ij}$ and $F_i = w_i^{\frac{1}{2}} f_i$. Here $\mathbf{X}$ and $\mathbf{F}$ are column matrices of order m and n respectively, and $\mathbf{A}$ is a rectangular matrix of order $m \times n$—i.e., with m columns and n rows. Note that the observational equations are multiplied by the square roots of the appropriate weights. From Eq. 4.12 it is evident that $\mathbf{N} = \mathbf{A'A}$ and $\mathbf{E} = \mathbf{A'F}$ where $\mathbf{A'}$ is the transpose of $\mathbf{A}$. Hence Eq. 4.11 can be rewritten

$$\mathbf{A'AX} = \mathbf{A'F}$$

with solution

$$\mathbf{X} = (\mathbf{A'A})^{-1}\,\mathbf{A'F} \tag{4.14}$$

for the vector of unknowns. This discussion may seem rather abstract to those unfamiliar with matrix notation. However, Eq. 4.14 is just an extremely concise recipe for finding the best values of the unknowns x_j in any given case; hence, it is instructive to write down the explicit arrays of numbers and carry out the actual calculation for a simple example. Say we have five observational equations (that we assume to be already weighted) in three unknowns.

$$\begin{aligned}
x_1 \qquad\qquad\qquad &= 2 \\
x_2 \qquad\quad &= 4 \\
x_3 &= 6 \\
2x_1 + 3x_2 + x_3 &= 21 \\
x_1 + 2x_2 + x_3 &= 17
\end{aligned}$$

[5]The order of multiplication is important. The number of columns m in $\mathbf{A}(m \times n)$ must equal the number of rows in $\mathbf{X}(p \times m)$. The product matrix $\mathbf{F}$ is of the order $p \times n$. Here $p = 1$.

The equations are obviously not self-consistent, and we seek the best estimates of x_1 x_2, x_3 according to the least-squares criterion. We write

$$\mathbf{A} = \begin{bmatrix} 1 & 0 & 0 \\ 0 & 1 & 0 \\ 0 & 0 & 1 \\ 2 & 3 & 1 \\ 1 & 2 & 1 \end{bmatrix} \qquad \mathbf{F} = \begin{bmatrix} 2 \\ 4 \\ 6 \\ 21 \\ 17 \end{bmatrix}$$

$$\mathbf{A'A} = \begin{bmatrix} 1 & 0 & 0 & 2 & 1 \\ 0 & 1 & 0 & 3 & 2 \\ 0 & 0 & 1 & 1 & 1 \end{bmatrix} \begin{bmatrix} 1 & 0 & 0 \\ 0 & 1 & 0 \\ 0 & 0 & 1 \\ 2 & 3 & 1 \\ 1 & 2 & 1 \end{bmatrix} = \begin{bmatrix} 6 & 8 & 3 \\ 8 & 14 & 5 \\ 3 & 5 & 3 \end{bmatrix}$$

$$\mathbf{A'F} = \begin{bmatrix} 1 & 0 & 0 & 2 & 1 \\ 0 & 1 & 0 & 3 & 2 \\ 0 & 0 & 1 & 1 & 1 \end{bmatrix} \begin{bmatrix} 2 \\ 4 \\ 6 \\ 21 \\ 17 \end{bmatrix} = \begin{bmatrix} 61 \\ 101 \\ 44 \end{bmatrix}$$

and the normal equations are

$$\mathbf{A'AX} = \mathbf{A'F}$$

$$\begin{bmatrix} 6 & 8 & 3 \\ 8 & 14 & 5 \\ 3 & 5 & 3 \end{bmatrix} \begin{bmatrix} x_1 \\ x_2 \\ x_3 \end{bmatrix} = \begin{bmatrix} 61 \\ 101 \\ 44 \end{bmatrix}$$

Inversion of $\mathbf{A'A}$ gives

$$(\mathbf{A'A})^{-1} = \frac{1}{24} \begin{bmatrix} 17 & -9 & -2 \\ -9 & 9 & -6 \\ -2 & -6 & 20 \end{bmatrix}$$

whence

$$\begin{bmatrix} x_1 \\ x_2 \\ x_3 \end{bmatrix} = \frac{1}{24} \begin{bmatrix} 17 & -9 & -2 \\ -9 & 9 & -6 \\ -2 & -6 & 20 \end{bmatrix} \begin{bmatrix} 61 \\ 101 \\ 44 \end{bmatrix} = \begin{bmatrix} 1.67 \\ 4.00 \\ 6.33 \end{bmatrix}$$

If the first three observational equations had been accepted as correct, then $\Sigma\Delta^2 = (0 + 0 + 0 + 1 + 1) = 2$. The least-squares estimates

$$x_1 = 1.67 \qquad x_2 = 4.00 \qquad x_3 = 6.33$$

give $\Sigma\Delta^2 = (0.33)^2 + 0 + (0.33)^2 + (0.67)^2 + 1 = 1.67$.

The matrix notation is not only very elegant compared with the cumbersome formulas that arise when the algebraic equations are written explicitly, it is also ideally suited for writing computer programs since standard procedures for matrix multiplication and inversion are usually available.

The least-squares method also allows us to estimate the precision of the derived parameters—their variances and covariances. The variance σ_j^2 of a single parameter x_j is the expectation value of $(x_j - \langle x_j \rangle)^2$; its square root σ_j is often called the standard deviation of x_j. In a multi-dimensional problem, we must take into account not only the variances of the individual parameters but also their covariances σ_{jk}, defined as the expectation values of $(x_j - \langle x_j \rangle)(x_k - \langle x_k \rangle)$. The joint normal probability distribution of a pair of parameters x_1 and x_2 with mean values of zero is

$$P(x_1,x_2) = \frac{\pi}{p_{11}p_{22}} \exp\left[-(p_{11}x_1^2 + p_{22}x_2^2 + 2p_{12}x_1x_2)\right]$$

and contours of equal probability are given by the set of ellipses

$$p_{11}x_1^2 + p_{22}x_2^2 + 2p_{12}x_1x_2 = \text{constant}$$

For the probability distribution of x_1 or x_2 alone, $\sigma_1^2 = 1/2p_{11}$ and $\sigma_2^2 = 1/2p_{22}$, but to describe the complete two-dimensional distribution the cross-term must also be taken into account. It is this cross-term that gives the covariance, $\sigma_{12} = 1/2p_{12}$, which allows for the possibility that errors in x_1 and x_2 may be correlated. For the multi-dimensional case, the array of variances σ_j^2 and covariances σ_{jk} form the variance–covariance matrix.

$$\mathbf{V} = \begin{bmatrix} \sigma_1^2 & \sigma_{12} & \sigma_{13} \cdots\cdot & \sigma_{1n} \\ \sigma_{12} & \sigma_2^2 & \sigma_{23} & \sigma_{2n} \\ & & & \\ \sigma_{1n} & \sigma_{2n} & \sigma_{3n} & \sigma_n^2 \end{bmatrix} \tag{4.15}$$

The corresponding matrix of correlation coefficients ρ_{ij} is obtained by dividing each element by $(\sigma_i^2\sigma_j^2)^{\frac{1}{2}}$ so that $\rho_{ii} = 1$.

The variances and covariances of the derived parameters x_j can be estimated if we are prepared to make certain assumptions about the deviations, which can be expressed as elements of a column vector $\mathbf{\Delta}$ of the same order $(1 \times n)$ as $\mathbf{F}$:

$$\mathbf{AX} + \mathbf{\Delta} = \mathbf{F} \tag{4.16}$$

If the weights have been properly assigned, inversely proportional to the variances of the individual observations, then the elements of $\mathbf{\Delta}$ can be regarded as having been randomly drawn from a population with mean zero and common variance σ^2. Indeed, one test for the suitability of the assigned weights is that the sum of the squares of the weighted deviations for any arbitrarily chosen sample of observations should be proportional to the size of the sample. If these conditions are fulfilled, the expectation values of the parameters are:

$$\begin{aligned}\langle \mathbf{X}\rangle &= \langle (\mathbf{A}'\mathbf{A})^{-1}\mathbf{A}'\mathbf{F}\rangle \\ &= \langle (\mathbf{A}'\mathbf{A})^{-1}\mathbf{A}'(\mathbf{A}\mathbf{X}+\mathbf{\Delta})\rangle \\ &= (\mathbf{A}'\mathbf{A})^{-1}\mathbf{A}'\mathbf{A}\mathbf{X} \\ &= \mathbf{X}\end{aligned}$$

as expected. The variance–covariance matrix of $\mathbf{X}$ is

$$\begin{aligned}\mathbf{V}(\mathbf{X}) &= \langle (\mathbf{X}-\langle \mathbf{X}\rangle)(\mathbf{X}'-\langle \mathbf{X}'\rangle)\rangle \\ &= (\mathbf{A}'\mathbf{A})^{-1}\mathbf{A}'\langle (\mathbf{F}-\langle \mathbf{F}\rangle)(\mathbf{F}'-\langle \mathbf{F}'\rangle)\rangle \mathbf{A}(\mathbf{A}'\mathbf{A})^{-1} \\ &= (\mathbf{A}'\mathbf{A})^{-1}\mathbf{A}'\langle \mathbf{\Delta}\mathbf{\Delta}'\rangle \mathbf{A}(\mathbf{A}'\mathbf{A})^{-1}\end{aligned}$$

Now $\langle \mathbf{\Delta}\mathbf{\Delta}'\rangle$ is a $n \times n$ matrix whose elements are the expectation values of $\Delta_i\Delta_j$. If the Δs are randomly drawn from a population with mean zero and common variance σ^2, this matrix reduces to one with every diagonal element equal to σ^2 and every nondiagonal element equal to zero. In other words

$$\langle \mathbf{\Delta}\mathbf{\Delta}'\rangle = \sigma^2 \mathbf{I}$$

when $\mathbf{I}$ is an $n \times n$ unit matrix. We then have

$$\begin{aligned}\mathbf{V}(\mathbf{X}) &= (\mathbf{A}'\mathbf{A})^{-1}\mathbf{A}'\mathbf{I}\mathbf{A}(\mathbf{A}'\mathbf{A})^{-1}\,\sigma^2 \\ &= (\mathbf{A}'\mathbf{A})^{-1}\,\sigma^2 \qquad (4.17)\end{aligned}$$

The matrix inversion required to determine the vector of unknowns (Eq. 4.14) thus also yields the variances and covariances of the derived quantities apart from a proportionality factor. If sufficient information is available to justify the assignment of variances and covariances to the individual observations—i.e., if the observations are correlated in any way—a more elaborate treatment is required.[4]

There remains the problem of estimating the variance of the Δ values, the proportionality factor σ^2 in Eq. 4.17. The best estimate turns out to be

$$\sigma^2 = \sum_i w_i\,\Delta_i^2/(n-m) \qquad (4.18)$$

the same as used in single parameter statistics (where $m = 1$).

For the numerical example on p. 193, assuming unit weight for the observational equations, $\sigma^2 = 1.67/(5-3) = 0.83$, thus, the variance-covariance matrix is

$$\begin{array}{c} \\ x_1 \\ x_2 \\ x_3 \end{array} \begin{array}{c} \begin{array}{ccc} x_1 & x_2 & x_3 \end{array} \\ \begin{bmatrix} 0.59 & -0.31 & -0.07 \\ -0.31 & 0.31 & -0.21 \\ -0.07 & -0.21 & 0.69 \end{bmatrix} \end{array}$$

and the correlation matrix is

$$\begin{bmatrix} 1.00 & -0.72 & -0.11 \\ -0.72 & 1.00 & -0.45 \\ -0.11 & -0.45 & 1.00 \end{bmatrix}$$

Thus, of the three parameters, x_2 is determined most precisely, and there is a large negative correlation between x_1 and x_2; this means that any positive error in x_1 can be nearly compensated by a negative error in x_2.

To summarize, the variance of the derived parameter x_j is

$$\sigma^2(x_j) = \frac{\mathrm{N}_{jj}^{-1} \sum w_i \Delta_i^2}{n - m} \tag{4.19}$$

This is the source of the statement frequently found in X-ray crystallographic papers that "standard deviations were estimated by inversion of the matrix of normal equations." If this matrix is approximately diagonal—i.e., if the nondiagonal terms are small compared with the diagonal ones, Eq. 4.19 can be approximated by

$$\sigma^2(x_j) = \frac{\sum w_i \Delta_i^2}{N_{jj}} \cdot \frac{1}{n - m} \tag{4.20}$$

In this case, the variance of the parameter x_j is inversely proportional to the corresponding diagonal element of the matrix of normal equations. Analogous to Eq. 4.19 the covariance between the derived parameters x_j and x_k is given by

$$\sigma(x_j,x_k) = \frac{N_{jk}^{-1} \sum w_i \Delta_i^2}{n - m} \tag{4.21}$$

Cases may arise where the matrix of normal equations is singular or nearly so—i.e., where its determinant is zero or nearly zero. A singular matrix cannot be inverted, and the inverse of a nearly singular matrix is said to be ill conditioned. Under these conditions, some or all of the parameters attain enormous variances and become more or less indeterminate.

As a simple example consider the observational equations

$$
\begin{aligned}
0.50\,x_1 + 0.99\,x_2 &= 2.50 \\
1.01\,x_1 + 1.99\,x_2 &= 4.99 \\
1.98\,x_1 + 4.02\,x_2 &= 10.01
\end{aligned}
\tag{4.22}
$$

Formally, there are three equations for two unknowns, but the three equations are nearly the same, apart from a multiplicative factor. It is as if, instead of making three independent measurements to estimate two quantities, we made the same measurement three times, obtaining slightly different values each time. However, assume we had not noticed this and were to set out blindly to determine x_1 and x_2 from the three equations by least-squares analysis. We would obtain

$$
\mathbf{N} = \mathbf{A'A} = \begin{bmatrix} 0.50 & 1.01 & 1.98 \\ 0.99 & 1.99 & 4.02 \end{bmatrix} \begin{bmatrix} 0.50 & 0.99 \\ 1.01 & 1.99 \\ 1.98 & 4.02 \end{bmatrix}
$$

$$
\mathbf{E} = \mathbf{A'F} = \begin{bmatrix} 0.50 & 1.01 & 1.98 \\ 0.99 & 1.99 & 4.02 \end{bmatrix} \begin{bmatrix} 2.50 \\ 4.99 \\ 10.01 \end{bmatrix}
$$

leading to the normal equations $\mathbf{NX} = \mathbf{E}$

$$
\begin{bmatrix} 5.191 & 10.465 \\ 10.465 & 21.101 \end{bmatrix} \begin{bmatrix} x_1 \\ x_2 \end{bmatrix} = \begin{bmatrix} 26.110 \\ 52.645 \end{bmatrix}
\tag{4.23}
$$

The determinant of $\mathbf{N}$ is 0.019066, and we obtain $\mathbf{X} = \mathbf{N}^{-1}\mathbf{E}$

$$
\begin{bmatrix} x_1 \\ x_2 \end{bmatrix} = \begin{bmatrix} 1106.7 & -548.9 \\ -548.9 & 272.3 \end{bmatrix} \begin{bmatrix} 26.110 \\ 52.645 \end{bmatrix}
$$

leading to $x_1 = -0.903$, $x_2 = 3.455$, which produce very large Δs in all three equations. Clearly, we would do much better with $x_1 \approx 1$, $x_2 \approx 2$, from simple inspection of the equations—or with $x_1 \approx 0$, $x_2 \approx 2.5$, or with an infinity of other solutions. Each equation defines a line in the parameter space (x_1, x_2), but the three lines thus defined are virtually the same, so that x_1 and x_2 cannot be individually determined; only some linear combination, something like $x_1 + 2x_2 \approx 5$, can be determined.

If the observational equations were more complicated, involving many unknowns, it might not be as easy to recognize such a case, and some of the parameters may be coupled and individually indeterminate. In such cases, however, the normal-equations matrix will always be nearly singular. The particular linear combinations of parameters whose values are well or poorly defined by the equations can be identified by diagonalizing the matrix of normal equations—i.e., by finding its eigenvalues and eigenvectors. Since, for a diagonal matrix, the variances are inversely proportional to the diagonal elements and covariances are null, components of the vector of unknowns along eigenvectors of large eigenvalues will be well defined while components along eigenvectors of zero eigenvalues will be indeterminate.

In matrix language, the original normal equations

$$\mathbf{NX} = \mathbf{E}$$

are transformed into another set of equations

$$\mathbf{DY} = \mathbf{G}$$

where $\mathbf{D} = \mathbf{S'NS}$ is a diagonal matrix whose elements are the eigenvalues of $\mathbf{N}$; the eigenvectors are the columns of $\mathbf{S}$. Since

$$(\mathbf{S'NS})\mathbf{Y} = \mathbf{G}$$

$$(\mathbf{SS'})\mathbf{N}(\mathbf{SY}) = \mathbf{SG}$$

so if $\mathbf{X} = \mathbf{SY}$, then $\mathbf{E} = \mathbf{SG}$. The components of the new vector of unknowns $\mathbf{Y}$ are normalized mutually orthogonal linear combinations of the components of $\mathbf{X}$, and the transformation ($\mathbf{G} = \mathbf{S'E}$) that takes $\mathbf{E}$ into $\mathbf{G}$ is the same as the transformation ($\mathbf{Y} = \mathbf{S'X}$) that takes $\mathbf{X}$ into $\mathbf{Y}$.

We can try this with our simple example. The eigenvalues of $\mathbf{N}$ (Eq. 4.23) are given by

$$\begin{vmatrix} 5.191-\lambda & 10.465 \\ 10.465 & 21.101-\lambda \end{vmatrix} = 0$$

$$(5.191-\lambda)(21.101-\lambda) - (10.465)^2 = 0$$

$$\lambda^2 - 26.292\lambda + 0.0191 = 0$$

$$\lambda_1 = 26.291 \qquad \lambda_2 = 0.000725$$

By definition, $\mathbf{NX} = \lambda_i\mathbf{X}$ or $(\mathbf{N}-\lambda_i\mathbf{I})\mathbf{X} = 0$ if $\mathbf{X}$ is an eigenvector of $\mathbf{N}$. Thus, the eigenvector corresponding to the larger eigenvalue is obtained from

$$(5.191-26.291)x_1 + 10.465x_2 = 0$$

$$-21.100x_1 + 10.465x_2 = 0$$

or
$$\frac{x_1}{x_2} = 0.4960$$

The unit vector in this direction has components of 0.444 and 0.896 on the original x_1- and x_2-axes, respectively, and the other eigenvector is orthogonal to this (Fig. 4.5). We now define a pair of normalized linear combinations:

$$X_1 = 0.444x_1 + 0.896x_2$$
$$X_2 = -0.896x_1 + 0.444x_2$$

and the normal equations (Eqs. 4.23) can be written in their diagonal form:

$$\begin{bmatrix} 26.291 & 0 \\ 0 & 0.001 \end{bmatrix} \begin{bmatrix} X_1 \\ X_2 \end{bmatrix} = \begin{bmatrix} 58.763 \\ 0.020 \end{bmatrix}$$

where the same transformations have been applied to the elements of

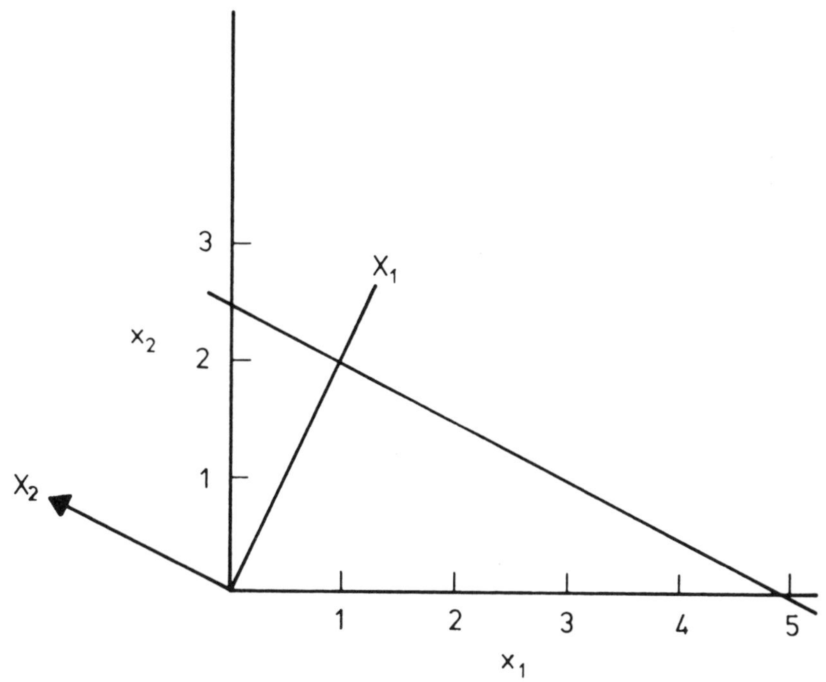

Figure 4.5. The linear combinations: $X_1 = 0.444x_1 + 0.896x_2$ and $X_2 = -0.896x_1 + 0.444x_2$ are normal and parallel to the "best line" $x_1 + 2.018x_2 = 5.034$ fitting the observational equations (Eq. 4.22). Although x_1 and x_2 are individually indeterminate, the value of the linear combination X_1 (the distance of the line from the origin) is well determined.

the **E** column vector as were applied to those of the **X** vector. One solution is

$$X_1 = 58.763/26.291 = 2.235$$

This is the distance of the line

$$x_1 + 2.018x_2 = 5.034,$$

from the origin (Fig. 4.5). The other solution is indeterminate, its variance being 10^4–10^5 times greater. In this case, the variance-covariance ellipsoid has degenerated practically into a line along the direction of X_2. Only the linear combination of parameters normal to this line can be determined from the information given by the observational equations (Eqs. 4.22).

Least-Squares Analysis in Crystal Structure Refinement

Although many observations—several hundred to several thousand observed $|F_{\mathbf{H}}|_o$ values, depending on the size and symmetry of the unit cell—are usually available, the observational equations are not linear. Under these circumstances, the least-squares procedure must be altered slightly so that the unknowns are the shifts to be applied to the parameters describing a postulated structural model rather than the parameters themselves. The observational equations can then be linearized by expanding $|F_{\mathbf{H}}|_c$ in a Taylor series about the trial values of the parameters p_j

$$|F_{\mathbf{H}}|_c = |F_{\mathbf{H}}(\mathbf{p})|_c + \sum_{j=1}^{m} \frac{\partial |F_{\mathbf{H}}|_c}{\partial p_j} \Delta p_j \qquad (4.24)$$

These parameters include fractional positional coordinates of all the atoms (usually three per atom), isotropic or anisotropic temperature factor parameters (up to six per atom), an overall scale factor, and possibly fractional occupation numbers if some atoms are statistically distributed over several sites. The full least-squares minimization process thus involves calculation of the elements of a normal-equations matrix of very large dimensionality and also the inversion of such a matrix—calculations that would be out of question without electronic computers. Moreover, since the Taylor series expansion above applies only for infinitesimally small shifts, convergence to the optimal parameters cannot be expected to occur in a single step. The calculated shifts must be applied to the parameters and the adjusted parameters used as trial values in a further least-squares cycle; this process is continued until the shift terms become negligible.

The function usually minimized is

$$R_1 = \sum_{\mathbf{H}} w_{\mathbf{H}} (|F_{\mathbf{H}}|_o - |F_{\mathbf{H}}|_c)^2 \tag{4.25}$$

The quantities directly observed are, of course, $|F_{\mathbf{H}}|^2$ rather than $|F_{\mathbf{H}}|$, and some people accordingly prefer to minimize

$$R_2 = \sum w'_{\mathbf{H}} (|F_{\mathbf{H}}|_o^2 - |F_{\mathbf{H}}|_c^2)^2 \tag{4.26}$$

Here we restrict the discussion mainly to the R_1 minimization. The unweighted observational equations are

$$|F_{\mathbf{H}}|_o - |F_{\mathbf{H}}|_c - \sum_{j=1}^{m} \frac{\partial |F_{\mathbf{H}}|_c}{\partial p_j} \Delta p_j = 0$$

The normal equations **NX** = **E** can now be developed as before. The appropriate matrix elements are (cf., Eq. 4.12):

$$N_{jk} = \sum_{\mathbf{H}} w_{\mathbf{H}} \left(\frac{\partial |F_{\mathbf{H}}|_c}{\partial p_j} \right) \left(\frac{\partial |F_{\mathbf{H}}|_c}{\partial p_k} \right) \tag{4.27}$$

$$X_j = \Delta p_j \tag{4.28}$$

$$E_k = \sum w_{\mathbf{H}} \left(\frac{\partial |F_{\mathbf{H}}|_c}{\partial p_k} \right) \Delta_{\mathbf{H}} \tag{4.29}$$

where $\Delta_{\mathbf{H}} = |F_{\mathbf{H}}|_o - |F_{\mathbf{H}}|_c$, and the vector of parameter shifts is obtained as

$$\mathbf{X} = \mathbf{N}^{-1}\mathbf{E}$$

as described previously for the linear case. The explicit expressions for the derivatives depend on the type of parameter involved.

In general, we have

$$(F_{\mathbf{H}})_c = A_{\mathbf{H}} + iB_{\mathbf{H}} = \sum f_k^0 T_k G(h,k,l,x_k,y_k,z_k) \tag{4.30}$$

where f_k^0 is the atomic scattering factor of atom k at rest, T_k is the temperature factor of atom k, and G consists of space-group dependent trigonometric expressions involving the fractional coordinates of atom k. The atomic form factors are usually accepted as constants of the structural model and are not refined. We now consider the various types of derivatives.

(a) Positional coordinates; x_k, y_k, $z_k = x_{ki}$ ($i = 1,2,3$).

$$|F|_c = (A^2+B^2)^{\frac{1}{2}} = \left\{\left(\sum_k A_k\right)^2 + \left(\sum_k B_k\right)^2\right\}^{\frac{1}{2}}$$

where A_k and B_k are the contributions of atom k to A and B, respectively. The derivatives are then:

$$\frac{\partial|F|_c}{\partial x_{ki}} = \frac{1}{2|F|_c}\left(2A\frac{\partial A}{\partial x_{ki}} + 2B\frac{\partial B}{\partial x_{ki}}\right)$$

$$= \frac{\partial A_k}{\partial x_{ki}}\cos\alpha + \frac{\partial B_k}{\partial x_{ki}}\sin\alpha$$

For example, in the space group $P1$,

$$A = \sum_k f_k^0 T_k \cos 2\pi(hx_k+ky_k+lz_k)$$

$$B = \sum_k f_k^0 T_k \sin 2\pi(hx_k+ky_k+lz_k)$$

$$\frac{\partial A}{\partial x_k} = -2\pi h f_k^0 T_k \sin 2\pi(hx_k+ky_k+lz_k)$$

$$\frac{\partial B}{\partial x_k} = 2\pi h f_k^0 T_k \cos 2\pi(hx_k+ky_k+lz_k)$$

For centrosymmetric space groups, $\cos\alpha = 1$, $\sin\alpha = 0$.

(b) Isotropic temperature-factor parameter Q common to all atoms:[6]

$$T_k = \exp(-Q\sin^2\theta/\lambda^2)$$

$$F_c = \exp(-Q\sin^2\theta/\lambda^2)\sum_k f_k^0 G_k$$

$$\frac{\partial|F|_c}{\partial Q} = -(\sin^2\theta/\lambda^2)|F|_c$$

where F_c is calculated with the trial value of Q.

(c) Individual isotropic temperature factors Q_k:

$$T_k = \exp(-Q_k\sin^2\theta/\lambda^2)$$

$$\frac{\partial|F|_c}{\partial Q_k} = \frac{\partial A_k}{\partial Q_k}\cos\alpha + \frac{\partial B_k}{\partial Q_k}\sin\alpha$$

$$= -(\sin^2\theta/\lambda^2)[A_k\cos\alpha + B_k\sin\alpha]$$

[6]Here we use the symbol Q instead of the more usual B to avoid using the same symbol for temperature-factor parameter and for the imaginary part of the structure factor.

where A_k and B_k are calculated with the trial value of Q_k.

(d) Individual anisotropic temperature factors; the most convenient form for the actual computations is

$$T = \exp\left[-(\beta_{11}h^2+\beta_{22}k^2+\beta_{33}l^2+\beta_{12}hk+\beta_{13}hl+\beta_{23}kl)\right]$$

where it is understood that the parameters β_{ij} are to be determined separately for each atom k.

$$\frac{\partial|F|_c}{\partial(\beta_{11})_k} = -h^2(A_k\cos\alpha + B_k\sin\alpha), \quad \text{etc.}$$

$$\frac{\partial|F|_c}{\partial(\beta_{12})_k} = -hk(A_k\cos\alpha + B_k\sin\alpha), \quad \text{etc.}$$

where A_k and B_k are calculated with the trial values of the parameters β_{ij}. In all space groups except $P\bar{1}$, symmetry relationships between the β_{ij} parameters of symmetry-equivalent atoms have to be taken into account;[7] the appropriate derivatives may be obtained by summing over contributions from equivalent atoms using the trigonometric expressions for $P1$ or $P\bar{1}$.

An alternative form of T is

$$T = \exp\left[-2\pi^2(U_{11}h^2a^{*2}+\ldots\ldots+2U_{12}hka^*b^*\ \ldots\ldots)\right]$$

where the parameters U_{ij} are more directly related to the thermal vibrational amplitudes. Comparison of the two forms gives:

$$U_{11} = \frac{\beta_{11}}{2\pi^2a^{*2}}, \text{ etc.} \qquad U_{12} = \frac{\beta_{12}}{4\pi^2a^*b^*}, \text{ etc.}$$

Sometimes the temperature-factor parameters are given as B_{ij} values ($B = 8\pi^2U_{ij}$). They are then on the same scale as the Debye factor B in the isotropic temperature-factor expression:

$$T = \exp(-B\sin^2\theta/\lambda^2)$$

(e) Scale factor. In principle, the $|F|_c$ values are on an absolute scale, and a scale factor should be applied to the $|F|_o$ values. However, during a round of least-squares refinement, the structural model is adjusted to fit the observations, and not vice-versa. From this standpoint, the $|F|_o$ values should be left as they are, and it is the scale of the $|F|_c$ values that has to be adjusted. Of course, once this scale factor g has been determined, the $|F|_o$ values have to be multiplied by $1/g$ to bring them to a more nearly absolute scale. The rescaled $|F|_o$ values can then be regarded as the

[7] K. N. Trueblood, *Acta Crystallogr.* **9**, 359 (1956).

observational data in a subsequent round of least-squares calculations. Thus the function to be minimized in any given cycle is

$$R = \sum_{\mathbf{H}} w_{\mathbf{H}}(|F_{\mathbf{H}}|_o - g|F_{\mathbf{H}}|_c)^2$$

This can be rewritten in terms of the reciprocal scale factor $k = 1/g$

$$R = \left(\frac{1}{k}\right)^2 \sum_{\mathbf{H}} w_{\mathbf{H}}(k|F_{\mathbf{H}}|_o - |F_{\mathbf{H}}|_c)^2$$

$$= \left(\frac{1}{k}\right)^2 \sum_{\mathbf{H}} w_{\mathbf{H}}\, \Delta_{\mathbf{H}}^2$$

where $\Delta_{\mathbf{H}}$ involves scaled $|F|_o$ values and unscaled $|F|_c$ values. The procedure usually followed is first to determine g and then to apply the inverse scale factor to the $|F|_o$ values, leaving all other derivatives unscaled, and to calculate the other parameter shifts. The effect of this is that although R must decrease from cycle to cycle, the quantity $\Sigma w_{\mathbf{H}}\Delta_{\mathbf{H}}^2$ may actually increase slightly if $k > 1$. However, the quantity

$$R' = \frac{\sum w\Delta^2}{k^2\sum|F_o|^2} = \frac{\sum w\Delta^2}{\sum|kF_o|^2}$$

must decrease in the same ratio as R. The value of $(R')^{\frac{1}{2}}$ is sometimes used as an alternative to the usual R (reliability) factor.

Errors in the assumed temperature-factor parameters can cause the general level of the $|F|_c$ values for high-order reflections to be too large or too small and can thus affect the determination of the scale factor g. This correlation is sometimes allowed for by estimating g together with an overall isotropic temperature-factor parameter, the corresponding pair of normal equations being solved separately and before the main set of equations. A common procedure is to write $g = 1 + g'$; the quantity to be minimized is then

$$\sum w_{\mathbf{H}}\{|F_{\mathbf{H}}|_o - |F_{\mathbf{H}}|_c - g'|F_{\mathbf{H}}|_c - \frac{\partial|F_{\mathbf{H}}|_c}{\partial Q}\,\Delta Q\}^2 = 0$$

and the 2×2 matrix of normal equations is

$$\begin{bmatrix} \sum w_{\mathbf{H}}|F_{\mathbf{H}}|_c^2 & -\sum w_{\mathbf{H}}s^2|F_{\mathbf{H}}|_c^2 \\ -\sum w_{\mathbf{H}}s^2|F_{\mathbf{H}}|_c^2 & \sum w_{\mathbf{H}}s^4|F_{\mathbf{H}}|_c^2 \end{bmatrix} \begin{bmatrix} g' \\ \Delta Q \end{bmatrix} = \begin{bmatrix} \sum w_{\mathbf{H}}\Delta_{\mathbf{H}}|F_{\mathbf{H}}|_c \\ \sum w_{\mathbf{H}}\Delta_{\mathbf{H}}s^2|F_{\mathbf{H}}|_c \end{bmatrix}$$

since $\partial|F_{\mathbf{H}}|_c/\partial Q = -s^2|F_{\mathbf{H}}|_c$ where $s = \sin\theta/\lambda$. One difficulty, however, is that the nondiagonal elements, being weighted sums of squares, can be quite large, and unless $(N_{11}N_{22}) \gg N_{12}^2$, the matrix may become nearly singular with very high correlation between the unknowns g' and ΔQ. This may be the case particularly when the overall temperature factor is so large that reflections with high s values are unobservable. In such circumstances, the scale factor should be estimated separately by comparison of $|F_o|$ and $|F_c|$ values for reflections exclusively in the low s range (say $s < 0.2$) where an error in Q has virtually no effect. The overall temperature factor can then be estimated, if desired, by fixing the scale factor and solving separately for ΔQ.

The calculation and storage of all the elements of the matrix **N** for a structure of even moderate complexity may stretch the capacity of all but the most powerful computers. With 30 atoms in the asymmetric unit, there are 90 positional coordinates and 180 anisotropic temperature-factor parameters to be refined, plus a scale factor, leading to a symmetric matrix of order 271, containing $271 \times 272/2 = 36856$ distinct elements, each involving a summation over a few thousand terms. For purely practical reasons, therefore, it is often advisable and sometimes essential to factorize the N-dimensional matrix into smaller block matrices by setting appropriate off-diagonal terms to zero. Setting any particular off-diagonal term to zero is equivalent to assuming that the corresponding pair of parameters is uncorrelated—i.e., that an error in one has no effect on the other. The most extreme procedure would be to neglect *all* off-diagonal terms, assuming complete independence of all the parameters. Such an assumption is far from reality in most cases. For example, if the crystal axes are oblique or if the vibration ellipsoids are markedly anisotropic, an error, say in the x-coordinate of a given atom, will obviously tend to introduce a compensating error in another coordinate (Fig. 4.6).

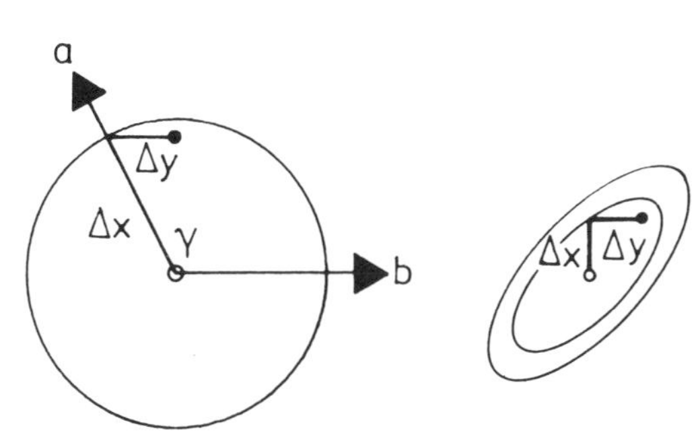

Figure 4.6. Correlation of errors in positional coordinates. Left: cross term caused by oblique coordinate system. Once the error Δx is introduced, the "best" y-coordinate, obtained by minimizing the distance from the true center (x_o, y_o) along the line Δx = constant, differs from y_o by $-\Delta x \cos \gamma$. Right: cross term caused by anisotropic thermal motion. If the electron density contours are anisotropic and one positional coordinate is in error, the "best" value of the other corresponds to the position of maximum electron density along the line Δx = constant.

Some insight into the source of this kind of correlation can be obtained by considering a particular off-diagonal element of the type

$$\sum_{\mathbf{H}} w_{\mathbf{H}} \left(\frac{\partial|F_{\mathbf{H}}|}{\partial x_j}\right) \left(\frac{\partial|F_{\mathbf{H}}|}{\partial y_j}\right)$$

for an oblique unit cell. Assuming space group $P\bar{1}$ and considering only ($hk0$) reflections

$$\frac{\partial|F_{\mathbf{H}}|}{\partial x_j} = -4\pi h f_j^0 T_j \sin 2\pi(hx_j+ky_j)$$

$$\frac{\partial|F_{\mathbf{H}}|}{\partial y_j} = -4\pi k f_j^0 T_j \sin 2\pi(hx_j+ky_j)$$

$$\left(\frac{\partial|F_{\mathbf{H}}|}{\partial x_i}\right) \left(\frac{\partial|F_{\mathbf{H}}|}{\partial y_i}\right) = 16\pi^2 hk(f_j^0)^2 T_j^2 \sin^2 2\pi(hx_j+ky_j)$$

The sign of each term in the summation depends only on the sign of the product hk. If the coordinate axes are at right angles, the number of positive terms will be approximately equal to the number of negative terms and the off-diagonal element will be nearly zero. However, if the interaxial angle differs appreciably from 90°, the positive and negative terms will not balance. For example, in Fig. 4.7 the "positive" region of reciprocal space is proportional to γ^*, the "negative" region to $\pi - \gamma^*$, and the net positive balance to $2\gamma^* - \pi$, which is zero only if $\gamma^* = \pi/2$. An analogous, but somewhat more complicated argument can be given when the overall temperature factor is anisotropic. Even for an orthogonal axial system, anisotropy in T will cause the terms in the summation to be systematically larger in one direction of reciprocal space than in the perpendicular direction, leading to off-diagonal terms that are not negligible.

Because of neglect of parameter correlation in the diagonal approximation, convergence to the least-squares minimum may be extremely slow, requiring many cycles of iteration, so that the intended economy in

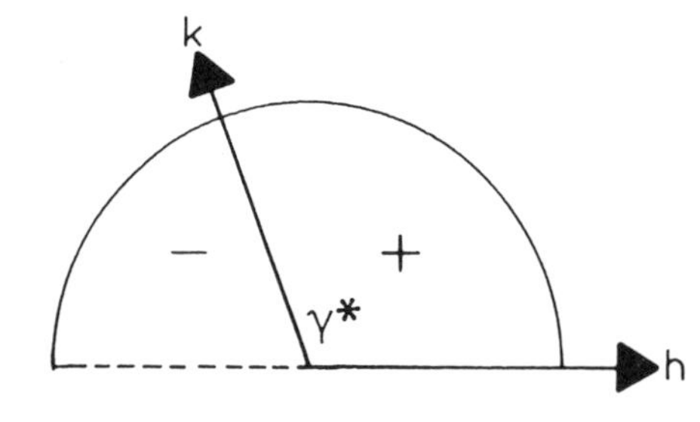

Figure 4.7. Origin of large off-diagonal term between x- and y-coordinates of the same atom. If $\gamma^* \neq 90°$, the region of positive hk products does not balance the region of negative hk products in the summation, leading to a nonzero off-diagonal term.

computational effort may be illusory. Also, the variances in the final parameter values obtained by consideration of only diagonal elements may be seriously underestimated.

The "block-diagonal" approximation often represents a reasonable compromise. Here each block consists of a 9×9 matrix comprising positional and anisotropic temperature-factor parameters of a single atom. Thus, correlations among these parameters are taken care of, but correlations among parameters of different atoms are ignored. The saving is considerable since only $45N$ matrix elements have to be computed and stored, in contrast to the $9N(9N+1)/2$ elements for the complete matrix. In general, the block-diagonal approximation works reasonably well, especially when the temperature factors are small enough so that no significant overlapping of the individual atomic peaks occurs. If overlapping is considerable or if the crystal structure is disordered, some of the interatomic correlations may be so large that full-matrix refinement is needed.

Intrinsic Parameter Correlation

Even with full-matrix refinement, large correlations among individual parameters may lead to ill-conditioning of the normal equations. Some situations where this might occur can be foreseen—e.g., in disordered structures where correlations between positional and temperature-factor parameters of a pair of nearly superimposed atoms, or between occupancy factor and temperature factor, are obviously close to unity.

Correlation problems also arise when the postulated model has a higher symmetry than that of the space group assumed for the least-squares analysis. For example, when a centrosymmetric and a noncentrosymmetric space group cannot be distinguished on the basis of systematically absent reflections (e.g., $P1$ or $P\bar{1}$), it is often convenient to start with the provisional assumption that the structure is centrosymmetric. Once a reasonable model has been found, the question arises as to whether or not the structure is really centrosymmetric. At this point, one might decide to expand the trial parameters over the questionable symmetry element and carry out least-squares refinement in the corresponding noncentrosymmetric space group. However, this procedure would lead to singularity of the normal-equations matrix.[8] The B parts of the structure factors are zero, derivatives of $|F_{\mathbf{H}}|$ with respect to pairs of centrosymmetrically related positional parameters are equal in magni-

[8] O. Ermer and J. D. Dunitz, *Acta Crystallogr.* **A26**, 163 (1970).

tude and opposite in sign, and derivatives with respect to corresponding pairs of temperature-factor parameters are equal in magnitude and sign. The elements of the normal-equations matrix are sums (over **H**) of products of pairs of these derivatives (Eq. 4.27), and the above relationships lead to a matrix **N** with the characteristic structure:

	X_1	X_2	$X_{1'}$	$X_{2'}$	B_1	$B_{1'}$
X_1	N_{11}	N_{12}	$-N_{11}$	$-N_{12}$	N_{13}	N_{13}
X_2	N_{12}	N_{22}	$-N_{12}$	$-N_{22}$	N_{23}	N_{23}
$X_{1'}$	$-N_{11}$	$-N_{12}$	N_{11}	N_{12}	$-N_{13}$	$-N_{13}$
$X_{2'}$	$-N_{12}$	$-N_{22}$	N_{12}	N_{22}	$-N_{23}$	$-N_{23}$
B_1	N_{13}	N_{23}	$-N_{13}$	$-N_{23}$	N_{33}	N_{33}
$B_{1'}$	N_{13}	N_{23}	$-N_{13}$	$-N_{23}$	N_{33}	N_{33}

Pairs of rows (and columns) are identical or opposite, and the matrix is obviously singular. The question cannot be answered by recourse to diagonal or block-diagonal refinements since these automatically preserve the problematic center of symmetry. Small, random shifts may be applied to the centrosymmetric set of parameters to make it only approximately centrosymmetric, but ill-conditioning of the normal-equations matrix has to be reckoned with.[9]

The best way to decide whether or not the structure is really or only nearly centrosymmetric is probably to scrutinize the few reflections that are most sensitive to the presence or absence of the questionable center of symmetry. If we look at the A and B parts of the structure factor (taking the origin of coordinates at the questionable center of symmetry), we see that A is relatively insensitive to small departures from centrosymmetry, while B is proportional to the magnitudes of such deviations, being zero for the exactly centrosymmetric structure. If Δ_j and Δ_j' are small displacements for a pair of related atoms from a postulated centrosymmetric structure, then

$$\begin{aligned} A_j &= \cos 2\pi h(x_j + \Delta_j) + \cos 2\pi h(-x_j + \Delta_j') \\ &= 2 \cos 2\pi h\left(\frac{\Delta_j + \Delta_j'}{2}\right) \cos 2\pi h\left(x_j + \frac{\Delta_j - \Delta_j'}{2}\right) \\ &\approx 2 \cos 2\pi h x_j \end{aligned}$$

[9]See S. Geller, *Acta Crystallogr.* **14,** 1026 (1961); A. I. M. Rae and E. N. Maslen, *ibid.* **16,** 703 (1963); R. Parthasarathy, J. G. Sime, and J. C. Speakman, *ibid.* **B25,** 1201 (1969) for further discussion of the inverse correlation problem.

$$B_j = \sin 2\pi h(x_j + \Delta_j) + \sin 2\pi h(-x_j + \Delta_j')$$

$$= 2 \sin 2\pi h\left(\frac{\Delta_j + \Delta_j'}{2}\right) \cos 2\pi h\left(x_j + \frac{\Delta_j - \Delta_j'}{2}\right)$$

$$\approx 2\pi h(\Delta_j + \Delta_j') \cos 2\pi h x_j$$

Now if $A = \Sigma A_j$ is large, then $F = (A^2+B^2)^{\frac{1}{2}} \approx A$, but if A is small, the value of F depends on the value of B as well, in particular whether it is zero or not. The sensitive reflections are then the ones, generally weak, for which the calculated value of A is close to zero. If, for these reflections, the centrosymmetric structure gives good agreement between $|F|_o$ and $|F|_c$, then no degradation of the symmetry is required by the experimental data. If, on the other hand, $|F|_o$ is systematically greater than $|F|_c$, then the inclusion of a B part of the structure factor is clearly called for and the structure can be regarded as noncentrosymmetric. In this way, it was possible to decide that the space group of cyclododecane is $C2/m$ (centrosymmetric, disordered) rather than $C2$ (noncentrosymmetric, ordered).[10] A more complicated case is illustrated by the structure of 1,5-diaza-6,10-cyclodecadione,[11] where analysis of the X-ray data in terms of an ordered structure in space group Pn led to an R factor of 0.084. The tests described above indicated that the structure must be disordered but still not exactly centrosymmetric. Of course, even if tests show that the symmetry of a proposed structure must be lowered, they provide no indication of the shifts that are to be applied to the atomic positions. Hints may sometimes be obtained from the Patterson function, calculated with the sensitive reflections only—those with small A and maximal B values.

However, it must be admitted that in the presence of very large parameter correlation, certain features of the structure may be indeterminable from the X-ray data alone. Diamond[12] has shown how the eigenvalue-eigenvector technique can be used to obtain the maximum amount of information in such cases.

Constrained Refinement

Apart from trivial constraints, such as fixing the values of certain parameters (e.g., atoms in special positions, origin specification in polar

[10] H. M. M. Shearer and J. D. Dunitz, *Helv. Chim. Acta* **43**, 18 (1960).
[11] T. Srikrishnan and J. D. Dunitz, *Acta Crystallogr.* **B31**, 1372 (1975).
[12] R. Diamond, *Acta Crystallogr.* **11**, 129 (1958).

space groups) or linear combinations of parameters (e.g., sum of occupation factors equals unity), it may sometimes be desirable to introduce more complicated constraints on the atomic parameters to make them satisfy some specific geometrical criterion. For example, if the asymmetric unit contains many atoms, or if light atoms are to be refined in the presence of heavy ones, or if the temperature factors are very high, or if the structure is disordered, or if the attainable atomic resolution is limited for any other reasons, then the individual atomic positions obtainable from straightforward least-squares refinement of the X-ray data may have such large uncertainties that molecular parameters derived from them are more or less meaningless. In such circumstances, it may be advantageous to force the atomic positions to satisfy certain geometric conditions by introducing appropriate constraints.

For structures that contain more-or-less rigid molecular fragments of well-established shape and dimensions (e.g., benzene rings), the number of independent parameters to be refined in the least-squares analysis can be considerably reduced, and the convergence can be accelerated by rigid-body group refinement—i.e., by fixing the geometries of these atomic groupings and varying only their positions and orientations in the unit cell.[13] For each such grouping consisting of N atoms whose relative positions are fixed in this way, the number of independent parameters is reduced from $3N$ to 6—three, x_o, y_o, z_o, defining the position of some reference point in the group, and three, ϕ, θ, ρ, defining the orientation of the group.

The derivatives of $|F_{\mathbf{H}}|$ with respect to these parameters can be calculated either numerically or analytically, using expressions of the type:

$$\frac{\partial|F|}{\partial x_o} = \sum_j^N \frac{\partial|F|}{\partial x_j}$$

$$\frac{\partial|F|}{\partial \theta} = \sum_j^N \left(\frac{\partial|F|}{\partial x_j}\frac{\partial x_j}{\partial \theta} + \frac{\partial|F|}{\partial y_j}\frac{\partial y_j}{\partial \theta} + \frac{\partial|F|}{\partial z_j}\frac{\partial z_j}{\partial \theta} \right)$$

where the summations extend over the N atoms of the group. For ease of computation and to keep the number of parameters small, isotropic temperature factors are usually assumed in group refinement.

It is, of course, essential that the fixed geometries assumed for mo-

[13]C. Scheringer, *Acta Crystallogr.* **16,** 546 (1963). See also R. J. Doedens in *Crystallographic Computing,* F. R. Ahmed, S. R. Hall, and C. P. Huber, Eds., Munksgaard, Copenhagen, 1970.

lecular fragments correspond to the true geometries; otherwise, systematic errors in the other molecular parameters will be introduced. For example, phenyl groups may deviate markedly from the full D_{6h} symmetry of benzene, and systematic errors are introduced if this symmetry is assumed in rigid-body refinement.[14]

The rigid-body assumption is tenable only for a few special atomic groupings. However, one may wish to fix certain molecular parameters—e.g., some of the interatomic distances or angles—without fixing them all in a given molecular fragment. This can be done by minimizing (Eq. 4.25) under the condition that the atomic parameters have to satisfy certain equations—equations of constraint. Problems of this kind are best solved by the method of undetermined multipliers (or Lagrangian multipliers). Say, we have N such constraints, each expressed as an equation that has to be satisfied by the atomic parameters p_j, i.e.,

$$G_n(p_1,p_2,\ldots p_m) = 0 \qquad n = 1,2,3,\ldots N$$

In general, the trial values p_j^o will not satisfy these equations exactly, so we may write

$$G_n^o(p_1^o,p_2^o,\ldots p_m^o) + \sum_j \frac{\partial G_n^o}{\partial p_j}\,\Delta p_j = 0$$

Now, instead of minimizing R_1 (Eq. 4.25), we minimize

$$\begin{aligned} R' &= \sum_{\mathbf{H}} w_{\mathbf{H}}(|F_{\mathbf{H}}|_o - |F_{\mathbf{H}}|_c)^2 + \sum_n \lambda_n G_n \\ &= \sum_{\mathbf{H}} w_{\mathbf{H}}\left(\Delta_{\mathbf{H}} - \sum_j \frac{\partial |F_{\mathbf{H}}|_c}{\partial p_j}\,\Delta p_j\right)^2 + \sum_n \lambda_n\left(G_n^o + \sum_j \frac{\partial G_n^o}{\partial p_j}\,\Delta p_j\right) \end{aligned}$$

with respect to the Δp's and the λ's. This leads to equations of the type

$$-\frac{1}{2}\frac{\partial R'}{\partial \Delta p_k} = \sum_{\mathbf{H}} w_{\mathbf{H}}\left(\Delta_{\mathbf{H}} - \sum_j \frac{\partial |F_{\mathbf{H}}|_c}{\partial p_j}\,\Delta p_j\right)\left(\frac{\partial |F_{\mathbf{H}}|_c}{\partial p_k}\right) - \frac{1}{2}\sum_n \lambda_n \frac{\partial G_n^o}{\partial p_k} = 0$$

$$\frac{\partial R'}{\partial \lambda_n} = G_n^o + \sum_j \frac{\partial G_n^o}{\partial p_j}\,\Delta p_j = 0$$

The augmented normal equations now have the general form:

[14]See A. Domenicano and A. Vaciago, *Acta Crystallogr.* **B31,** 2553 (1975) for discussion of this point.

$$
\begin{aligned}
&N_{11}\Delta p_1 + \ldots + N_{1k}\Delta p_k + \ldots + M_{1q}\lambda_q + \ldots\ldots &&= E_1\\
&N_{21}\Delta p_1 + \ldots + N_{2k}\Delta p_k + \ldots + M_{2q}\lambda_q + \ldots\ldots &&= E_2\\
&\ldots\ldots\ldots\ldots\ldots\ldots &&\\
&N_{j1}\Delta p_1 + \ldots + N_{jk}\Delta p_k + \ldots + M_{jq}\lambda_q + \ldots &&= E_j\\
&M_{11}\Delta p_1 + \ldots + M_{1k}\Delta p_k &&= G_1^o/2\\
&\ldots\ldots\ldots\ldots\ldots\ldots &&\\
&M_{q1}\Delta p_1 + \ldots + M_{qk}\Delta p_k &&= G_q^o/2
\end{aligned}
$$

They can be expressed symbolically as

$$
\begin{bmatrix} N_{jk} & M_{jq} \\ M_{qj} & 0 \end{bmatrix} \begin{bmatrix} \Delta p_k \\ \lambda_q \end{bmatrix} = \begin{bmatrix} E_j \\ \frac{1}{2} G_q^o \end{bmatrix}
$$

where the partitioned matrix elements N_{jk}, Δp_k, and E_j are defined as in Eqs. 4.27–4.29, and

$$
M_{jq} = -\frac{1}{2}\frac{\partial G_q^o}{\partial p_j}
$$

Constrained minimization using Lagrangian multipliers can be applied more generally than rigid-body group refinement. For example, a group of atoms may be held in a coplanar arrangement while bond lengths and angles are permitted to vary. This method has the minor disadvantage, however, that the order of the normal equations matrix is increased by one for every added equation of constraint.

Instead of exact constraints, more elastic ones may be imposed—e.g., that bond distances and angles derived from the atomic parameters are permitted to deviate from standard values but not by much, or that certain groups of atoms should be approximately coplanar. For this purpose the function to be minimized may be modified[15] to

$$
M = \sum w_{\mathbf{H}}(|F_{\mathbf{H}}|_o - |F_{\mathbf{H}}|_c)^2 + \sum w_k[g_k(p_1, p_2, \ldots p_n) - g_k^o]^2 \qquad (4.31)
$$

where g_k is the value of some quantity that depends on the atomic coordinates, and g_k^o is the specified value of this quantity; w_k can be regarded as being the weight assigned to the constraint in question. This way of including the constraints is equivalent to adding a set of further observational equations, and the order of the matrix of normal equations

[15] J. Waser, *Acta Crystallogr.* **16,** 1091 (1963).

is thereby neither increased or reduced. The method of conditional or slack least-squares refinement is applicable when strong correlation among some of the positional parameters is present or when the quality of the available X-ray data is insufficient to determine derived molecular parameters as reliably as they can be guessed on the basis of previous knowledge. Under these circumstances, the "best" structure is not likely to be the one that can be fitted optimally to the X-ray data alone. We skim over the problem of deciding the relative weights of the actual observations and of the added constraints. The basic difficulty is, of course, that the first sum in Eq. 4.31 is in electrons squared, while the second can be regarded as being an energy quantity, w_k being an effective force constant. Waser suggested that a ratio of about 1:100 for the relative weights of the two sums may be appropriate. In any case, it is clear that the down-weighting of the observations must not be carried too far. As Rollett has pointed out,[16] a normal-equations matrix based exclusively on constraint equations alone becomes singular—i.e., it yields no unique solution.

Least-Squares Weights

A. I'm planning to do a least-squares refinement of a crystal structure and I've been wondering if you, as a statistician, could give me some advice.

B. What's the problem?

A. Well, I don't really know how to assign weights to the $|F_o|$ values.

B. Oh, that's quite simple. You just have to write down the variance-covariance matrix of the observations and invert it. There are computer programs for doing this kind of thing.

A. Yes, but the trouble is that I don't know the variance-covariance matrix.

B. But you must have some idea, for example, whether the errors in your observations are correlated or not.

A. I should imagine that there may be some systematic errors that are highly correlated—absorption errors and so on—but as far as the random errors are concerned, each observation should be more or less independent of the others.

B. I'm afraid I couldn't help you with the systematic errors—you would have to analyze the details of the measurement process, and that's

[16] J. S. Rollett, *Crystallographic Computing*, F. R. Ahmed, S. R. Hall, and C. P. Huber, Eds., Munksgaard, Copenhagen, 1970, p. 169 ff.

not my department. If you think the random errors aren't correlated seriously, then the weights are just the reciprocals of your variances—the off-diagonal elements of the matrix are small and can be neglected in a first approximation, which is all you're interested in anyway.

A. But I don't even know the variances! I had to measure the intensities of 3500 reflections and you can't expect me to repeat the whole thing 20 times just to be able to get some reasonable statistical estimates of the errors.

B. No, I agree, that would be asking a bit too much. But there is always counting statistics to fall back on. If $\bar{N}$ is the average of N, the number of counts in a given time t, then the distribution of N for large $\bar{N}$ should be Gaussian with standard deviation equal to $N^{\frac{1}{2}}$, and . . .

A. Yes, I know all that, of course, but there are all sorts of other errors besides the kind of fluctuation you refer to.

B. Well, in that case you might appeal to the Central Limit theorem, which says essentially that if you have many sources of random error, the sum of the errors tends to the normal error distribution, regardless of the distributions of the individual errors.

A. That's all very well in theory, but it's my impression that many actual error distributions that are supposed to be Gaussian are not quite Gaussian—the tails are too large. It seems to me that this might happen if the observations were drawn from two normal distributions with the same mean but different variances, for example. If the two sets of observations were mixed, the result would be the sum of two Gaussians, which isn't a Gaussian.

B. Yes, it is! No, excuse me, you're quite right, it's the product of two Gaussians that's a Gaussian, of course. But even if your distribution isn't Gaussian, one would still expect it to have a finite variance,[17] and that is all that matters.

A. Even if I don't know what the variance is?

B. Yes, if you don't mind my saying so, I think you may be worrying too much about this weighting problem. My advice is to make some reasonable guesses and go ahead with your analysis. Even if your

[17]Would one? A skeptic might claim, with some justification, that most experimental error distributions can be fitted to a Lorentzian function, $f(x) = (1+a^2x^2)^{-1}$, just about as well as to a Gaussian function, $g(x) = e^{-a^2x^2}$. The odd moments of $(1+a^2x^2)^{-1}$ are all zero, like those of the Gaussian, but the even moments are all infinite, quite unlike those of the Gaussian. As far as the law of errors is concerned, "All the world believes it firmly, because the mathematicians imagine that it is a fact of observation, and the observers that it is a theorem of mathematics" [an "eminent physicist" cited in H. Poincaré, *The Foundations of Science. Science and Hypothesis,* Science Press, Lancaster, Pa., 1946, p. 119 (translated by G. B. Halsted)].

weights are wrong, the estimates of the parameters themselves will be unbiased, provided only that the observations are unbiased. The wrong weights only introduce bias into the estimates of the variances of the parameters.

A. Thank you very much, it has been most helpful to talk to you.

B. A pleasure, don't mention it.

It is true that we know very little about the statistical distribution of experimental errors to be expected among the observations because hardly anyone is prepared to measure a set of $|F_{\mathbf{H}}|$ values sufficiently often and under sufficiently varied conditions to obtain the necessary statistics. In one project initiated by the International Union of Crystallography several years ago, 17 sets of measurements were made by different participants for crystals of D(+)-tartaric acid.[18] The mutual consistency of different data sets i and j was estimated by evaluating

$$R_{ij} = \frac{\sum_{\mathbf{H}} |F_i(\mathbf{H}) - F_j(\mathbf{H})|}{\frac{1}{2}\sum_{\mathbf{H}} |F_i(\mathbf{H}) + F_j(\mathbf{H})|}$$

and the consistency between set i and the mean was estimated by evaluating

$$R_{i\mu} = \frac{\sum_{\mathbf{H}} |F_i(\mathbf{H}) - \langle F(\mathbf{H})\rangle|}{\sum_{\mathbf{H}} \langle F(\mathbf{H})\rangle}$$

The distribution of R_{ij} for the complete group of data sets is shown in Fig. 4.8a. The broad peak near $R = 0.45$ is caused by one egregious data set, the peak at $R = 0.14$ is caused mainly by another, and the shoulder at $R = 0.09$ is caused mainly by two others. After these data sets are removed, the distribution of the remaining R_{ij} values is more nearly single peaked (Fig. 4.8b) and may be interpreted as arising from a broad spread of error sources. The results seem to show that if two crystallographers measure different crystals of the same low-absorption compound on different diffractometers, their results are most likely to differ by 6% in R_{ij} and that they are unlikely to differ by less than 3% or more than about 10%, although they may differ by as much as 50% in extreme cases where some systematic error is probably involved. When the data

[18] S. C. Abrahams, W. C. Hamilton, and A. Mc L. Mathieson, *Acta Crystallogr.* **A26**, 1 (1970).

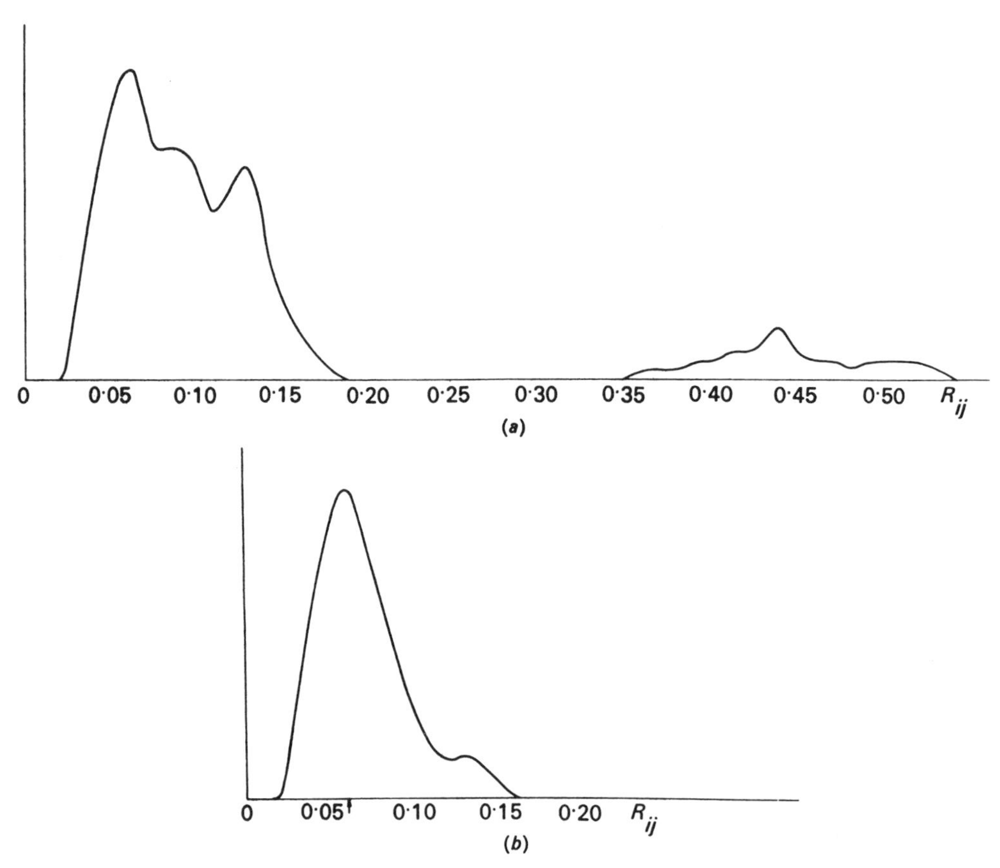

Figure 4.8. (a) R_{ij} curve for all 17 sets of data. (b) R_{ij} curve excluding four egregious data sets. From Abrahams, Hamilton, and Mathieson, *Acta Crystallogr.* **A26,** 1 (1970).

sets were subjected to least-squares refinement[19] of positional and anisotropic temperature factors of the C and O atoms, with weights inversely proportional to $|F_o|^2$, they all produced conventional R factors in the range 0.034–0.112 and gave estimated standard deviations in positional coordinates of less than 0.01 Å (several gave much less). However, with some of the data sets, it was not possible to refine the hydrogen atom positions.[20] From the internal consistencies obtained, it was concluded that the estimated standard deviations obtained by least-squares refinement could be too small by a factor of about 2 for positional parameters, for

[19]W. C. Hamilton and S. C. Abrahams, *Acta Crystallogr.* **A26,** 18 (1970).

[20]Criteria for the feasibility of refining light-atom parameters in the presence of heavy atoms (e.g., hydrogen in presence of carbon, nitrogen, oxygen, and so forth, or carbon in presence of uranium) have been discussed by E. Huber-Buser, *Z. Kristallogr.* **133,** 150 (1971).

thermal parameters even more. The weights that are commonly applied are therefore unrealistic even when they reflect the correct relative variances in the observations.

The weights assigned to the X-ray data are usually based on a combination of theoretical assumptions and practical experience. In my laboratory we take the standard deviation of an individual intensity measurement as

$$\sigma(I) = D_1(P+k^2B)^{\frac{1}{2}} + D_2P + D_3 \tag{4.32}$$

where $I = P-kB$, the difference between a peak count and a suitably scaled background count. The first term thus allows for the statistical error in the total count. The second term, proportional to the peak count, is supposed to allow for possible errors arising from instrumental instability, incorrect settings, and so forth, and the proportionality factor D_2 is typically taken to be about 0.03. The third term allows for all errors, whatever their nature, that might affect all observations to the same extent. Usually we take $D_1 = 1$, $D_2 \approx 0.03$, D_3 in the range 10–30, but other choices may be made.

For multiple measurements of the same reflection intensity I

$$\langle \mathrm{I} \rangle = \frac{\Sigma\omega_i \mathrm{I}_i}{\Sigma\omega_i} \quad \text{where} \quad \omega_i = \frac{1}{\sigma_i^2}$$

and the standard deviation of the mean intensity can be estimated two ways:

$$\sigma_\mathrm{A}\langle \mathrm{I} \rangle = \left[\Sigma\left(\frac{1}{\omega_i}\right)\right]^{\frac{1}{2}} \tag{4.33}$$

$$\sigma_B\langle I \rangle = (I_{\max}-I_{\min})T(N) \tag{4.34}$$

where $T(N)$ is a factor taken from tables of small-number statistics that depends on the sample range.[21] If σ_B is consistently larger (or smaller) than σ_A, the σ_i estimates used in Eq. 4.33 are probably unrealistic, and suitable corrective measures (i.e., altering D_2 or D_3 in Eq. 4.32) can be taken. Once the σ_A estimates are thought to be reasonable on the whole, we then choose the larger of σ_A and σ_B for each reflection as the more realistic (i.e., the more pessimistic) estimate. In applying the various correction factors (discussed in Chapter 5) to convert the raw intensities to F^2 values, we assume that $\sigma(F^2)/F^2$ is the same as $\sigma(I)/I$. Since

$$\frac{d(F^2)}{F^2} = \frac{2FdF}{F^2} = \frac{2\,dF}{F}$$

we can also assume

[21]See, for example, B. W. Lindgren and G. W. Mc Elrath, *Introduction to Probability and Statistics,* 2nd ed., Macmillan, New York, 1966, p. 185.

$$\frac{\sigma(F)}{F} = \frac{1}{2}\frac{\sigma(F^2)}{F^2}$$

as long as $\sigma(F^2)/F^2$ is small, in which case the error in F^2 can be expected to follow an approximately normal distribution. However, if the estimate of $\sigma(F^2)$ is about the same magnitude as F^2 itself, the error distribution cannot be symmetric about the mean since negative values of F^2 are physically meaningless. In such cases, all we can say is that if $\sigma(F^2) \approx F^2$, then $\sigma(F) \approx |F|$. The rule-of-thumb procedure actually adopted is to take

$$\sigma(F) = |F| - (F^2 - \sigma(F^2))^{\frac{1}{2}}$$

that is,

$$\frac{\sigma(F)}{|F|} = 1 - \left\{1 - \frac{\sigma(F^2)}{F^2}\right\}^{\frac{1}{2}} \quad \text{if} \quad F^2 \geq \sigma(F^2)$$

and

$$\sigma(F) = |F| \quad \text{if} \quad F^2 \leq \sigma(F^2).$$

The "experimental weights" are taken as being inversely proportional to the squares of the standard deviations estimated this way. The main thing is that observations that are believed to be unreliable get very small weights. To avoid extensive calculations on many low-quality data which hardly contribute to the normal equations because of their small weights, reflections for which $|F| < k\,\sigma(F)$ are sometimes thrown out of the data set altogether—the cut-out value of k depends on the amount of more reliable data that may be available.

Often the experimental weights assigned by this procedure or by other similar procedures are inadequate. They do not satisfy the criterion that $\langle w(\Delta F)^2\rangle$ is roughly constant for arbitrarily chosen subsets of the data. This result may be caused by the presence of systematic errors in the data (e.g., extinction errors, discussed in Chapter 5), but it may also be caused by inadequacies in the postulated model—e.g., the inability of the assumed atomic form factors to reproduce exactly the actual electron density in the crystal. One way out of this difficulty is to use an empirical weighting function whose parameters are chosen to enforce constancy of $\langle w(\Delta F)^2\rangle$ over selected ranges of $\sin\theta/\lambda$ or $|F_o|$. Weights chosen this way lead to estimates of precision for the parameters of the model that "allow for all random experimental errors, for such systematic experimental errors as cannot be paralleled in the calculated model, and for such defects of the model as are not paralleled in the experimental data."[22] The main weakness of this approach is that

[22]D. W. J. Cruickshank, *Crystallographic Computing,* F. R. Ahmed, S. R. Hall, and C. P. Huber, Eds., Munksgaard, Copenhagen, 1970, p. 195.

although the model has been tampered with, one does not know in what way and to what extent.

Once we admit that the most appropriate weights depend not only on the experimental data but also on possible inadequacies in the postulated model, we are led to ask which parameters of the model are of particular interest and what weighting scheme is most appropriate for obtaining optimal estimates of these parameters, possibly at the cost of some loss of precision in estimating the parameters that are of less interest. To answer these questions we must consider the relationship between least-squares refinement and the fit of the postulated model to the experimental electron density.

Suppose we had a set of experimental $|F_o|$ values of uniform quality and free from systematic error. It is easily shown from the convolution theorem (Chapter 1, p. 38) that minimization of

$$\sum w_{\mathbf{H}}(|F_{\mathbf{H}}|_o - |F_{\mathbf{H}}|_c)^2$$

with $w_{\mathbf{H}} = 1$ (unit weights) is equivalent to minimizing the integral of the difference density squared over the unit cell volume:

$$\sum_{\mathbf{H}} (|F_{\mathbf{H}}|_o - |F_{\mathbf{H}}|_c) \exp(-2\pi i \mathbf{H}\cdot\mathbf{r}) = \rho_o(\mathbf{r}) - \rho_c(\mathbf{r}) = \delta(\mathbf{r})$$

$$\sum_{\mathbf{H}} (|F_{\mathbf{H}}|_o - |F_{\mathbf{H}}|_c)^2 \cos 2\pi\mathbf{H}\cdot\mathbf{r}' = D(\mathbf{r}') = \int_V \delta(\mathbf{r})\, \delta(\mathbf{r}-\mathbf{r}')\, d\mathbf{r}$$

$$\sum_{\mathbf{H}} (|F_{\mathbf{H}}|_o - |F_{\mathbf{H}}|_c)^2 = D(0) = \int_V \delta^2(\mathbf{r})\, d\mathbf{r} \qquad (4.35)$$

In minimizing the integral on the right side of Eq. 4.35, contributions from all positions $\mathbf{r}$ in the unit cell have equal weight. This means that if the peaks of ρ_o were asymmetric and those of ρ_c were symmetric,[23] the best fit would be obtained by adjusting the peak centers of ρ_c to coincide with the centroids of the ρ_o peaks rather than with their maxima. Asymmetry of the ρ_o peaks can arise from several sources—i.e., from asymmetry in the bonding electron density or from anharmonicity in the thermal vibrations—and it has been known for some time that the use of spherically symmetric atomic form factors in unit-weight least-squares analysis can lead to spurious shifts in atomic positions and to apparent anisotropy in temperature factors.[24] We can reduce these effects by weighting the difference density so that more importance is ascribed to

[23]In this context, symmetric means centrosymmetric about the peak maximum, asymmetric means noncentrosymmetric or skew about the peak maximum.

[24]B. Dawson, *Acta Crystallogr.* **17**, 990 (1964).

fitting ρ_c to ρ_o at positions near the maxima than at positions further away from the maxima—i.e., by making the peaks of ρ_o and of ρ_c sharper than their "natural" shapes. Of course, we lose information in the process, information about the details of the deformation density that has been suppressed. However, this lost information is usually not obtained anyway in normal least-squares analysis designed to estimate the atomic positions and vibration parameters. Indeed, unless the deformation density is specifically included in the least-squares model by addition of appropriate parameters (Chapter 8), it only interferes with the estimates of the other parameters. Finally, the information suppressed in the least-squares analysis can always be recovered later in a subsequent difference synthesis designed to give it prominence.

The modification of the Fourier coefficients that is needed to accentuate the fit of ρ_c to ρ_o near peak maxima is formally equivalent to the introduction of appropriate weights in the least-squares analysis. One possibility is to assign zero weight to low-order reflections. Indeed, there is considerable evidence that least-squares refinements based only on high-order reflections yield atomic coordinates that are closer to the true nuclear positions (as determined by neutron diffraction) than those derived from refinements including all observed reflections.[25] The thermal parameters are also improved. However, such high-order refinements are implicitly based on discontinuous weighting schemes, with a sudden change from zero to unit or experimental weights at some arbitrary radius in reciprocal space. The corresponding peaks in the difference syntheses are Fourier transforms of step functions and contain high-frequency ripples of considerable amplitude at some distance from the atomic centers. A smoother, more flexible and more readily interpretable weighting system can be derived by sharpening the peaks of ρ_o and ρ_c by Gaussian functions.[26]

If $\rho_o(\mathbf{r})$ and $\rho_c(\mathbf{r})$ are sums of atomic peaks

$$\delta(\mathbf{r}) = \sum_j \{\rho_{oj}(\mathbf{r}-\mathbf{r}_j)-\rho_{cj}(\mathbf{r}-\mathbf{r}_{cj})\} \tag{4.36}$$

then the quantity to be minimized (cf., Eq. 4.35) could be taken as

$$D'(0) = \sum_j \int \delta_j^2(\mathbf{r}-\mathbf{r}_j) \exp\left[-2q(\mathbf{r}-\mathbf{r}_j)^2\right] d\mathbf{r}$$

[25] S. R. Hall and E. N. Maslen, *Acta Crystallogr.* **22,** 216 (1967); R. F. Stewart and L. H. Jensen, *Z. Kristallogr.* **128,** 133 (1969); A. M. O'Connell, *Acta Crystallogr.* **B25,** 1273 (1969); D. M. Collins and J. L. Hoard, *J. Am. Chem. Soc.* **92,** 3761 (1970); P. Coppens and A. Vos, *Acta Crystallogr.* **B27,** 146 (1971).

[26] J. D. Dunitz and P. Seiler, *Acta Crystallogr.* **B29,** 589 (1973).

This is equivalent to multiplying the peaks of $\rho_0(\mathbf{r})$ and $\rho_c(\mathbf{r})$ by Gaussian functions, $\exp[-q(\mathbf{r}-\mathbf{r}_j)^2]$, and the Fourier coefficients $|F_\mathbf{H}|_o$ and $|F_\mathbf{H}|_c$ have to be modified accordingly. Assume that the "natural" peaks are also roughly Gaussian in shape—not a bad approximation, as mentioned earlier:

$$\rho_{0j}(\mathbf{r}-\mathbf{r}_j) \approx \rho_c(\mathbf{r}-\mathbf{r}_j) \approx A \exp[-p(\mathbf{r}-\mathbf{r}_j)^2] \tag{4.37}$$

The atomic form factors (including isotropic temperature factor) are then (Chapter 1, p. 46) approximated by

$$f(H) \approx A\left(\frac{\pi}{p}\right)^{3/2} \exp\left(-\frac{\pi^2 H^2}{p}\right) \tag{4.38}$$

The sharpened atomic peak shapes are then given by

$$\rho'_{0j}(\mathbf{r}-\mathbf{r}_j) \approx \rho'_c(\mathbf{r}-\mathbf{r}_j) \approx A \exp[-(p+q)(\mathbf{r}-\mathbf{r}_j)^2] \tag{4.39}$$

corresponding to modified form factors

$$f'(H) = A\left(\frac{\pi}{p+q}\right)^{3/2} \exp\left(-\frac{\pi^2 H^2}{p+q}\right) \tag{4.40}$$

Thus, to the extent that all atoms can be assumed to have similar form factors the ratio of modified to unmodified Fourier coefficients is

$$\frac{F'(\mathbf{H})}{F(\mathbf{H})} = \frac{f'(H)}{f(H)} = \left(\frac{p}{p+q}\right)^{3/2} \exp\left[\frac{\pi^2 H^2 q}{p(p+q)}\right]$$

If the least-squares analysis is to correspond to minimization of $D'(0)$ rather than of $D(0)$, the experimental weights $w_\mathbf{H}$ must be modified to show a $\sin\theta/\lambda$ dependence,

$$w'_\mathbf{H} = w_\mathbf{H} \exp(t \sin^2\theta/\lambda^2) \tag{4.41}$$

where
$$t = \frac{8\pi^2 q}{p(p+q)}$$

since $|\mathbf{H}| = 2\sin\theta/\lambda$. The maximum permissable value of t, corresponding to infinitely sharp peaks in ρ'_0 and ρ'_c is then $t_{max} = 8\pi^2/p$.

In light-atom structures, the overall form factor, including isotropic thermal motion, can be approximated by

$$f(H) = A \exp(-B' \sin^2\theta/\lambda^2) = A \exp(-B' H^2/4)$$

corresponding to an average atom with density fall-off given by:

$$\exp(-pr^2)$$

with $p = 4\pi^2/B'$. As already mentioned, typical values of p are 4–6 Å^{-2}, smaller if thermal motion is very severe. The value of q can be chosen, depending on how much we wish to suppress the effect of the outer regions of the atomic peaks. With $q = 9$ Å^{-2}, $\delta^2(\mathbf{r})$ is reduced to 5% of its value at radial distance 0.4 Å from an atomic center. If q is set equal to p—i.e., if the modified peaks are made just twice as sharp as the natural ones, then $t = 4\pi^2/p = B'$ and

$$w'_{\mathbf{H}} = w_{\mathbf{H}} \exp (B' \sin^2 \theta/\lambda^2) \approx w_{\mathbf{H}}/f(\mathbf{H})$$

As Cochran showed many years ago,[27] the atomic positions obtained by least-squares minimization with weights inversely proportional to $f(H)$ coincide with those at which the gradient of the unweighted difference synthesis is zero. A suitable value of t can also be estimated from the criterion that $\langle w' \Delta^2 \rangle$ should be more or less constant in different $\sin \theta/\lambda$ ranges. This is often far from the case when unmodified experimental weights are used.

The kind of modified weighting system discussed here allows for inadequacies of the least-squares model as well as for experimental errors in the observations. In applying modified weights we are really altering the model to correspond to a different set of observations, which, it must be admitted, are not experimentally available. We replace the missing experimental data by mathematical operations on the existing data. Caution is obviously indicated. Nevertheless, experience shows that atomic coordinates and thermal parameters obtained by least-squares analysis can often be markedly improved—i.e., they are closer to the values obtained by neutron diffraction or by low-temperature studies if a suitably modified weighting scheme is adopted. In addition, the use of modified weights can have other advantages; one is faster convergence (if the input parameters are not too far from the final ones); another is that because of sharpening of the atomic peaks, off-diagonal elements corresponding to interactions between parameters of different atoms are reduced in magnitude so that block-diagonal refinement can be used in place of full-matrix refinement—a considerable saving in computational outlay.

Wrong Structures

The function R_1 (Eq. 4.25) has countless local maxima and minima in the polydimensional parameter space. An arbitrary trial structure

[27]W. Cochran, *Acta Crystallogr.* **1**, 138 (1948).

corresponds to an arbitrary point in this space. Starting from an arbitrary structure, the standard least-squares refinement shifts the point toward the nearest local turning point, which may be a maximum or a minimum of R_1. Thus R_1 may increase or decrease. If it increases, the starting model can be rejected, and even if it decreases, the refinement may converge to one of the local minima instead of to the deepest minimum, the true "least-squares" minimum. Convergence to the deepest minimum can occur only if the parameters of the trial structure are sufficiently close to those of the actual structure. How close must they be? From Eq. 4.35 a rough criterion might be that the atomic peaks of $\rho_c(\mathbf{r})$ should at least overlap appreciably with those of $\rho_0(\mathbf{r})$.

If the postulated model has serious shortcomings (grossly misplaced atoms, incorrectly identified atoms, missing or extra atoms) it may still refine to a reasonable R value if most of the interatomic vectors of the postulated structure are compatible with the Patterson function. Gross errors in the postulated model should be manifested in one way or another. One or two grossly misplaced or misidentified atoms can often be recognized by erratic behavior of the associated temperature-factor parameters. If the model puts electron density at positions where none is present, the refinement will tend to smear the postulated density peak over an infinitely large volume, leading to unreasonably large temperature-factor parameters. If the model puts too little density at a given site, the refinement will tend to sharpen the postulated density, leading often to an apparently negative temperature factor. An incorrect postulated model corresponding to an approximately correct relative arrangement of atoms, each displaced by the same amount from an actual position, might be more difficult to diagnose, but it should be revealed by a few major discrepancies between observed and calculated $|F|$ values for low-order reflections.

At the close of a refinement with modified or experimental weights, it is essential to calculate a final difference synthesis to check that no significant residual density remains after subtraction of the density corresponding to the least-squares model. If this condition is met, if the R factor is close to the expectation value, and if there are no individual discrepancies between $|F_{\mathbf{H}}|_c$ and $|F_{\mathbf{H}}|_0$ that are larger than say 3 $\sigma(F)$, then the postulated model may be regarded as being essentially correct or at least in full agreement with the experimental data. It is not the only model consistent with these data, but it is likely to be the only chemically reasonable one—assuming, of course, that it *is* chemically reasonable.

A refined structure that contains "unusual" features is automatically suspect, which is not to say that it is necessarily wrong. Abnormal bond lengths, bond angles, or intermolecular contacts are highly interesting, but there are too many examples where they have been shown to be spurious—artifacts of refinements based on an incorrect model or faulty experimental data. Unusual features call for careful scrutiny of the data and a search for alternative models that might fit the data as well as the suspect one. In particular, the space-group assignment should be checked for possible alternatives. Some of the pitfalls and ways of avoiding them have been discussed in recent articles by Ibers[28] and Donohue.[29]

[28]J. A. Ibers, *Critical Evaluation of Chemical and Physical Structure Information,* D. R. Lide and M. A. Paul, Eds., National Academy of Sciences, Washington, D.C., 1974, p. 186.
[29]J. Donohue, *Critical Evaluation of Chemical and Physical Structure Information,* D. R. Lide and M. A. Paul, Eds., National Academy of Sciences, Washington, D.C., 1974, p. 199.

5. Treatment of the Results

"Speak English!" said the Eaglet. "I don't know the meaning of half those long words, and, what's more, I don't believe you do either."

Long Lists of Numbers

The results of a crystal structure analysis are usually presented as numerical tables of positional coordinates and vibrational parameters for the atoms contained in an asymmetric unit of structure. This is the unit that is repeated by the appropriate combination of space-group symmetry operations and lattice translations to give the complete crystal structure. The choice of any particular asymmetric unit in a symmetric, repeating pattern is, of course, arbitrary. The unit chosen may encompass an entire molecule or only a part of one (if the molecule has crystallographic symmetry), but it may also comprise nonequivalent atoms that belong to different molecules.

Atomic positions are usually expressed as fractional coordinates x_i, y_i, z_i, fractional scalar components along the **a**, **b**, **c** crystal axes respectively. The origin should be explicitly specified unless no ambiguity is possible. The equivalent positions at which symmetry-related atoms occur are tabulated with respect to standard origins for all 230 space groups in Volume I of *International Tables for X-ray Crystallography.*

Most authors of papers that describe crystal structures also provide tables of selected interatomic distances and angles, together with other metrical information that may be of interest in connection with molecular structure. Short distances between atoms in different molecules are also often tabulated or at least mentioned in the text. In addition, much information is often given in pictorial form—e.g., figures illustrating aspects of molecular conformation and packing. Stereoviews[1] are especially valu-

[1]A computer program (ORTEP, C. K. Johnson, Oak Ridge National Laboratory, Report ORNL-3794) for preparing stereo pictures from three-dimensional positional coordinates along any desired viewing direction has been available for some time and is in use in many crystal structure laboratories. The program also depicts thermal vibration ellipsoids from the appropriate thermal parameters. Stereoscopes for viewing these pic-

able since they convey the three-dimensional impression that is so essential in problems of molecular conformation and packing but that is often lost in planar representation.

Vibrational parameters are expressed in various ways, depending on the form chosen for the temperature factor. The most convenient form for the least-squares computations is:

$$T = \exp\left[-(\beta_{11}h^2+\beta_{22}k^2+\beta_{33}l^2+\beta_{12}hk+\beta_{13}hl+\beta_{23}kl)\right]$$

and many authors simply tabulate the β_{ij} coefficients for each atom in the asymmetric unit. One confusing point is that the cross terms cited are sometimes those corresponding to the above quadratic form, sometimes those corresponding to the expression with all cross terms multiplied by 2. The symbol β_{ij} seems to be used indiscriminately to cover both cases. There are strong reasons for preferring the alternative form

$$T = \exp\left[-2\pi^2(U_{11}h^2\mathbf{a}^{*2}+ \ldots +2U_{12}hk\mathbf{a}^*\mathbf{b}^* \ldots .)\right]$$

because the coefficients are more directly related to vibrational amplitudes (see Chapter 1, p. 48). It is an easy matter to transform one set of coefficients into the other:

$$U_{11} = \frac{\beta_{11}}{2\pi^2 a^{*2}}, \text{ etc.} \qquad U_{12} = \frac{\beta_{12}}{4\pi^2 a^* b^*}, \text{ etc.}$$

Again, the coefficients are sometimes given as B_{ij} values ($B_{ij} = 8\pi^2 U_{ij}$), to put them on the same scale as the Debye factor B in the isotropic temperature factor expression:

$$T = \exp(-B \sin^2\theta/\lambda^2)$$

If vibrational parameters of symmetry-equivalent atoms are required, the necessary transformations have to be carried out.

Readers of crystallographic papers may wish to check the values of derived quantities or to compute values of quantities not explicitly listed by the authors. It is worthwhile here to add a few remarks concerning the calculations involved.

Hazards of Oblique Coordinate Systems

The fractional scalar components can always be converted to distance components by multiplying them by the axial lengths, but the resulting

---

tures are available from the Hubbard Scientific Co., Northbrook, Ill. With practice, many people are capable of naked-eye stereopsis; recommendations for developing this faculty are given by J. C. Speakman, *Chem. Brit.* **14,** 107 (1978).

coordinate system is Cartesian only if the crystal axes are mutually orthogonal (orthorhomic, tetragonal, and cubic systems). It is hardly necessary to mention that the elementary formulas for calculating distances and angles

$$|\mathbf{V}| = (x^2 + y^2 + z^2)^{\frac{1}{2}} \tag{5.1}$$

$$\cos(\mathbf{V}_1,\mathbf{V}_2) = \frac{x_1x_2 + y_1y_2 + z_1z_2}{|\mathbf{V}_1\ \mathbf{V}_2|} \tag{5.2}$$

apply only in Cartesian systems. In an oblique system the length of a vector **V** with scalar components x, y, z along **a**, **b**, **c** respectively is

$$|\mathbf{V}| = (\mathbf{V}\cdot\mathbf{V})^{\frac{1}{2}} = [(x\mathbf{a}+y\mathbf{b}+z\mathbf{c})\cdot(x\mathbf{a}+y\mathbf{b}+z\mathbf{c})]^{\frac{1}{2}} \tag{5.3}$$

$$= (x^2a^2+y^2b^2+z^2c^2+2xy\,ab\cos\gamma+2xz\,ac\cos\beta+2yz\,bc\cos\alpha)^{\frac{1}{2}}$$

The expression within the parentheses may be written as a matrix product:

$$[x\quad y\quad z]\begin{bmatrix} a^2 & ab\cos\gamma & ac\cos\beta \\ ab\cos\gamma & b^2 & bc\cos\alpha \\ ac\cos\beta & bc\cos\alpha & c^2 \end{bmatrix}\begin{bmatrix} x \\ y \\ z \end{bmatrix}$$

$$= [x\quad y\quad z]\begin{bmatrix} \mathbf{a}\cdot\mathbf{a} & \mathbf{a}\cdot\mathbf{b} & \mathbf{a}\cdot\mathbf{c} \\ \mathbf{a}\cdot\mathbf{b} & \mathbf{b}\cdot\mathbf{b} & \mathbf{b}\cdot\mathbf{c} \\ \mathbf{a}\cdot\mathbf{c} & \mathbf{b}\cdot\mathbf{c} & \mathbf{c}\cdot\mathbf{c} \end{bmatrix}\begin{bmatrix} x \\ y \\ z \end{bmatrix}$$

$$= \mathbf{x}^T\,\mathbf{G}\,\mathbf{x}$$

so that

$$|\mathbf{V}| = (\mathbf{x}^T\mathbf{G}\mathbf{x})^{\frac{1}{2}} \tag{5.4}$$

where **x** in Eq. 5.4 symbolizes the components of **V** along the three axes **a**, **b**, **c** (column vector), and $\mathbf{x}^T$ symbolizes the same information expressed as a row vector. The distinction between row and column vectors is a formality required by the rules of matrix multiplication. The matrix **G** is known as the metric matrix. Its determinant equals the square of the volume of the parallelepiped formed by the three basis vectors

$$|\mathbf{G}| = a^2b^2c^2(1-\cos^2\alpha-\cos^2\beta-\cos^2\gamma+2\cos\alpha\cos\beta\cos\gamma) \tag{5.5}$$

as can easily be checked by multiplication.

If the vector **V** is written

$$\mathbf{V} = x^i\mathbf{a}_i \qquad i = 1,2,3 \tag{5.6}$$

as in tensor notation,[2] then

$$|\mathbf{V}|^2 = g_{ij}x^i x^j \tag{5.7}$$

a summation being implied over any index that is repeated. Here g_{ij} is the metric tensor corresponding to **G**. Superscripts imply that the components of **V** are contravariant—i.e., under a change of axes they do not transform as the base vectors $\mathbf{a}_i$ but rather as the reciprocal base vectors $\mathbf{b}^i$ (see next section, Linear Transformations).

The angle between two vectors $\mathbf{V}_1$ and $\mathbf{V}_2$ is given by

$$\cos\theta_{12} = \frac{\mathbf{V}_1\cdot\mathbf{V}_2}{|V_1V_2|}$$

$$= \frac{g_{ij}x_1^i x_2^j}{[(g_{ij}x_1^i x_1^j)(g_{ij}x_2^i x_2^j)]^{\frac{1}{2}}} \tag{5.8}$$

For the cross-product of two vectors with components along oblique axes, we have

$$\begin{aligned}
&\mathbf{V}_1(x_1,y_1,z_1) \times \mathbf{V}_2(x_2,y_2,z_2)\\
&= (x_1\mathbf{a}+y_1\mathbf{b}+z_1\mathbf{c}) \times (x_2\mathbf{a}+y_2\mathbf{b}+z_2\mathbf{c})\\
&= (x_1y_2-x_2y_1)(\mathbf{a}\times\mathbf{b}) + (y_1z_2-y_2z_1)(\mathbf{b}\times\mathbf{c}) + (z_1x_2-z_2x_1)(\mathbf{c}\times\mathbf{a})\\
&= [\mathbf{abc}]\begin{vmatrix} \mathbf{a}^* & \mathbf{b}^* & \mathbf{c}^* \\ x_1 & y_1 & z_1 \\ x_2 & y_2 & z_2 \end{vmatrix}
\end{aligned} \tag{5.9}$$

using the relationships between direct and reciprocal base vectors (p. 33)

$$\mathbf{a}^* = \frac{\mathbf{b}\times\mathbf{c}}{[\mathbf{abc}]} \qquad \mathbf{b}^* = \frac{\mathbf{c}\times\mathbf{a}}{[\mathbf{abc}]} \qquad \mathbf{c}^* = \frac{\mathbf{a}\times\mathbf{b}}{[\mathbf{abc}]}$$

The quantity $[\mathbf{abc}]$ is the triple product $\mathbf{a}\cdot\mathbf{b}\times\mathbf{c} = |\mathbf{G}|^{\frac{1}{2}}$. The triple product of three vectors $\mathbf{V}_1$, $\mathbf{V}_2$, $\mathbf{V}_3$ is

$$[\mathbf{V}_1\mathbf{V}_2\mathbf{V}_3] = [\mathbf{abc}]\begin{vmatrix} x_1 & y_1 & z_1 \\ x_2 & y_2 & z_2 \\ x_3 & y_3 & z_3 \end{vmatrix} \tag{5.10}$$

[2]For an admirably clear and concise introduction to vector and tensor analysis, see *International Tables for X-ray Crystallography*, Vol. II, Kynoch Press, Birmingham, 1959, section 2.4 (contributed by A. L. Patterson).

the determinant being zero if and only if $\mathbf{V}_1$, $\mathbf{V}_2$, $\mathbf{V}_3$ lie in a common plane. It is only in an orthogonal coordinate system that

$$\mathbf{a}^* = \frac{\mathbf{a}}{|\mathbf{a}^2|} \qquad \mathbf{b}^* = \frac{\mathbf{b}}{|\mathbf{b}^2|} \qquad \mathbf{c}^* = \frac{\mathbf{c}}{|\mathbf{c}^2|}$$

and only in a Cartesian (orthonormal) system that direct and reciprocal base vectors are equal in magnitude and direction.

In an oblique coordinate system, the components of a vector along the **a**, **b**, **c** axes are not equal to the projections on these axes. This is immediately clear from the two-dimensional example (Fig. 5.1), where $\mathbf{V} = x\mathbf{a} + y\mathbf{b}$. The x component is equal to $\mathbf{V}\cdot\mathbf{a}^*$:

$$\begin{aligned}\mathbf{V}\cdot\mathbf{a}^* &= (\text{projection of } \mathbf{V} \text{ on direction of } \mathbf{a}^*)(\text{length of } \mathbf{a}^*)\\ &= ax\cos(\gamma-90°)\cdot\frac{1}{a\sin\gamma}\\ &= x\end{aligned}$$

and similarly $\mathbf{V}\cdot\mathbf{b}^* = y$. Thus the vector can be written:

$$\mathbf{V} = (\mathbf{V}\cdot\mathbf{a}^*)\mathbf{a}+(\mathbf{V}\cdot\mathbf{b}^*)\mathbf{b}$$

Similarly, in three dimensions

$$\mathbf{V} = (\mathbf{V}\cdot\mathbf{a}^*)\mathbf{a}+(\mathbf{V}\cdot\mathbf{b}^*)\mathbf{b}+(\mathbf{V}\cdot\mathbf{c}^*)\mathbf{c} \tag{5.11}$$

as may easily be checked by dot multiplication of **V** with $\mathbf{a}^*$, $\mathbf{b}^*$, $\mathbf{c}^*$, respectively. Also, for the components along the reciprocal axes

$$\mathbf{V} = (\mathbf{V}\cdot\mathbf{a})\mathbf{a}^*+(\mathbf{V}\cdot\mathbf{b})\mathbf{b}^*+(\mathbf{V}\cdot\mathbf{c})\mathbf{c}^* \tag{5.12}$$

Components of a vector along direct axes are thus equal to projections on reciprocal axes times the lengths of reciprocal axes, and components

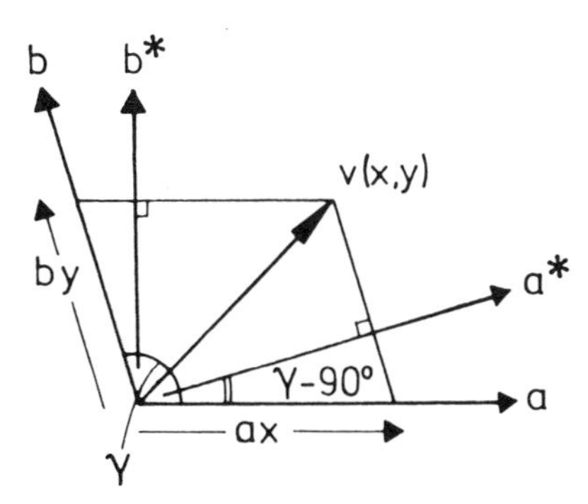

Figure 5.1. The components of **V** along **a** and **b** are given by $\mathbf{V}\cdot\mathbf{a}^*$ and $\mathbf{V}\cdot\mathbf{b}^*$ respectively.

along reciprocal axes are equal to projections on direct axes times the lengths of these axes.

Note that direct base vectors may be written in terms of their components along reciprocal axes:

$$\mathbf{a} = (\mathbf{a}\cdot\mathbf{a})\mathbf{a}^* + (\mathbf{a}\cdot\mathbf{b})\mathbf{b}^* + (\mathbf{a}\cdot\mathbf{c})\mathbf{c}^*$$

$$\mathbf{b} = (\mathbf{b}\cdot\mathbf{a})\mathbf{a}^* + (\mathbf{b}\cdot\mathbf{b})\mathbf{b}^* + (\mathbf{b}\cdot\mathbf{c})\mathbf{c}^*$$

$$\mathbf{c} = (\mathbf{c}\cdot\mathbf{a})\mathbf{a}^* + (\mathbf{c}\cdot\mathbf{b})\mathbf{b}^* + (\mathbf{c}\cdot\mathbf{c})\mathbf{c}^*$$

or in tensor notation

$$\mathbf{a}_i = (\mathbf{a}_i\cdot\mathbf{a}_j)\mathbf{b}^j = g_{ij}\mathbf{b}^j \tag{5.13}$$

similarly,

$$\mathbf{b}^i = (\mathbf{b}^i\cdot\mathbf{b}^j)\mathbf{a}_j = g^{ij}\mathbf{a}_j \tag{5.14}$$

In an oblique coordinate system the direction cosines of a line (unit vector) no longer satisfy the relationship

$$l^2 + m^2 + n^2 = 1$$

A relationship they do satisfy is given by the determinantal equation

$$\begin{vmatrix} 1 & \cos\gamma & \cos\beta & l \\ \cos\gamma & 1 & \cos\alpha & m \\ \cos\beta & \cos\alpha & 1 & n \\ l & m & n & 1 \end{vmatrix} = 0 \tag{5.15}$$

This is the condition satisfied by the six angles between four vectors emanating from a common point (the six angles at the center of a tetrahedron). To see this, consider a unit vector $\mathbf{v}$ with components x, y, z along $\mathbf{a}$, $\mathbf{b}$, $\mathbf{c}$ (here taken also as unit vectors):

$$\mathbf{v} = x\mathbf{a} + y\mathbf{b} + z\mathbf{c}$$

From Eq. 5.12 the direction cosines of $\mathbf{v}$ are the components of this vector along the reciprocal axes $\mathbf{a}^*$, $\mathbf{b}^*$, $\mathbf{c}^*$:

$$l = \mathbf{v}\cdot\mathbf{a} = x + y\cos\gamma + z\cos\beta$$

$$m = \mathbf{v}\cdot\mathbf{b} = x\cos\gamma + y + z\cos\alpha$$

$$n = \mathbf{v}\cdot\mathbf{c} = x\cos\beta + y\cos\alpha + z$$

but we also have

$$\mathbf{v}\cdot\mathbf{v} = [(\mathbf{v}\cdot\mathbf{a})\mathbf{a}^* + (\mathbf{v}\cdot\mathbf{b})\mathbf{b}^* + (\mathbf{v}\cdot\mathbf{c})\mathbf{c}^*]\cdot[x\mathbf{a}+y\mathbf{b}+z\mathbf{c}]$$

or, since $\mathbf{v}$ is a unit vector,

$$1 = lx + my + nz$$

For self-consistency of the four equations in x, y, z, the coefficients must satisfy the determinantal equation Eq. 5.15.

Note that if, as in the above example, the axes **a**, **b**, **c** of an oblique coordinate system are taken to be unit vectors, the lengths of the reciprocal axes are not equal to unity since

$$\mathbf{a}^* = \frac{\mathbf{b} \times \mathbf{c}}{[\mathbf{abc}]} = \frac{\sin \alpha}{G^{\frac{1}{2}}} = \frac{1}{\sin \beta \sin \gamma^*}$$

The reciprocal axes are thus longer than the unit-length direct axes. (The components of **v** along unit vectors in the directions of the reciprocal axes would be $l/\sin \beta \sin \gamma^*$, etc.).

Another problem that requires special treatment in a non-Cartesian coordinate system is that of finding principal axes of a quadratic form, such as

$$\begin{aligned} q(x,y,z) &\equiv q_{11}x^2 + q_{22}y^2 + q_{33}z^2 + 2q_{12}xy + 2q_{13}xz + 2q_{23}yz \\ &\equiv q_{ij}x^i x^j \end{aligned} \tag{5.16}$$

When x, y, z are referred to Cartesian axes, this is equivalent to the well-known problem of finding the eigenvalues and eigenvectors of the matrix $\mathbf{Q}$ with elements q_{ij}. This problem is usually handled by seeking an orthogonal transformation of the Cartesian axes that diagonalizes $\mathbf{Q}$. If the matrix product $\mathbf{Q}' = \boldsymbol{\beta}^T\mathbf{Q}\boldsymbol{\beta}$ is to be diagonal, then the columns of the transformation matrix $\boldsymbol{\beta}$ are the components of the normalized eigenvectors along the original axes, and these can be found by solving the characteristic equation $|\mathbf{Q} - \lambda\mathbf{I}| = 0$ to obtain the eigenvalues λ_r (latent roots) and substituting these in turn into $\mathbf{Q}\mathbf{x}_r = \lambda\mathbf{x}_r$. In a non-Cartesian coordinate system, the vectors found by this procedure would not be mutually orthogonal and hence cannot possibly correspond to principal axes of the quadratic form.

More generally, the problem of finding principal axes of q is equivalent to that of finding the directions in which $q(\mathbf{x})$ has a maximum or minimum value (or a stationary intermediate value) as the vector $\mathbf{x}$ of fixed length, say unity, is rotated about the origin. From Eq. 5.7 the components of such a vector have to satisfy the condition

$$g_{ij}x^i x^j = 1$$

where g_{ij} is the metric tensor for the coordinate system in question. Problems of this kind are conveniently treated by introducing Lagrangian multipliers,[3]—i.e., we seek the vectors $\mathbf{x}$ for which the modified function

$$q' = q_{ij}x^i x^j - \lambda(g_{ij}x^i x^j - 1) \tag{5.17}$$

has stationary values. Differentiation with respect to $x, y, z(x^1,x^2,x^3)$ yields the three conditions ($i = 1,2,3$)

$$\frac{1}{2}\frac{\partial q'}{\partial x^i} = (q_{ij} - \lambda\, g_{ij})\, x^j = 0 \tag{5.18}$$

which have a common solution only if the determinantal equation

$$|\mathbf{Q} - \lambda\,\mathbf{G}| = 0 \tag{5.19}$$

is satisfied. Substitution of the three roots λ_r into Eq. 5.18 then leads to the principal axis vectors $\mathbf{x}_r$. For Cartesian axes, $\mathbf{G}$ reduces to the identity matrix $\mathbf{I}$ with

$$g_{ij} = 1\ (i=j) \quad \text{and} \quad g_{ij} = 0\ (i \neq j)$$

Note that if the quadratic form is referred to components along reciprocal base vectors (as in the vibration ellipsoid $\mathbf{h}^T\mathbf{U}\mathbf{h}$), then the metric tensor $g_{ij} \equiv (\mathbf{a}_i\cdot\mathbf{a}_j)$ has to be replaced by the reciprocal metric tensor $g^{ij} \equiv (\mathbf{b}^i\cdot\mathbf{b}^j)$. Otherwise the treatment is identical.

The foregoing indicates that geometrical calculations are much simpler in a Cartesian coordinate system than in an oblique system. Thus, it is often advisable to transform the fractional coordinates based on the crystal axes into Cartesian coordinates based on orthonormal axes. All subsequent calculations can then be carried out in the transformed coordinate system. In addition, we may sometimes wish to transform from one set of crystal axes to another. After all, the choice of crystal axes is somewhat arbitrary, and what is convenient for one purpose may be less convenient for another. It is very useful to be able to move back and forth from one coordinate system to another, which brings us to the description of linear transformations in general.

Linear Transformations

In three-dimensional space, a coordinate system is usually based on three noncoplanar vectors (base vectors). A second set of base vectors may be defined in terms of the first set by three equations:

[3]This treatment follows essentially that described by J. Waser, *Acta Crystallogr.* **8**, 731 (1955).

$$\mathbf{A}_1 = \alpha_{11}\mathbf{a}_1 + \alpha_{12}\mathbf{a}_2 + \alpha_{13}\mathbf{a}_3$$
$$\mathbf{A}_2 = \alpha_{21}\mathbf{a}_1 + \alpha_{22}\mathbf{a}_2 + \alpha_{23}\mathbf{a}_3 \qquad (5.20)$$
$$\mathbf{A}_3 = \alpha_{31}\mathbf{a}_1 + \alpha_{32}\mathbf{a}_2 + \alpha_{33}\mathbf{a}_3$$

where the coefficients α_{ij} are any real numbers. In matrix notation

$$\begin{bmatrix} \mathbf{A}_1 \\ \mathbf{A}_2 \\ \mathbf{A}_3 \end{bmatrix} = \begin{bmatrix} \alpha_{11} & \alpha_{12} & \alpha_{13} \\ \alpha_{21} & \alpha_{22} & \alpha_{23} \\ \alpha_{31} & \alpha_{32} & \alpha_{33} \end{bmatrix} \begin{bmatrix} \mathbf{a}_1 \\ \mathbf{a}_2 \\ \mathbf{a}_3 \end{bmatrix} \qquad (5.21a)$$

often condensed to $\mathbf{A} = \boldsymbol{\alpha}\mathbf{a}$ (5.21b)

or in tensor notation $\mathbf{A}_i = \alpha_{ij}\mathbf{a}_j$ (5.21c)

The reverse transformation is simply

$$\mathbf{a} = \boldsymbol{\alpha}^{-1}\mathbf{A} \qquad \text{or} \qquad \mathbf{a}_i = \boldsymbol{\beta}_{ij}\mathbf{A}_j \qquad (5.22)$$

where $\boldsymbol{\beta}$ is the inverse matrix of $\boldsymbol{\alpha}$. To calculate the inverse matrix, the elements of $\boldsymbol{\beta}$ are most conveniently expressed as

$$\beta_{ij} = \frac{(-1)^{i+j}}{|\boldsymbol{\alpha}|} \cdot (\text{minor of } \alpha_{ji})$$

A vector is unaffected by a change of axes, but its components are altered. Let $\mathbf{V}(x_1, x_2, x_3)$ be the vector in coordinate system $\mathbf{a}$, and let $\mathbf{V}(X_1, X_2, X_3)$ be the same vector in coordinate system $\mathbf{A}$:

$$\begin{aligned} \mathbf{V} &= X_1\mathbf{A}_1 + X_2\mathbf{A}_2 + X_3\mathbf{A}_3 = x_1\mathbf{a}_1 + x_2\mathbf{a}_2 + x_3\mathbf{a}_3 \\ &= x_1(\beta_{11}\mathbf{A}_1 + \beta_{12}\mathbf{A}_2 + \beta_{13}\mathbf{A}_3) \\ &+ x_2(\beta_{21}\mathbf{A}_1 + \beta_{22}\mathbf{A}_2 + \beta_{23}\mathbf{A}_3) \\ &+ x_3(\beta_{31}\mathbf{A}_1 + \beta_{32}\mathbf{A}_2 + \beta_{33}\mathbf{A}_3) \\ &= \quad (\beta_{11}x_1 + \beta_{21}x_2 + \beta_{31}x_3)\mathbf{A}_1 \\ &+ \quad (\beta_{12}x_1 + \beta_{22}x_2 + \beta_{32}x_3)\mathbf{A}_2 \\ &+ \quad (\beta_{13}x_1 + \beta_{23}x_2 + \beta_{33}x_3)\mathbf{A}_3 \end{aligned}$$

Equating coefficients,

$$X_1 = \beta_{11}x_1 + \beta_{21}x_2 + \beta_{31}x_3$$
$$X_2 = \beta_{12}x_1 + \beta_{22}x_2 + \beta_{32}x_3 \qquad (5.23)$$
$$X_3 = \beta_{13}x_1 + \beta_{23}x_3 + \beta_{33}x_3$$

or, $$X_i = \beta_{ji} x_j \tag{5.23a}$$

We see that the transformation matrix for vector components has elements β_{ji} and is the transpose of the matrix $\boldsymbol{\beta} = \boldsymbol{\alpha}^{-1}$. The above derivation is straightforward but cumbersome, and to show the advantages of tensor notation, we repeat it in condensed form:

$$\mathbf{V} = X_i \mathbf{A}_i = x_i \mathbf{a}_i = x_i \beta_{ij} \mathbf{A}_j = x_j \beta_{ji} \mathbf{A}_i$$

hence $$X_i = x_j \beta_{ji}$$

as in Eq. 5.23a. In the same way we can show that

$$x_i = X_j \alpha_{ji} \tag{5.24}$$

Corresponding to the set of base vectors $\mathbf{a}_1$, $\mathbf{a}_2$, $\mathbf{a}_3$, a set of reciprocal base vectors $\mathbf{b}_1$, $\mathbf{b}_2$, $\mathbf{b}_3$ can be defined such that

$$\mathbf{a}_i \cdot \mathbf{b}_i = 1$$

$$\mathbf{a}_i \cdot \mathbf{b}_j = 0$$

Similarly, the vectors $\mathbf{B}_1$, $\mathbf{B}_2$, $\mathbf{B}_3$ are defined as the set reciprocal to $\mathbf{A}_1$, $\mathbf{A}_2$, $\mathbf{A}_3$. How do the reciprocal base vectors transform? From Eq. 5.12 we have:

$$\mathbf{B}_i = (\mathbf{B}_i \cdot \mathbf{a}_j)\mathbf{b}_j$$

for the components of the **B** vectors along the **b** axes and from Eq. 5.11

$$\mathbf{a}_i = (\mathbf{a}_i \cdot \mathbf{B}_j)\mathbf{A}_j$$

for the components of the **a** vectors along the **A** axes. However, we already know that

$$\mathbf{a}_i = \beta_{ij} \mathbf{A}_j$$

so $$\mathbf{a}_i \cdot \mathbf{B}_j = \beta_{ij}$$

and $$\mathbf{B}_i \cdot \mathbf{a}_j = \beta_{ji}$$

leading to $$\mathbf{B}_i = \beta_{ji} \mathbf{b}_j \tag{5.25}$$

Thus, reciprocal base vectors transform in the same way as *components* along direct base vectors. Similarly, it is easily shown that components along reciprocal base vectors—e.g., $\mathbf{H}(h_1 h_2 h_3)$ transform in the same way as direct base vectors. Quantities that transform like direct base vectors are called covariant; those that transform like reciprocal base vectors are called contravariant. The distinction is often made by writing symbols for the former with a subscript (e.g., $\mathbf{a}_i$, h_i) and for the latter with a superscript (e.g., x^i, $\mathbf{b}^i$), although this will not be necessary for most of the examples discussed here. We may summarize the above results:

$$\begin{array}{ll} \mathbf{A}_i = \alpha_{ij}\mathbf{a}_j & \mathbf{a}_i = \beta_{ij}\mathbf{A}_j \\ H_i = \alpha_{ij}h_j & h_i = \beta_{ij}H_j \\ \mathbf{B}^i = \mathbf{b}^j\beta_{ji} & \mathbf{b}^i = \mathbf{B}^j\alpha_{ji} \\ X^i = x^j\beta_{ji} & x^i = X^j\alpha_{ji} \end{array} \tag{5.26}$$

or, in matrix notation

$$\begin{array}{ll} \mathbf{A} = \boldsymbol{\alpha}\mathbf{a} & \mathbf{a} = (\boldsymbol{\alpha}^{-1})\mathbf{A} \\ \mathbf{H} = \boldsymbol{\alpha}\mathbf{h} & \mathbf{h} = (\boldsymbol{\alpha}^{-1})\mathbf{H} \\ \mathbf{B} = (\boldsymbol{\alpha}^{-1})^T\mathbf{b} & \mathbf{b} = \boldsymbol{\alpha}^T\mathbf{B} \\ \mathbf{X} = (\boldsymbol{\alpha}^{-1})^T\mathbf{x} & \mathbf{x} = \boldsymbol{\alpha}^T\mathbf{X} \end{array} \tag{5.26a}$$

The transformation of a quadratic form, such as

$$u_{ij}h_ih_j = \mathbf{h}^T\mathbf{u}\mathbf{h} \tag{5.27}$$

for a vibrational ellipsoid may also be needed. Under the axial transformation $\mathbf{A} = \boldsymbol{\alpha}\mathbf{a}$, components along reciprocal base vectors transform as $\mathbf{h} = (\boldsymbol{\alpha}^{-1})\mathbf{H} = \boldsymbol{\beta}\mathbf{H}$ so that the quadratic form (Eq. 5.27) becomes

$$\mathbf{h}^T\mathbf{u}\mathbf{h} = \mathbf{H}^T\boldsymbol{\beta}^T\mathbf{u}\boldsymbol{\beta}\mathbf{H} \tag{5.28}$$

or in the tensor notation (ignoring the superscript-subscript distinction)

$$u_{ij}h_ih_j = \beta_{ik}u_{ij}\beta_{jl}H_kH_l \tag{5.28a}$$

Transformation from Triclinic to Orthonormal Axes

To begin, we list (without proof) some of the more important relationships between direct and reciprocal cell constants:[4]

$$a^* = \frac{bc \sin \alpha}{V} \qquad b^* = \frac{ac \sin \beta}{V} \qquad c^* = \frac{ab \sin \gamma}{V}$$

$$\cos \alpha^* = \frac{\cos \beta \cos \gamma - \cos \alpha}{\sin \beta \sin \gamma} \qquad \cos \beta^* = \frac{\cos \alpha \cos \gamma - \cos \beta}{\sin \alpha \sin \gamma}$$

$$\cos \gamma^* = \frac{\cos \alpha \cos \beta - \cos \gamma}{\sin \alpha \sin \beta}$$

$$V = abc(1-\cos^2\alpha-\cos^2\beta-\cos^2\gamma+2\cos \alpha \cos \beta \cos \gamma)^{\frac{1}{2}}$$

$$= abc \sin \alpha \sin \beta \sin \gamma^* = abc \sin \alpha \sin \beta^* \sin \gamma = abc \sin \alpha^* \sin \beta \sin \gamma$$

[4]For a complete list, see M. J. Buerger, *X-Ray Crystallography,* Wiley, New York, 1942, pp. 360–361.

The inverse relationships are obtained by interchanging starred and unstarred quantities—e.g., $V^* = a^*b^*c^* \sin\alpha^* \sin\beta^* \sin\gamma$, etc. We shall need some of these relationships in the following discussion.

For the transformation from triclinic axes **a**, **b**, **c** to orthonormal axes **A**, **B**, **C**, we choose **A** as unit vector along **a**, **B** as unit vector normal to **a** in the **ab** plane, and **C** normal to **A** and **B**. Instead of finding the transformation matrix $\boldsymbol{\alpha}$ ($\mathbf{A} = \boldsymbol{\alpha}\mathbf{a}$) directly, it is easier to proceed by first deriving the elements of the inverse matrix $\boldsymbol{\alpha}^{-1}$ ($\mathbf{a} = \boldsymbol{\alpha}^{-1}\mathbf{A}$) since the components of $\mathbf{a}/a$, etc., along the Cartesian axes are direction cosines. Thus, we have

$$\begin{bmatrix} \mathbf{a}/a \\ \mathbf{b}/b \\ \mathbf{c}/c \end{bmatrix} = \begin{bmatrix} l_1 & m_1 & n_1 \\ l_2 & m_2 & n_2 \\ l_3 & m_3 & n_3 \end{bmatrix} \begin{bmatrix} \mathbf{A} \\ \mathbf{B} \\ \mathbf{C} \end{bmatrix}$$

where $l_i^2 + m_i^2 + n_i^2 = 1$. We can write down some of the coefficients by inspection (Fig. 5.2), e.g.,

$$\begin{bmatrix} 1 & 0 & 0 \\ \cos\gamma & \sin\gamma & 0 \\ \cos\beta & m_3 & n_3 \end{bmatrix}$$

and determine m_3 from

$$\begin{aligned} \cos\alpha &= l_2 l_3 + m_2 m_3 + n_2 n_3 \\ &= \cos\gamma\cos\beta + m_3 \sin\gamma \end{aligned}$$

so
$$m_3 = \frac{\cos\alpha - \cos\beta\cos\gamma}{\sin\gamma} = -\sin\beta\cos\alpha^*$$

The remaining coefficient n_3 is obtained from $\cos^2\beta + m_3^2 + n_3^2 = 1$:

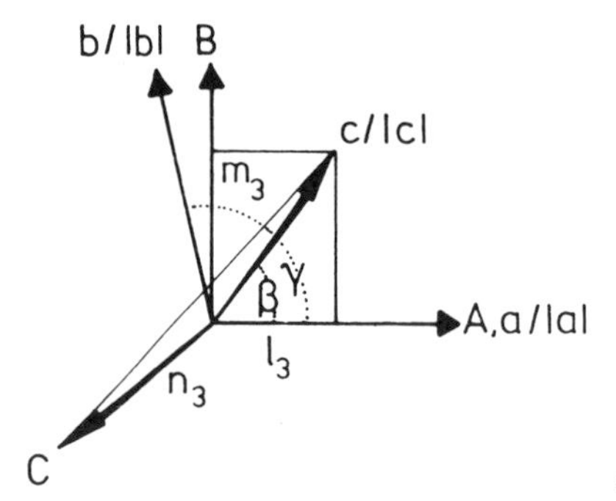

Figure 5.2. Components of **a**, **b**, **c** on **A**, **B**, **C** respectively.

$$n_3^2 = 1 - \cos^2\beta - \sin^2\beta\cos^2\alpha^*$$
$$= \sin^2\beta\sin^2\alpha^*$$

$$n_3 = \sin\beta\sin\alpha^* = \frac{v}{\sin\gamma}$$

where v is the volume of the parallelepiped formed by the unit vectors **a**/a, **b**/b, **c**/c. The transformation is therefore

$$\begin{bmatrix} \mathbf{a}/a \\ \mathbf{b}/b \\ \mathbf{c}/c \end{bmatrix} = \begin{bmatrix} 1 & 0 & 0 \\ \cos\gamma & \sin\gamma & 0 \\ \cos\beta & \dfrac{\cos\alpha - \cos\beta\cos\gamma}{\sin\gamma} & \dfrac{v}{\sin\gamma} \end{bmatrix} \begin{bmatrix} \mathbf{A} \\ \mathbf{B} \\ \mathbf{C} \end{bmatrix}$$

or

$$\begin{bmatrix} \mathbf{a} \\ \mathbf{b} \\ \mathbf{c} \end{bmatrix} = \begin{bmatrix} a & 0 & 0 \\ b\cos\gamma & b\sin\gamma & 0 \\ c\cos\beta & \dfrac{c(\cos\alpha - \cos\beta\cos\gamma)}{\sin\gamma} & \dfrac{cv}{\sin\gamma} \end{bmatrix} \begin{bmatrix} \mathbf{A} \\ \mathbf{B} \\ \mathbf{C} \end{bmatrix} \quad (5.29)$$

which we write $\mathbf{a} = \boldsymbol{\beta}\mathbf{A} = (\boldsymbol{\alpha}^{-1})\mathbf{A}$

We then have $\mathbf{X} = \boldsymbol{\beta}^T\mathbf{x} = (\boldsymbol{\alpha}^{-1})^T\mathbf{x}$

or

$$\begin{bmatrix} X \\ Y \\ Z \end{bmatrix} = \begin{bmatrix} a & b\cos\gamma & c\cos\beta \\ 0 & b\sin\gamma & \dfrac{c(\cos\alpha - \cos\beta\cos\gamma)}{\sin\gamma} \\ 0 & 0 & \dfrac{cv}{\sin\gamma} \end{bmatrix} \begin{bmatrix} x \\ y \\ z \end{bmatrix} \quad (5.30)$$

for the transformation from crystal fractional coordinates (dimensionless) to Cartesian coordinates with the dimension of length. The matrix $\boldsymbol{\alpha}$ is obtained by inversion of $\boldsymbol{\beta} = \boldsymbol{\alpha}^{-1}$; the determinant $|\boldsymbol{\alpha}^{-1}| = abc\,v$, and the individual matrix elements α_{ij} are easily evaluated from

$$\alpha_{ij} = \frac{(-1)^{i+j}}{|\boldsymbol{\beta}|} \cdot (\text{minor of } \beta_{ji}):$$

$$\alpha_{11} = \frac{1}{a} \qquad \alpha_{12} = 0 \qquad \alpha_{13} = 0$$

$$\alpha_{21} = \frac{-\cos\gamma}{a\sin\gamma} \qquad \alpha_{22} = \frac{1}{b\sin\gamma} \qquad \alpha_{23} = 0$$

$$\alpha_{31} = \frac{1}{av}\left(\frac{\cos\gamma}{\sin\gamma}(\cos\alpha - \cos\beta\cos\gamma) - \cos\beta\sin\gamma\right)$$

$$= \frac{\cos\gamma\cos\alpha - \cos\beta\cos^2\gamma - \cos\beta\sin^2\gamma}{a\,v\sin\gamma}$$

$$= \frac{\cos\gamma\cos\alpha - \cos\beta}{a\,v\sin\gamma}$$

$$\alpha_{32} = \frac{\cos\gamma\cos\beta - \cos\alpha}{b\,v\sin\gamma}$$

$$\alpha_{33} = \frac{\sin\gamma}{c\,v}$$

$$\begin{bmatrix} \mathbf{A} \\ \mathbf{B} \\ \mathbf{C} \end{bmatrix} = \begin{bmatrix} \frac{1}{a} & 0 & 0 \\ \frac{-\cos\gamma}{a\sin\gamma} & \frac{1}{b\sin\gamma} & 0 \\ \frac{\cos\gamma\cos\alpha - \cos\beta}{a\,v\sin\gamma} & \frac{\cos\gamma\cos\beta - \cos\alpha}{b\,v\sin\gamma} & \frac{\sin\gamma}{c\,v} \end{bmatrix} \begin{bmatrix} \mathbf{a} \\ \mathbf{b} \\ \mathbf{c} \end{bmatrix} \tag{5.31}$$

which condenses to $\mathbf{A} = \boldsymbol{\alpha}\mathbf{a}$.

Finally, for the reverse transformation from orthogonal coordinates to crystal coordinates,

$$\begin{bmatrix} x \\ y \\ z \end{bmatrix} = \begin{bmatrix} \frac{1}{a} & \frac{-\cos\gamma}{a\sin\gamma} & \frac{\cos\gamma\cos\alpha - \cos\beta}{a\,v\sin\gamma} \\ 0 & \frac{1}{b\sin\gamma} & \frac{\cos\gamma\cos\beta - \cos\alpha}{b\,v\sin\gamma} \\ 0 & 0 & \frac{\sin\gamma}{c\,v} \end{bmatrix} \begin{bmatrix} X \\ Y \\ Z \end{bmatrix} \tag{5.32}$$

which condenses to $\mathbf{x} = \boldsymbol{\alpha}^T\mathbf{X}$

The transformations among the reciprocal base vectors and among components along these vectors follow the rules given in the previous section:

$$\begin{array}{ll} \mathbf{b} \to \mathbf{B} & \mathbf{B} = \boldsymbol{\beta}^T\mathbf{b} \\ \mathbf{B} \to \mathbf{b} & \mathbf{b} = \boldsymbol{\alpha}^T\mathbf{B} \\ \mathbf{h} \to \mathbf{H} & \mathbf{H} = \boldsymbol{\alpha}\mathbf{h} \\ \mathbf{H} \to \mathbf{h} & \mathbf{h} = \boldsymbol{\beta}\mathbf{H} \end{array} \tag{5.33}$$

The quadratic form $U_{11}h^2\mathbf{a}^{*2}+\ldots+2U_{12}hk\,\mathbf{a}^*\mathbf{b}^*\ldots$ for the mean-square vibrational amplitude in the direction of the vector with components h, k, l along the reciprocal base vectors $\mathbf{a}^*$, $\mathbf{b}^*$, $\mathbf{c}^*$, can be written:

$$\mathbf{h}^T\mathbf{DUDh}$$

where $\mathbf{D}$ is the diagonal matrix

$$\mathbf{D} = \begin{bmatrix} a^* & 0 & 0 \\ 0 & b^* & 0 \\ 0 & 0 & c^* \end{bmatrix}$$

On transforming to the above set of Cartesian axes,

$$\mathbf{h} = \boldsymbol{\beta}\mathbf{H}$$

so the quadratic form becomes

$$\mathbf{H}^T\boldsymbol{\beta}^T\mathbf{DUD}\boldsymbol{\beta}\mathbf{H} = \mathbf{H}^T\mathbf{VH} \tag{5.34}$$

where $$\mathbf{V} = (\boldsymbol{\beta}^T\mathbf{D})\,\mathbf{U}\,(\mathbf{D}\boldsymbol{\beta})$$

Example: Hexaethylidenecyclohexane (space group $P\bar{1}$)[5]

$$a = 7.877\ \text{Å} \qquad \alpha = 105.563°$$

$$b = 7.210\ \text{Å} \qquad \beta = 116.245°$$

$$c = 7.891\ \text{Å} \qquad \gamma = 79.836°$$

$$v = 0.86209 \qquad \text{cell volume} = abc\,v = 386.35\ \text{Å}^3$$

$$a^* = \sin\alpha/a\,v = 0.14186\ \text{Å}^{-1}$$

$$b^* = \sin\beta/b\,v = 0.14430\ \text{Å}^{-1}$$

$$c^* = \sin\gamma/c\,v = 0.14469\ \text{Å}^{-1}$$

$$\alpha^* = \cos^{-1}\left(\frac{\cos\beta\cos\gamma - \cos\alpha}{\sin\beta\sin\gamma}\right) = 77.554°$$

$$\beta^* = \cos^{-1}\left(\frac{\cos\alpha\cos\gamma - \cos\beta}{\sin\alpha\sin\gamma}\right) = 65.391°$$

$$\gamma^* = \cos^{-1}\left(\frac{\cos\alpha\cos\beta - \cos\gamma}{\sin\alpha\sin\beta}\right) = 93.837°$$

For the transformation of fractional crystal coordinates to Cartesian distance coordinates: $\mathbf{X} = \boldsymbol{\beta}^T\mathbf{x}$

[5]W. Marsh and J. D. Dunitz, *Helv. Chim. Acta* **58,** 707 (1975).

$$\boldsymbol{\beta}^T = \begin{bmatrix} 7.87700 & 1.27232 & -3.48948 \\ 0 & 7.09685 & -1.52530 \\ 0 & 0 & 6.91121 \end{bmatrix}$$

Atom	x	y	z	X(Å)	Y(Å)	Z(Å)
C1	−0.13614	0.15714	−0.07165	−0.62242	1.22449	−0.49519
C2	−0.09631	0.11567	0.11837	−1.02451	0.64034	0.81808
C3	0.09193	0.01092	0.20827	0.01127	−0.24018	1.43940
C4	−0.19555	0.33436	−0.10572	−0.74603	2.53416	−0.73065
C5	−0.22720	0.40451	−0.28055	−0.29601	3.29867	−1.93894
C6	−0.21823	0.15538	0.20024	−2.22004	0.79728	1.38390
C7	−0.41675	0.24447	0.12346	−3.40251	1.54665	0.85326
C8	0.21739	0.06558	0.39063	0.43272	−0.13042	2.69973
C9	0.20332	0.23919	0.53948	0.02337	0.87463	3.72846

Coordinates of the other nine carbon atoms are obtained by inversion across the center of symmetry at (0,0,0).

For the transformation of vibrational tensors from oblique crystal coordinate system to the Cartesian system:

$$\mathbf{V} = (\boldsymbol{\beta}^T\mathbf{D})\,\mathbf{U}\,(\mathbf{D}\boldsymbol{\beta})$$

$$\boldsymbol{\beta}^T\mathbf{D} = \begin{bmatrix} 7.87700 & 1.27232 & -3.48948 \\ 0 & 7.09685 & -1.52530 \\ 0 & 0 & 6.91121 \end{bmatrix} \begin{bmatrix} 0.14186 & 0 & 0 \\ 0 & 0.14430 & 0 \\ 0 & & 0.14469 \end{bmatrix}$$

$$= \begin{bmatrix} 1.11743 & 0.18360 & -0.50489 \\ 0 & 1.02408 & -0.22070 \\ 0 & 0 & 1.00000 \end{bmatrix}$$

Anyone who wishes to carry out sequences of matrix multiplications is strongly advised to use a programmable electronic computer. Experience shows that the results of hand calculations are untrustworthy.

Calculations in Cartesian Coordinates

From now on we assume that atomic positions and vibrational parameters are expressed in a Cartesian coordinate system. Interatomic distances and angles may then be calculated easily.

A few words are required about the calculation of torsion angles. For a sequence of four atoms A, B, C, D, the torsion angle ω(ABCD) (Fig. 5.3) is defined as the angle between the directions BA and CD in projection down BC or as the angle between the normals to the planes ABC

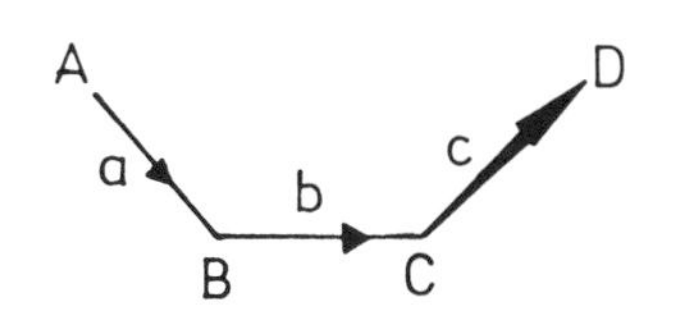

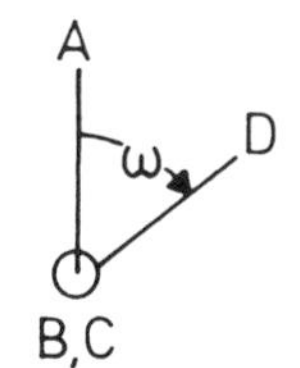

Figure 5.3. Definition of a torsion angle.

and BCD. By convention,[6] ω is considered positive if the sense of rotation from BA to CD, viewed down BC, is clockwise; it is negative if this sense is counterclockwise. A positive value of ω means that the sequence of atoms ABCD forms a right-handed screw. Note that if ω(ABCD) is positive, then so is ω(DCBA)—a right-handed corkscrew remains right-handed when it is turned from back to front. From the definition of a torsion angle

$$\cos \omega = \frac{(\overline{AB} \times \overline{BC}) \cdot (\overline{BC} \times \overline{CD})}{AB(BC)^2\, CD \sin \theta_{ABC} \sin \theta_{BCD}} \tag{5.35}$$

$$\frac{(\overline{BC})}{BC} \sin \omega = \frac{(\overline{AB} \times \overline{BC}) \times (\overline{BC} \times \overline{CD})}{AB(BC)^2\, CD \sin \theta_{ABC} \sin \theta_{BCD}} \tag{5.36}$$

The torsion angle is not defined when one or both bond angles is 0° or 180°. These vector formulas are valid in any coordinate system, but the usual expressions for vector dot and cross-products in terms of vector components:

$$\mathbf{V}_1(x_1y_1z_1) \cdot \mathbf{V}_2(x_2y_2z_2) = x_1x_2 + y_1y_2 + z_1z_2$$

$$\mathbf{V}_1(x_1y_1z_1) \times \mathbf{V}_2(x_2y_2z_2) = \begin{vmatrix} \mathbf{i} & \mathbf{j} & \mathbf{k} \\ x_1 & y_1 & z_1 \\ x_2 & y_2 & z_2 \end{vmatrix}$$

hold only in a Cartesian system (cf., Eqs. 5.8 and 5.9 for analogous expressions in other coordinate systems). The sign of a torsion angle is unchanged by rotation or translation but is reversed by reflection or inversion. The magnitude, but not the sign, of a torsion angle can also be calculated from interatomic distances and angles (Chapter 9).

Crystallographers will have noticed that a torsion angle is just the angle between two reciprocal vectors. Denoting the vectors $\overline{AB}$, $\overline{BC}$, $\overline{CD}$ as **a**, **b**, **c**, respectively, we have $\mathbf{c}^* = \mathbf{a} \times \mathbf{b}/V$ and $\mathbf{a}^* = \mathbf{b} \times \mathbf{c}/V$; thus, the required torsion angle is β^*. In terms of direct and reciprocal vectors, Eq. 5.35 becomes

[6]W. Klyne and V. Prelog, *Experientia* **16,** 521 (1960).

$$\cos \omega = \frac{(\mathbf{a} \times \mathbf{b}) \cdot (\mathbf{b} \times \mathbf{c})}{ab^2c \sin \alpha \sin \gamma} = \frac{\mathbf{c}^* \cdot \mathbf{a}^* V^2}{ab^2c \sin \alpha \sin \gamma} = \frac{a^*c^* \cos \beta^* V^2}{ab^2c \sin \alpha \sin \gamma}$$

but $a^* = bc \sin \alpha/V$, $c^* = ab \sin \gamma/V$, so $\cos \omega = \cos \beta^*$. Similarly,

$$\frac{\mathbf{b}}{b} \sin \omega = \frac{(\mathbf{a} \times \mathbf{b}) \times (\mathbf{b} \times \mathbf{c})}{ab^2c \sin \alpha \sin \gamma} = \frac{\mathbf{c}^* \times \mathbf{a}^* V^2}{ab^2c \sin \alpha \sin \gamma} = \frac{\mathbf{b}\, V}{ab^2c \sin \alpha \sin \gamma}$$

but $V = abc \sin \alpha \sin \beta^* \sin \gamma$, so $\sin \omega = \sin \beta^*$. The sign of ω is the sign of the triple product,

$$[\mathbf{abc}] = \begin{vmatrix} x_a & y_a & z_a \\ x_b & y_b & z_b \\ x_c & y_c & z_c \end{vmatrix}$$

Given the values of the direct angles α, β, γ (which can be calculated from the direction cosines of **a**, **b**, **c**), then

$$\cos \beta^* = \frac{\cos \beta - \cos \alpha \cos \gamma}{\sin \alpha \sin \gamma}$$

A BASIC computer program for calculating interatomic distances and angles and torsion angles from crystal coordinates is listed in Appendix I.

The vector equation of a plane is $\mathbf{x} \cdot \mathbf{N} = d$ where $\mathbf{N}(l,m,n)$ is a unit vector along the normal, and d is the distance from the origin to the plane. This equation can always be expressed in the form

$$Ax + By + Cz = d \tag{5.37}$$

but it is only in a Cartesian system that the coefficients A, B, C are equal to the direction cosines of the plane normal. The equation then becomes

$$x \cos \alpha + y \cos \beta + z \cos \gamma = d \tag{5.37a}$$

The distance from any other point (x', y', z') to the plane is

$$\pm (x' \cos \alpha + y' \cos \beta + z' \cos \gamma - d)$$

The sign is positive if the point (x', y', z') and the origin are on opposite sides of the plane; it is negative if they are on the same side. The vector equation of the plane passing through three given points $\mathbf{p}_1$, $\mathbf{p}_2$, $\mathbf{p}_3$ is $[(\mathbf{x}-\mathbf{p}_1)(\mathbf{p}_2-\mathbf{p}_1)(\mathbf{p}_3-\mathbf{p}_1)] = 0$; in terms of components this becomes

$$\begin{vmatrix} x - x_1 & y - y_1 & z - z_1 \\ x_2 - x_1 & y_2 - y_1 & z_2 - z_1 \\ x_3 - x_1 & y_3 - y_1 & z_3 - z_1 \end{vmatrix} = 0$$

The best plane through a set of points in a Cartesian system may be found by minimizing

$$\sum_i \omega_i(x_i l + y_i m + z_i n - d)^2$$

with respect to l, m, n, and d under the condition that $l^2 + m^2 + n^2 = 1$. The weights ω_i should be taken as being inversely proportional to the variances of the atomic positions in the direction normal to the desired plane, but they are often simply set to unity. If the weights are regarded as being point masses, the problem of finding the best plane is equivalent to that of finding the plane perpendicular to the axis of maximum inertia for the set of weighted points. The best procedure[7] is to carry out a principal axis transformation of the symmetric matrix

$$\begin{bmatrix} \sum\omega_i X_i^2 & \sum\omega_i X_i Y_i & \sum\omega_i X_i Z_i \\ \sum\omega_i X_i Y_i & \sum\omega_i Y_i^2 & \sum\omega_i Y_i Z_i \\ \sum\omega_i X_i Z_i & \sum\omega_i Y_i Zi & \sum\omega_i Z_i^2 \end{bmatrix}$$

where the origin has been taken at the weighted center of mass. The eigenvalues $\lambda^{(1)}$, $\lambda^{(2)}$, $\lambda^{(3)}$ of this matrix are the weighted sums of squares of distances from the best plane, an intermediate plane, and the worst plane, all at right angles to one another. The corresponding eigenvectors are the normals to these planes. If we write

$$\lambda^{(1)} = \sum\omega_i X_i'^2 \qquad \lambda^{(2)} = \sum\omega_i Y_i'^2 \qquad \lambda^{(3)} = \sum\omega_i Z_i'^2$$

then the new coordinates X', Y', Z' are related to the old ones by

$$\begin{bmatrix} X' \\ Y' \\ Z' \end{bmatrix} = \begin{bmatrix} l_1 & m_1 & n_1 \\ l_2 & m_2 & n_2 \\ l_3 & m_3 & n_3 \end{bmatrix} \begin{bmatrix} X \\ Y \\ Z \end{bmatrix}$$

where l_j, m_j, n_j are the direction cosines of the eigenvector corresponding to $\lambda^{(j)}$.

We would generally be interested in the smallest eigenvalue $\lambda^{(1)}$ and its eigenvector, normal to the plane $X' = l_1X + m_1Y + n_1Z = 0$, but the other two eigenvectors are useful in completing a convenient set of orthonormal basis vectors for a "molecular" coordinate system. If the primed coordinate system is to have the same chirality sense as the original unprimed one, the eigenvectors must be chosen so that the determinant of

[7] V. Schomaker, J. Waser, R. E. Marsh, and G. Bergman, *Acta Crystallogr.* **12**, 600 (1959).

the transformation matrix involving the direction cosines is +1. Caution is required if two or more of the eigenvalues are equal or approximately so because in such cases (degeneracy or near degeneracy) the eigenvectors obtained in the usual way may not be mutually perpendicular. They can be made so by the following procedure.

Suppose that the normalized (but not necessarily orthonormal) eigenvectors are $\mathbf{v}_1$, $\mathbf{v}_2$, $\mathbf{v}_3$. Choose $\mathbf{v}_1' = \mathbf{v}_1$ and $\mathbf{v}_2' = \mathbf{v}_2 - \mathbf{v}_1 \cos(\mathbf{v}_1\mathbf{v}_2)$, so that $\mathbf{v}_1' \cdot \mathbf{v}_2' = 0$. Now take $\mathbf{v}_3' = \mathbf{v}_1' \times \mathbf{v}_2'$, to complete a right-handed Cartesian system.

Thermal Motion Analysis

As mentioned earlier, many modern crystal structure analyses include the determination of anisotropic vibrational parameters U_{ij} for the individual atoms in the structure. These parameters are affected by absorption errors and by improper weighting in the least-squares refinement process. In addition, any smearing of the electron density peaks caused by chemical bonding tends to be absorbed in X-ray vibrational parameters, although not in neutron parameters. Nevertheless, if the appropriate precautions are taken, X-ray vibrational parameters are often accurate enough to provide a basis for a partial analysis of the overall thermal motion into contributions from various sources. Since bond-stretching vibrations have a much smaller amplitude than other vibrations (bond bending, torsional, rigid-body translational and rotational oscillations), the mean vibrational amplitudes of pairs of bonded atoms should be approximately equal in the bond direction, even though they may be widely different in other directions. As Hirshfeld[8] has pointed out, this provides a necessary (although by no means sufficient) condition that observed thermal ellipsoids represent genuine vibrational ellipsoids. If the condition is seriously violated, the U_{ij} values may be suspected of contamination by charge-density deformation contributions or absorption or other systematic errors.

In the least-squares refinement, the U_{ij} parameters are usually treated as independent variables, any correlation that may exist between the U values of different atoms in the asymmetric unit being ignored (except in certain kinds of constrained refinement). However, if the crystal contains more-or-less rigid groupings of atoms (e.g., molecules), the translational and librational oscillations of these groupings introduce very strong correlations among the U_{ij} parameters of different atoms.

[8] F. L. Hirshfeld, *Acta Crystallogr.* **A32,** 239 (1976).

Indeed, if the grouping were truly rigid, the U_{ij} values would be completely determined by these translational and librational oscillations. Of course, molecules are not rigid, but the contributions from internal vibrations are generally much smaller than those from translational and librational motions. It is therefore reasonable to use the rigid-body model, at least as a first approximation, and attempt to fit the parameters that describe its translational and librational motions to the observed U_{ij} values.

For the subsequent discussion we assume that the U_{ij} values are expressed in a Cartesian coordinate system, in which case the scalar quantity

$$\begin{aligned} U^n(l_1,l_2,l_3) &= U^n_{11}l_1^2+U^n_{22}l_2^2+U^n_{33}l_3^2+2U^n_{12}l_1l_2+2U^n_{13}l_1l_3+2U^n_{23}l_2l_3 \\ &= U^n_{ij}l_il_j \end{aligned}$$

represents the mean-square vibration amplitude of atom n in the direction of the unit vector $\boldsymbol{l}$ with components l_1, l_2, l_3 along the Cartesian axes. For some purposes, it is advantageous to choose the Cartesian system with origin at the centroid of the molecule and with axes along the principal inertial axes of the molecule. We assume that the atomic positions $\mathbf{r}^n(r_1,r_2,r_3)$ and U^n_{ij} values are given in such a system. The mean-square amplitude for translational vibration of the molecule as a whole can be written $T_{ij}l_il_j$; the translational contribution is equal for all the atoms. The librational motion is more complicated.

The Cruickshank Model

In his original analysis of thermal motion in terms of rigid-body motions, Cruickshank[9] assumed that the librational axes intersect at the origin of the coordinate system (at the molecular centroid in the molecular coordinate system). This assumption is not necessarily correct, as was shown later,[10] unless the origin happens to coincide with a crystallographic center of inversion or $\bar{6}(D_{3h})$ symmetry. Nevertheless, we adopt it for the time being since it simplifies the analysis considerably. Later, we shall mention some complications that arise in the more general case of nonintersecting libration axes. Consider a small rotation of a point $\mathbf{r}$ about some central axis $\boldsymbol{\lambda}(\lambda_1, \lambda_2, \lambda_3)$. The length of the axial vector $|\lambda|$ can be identified with the rotation angle. The displacement $\delta\mathbf{r}$ is given by

$$\delta\mathbf{r} = (\boldsymbol{\lambda} \times \mathbf{r}) \tag{5.38}$$

or, written out,

[9] D. W. J. Cruickshank, *Acta Crystallogr.* **9,** 754 (1956).
[10] V. Schomaker and K. N. Trueblood, *Acta Crystallogr.* **B24,** 63 (1968).

$$\begin{aligned}\delta r_1 &= \lambda_2 r_3 - \lambda_3 r_2 \\ \delta r_2 &= \lambda_3 r_1 - \lambda_1 r_3 \\ \delta r_3 &= \lambda_1 r_2 - \lambda_2 r_1\end{aligned} \tag{5.38a}$$

so that

$$\delta \mathbf{r} = \mathbf{D}\mathbf{r} \tag{5.38b}$$

where the transformation matrix $\mathbf{D}$ is skew-symmetric about its main diagonal

$$\mathbf{D} = \begin{bmatrix} 0 & -\lambda_3 & \lambda_2 \\ \lambda_3 & 0 & -\lambda_1 \\ -\lambda_2 & \lambda_1 & 0 \end{bmatrix}$$

It is also useful to have these relationships in tensor notation:

$$\delta r_i = D_{ij} r_j \tag{5.38c}$$

$$= -\epsilon_{ijk} \lambda_k r_j \tag{5.38d}$$

where ϵ_{ijk} equals $+1$ for i, j, k, a cyclic permutation of the indices 1, 2, 3, or -1 for a noncyclic permutation, or zero otherwise. Note that the result of the rotation $\boldsymbol{\lambda}(\lambda_1, \lambda_2, \lambda_3)$ is the same as that obtained by rotations of $\lambda_1, \lambda_2, \lambda_3$ about the coordinate axes in any order. (This holds only for infinitesimal rotations.)

If $\boldsymbol{\lambda}$ is supposed to vary with time (or with lattice position) the mean-square displacements caused by libration are obtained by forming the respective products and taking the appropriate averages. With the assumption that the librational and translational motions are independent, the elements of U^n_{ij} can be written

$$\begin{aligned}
U^n_{11} &= T_{11} + L_{22} r_3^2 + L_{33} r_2^2 - 2L_{23} r_2 r_3 \\
U^n_{22} &= T_{22} + L_{33} r_1^2 + L_{11} r_3^2 - 2L_{13} r_1 r_3 \\
U^n_{33} &= T_{33} + L_{11} r_2^2 + L_{22} r_1^2 - 2L_{12} r_1 r_2 \\
U^n_{12} &= T_{12} - L_{33} r_1 r_2 - L_{12} r_3^2 + L_{13} r_2 r_3 + L_{23} r_1 r_3 \\
U^n_{13} &= T_{13} - L_{22} r_1 r_3 + L_{12} r_2 r_3 - L_{13} r_2^2 + L_{23} r_1 r_2 \\
U^n_{23} &= T_{23} - L_{11} r_2 r_3 + L_{12} r_1 r_3 - L_{13} r_1 r_2 - L_{23} r_1^2
\end{aligned} \tag{5.39}$$

where the coefficients $L_{ij} = \langle \lambda_i \lambda_j \rangle$ and T_{ij} form the elements of two tensors, known as the libration tensor $\mathbf{L}$ and the translation tensor $\mathbf{T}$ respectively.

The twelve coefficients T_{ij} and L_{ij}, the elements of the $\mathbf{T}$ and $\mathbf{L}$ tensors, are to be determined, and this can be done by a least-squares fit of the unknown quantities to the experimental U^n_{ij} values. There are thus $6N$ observational equations (for N crystallographically independent atoms in general positions) and 12 unknowns. If the standard deviations of the experimental U^n_{ij} values are known, as is usually the case, the standard deviations of the derived tensor elements can be determined in the usual way, by inversion of the normal-equations matrix (Chapter 4).

Once the **T** and **L** tensors have been found, they may be diagonalized to yield the magnitudes and directions of their principal axes. The eigenvectors of the **L** tensor would represent the three principal libration axes, and its eigenvalues would represent the mean-square librational amplitudes about these axes. We might hope to relate these to the inertial tensor of the molecule, or, after suitable transformation, back to the crystal coordinate system, to the intermolecular packing.

One of the first applications[11] of thermal motion analysis was to the anthracene crystal structure. Although the experimental data (original three-dimensional $|F_H|$ values of Mathieson, Sinclair, and Robertson,[12] based on photographic intensity measurements) were of poor quality by today's standards, Cruickshank was able to derive useful information about the molecular rigid-body vibrations. Within rather large experimental uncertainty, the principal axes of the **T** and **L** tensors were found to coincide with the principal inertial axes of the molecule; both the largest translational vibration ($U \approx 0.4$ Å^2) and the major librational axis ($L_I \approx 16$ degrees^2) are roughly parallel to the long axis of the molecule. The root-mean-square librational amplitudes of 3.9, 3.0, and 2.2° also agreed reasonably well with those derived from Raman spectra.

As an example of the good fit between observed and calculated U_{ij} values that can be obtained in favorable cases, we cite results from a recent crystal structure analysis of the centrosymmetric isomer of tetramethyltetraasterane:[13]

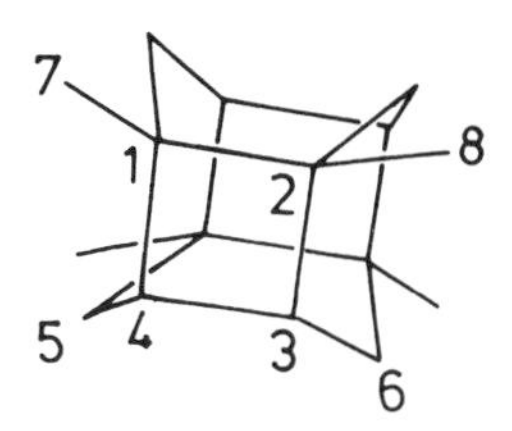

[11]D. W. J. Cruickshank, *Acta Crystallogr.* **9,** 915 (1956).

[12]A. Mc L. Mathieson, J. M. Robertson, and V. C. Sinclair, *Acta Crystallogr.* **3,** 245 (1950).

[13]J. P. Chesick, J. D. Dunitz, U. v. Gizycki, and H. Musso, *Chem. Ber.* **106,** 150 (1973).

	U_{11}	U_{22}	U_{33}	U_{12}	U_{13}	U_{23}
C1	363(12)	486(16)	489(14)	23(11)	9(10)	− 13(12)
	370	493	480	27	22	− 14
C2	377(12)	543(17)	424(14)	− 9(12)	− 6(10)	− 65(13)
	376	570	447	− 29	− 10	− 86
C3	430(13)	587(17)	371(14)	2(12)	32(10)	8(13)
	439	581	383	0	40	− 9
C4	457(14)	444(16)	518(17)	16(12)	53(12)	36(13)
	455	458	514	8	70	52
C5	418(13)	476(16)	619(18)	− 89(12)	95(12)	− 97(14)
	434	453	621	− 80	83	− 82
C6	389(12)	547(17)	478(15)	− 6(12)	101(10)	− 75(13)
	401	533	463	− 7	85	− 69
C7	496(16)	716(23)	723(22)	146(15)	2(14)	42(19)
	462	706	740	160	5	34
C8	483(16)	992(29)	622(20)	− 38(17)	−100(15)	−240(21)
	476	1000	598	− 35	−110	−230

For each carbon atom the upper row of numbers gives observed U_{ij} values ($Å^2 \times 10^4$) with their standard deviations $\sigma(U_{ij})$ in parentheses. The lower row of numbers has been calculated on the basis of the rigid-body model,[14] and a better level of agreement could hardly have been expected. The eigenvalues of the **T** tensors are 0.0427, 0.0380, and 0.0329 $Å^2$; those of the **L** tensor are 23.3, 11.8, and 8.1 degrees^2, but neither set of eigenvectors has any obvious relationship to the inertial axes of the molecule or to the crystal packing.

Effect of Libration on Apparent Intramolecular Distances

As first noted by Cruickshank,[15] rotational oscillations of molecules cause the apparent atomic positions to be slightly displaced from the true positions towards the rotation axes. If the root-mean-square amplitude of libration about the axis in question is ω, the radial displacement is given approximately by $d\omega^2/2$, where d is the distance of the atom from the axis (Fig. 5.4). Errors arising from librational motions about orthogonal axes are additive.

More generally, X-ray analysis locates the centroids of atomic distributions that are undergoing vibrations, and separations computed from

[14]In assessing the goodness of fit in problems of this kind, we have to keep in mind that a poor statistical fit can nevertheless correspond to a reasonably good fit of the physical model. The rigid-body model is, after all, at best only a reasonable approximation. In the above example, the residual $R = \Sigma|\Delta U_{ij}|/\Sigma|U_{ij}|$ is 0.039, the root-mean-square value of ΔU_{ij} is 14×10^{-4} $Å^2$, and that of $\sigma(U_{ij})$ is 15×10^{-4} $Å^2$. The value of $\chi^2 = \Sigma(\Delta/\sigma)^2$ is 43 for 36 degrees of freedom. The χ^2 = test (p. 263) suggests that there is about a one in five chance that the discrepancies ΔU_{ij} could arise by mere chance. In this case there would be little point in trying to improve the model.

[15]D. W. J. Cruickshank, *Acta Crystallogr.* **9**, 757 (1956).

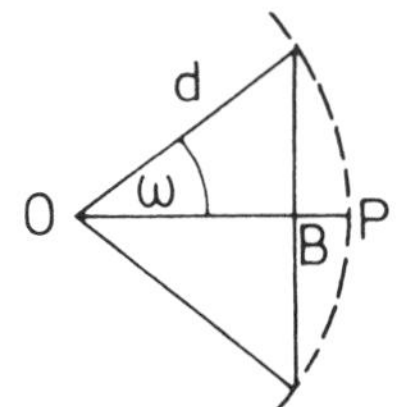

Figure 5.4. Displacement PB $= d - d\cos\omega \approx dw^2/2$.

these positions cannot be interpreted directly as interatomic distances.[16] To determine these distances, the correlation between the atomic motions would have to be known completely, and this information is seldom available. In the rigid-body model, molecular libration corresponds to a correlated motion that makes the molecule appear to shrink slightly.

Once the magnitudes and directions of the principal axes of the **L** tensor have been obtained, it is easy to calculate the approximate corrections and apply them to the observed atomic positions. The main effect is to dilate the uncorrected molecule, leading to slightly larger bond distances. For the tetramethyltetraasterane example discussed above, the corrections to C–C bond distances amount to 0.005–0.008 Å, but sometimes the corrections may be appreciably larger. For example, a root-mean-square librational amplitude of 10° (L = 100 degrees^2 = 0.03 radian^2) could lead to an apparent contraction of up to 0.025 Å for a typical bond distances of 1.5 Å. Note that the correct atomic positions may be used only for estimating *intra*molecular distances and angles since it is only for atoms within the rigid body that the motions can be regarded as being correlated. The rigid-body model provides no information about correlations between atomic motions in different molecules.

General Treatment of Rigid-Body Vibrations

One of the main assumptions of the Cruickshank model is that the three principal libration axes intersect at the molecular center. This assumption cannot be expected to hold unless the libration center is fixed by symmetry considerations, as was recognized by several subsequent authors who assumed that the libration axes intersect somewhere and sought to locate the intersection point by one means or another. It remained for Schomaker and Trueblood[10] to show that the search for a common intersection point was illusory because unless the existence of such a point was a consequence of crystal symmetry, there was no reason to assume that the libration axes intersected at all. The

[16] W. R. Busing and H. A. Levy, *Acta Crystallogr.* **17**, 142 (1964).

general treatment of rigid-body motion, as developed by Schomaker and Trueblood, is complicated, but it is worth describing its main features. For further details, consult the original paper[10] or the account by Johnson.[17]

The most general motion of a rigid body is a rotation about an axis, coupled with a translation parallel to this axis—a screw rotation. Suppose that the screw axis is correctly oriented but incorrectly positioned. The error in position introduces an additional translation component perpendicular to the rotation axis (Fig. 5.5). The rotation angle and the parallel component of the translation are invariant to the position of the axis, but the perpendicular component is not. This implies that the **L** tensor is unaffected by any assumptions about the positions of the libration axes, whereas the **T** tensor will be different, depending on what assumptions are made. Since **L** is independent of the origin and **T** varies with it, the fit of observed and calculated U_{ij} values will depend on the choice of origin in the two-tensor analysis.

The way around this difficulty is to introduce an additional tensor (called **S**) to account for the correlation between libration and translation. When this tensor is included in the analysis, the least-squares fit of calculated and observed U_{ij} values is independent of the origin chosen although the components of **S** and **T** vary with it. There is thus, in general, an unavoidable element of arbitrariness in the description of the rigid-body motion leading to an observed set of mean-square vibrational amplitudes. If the molecular center is situated at a crystallographic inversion center, however, **S** is identically zero, and the analysis reduces to Cruickshank's treatment.

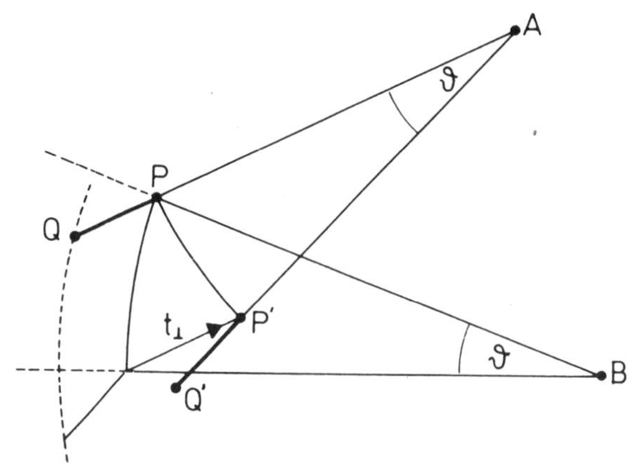

Figure 5.5. The motion PQ to P′Q′ consists of a rotation through θ about axis A coupled with a translation $\mathbf{t}_{\parallel}$ normal to the plane of the paper (parallel to A). If the rotation axis is correctly oriented but incorrectly positioned (B), an additional translation component $\mathbf{t}_{\perp}$ must be introduced to reproduce the same motion.

[17] C. K. Johnson, *Crystallographic Computing,* F. R. Ahmed, S. R. Hall, and C. P. Huber, Eds., Munksgaard, Copenhagen, 1970, pp. 209–219.

The quadratic correlation between librational and translational motions can be allowed for by rewriting Eq. 5.38c to include the translational displacement

$$\delta r_i = D_{ij} r_j + t_i \qquad (5.40)$$

before taking products and averaging.

The mean-square displacements are then

$$\begin{aligned} U_{ij} &= \langle D_{ik} D_{jl} \rangle r_k r_l + \langle D_{ik} t_j + D_{jk} t_i \rangle r_k + \langle t_i t_j \rangle \\ &= A_{ijkl} r_k r_l + B_{ijk} r_k + \langle t_i t_j \rangle \end{aligned}$$

It may be helpful to write out some of the expressions explicitly:

$$\begin{aligned} U_{11} &= \langle \delta r_1 \rangle^2 = \langle \lambda_3^2 \rangle r_2^2 + \langle \lambda_2^2 \rangle r_3^2 - 2\langle \lambda_2 \lambda_3 \rangle r_2 r_3 - 2\langle \lambda_3 t_1 \rangle r_2 + 2\langle \lambda_2 t_1 \rangle r_3 + \langle t_1^2 \rangle \\ U_{12} &= \langle \delta r_1 \delta r_2 \rangle = -\langle \lambda_3^2 \rangle r_1 r_2 + \langle \lambda_1 \lambda_3 \rangle r_2 r_3 + \langle \lambda_2 \lambda_3 \rangle r_1 r_3 - \langle \lambda_1 \lambda_2 \rangle r_3^2 \qquad (5.41) \\ &\quad + \langle \lambda_3 t_1 \rangle r_1 - \langle \lambda_1 t_1 \rangle r_3 - \langle \lambda_3 t_2 \rangle r_2 + \langle \lambda_2 t_2 \rangle r_3 + \langle t_1 t_2 \rangle \end{aligned}$$

Comparison with Eqs. 5.39 shows that products of the type $\langle t_i t_j \rangle$ and $\langle \lambda_i \lambda_j \rangle$ are just the elements of the previously defined **T** and **L** tensors, respectively. Since pairs of terms such as $\langle t_i t_j \rangle$ and $\langle t_j t_i \rangle$ correspond to averages over the same two scalar quantities, these tensors are symmetrical.

The new feature is the appearance of products of the type $\langle \lambda_i t_j \rangle$ that form the components of a new tensor **S**, which is unsymmetrical, since $\langle \lambda_i t_j \rangle$ is different from $\langle \lambda_j t_i \rangle$. Inspection of Eq. 5.41 shows that terms involving elements of **S** may be grouped as

$$\langle \lambda_3 t_1 \rangle r_1 - \langle \lambda_3 t_2 \rangle r_2 + (\langle \lambda_2 t_2 \rangle - \langle \lambda_1 t_1 \rangle) r_3$$

or

$$S_{31} r_1 - S_{32} r_2 + (S_{22} - S_{11}) r_3$$

The diagonal elements occur as differences, and this is typical. It means that a constant may be added to each of the diagonal terms without changing the observational equations. In other words, the trace of **S** is indeterminate. In terms of the **L**, **T**, and **S** tensors, the observational equations can be written:

$$U_{ij} = G_{ijkl} L_{kl} + H_{ijkl} S_{kl} + T_{ij} \qquad (5.42)$$

where the arrays G_{ijkl} and H_{ijkl}, involving the atomic coordinates $(x,y,z) = (r_1,r_2,r_3)$, are found by straightforward algebra (Table 5.1).

If the origin is placed at a center of symmetry, then for each atom at **r** with vibration tensor $\mathbf{U}^n$ thre is an equivalent atom at $-\mathbf{r}$ with the same

Table 5.1. The arrays G_{ijkl} and H_{ijkl} to be used in observational equations:
$U_{ij} = G_{ijkl}L_{kl} + H_{ijkl}S_{kl} + T_{ij}$

ij \ kl	11	22	33	23	31	12	
11	0	z^2	y^2	$-2yz$	0	0	
22	z^2	0	x^2	0	$-2xz$	0	
33	y^2	x^2	0	0	0	$-2xy$	G_{ijkl}
23	$-yz$	0	0	$-x^2$	xy	xz	
31	0	$-xz$	0	xy	$-y^2$	yz	
12	0	0	$-xy$	xz	yz	$-z^2$	

ij \ kl	11	22	33	23	31	12	32	13	21	
11	0	0	0	0	$-2y$	0	0	0	$2z$	
22	0	0	0	0	0	$-2z$	$2x$	0	0	
33	0	0	0	$-2x$	0	0	0	$2y$	0	
23	0	$-x$	x	0	0	y	0	$-z$	0	H_{ijkl}
31	y	0	$-y$	z	0	0	0	0	$-x$	
12	$-z$	z	0	0	x	0	$-y$	0	0	

vibration tensor. When the observational equations for these two atoms are added, the terms involving elements of **S** disappear since they are linear in the components of **r**. The other terms, involving elements of the **T** and **L** tensors, are simply doubled, like the $\mathbf{U}^n$ components. The analysis then reduces to the Cruickshank treatment where **S** does not appear.

In the general case the elements of the **T**, **L**, and **S** tensors can be found by a least-squares fit to the observed U^n_{ij} values for any arbitrary origin. One of the diagonal elements of **S** must be fixed in advance or some other suitable constraint applied because of the indeterminancy of Tr(**S**). Usually, Tr(**S**) is set equal to zero. There are thus eight elements of **S** to be determined, as well as the six each of **L** and **T**, making 20 in all. A shift of origin leaves **L** invariant, but it intermixes **T** and **S**.

The physical meaning of the **T** and **L** tensors is the same as in Cruickshank's treatment—i.e., $T_{ij}l_il_j$ is the mean-square amplitude of translational vibration in the direction of the unit vector $\boldsymbol{l}$ with components l_1, l_2, l_3 along the Cartesian axes, and $L_{ij}l_il_j$ is the mean-square amplitude of libration about an axis in this direction. The quantity $S_{ij}l_il_j$ represents the mean correlation between libration about the axis $\boldsymbol{l}$ and translation parallel to this axis. This quantity, like $T_{ij}l_il_j$, depends on the choice of origin although the sum of the two quantities is independent of the origin.

Some insight into this correlation problem can be obtained if the unsymmetrical tensor **S** is regarded as being the sum of a symmetric tensor with elements $S^S_{ij} = (S_{ij} + S_{ji})/2$ and a skew-symmetric tensor with elements $S^A_{ij} = (S_{ij} - S_{ji})/2$. The symmetric part $\mathbf{S}^S$ transforms like an ordinary quadratic form under rotation of the coordinate system. The indeterminacy of tr(**S**) = tr($\mathbf{S}^S$) would actually produce a whole family of quadratic forms, all with a common set of principal axis directions. Setting tr(**S**) = 0 corresponds to an arbitrary selection of one of these quadratic forms. Expressed in terms of principal axes, $\mathbf{S}^S$ consists of three principal screw correlations $\langle \lambda_I t_I \rangle$. Positive and negative screw correlations would correspond to opposite senses of helicity, but since an arbitrary constant may be added to all three correlation terms, only the differences between them can be determined from the data.

The skew-symmetric part $\mathbf{S}^A$ is equivalent to a vector $(\boldsymbol{\lambda} \times \mathbf{t})/2$ with components $(\boldsymbol{\lambda} \times \mathbf{t})_i/2 = (\lambda_j t_k - \lambda_k t_j)/2$, involving correlations between a libration and a perpendicular translation. These components can be reduced to zero—i.e., **S** can be made symmetric—by a change of origin. It can be shown that the origin shift that symmetrizes **S** also minimizes the trace of **T**. The required origin shift ρ is expressed most simply in a coordinate system based on the principal axes of **L**. In terms of this system (denoted by carets)

$$\hat{\rho}_1 = \frac{\hat{S}_{23} - \hat{S}_{32}}{\hat{L}_{22} + \hat{L}_{33}} \qquad \hat{\rho}_2 = \frac{\hat{S}_{31} - \hat{S}_{13}}{\hat{L}_{11} + \hat{L}_{33}} \qquad \hat{\rho}_3 = \frac{\hat{S}_{12} - \hat{S}_{21}}{\hat{L}_{11} + \hat{L}_{22}}$$

The description of the averaged motion can be simplified further by shifting to three generally nonintersecting libration axes (one for each of the principal axes of **L**). It can be shown[10,16] that a shift of the $\mathbf{L}_1$ axis (eigenvector of **L**) by

$$^1\hat{\rho}_2 = -\hat{S}_{13}/\hat{L}_{11} \qquad ^1\hat{\rho}_3 = \hat{S}_{12}/\hat{L}_{11}$$

annihilates the S_{12} and S_{13} terms of the symmetrized **S** tensor and simultaneously effects a further reduction in Tr(**T**) (the presuperscript in the above equations denotes that axis $\mathbf{L}_1$ is shifted, and the subscripts denote shift components parallel to the other axes). Analogous equations apply for displacements of the $\mathbf{L}_2$ and $\mathbf{L}_3$ axes. If all three axes are appropriately displaced, all the off-diagonal terms of **S** are eliminated. The remaining diagonal terms (referred to the principal axes of **L**) represent screw correlations along these axes and are independent of origin shifts.

The elements of the reduced **T** are:

$$^rT_{II} = \hat{T}_{II} - \sum_{K \neq I} (\hat{S}_{KI})^2/\hat{L}_{KK}$$

$$^rT_{IJ} = \hat{T}_{IJ} - \sum_{K} \hat{S}_{KI}\hat{S}_{KJ}/\hat{L}_{KK} \qquad J \neq I$$

These origin shifts lead to a description of the average rigid-body motion in terms of six independently distributed instantaneous motions—three screw librations about nonintersecting axes (with screw pitches given by $\hat{S}_{11}/\hat{L}_{11}$, etc.) and three translations. All 21 parameters are involved; three libration and three translation amplitudes, six angles of orientation for the principal axes of **L** and **T**, six coordinates of axis displacement, and three screw pitches, one of which has to be chosen arbitrarily.

Since diagonal elements of **S** enter into the expression for $^rT_{IJ}$, the indeterminacy of Tr(**S**) introduces a corresponding indeterminacy in $^r\mathbf{T}$. The reason for setting Tr(**S**) = 0 is that this constraint is unaffected by the various rotations and translations of the coordinate systems used in the course of the analysis. Nevertheless, the element of arbitrariness in **S** and $^r\mathbf{T}$, and hence in the description of the overall motion, remains.

Example: 11,13-Dioxo-12-methyl-12-aza[4.4.3]propella-3,8-diene.[18]

The molecule (Fig. 5.6), which occupies a general position, is significantly distorted from the C_{2v} symmetry implied by its structural formula. The thermal motion analysis of the observed U_{ij} values (Table 5.2) leads to the following results (inertial frame):

X along long axis of the molecule
Y normal to mean plane of the two six-membered rings
Z along short axis of the molecule

L	0.00386	−0.00076	−0.00032	radian²
		0.00701	0.00131	
			0.00353	
T	0.03574	0.00324	−0.00281	Å²
		0.02930	−0.00207	
			0.03031	
S	−0.00058	0.00090	−0.00127	radian Å
	−0.00169	−0.00035	0.00442	
	−0.00080	−0.00157	0.00092	

[18]M. Kaftory and J. D. Dunitz, *Acta Crystallogr.* **B32**, 617 (1976).

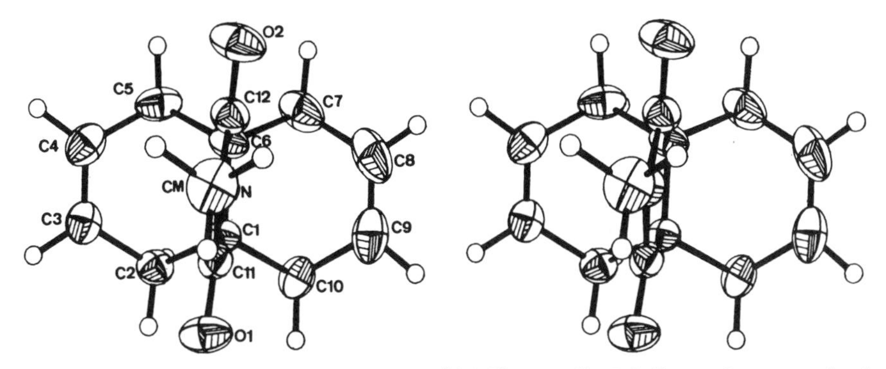

Figure 5.6. 11,13-Dioxo-12-methyl-12-aza[4,4,3] propella-3,8-diene. Stereoscopic view of molecule, showing vibration ellipsoids at the 50% probability level. From Kaftory and Dunitz, *Acta Crystallogr.* **B32,** 617 (1976).

Eigenvalues and eigenvectors of **L** and **T**:

*L*1	0.007629 radian²	0.2141	0.9712	0.1047
*L*2	0.003686	−0.9254	0.2360	−0.2965
*L*3	0.003085	−0.3127	−0.0339	0.9493
*T*1	0.038507 Å²	0.8401	−0.5096	−0.1861
*T*2	0.029335	0.3821	0.3124	0.8697
*T*3	0.027517	−0.3850	−0.8017	0.4571

The main libration motion (*L*1) is roughly along the *Y*-axis. The results that follow are expressed in a coordinate system based on the *L*1-, *L*2- and *L*3-axes (*L*-frame).

Shift of origin required to symmetrize **S**:

$$\rho_1 = 0.006 \text{ Å}, \qquad \rho_2 = 0.546 \text{ Å}, \qquad \rho_3 = 0.261 \text{ Å}$$

The largest shifts are along *L*2 and *L*3 (nearly parallel to *X* and *Z*, respectively)

Old **S** (rad. Å)

0.000887	0.002036	−0.004539
−0.000913	−0.000700	−0.000557
0.001308	−0.000600	−0.000187

New **S** (rad. Å)

0.000887	0.000048	−0.000375
	−0.000700	−0.000580
		−0.000187

Old **T** (Å²)

0.027588	−0.000433	0.000472
	0.037079	−0.003297
		0.030692

New **T** (Å²)

0.026853	−0.000522	0.000913
	0.036529	−0.002085
		0.028017

Table 5.2. Comparison of observed (upper rows) and calculated (lower rows) U_{ij} values for example discussed in text[a]

$\times 10^3$ (Å^2)	U_{11}	U_{22}	U_{33}	U_{12}	U_{13}	U_{23}
C(1)	32	37	28	−2	−2	2
	32	37	29	−3	−2	4
C(2)	48	46	30	−1	4	1
	42	49	29	−2	3	1
C(3)	53	44	40	6	−1	−6
	55	48	40	10	2	−5
C(4)	68	38	43	1	−3	−2
	72	35	48	1	−4	0
C(5)	65	46	48	−21	6	1
	59	44	46	−21	−2	5
C(6)	32	47	33	−9	0	1
	32	48	33	−8	1	3
C(7)	33	86	50	−3	−1	5
	30	92	52	−6	1	2
C(8)	43	104	57	27	−2	0
	42	99	61	25	−3	−4
C(9)	64	61	64	24	−16	0
	61	66	59	26	−13	1
C(10)	46	49	41	4	−10	7
	51	48	40	4	−10	10
C(11)	34	33	37	−1	−2	3
	36	36	35	−6	−4	3
C(12)	37	48	33	2	3	4
	39	53	29	−2	4	3
N	37	41	28	−2	−5	−2
	40	42	31	0	−4	−2
CM	62	60	37	−2	−2	−9
	62	61	39	−1	−1	−9
O(1)	48	61	53	−23	−5	11
	48	50	52	−21	−6	9
O(2)	57	92	34	−9	10	11
	56	85	34	−5	12	8

$\langle(\Delta U_{ij})^2\rangle^{\frac{1}{2}} \approx 3 \times 10^{-3}$ Å^2

$R = \sum|\Delta U_{ij}|/\sum|U_{ij}| = 0.085$

[a]Crystal coordinate system. Standard deviations are about 1–2 $\times 10^{-3}$ Å^2.

Displacements (Å) of libration axes from origin of inertial frame and screw pitches along these axes:

$L1$	0	0.595	0.267	0.116 Å/radian
$L2$	−0.151	0	0.248	−0.190
$L3$	0.195	0.424	0	−0.061

The **T** tensor is not too far from isotropic, but the **L** tensor is, with the largest libration amplitude (0.0076 radian^2 = 25.0 degrees^2) about an axis that is roughly parallel to the Y-axis of the inertial frame. This corresponds to the qualitative impression gained from the shapes of the vibration ellipsoids shown in Fig. 5.6.

The agreement between observed and calculated U_{ij} values (Table 5.2, $R = 0.085$) is not as good as in the tetramethyltetraasterane example cited earlier, but then the rigid-body model cannot be expected to hold as well as for that molecule.

Nonrigid-Body Thermal Motion Analysis

Although the analysis of molecular thermal motion in terms of a rigid-body model often produces fair to excellent agreement between observed and calculated U_{ij} values ($\langle(\Delta U_{ij})^2\rangle^{\frac{1}{2}} \approx 1-4\times10^{-3}\text{Å}^2$), there are many examples where the agreement is much worse ($\langle(\Delta U_{ij})^2\rangle^{\frac{1}{2}} \approx 10^{-2}\text{Å}^2$ or more). In some of these cases, the poor agreement can plausibly be ascribed to internal molecular motions that invalidate the rigid-body assumption. The presence of such internal motions can sometimes be inferred, or at least suspected, from visual inspection of a stereoview of the molecule, showing the shapes of the thermal ellipsoids. In Fig. 5.7 for example, which shows the molecule of a dinitro compound (1,5-dinitro-3-methyl-3-azabicyclo[3.3.1]nonane-7-one)[19] one gets the distinct impression that one of the nitro groups ($N_1O_1O_2$) is carrying out a torsional vibration about its N–C bond, independent of the motion of the rest of the molecule.

The general treatment of internal motions, given only the vibration tensors, which are sums of mean-square displacements, is impracticable for most molecules. For some molecules, however, such as the example mentioned, we may be able to postulate the existence of specific kinds of motion and estimate their magnitudes by including appropriate parameters in the least-squares equations.[20,21]

Assume, for example, that atom C librates about an axis passing through the bond AB (Fig. 5.8). Its motion is along the unit vector $\mathbf{n} = \mathbf{m} \times \mathbf{a}/|\mathbf{m} \times \mathbf{a}|$, and the magnitude of the displacement is the product of the libration angle Ω and the perpendicular distance $R = |\mathbf{m} \times \mathbf{a}|/|\mathbf{a}|$ of C from the libration axis. We can now describe the motion of C in terms of the usual rigid-body contributions plus a nonrigid-body contribution and rewrite the observational equations (Eqs. 5.42) involving atom C as

[19] M. Kaftory and J. D. Dunitz, *Acta Crystallogr.* **B32,** 1 (1976).
[20] J. D. Dunitz and D. N. J. White, *Acta Crystallogr.* **A29,** 93 (1973).
[21] C. K. Johnson, Program and Abstracts, American Crystallographic Association, Summer Meeting, Minneapolis, 1967, p. 82.

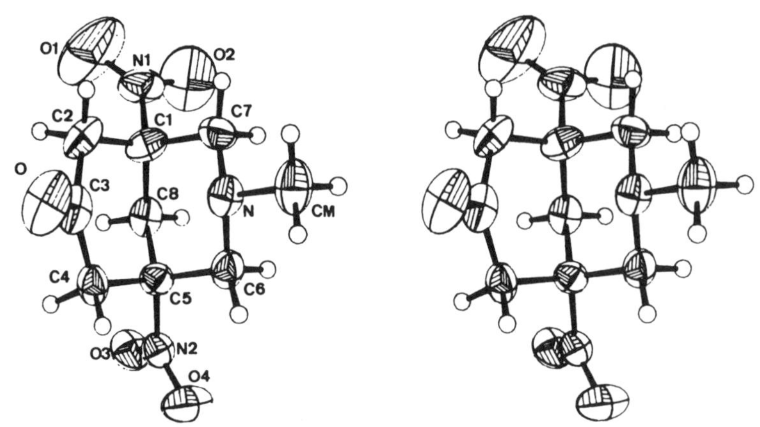

Figure 5.7. 1,5-Dinitro-3-methyl-3-azabicyclo[3,3,1] nonan-7-one: stereoscopic view of molecule, showing vibration ellipsoids at the 50% probability level. From Kaftory and Dunitz, *Acta Crystallogr.* **B32**, 1 (1976).

$$U_{ij} = G_{ijkl}L_{kl} + H_{ijkl}S_{kl} + T_{ij} + \Omega^2R^2n_in_j \tag{5.43}$$

More generally, several intramolecular libration axes $\mathbf{a}_j$ can be postulated, each acting on one or more atoms in the molecule. The last term in Eq. 5.43 is then replaced by the appropriate sum over the various axes $\mathbf{a}_j$ that contribute to the motion of the atom in question. This treatment assumes that the resultant mean-square amplitude can be approximated by compounding individual librational motions about prescribed axes. The axes have to be chosen by chemical intuition or guesswork; visual inspection of the thermal ellipsoid pattern can sometimes be helpful, as in the dinitro compound mentioned (Fig. 5.7).

For the dinitro compound,[19] the usual rigid-body analysis gave very poor agreement between observed and calculated U_{ij} values ($\langle(\Delta U_{ij})^2\rangle^{\frac{1}{2}} =$ 0.019 Å^2), even allowing for the relatively low quality of the observed

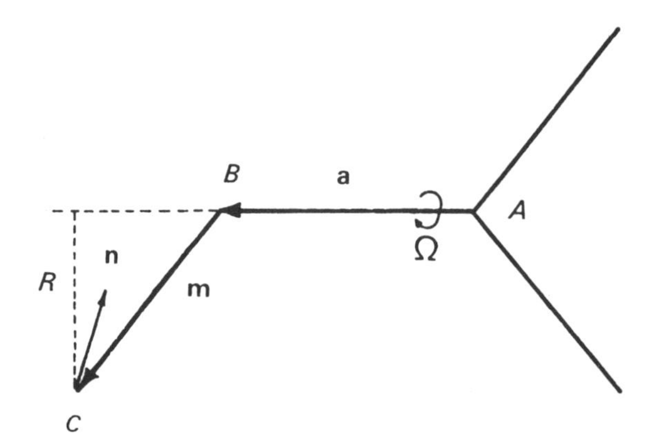

Figure 5.8. Atom C is supposed to librate about the bond AB. From Dunitz and White, *Acta Crystallogr.* **A29**, 94 (1973).

values ($\langle\sigma^2(U_{ij})\rangle^{1/2} \approx 0.008$ Å^2). A model in which both nitro groups were allowed to undergo torsional librations about their respective C–N bonds led to a much improved fit ($\langle(\Delta U_{ij})^2\rangle^{1/2} = 0.009$ Å^2). The root-mean-square amplitudes of torsional libration Ω_1 and Ω_2 were 25.2° and 8.7° for the two nitro groups, with estimated standard deviations of 0.7° and 2.1°, respectively.

In the crystal structure analysis of lithium purpurate dihydrate,[22] the usual rigid-body treatment of the complexing anion

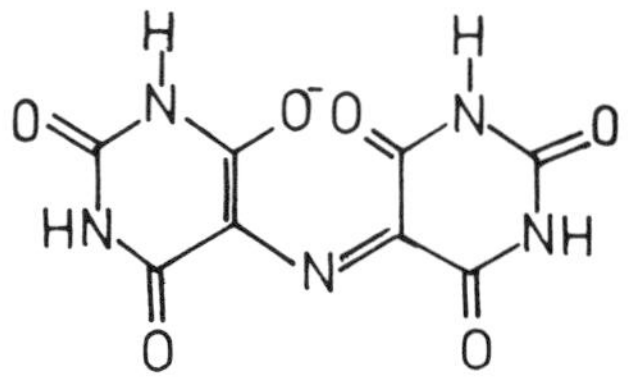

gave $\langle(\Delta U_{ij})^2\rangle^{1/2} = 0.0022$ Å^2, compared with $\langle\sigma^2(U_{ij})\rangle^{1/2} \approx 0.0015$ Å^2. Inclusion of a nonrigid-body model that incorporated the possibility of libration of both rings (with their substituents) about their respective central bonds gave a negligible reduction of the residual and led to values of Ω that were zero within one standard deviation. In an alternative model, the oxygen atoms were permitted to move perpendicular to the planes of their respective rings (such out-of-plane motions can be described as restricted librations about the adjacent ring bonds, the motions being imparted only to the oxygens). This model reduced $\langle(\Delta U_{ij})^2\rangle^{1/2}$ to 0.0017 Å^2 and yielded Ω values in the range 3°–8°, significantly different from zero. It was thus concluded that the internal motions of largest amplitude involve out-of-plane vibrations of the oxygen atoms rather than torsional vibrations about the central C–N bonds.

Several objections can be raised against this kind of procedure; the model is oversimplified, and the internal motions introduced are, to some extent, arbitrary. Nevertheless, if applied with caution, the procedure seems capable of providing a semi-quantitative description of the more important internal motions of nonrigid molecules. [A recent study by K. N. Trueblood, *Acta Crystallogr.* **A34,** 950 (1978), shows that the simple model described above yields results not significantly different from those of more elaborate models for librational motions about bonds, even when the root-mean-square amplitudes of these librations are as great as 24°.]

[22]H. B. Bürgi, S. Djuric, M. Dobler, and J. D. Dunitz, *Helv. Chim. Acta* **55,** 1771 (1972).

Lattice-Dynamical Treatment of Rigid-Body Motion

The **T**, **L**, and **S** tensors represent averages over lattice vibrations and thus contain information about intermolecular interactions in the crystal, albeit in a hopelessly scrambled form. The best that can be done is to make certain assumptions about the nature of the intermolecular force field, to estimate the potential energy increases associated with individual modes of displacement from equilibrium, and to derive the corresponding mean-square vibrational amplitudes for comparison with the observed tensor components.[23,24] This procedure at least allows a check to be made on whether or not the assumptions are compatible with the observations.

One model (analogous to the Einstein model for estimating specific heats) assumes that each molecule vibrates independently of all others; the corresponding potential is obtained by summing over all relevant intermolecular interactions, assuming one molecule to be displaced from its equilibrium position and all others to be fixed at theirs. Once the potential for a given kind of displacement has been calculated, the corresponding mean-square vibrational amplitude $\langle\phi^2\rangle$ can be estimated in various ways—it is approximately equal to $kT/(\partial^2 V/\partial\phi^2)_{\phi=0}$ for the displacement parameter ϕ. Whereas Shmueli and Goldberg find a reasonably satisfactory proportionality between observed and calculated librational amplitudes for several molecular crystals using this model,[25] Cerrini and Pawley find that the calculated potentials are, in general, too hard.[24] The latter result is qualitatively what one might have expected from a model in which correlations between the motions of neighboring molecules are neglected. The Einstein model has the advantages of simplicity and easy visualization, but it can hardly be expected to reproduce the observed tensors, even if it were based on a realistic force field.

The Einstein model tends to underestimate the observed tensor components. Cerrini and Pawley obtained somewhat better agreement for pyrene with the Born-van Kármán model, in which the potential is set up in terms of simultaneous displacements of all the molecules (if anything, the calculation tends to overestimate the observed tensor components). Other calculations on the same basis for anthracene give excellent agreement with observed Raman frequencies.[26] Here too, the calculated values of diagonal elements of the **U** tensors for the individual

[23]C. Scheringer, *Acta Crystallogr.* **A29**, 554 (1973).
[24]S. Cerrini and G. S. Pawley, *Acta Crystallogr.* **A29**, 660 (1973).
[25]U. Shmueli and I. Goldberg, *Acta Crystallogr.* **B29**, 2466 (1973).
[26]G. Filippini, C. M. Gramaccioli, M. Simonetta, and G. B. Suffritti, Proceedings of the First European Crystallographic Meeting, Bordeaux, 1973.

atoms are consistently larger than the experimental values. The authors raise the possibility that the calculated values are better than the experimental ones.

Accuracy of Derived Parameters

Published tables of positional and vibrational parameters are usually supplemented by a list of estimated standard deviations. For convenience and conciseness, these quantities are expressed in units of the least significant digit of the parameter value and are enclosed in parentheses—e.g., 2.1121(4) means that the parameter value is 2.1121 with estimated standard deviation, $\sigma = 0.0004$. The covariances between the individual parameters are usually not listed, even though they are often calculated toward the end of the analysis;[27] thus, we usually have to ignore any correlation that may be present between the parameters.

We are likely to be more interested in the accuracy of derived parameters (such as interatomic distances and angles) than in the atomic positions themselves. The complete expression for the variance of a derived parameter f, in terms of the variances and covariances of the directly determined parameters p_i, is

$$\sigma^2(f) = \sum_i \sum_j \left(\frac{\partial f}{\partial p_i}\right)\left(\frac{\partial f}{\partial p_j}\right) v(p_i,p_j) \tag{5.44}$$

where $v(p_i,p_j)$ is the appropriate element of the variance–covariance matrix. If the parameters p_i are regarded as being uncorrelated, this reduces to

$$\sigma^2(f) = \sum_i \left(\frac{\partial f}{\partial p_i}\right)^2 \sigma^2(p_i)$$

involving only the diagonal matrix elements, the squares of the standard deviations. For the general case of a triclinic crystal, this equation expands to a lengthy expression, even for the simplest kind of derived parameter—a distance between atoms i and j:

$$\begin{aligned}\sigma(d) = \frac{1}{d}\,[&(\Delta x + \Delta y \cos\gamma + \Delta z \cos\beta)^2\,(\sigma^2(x_i) + \sigma^2(x_j))\\ &+ (\Delta y + \Delta x \cos\gamma + \Delta z \cos\alpha)^2\,(\sigma^2(y_i) + \sigma^2(y_j))\\ &+ (\Delta z + \Delta x \cos\beta + \Delta y \cos\alpha)^2\,(\sigma^2(z_i) + \sigma^2(z_j))]^{\frac{1}{2}}\end{aligned}$$

[27]For understandable reasons, editors do not encourage publication of the complete error matrix, which would typically require several additional journal pages. It has to be admitted that such matrices are unlikely to make fascinating reading.

where $\Delta x = a(x_j - x_i)$, etc. This treatment assumes that the errors in unit cell dimensions can be neglected; if this is not the case, an even lengthier expression is required.[28] If the σ values in ax_i, by_i, and cz_i are not too different, we may be content to regard σ_i as being isotropic, in which case the expression for $\sigma(d)$ is greatly simplified:

$$\sigma(d) = (\sigma_i^2 + \sigma_j^2)^{\frac{1}{2}}$$

again assuming that the atomic positions are uncorrelated. If the atoms are symmetry-related, then any error in one position is completely correlated with the error in the other. For a pair of atoms related by a center of symmetry

$$\sigma(d) = 2\sigma_i$$

The expressions for the standard deviation of a bond angle[29] θ or a torsion angle[30,31] ω only become manageable in the isotropic error approximation.

$$\sigma_\theta^2(\mathrm{ABC}) = \frac{\sigma_\mathrm{A}^2}{d_\mathrm{AB}^2} + \frac{\sigma_\mathrm{B}^2 d_\mathrm{AC}^2}{d_\mathrm{AB}^2\, d_\mathrm{BC}^2} + \frac{\sigma_\mathrm{C}^2}{d_\mathrm{BC}^2}$$

$$\sigma_\omega^2(\mathrm{ABCD}) = \frac{\sigma_\mathrm{A}^2}{d_\mathrm{AB}^2 \sin^2(\mathrm{ABC})} + \frac{\sigma_\mathrm{D}^2}{d_\mathrm{CD}^2 \sin^2(\mathrm{BCD})} + \frac{\sigma_\mathrm{B}^2}{d_\mathrm{BC}^2}\Bigg\{ \cot^2(\mathrm{BCD})$$

$$+ \left(\frac{d_\mathrm{BC} - d_\mathrm{AB}\cos(\mathrm{ABC})}{d_\mathrm{AB}\sin(\mathrm{ABC})}\right)^2 - 2\cos\omega\cot(\mathrm{BCD}) \left(\frac{d_\mathrm{BC} - d_\mathrm{AB}\cos(\mathrm{ABC})}{d_\mathrm{AB}\sin(\mathrm{ABC})}\right)\Bigg\}$$

$$+ \frac{\sigma_\mathrm{C}^2}{d_\mathrm{BC}^2}\Bigg\{ \cot^2(\mathrm{ABC}) + \left(\frac{d_\mathrm{BC} - d_\mathrm{CD}\cos(\mathrm{BCD})}{d_\mathrm{CD}\sin(\mathrm{BCD})}\right)^2$$

$$-2\cos\omega\cot(\mathrm{ABC}) \left(\frac{d_\mathrm{BC} - d_\mathrm{CD}\cos(\mathrm{BCD})}{d_\mathrm{CD}\sin(\mathrm{BCD})}\right)\Bigg\}$$

If the bond distances and bond angles have the common values d and θ, respectively, and if a common σ_i^2 is assumed, these expressions simplify to

[28]For a detailed analysis of the accuracy of bond distances in oblique coordinate systems see D. H. Templeton, *Acta Crystallogr.* **12,** 771 (1959).

[29]D. W. J. Cruickshank and A. P. Robertson, *Acta Crystallogr.* **6,** 698 (1953).

[30]P. J. Huber in an appendix to E. Huber-Buser and J. D. Dunitz, *Helv. Chim. Acta* **44,** 2027 (1961).

[31]R. H. Stanford, Jr. and J. Waser, *Acta Crystallogr.* **A28,** 213 (1972).

$$\sigma^2(\theta) = \frac{2\sigma^2(2 - \cos\theta)}{d^2}$$

$$\sigma^2(\omega) = \frac{4\sigma^2}{d^2 \sin^2\theta}\,[1 - \cos\theta\,(1 - \cos\theta)\,(1 + \cos\omega)]$$

In transforming from one coordinate system to another, it may be desirable to transform the standard deviations of the initial positional and vibrational parameters. Since the transformation equations are linear in X_i and in U_{ij}, this can be done readily by the appropriate expansion of the general expression, Eq. 5.44.

To assess the significance of deviations among individual parameters, say bond distances, the standard statistical criteria may be used. For example, we may be interested in the question of whether or not it is justified to regard a sample of observed distances d_i as being drawn from a common population, normally distributed about the mean, $\bar{d}$. We should then evaluate the quantity

$$\chi^2 = \sum_i (\bar{d} - d_i)^2/\sigma_i^2$$

and look up tables of the χ^2 distribution.[32] There is roughly a 50% probability that a random sample would give $\chi^2 \geq N$, the size of the sample. For large N (say, greater than 20) the probability that random sampling would give $\chi^2 \geq 2N$ is virtually zero.

When more than one set of independent measurements of the same quantities is available, a convenient way of estimating the statistical significance of the discrepancies is to construct a normal probability plot.[33] Any two sets of measured or derived quantities (on the same scale) can be analyzed with respect to their errors by calculating the weighted differences between corresponding pairs

$$\delta m_i = \frac{p_i(1) - p_i(2)}{\{\sigma^2[p_i(1)] + \sigma^2[p_i(2)]\}^{\frac{1}{2}}}$$

The δm_i values are then rearranged in order of increasing magnitude and plotted against the values expected for a normal error distribution. If the p_i values contain only random error and the σ_i values have been correctly evaluated, the resulting plot will be scattered about a straight line of unit slope that passes through the origin.

Comparison of results obtained in two independent analyses (CSS and

[32]To be found in any modern book on statistics.
[33]S. C. Abrahams and E. T. Keve, *Acta Crystallogr.* **A27**, 157 (1971).

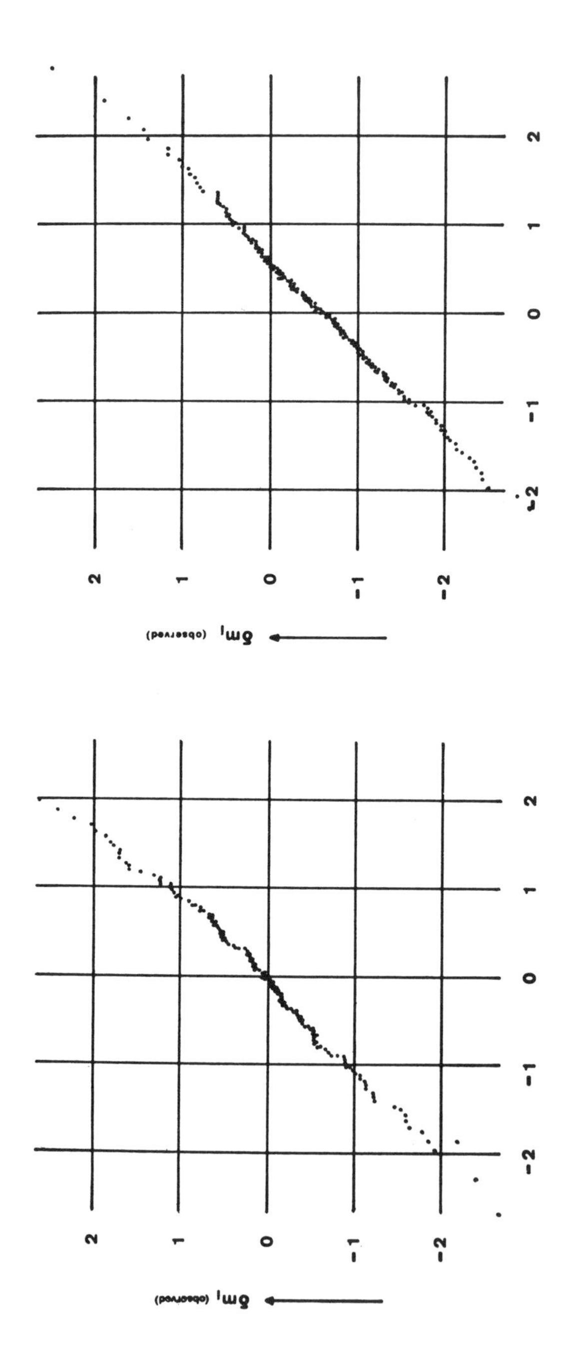

Figure 5.9. Normal probability plots (left) of δm_i from 146 positional coordinates, and (right) of δm_i from 179 vibrational parameters. From Kratky and Dunitz, *Acta Crystallogr.* **B31**, 1586 (1975).

KD) of the ethylchlorophyllide A dihydrate crystal structure[34,35] shows that these conditions are well fulfilled as far as the positional coordinates are concerned (Fig. 5.9, left). On the other hand, there is a systematic difference between the two sets of vibrational parameters since this plot (Fig. 5.9, right) does not pass through the origin. The KD vibrational parameters are systematically smaller, by about $\Delta U = 0.004$ Å^2 or about half a standard deviation, than the CSS parameters. This difference might result from various sources: different weighting schemes, neglect of absorption corrections in the KD analysis, overcompensation for absorption errors in the CSS analysis, or possibly from a genuine difference in temperature factors depending on choice of crystal specimen and experimental conditions. At any rate, it is interesting that such small differences can be detected easily by suitable comparison of the results.

[34] H. C. Chow, R. S. Serlin, and C. E. Strouse, *J. Am. Chem. Soc.* **97,** 7230 (1975).
[35] C. Kratky and J. D. Dunitz, *Acta Crystallogr.* **B31,** 1586 (1975).

6. Experimental Aspects of X-Ray Analysis

"Explain all that," said the Mock Turtle.
"No, no! The adventures first!" said the Gryphon in an impatient tone: "explanations take such a dreadful time."

Introductory Remarks

Experimental aspects of crystal structure analysis have been more or less ignored in the previous chapters. However, the diffraction pattern of a crystal is not presented to the investigator on a plate, instantly available on order, ready to solve. Progress in X-ray crystallography over the last quarter century has been due just as much to advances in instrumentation as to those in structure-solving techniques. After all, the results of an X-ray analysis can be no better than the experimental data on which they are based, although they can well be worse. In this chapter we describe the main methods used for recording and measuring diffraction patterns produced by crystals, and we also discuss various structure-independent factors that affect the raw intensities and have to be allowed for in converting these to structure amplitudes.

The geometric conditions that must be satisfied for producing a diffracted beam from a three-dimensionally periodic crystal are described in Chapter 1 and can be expressed in various equivalent ways—as Laue conditions, Bragg conditions, or, perhaps most conveniently, in terms of the Ewald construction (Fig. 1.6). In general, for an arbitrary orientation of a crystal to an incident beam and for a fixed wavelength, these conditions are not satisfied, and no diffraction occurs. However, if the mutual orientation of crystal and incident beam is varied, each reciprocal lattice point $\mathbf{H}(hkl)$ within a limiting radius of $2/\lambda$ can be made to sweep through the surface of the Ewald sphere (Fig. 6.1) and to produce a diffracted beam. The number of reciprocal lattice (R. L.) points that lie within the limiting sphere is the ratio of its volume to $V^* = 1/V$, where V is the volume of the primitive unit cell. Thus

$$N = V \cdot \frac{4\pi}{3} \left(\frac{2}{\lambda}\right)^3 = \frac{32\pi V}{3\lambda^3}$$

and V can be estimated as roughly 20 Å^3 for each nonhydrogen atom

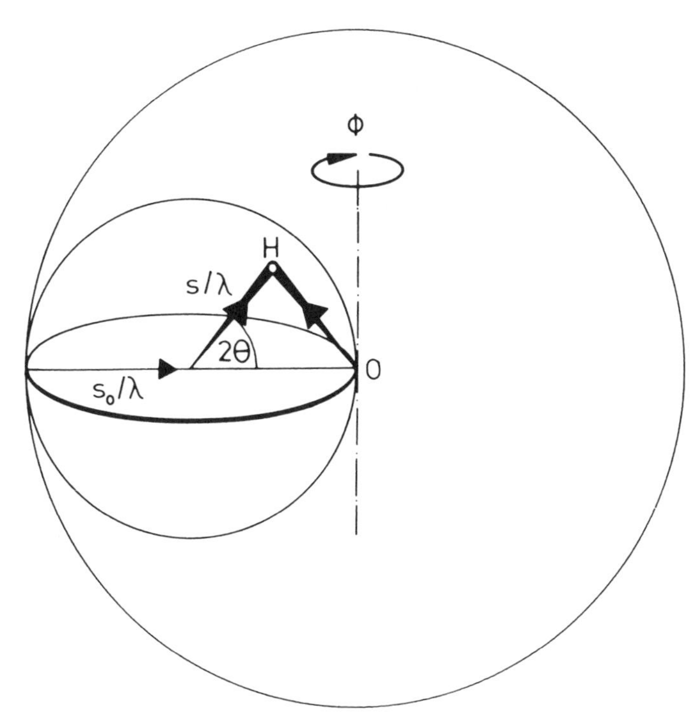

Figure 6.1. As the crystal is rotated about ϕ, each reciprocal lattice vector $\mathbf{H}(hkl)$ intersects the surface of the Ewald sphere, the smaller sphere of radius $1/\lambda$ centered at $-\mathbf{s}_0/\lambda$ from the R. L. origin. For each such intersection, the diffraction condition is satisfied and a diffracted beam is produced. All R. L. points within the limiting sphere, radius $2/\lambda$, centered at O can cut the Ewald sphere for some orientation of the crystal.

present. For a unit cell that contains 100 atoms, this gives about 18,000 R. L. points within the limiting sphere for Cu*K*α radiation ($\lambda = 1.542$ Å). If symmetry-independent reflections alone are to be considered, this number is to be divided by the order of the Laue symmetry group. In any case, for crystals of even moderate complexity, a large number of diffracted beams must be recorded.

It is not necessary to use a large crystal to record a diffraction diagram. Indeed, for accurate intensity measurements the crystal should be small enough (ca. 0.2–0.3 mm on edge) to be completely bathed in the incident beam and should preferably be roughly isotropic in linear dimension to reduce absorption errors (see later section on Absorption).

Photographic Methods

There are two main classes of methods for recording diffraction patterns of crystals and measuring the diffracted intensities—photographic methods and direct-counting diffractometer methods. For rapid, automatic, accurate intensity data collection, diffractometer methods offer many advantages, but it is only within the past 10 or 15 years that they have come into general use. Photographic methods, which were used almost universally before that time, are by no means obsolete, nor are they ever likely to be. A great deal of information can be recorded and stored in semi-permanent form on a single photographic film. Also, it is much easier to recognize significant features of the diffraction pattern—symmetry, slight deviations from symmetry, relative sharpness or diffuseness of spots, systematic variations in intensity—by inspection of a film than by scrutiny of information in digital form. The trained eye takes in the visual pattern as a whole, and the trained brain selects what is significant and filters out what is unimportant for the purpose in mind.

We review briefly some of the main methods for photographic recording of X-ray diffraction patterns of single crystals, for determining lattice constants, and for indexing the patterns—i.e., assigning reflection indices $(h_1h_2h_3)$ to the diffracted beams.

Rotation and Oscillation Photographs

The crystal is set to rotate about some prominent direction, say the **a** axis. The incident beam is normal to the rotation axis; and the diffraction pattern is usually recorded on a cylindrical film coaxial with this axis. In this arrangement, the reciprocal lattice points lie on a set of parallel planes normal to **a**; the equatorial plane contains the points $(0kl)$, and the others contain $(\pm 1kl)$, $(\pm 2kl)$, and so forth. As the R. L. rotates with the crystal, each such plane intersects the Ewald sphere in a circle, and hence the diffracted beams also intersect the cylindrical film in a set of circles that appear as straight lines (layer lines) when the film is flattened out (Fig. 6.2). The separation d_n between the equator and the nth layer line gives the periodicity along **a**, assuming that the crystal-to-film distance R and the wavelength λ are known:

$$\frac{d_n}{R} = \tan \alpha_n$$

$$\frac{n\lambda}{a_1} = \sin \alpha_n = \sin \tan^{-1}\left(\frac{d_n}{R}\right)$$

To each reflection, one index h can be assigned by inspection of the film, but it is not as easy to assign the other two indices because we do not know the instantaneous orientation of the R. L. as the individual reflections were produced. This information is lost by rotating the crystal

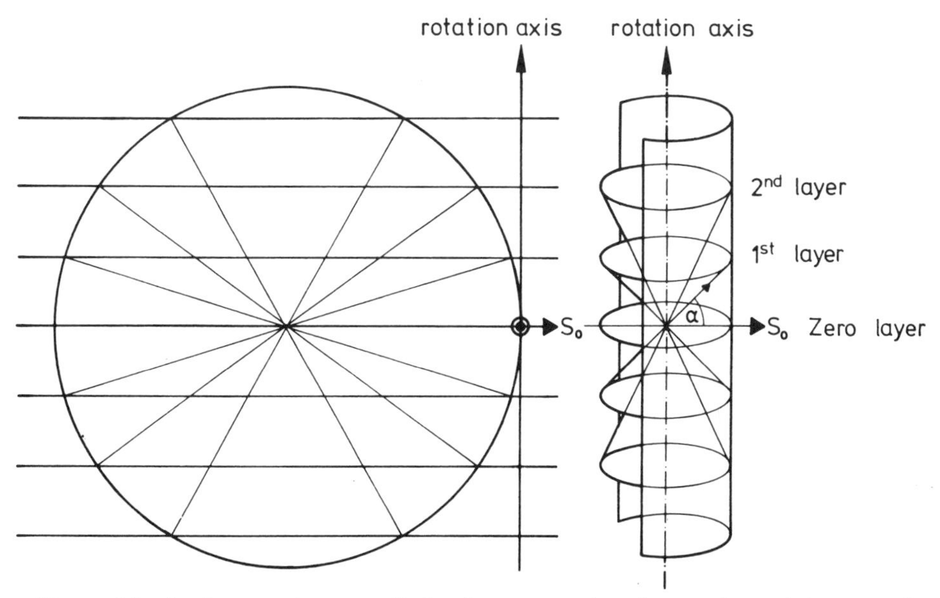

Figure 6.2. As the crystal rotates, R. L. planes normal to the rotation axis intersect the Ewald sphere in circles, recorded on a cylindrical film. These appear as straight lines when the film is flattened out.

during the exposure. Along each layer line we can measure only one quantity—the distance from the meridian—and this is usually insufficient to determine the two missing indices unequivocally. Besides, for crystals with large unit cells, the number of reflections in each layer line is so great that individual reflections are often superimposed.

For a given direction of the rotation axis, only a part of the R. L. can be observed, corresponding to those R. L. points that can intersect the Ewald sphere; these points lie within the torus obtained by rotating the sphere about the rotation axis (Fig. 6.2). Also, for a cylindrical film of finite height, some of the R. L. points theoretically attainable are lost.

More generally, if the rotation axis is any crystal direction

$$\mathbf{u} = u_1\mathbf{a}_1 + u_2\mathbf{a}_2 + u_3\mathbf{a}_3 \qquad u_1,\ u_2,\ u_3 \text{ integers}$$

the R. L. points $\mathbf{H}(h_1,h_2,h_3)$ that lie in planes normal to the rotation axis $\mathbf{u}$ are given by

$$\begin{aligned}\mathbf{H} \cdot \mathbf{u} &= (h_1\mathbf{b}_1+h_2\mathbf{b}_2+h_3\mathbf{b}_3) \cdot (u_1\mathbf{a}_1+u_2\mathbf{a}_2+u_3\mathbf{a}_3)\\ &= h_1u_1+h_2u_2+h_3u_3\\ &= n \qquad n = 0,\ \pm1,\ \pm2,\ldots\end{aligned}$$

For $\mathbf{u}(1,0,0)$, $h_1 = n$, as in the example mentioned. Details of indexing rotation photographs using the Bernal chart are given elsewhere.[1]

The indeterminacy of the rotation angle of the crystal during the exposure can be reduced by oscillating the crystal over a small angle rather than rotating it through the full 360° range. The full range is then covered in a series of exposures (oscillation photographs) with different angular limits. With this procedure, the density of spots on a given layer line is much less than in a rotation photograph. This means that the spots are usually resolved from one another, and hence their intensities can be estimated without much trouble.

Sometimes a flat film or plate rather than a cylindrical film is used to record the diffracted beams. Clearly, only the part of the reciprocal lattice near the origin can be recorded in this arrangement—a serious disadvantage unless the crystal has such a large temperature factor that its diffraction pattern is limited to the θ range observable with the flat plate.

Weissenberg Photographs

On a rotation photograph, particular nets of reciprocal lattice points are recorded as distinct layer lines. The Weissenberg camera (Fig. 6.3) involves a screen containing an annular slit that allows only one pre-selected layer line to pass at a time. A mechanism is also provided to move the film synchronously with the rotation of the crystal, so that the reflections of the selected layer line are spread over the entire film. This greatly simplifies the indexing problem. If the crystal is rotated about a main axis, one index, say h, depends on the particular layer line selected. Each reflection, of known h, is then described by two coordinates on the film: x, proportional to the glancing angle Υ (equals 2θ for the zero layer), and z, proportional to the angle ω through which the crystal has been rotated from some reference orientation (Fig. 6.3). The proportionality factors depend on the design of the instrument. Usually, the camera diameter is taken as 57.296 mm so that

$$\Upsilon\ (\text{degrees}) = 2\,x\ (\text{mm})$$

and the second instrumental constant is usually chosen to make

$$\omega\ (\text{degrees}) = 2\,z\ (\text{mm})$$

The picture of a reciprocal lattice net obtained on a Weissenberg

[1] M. J. Buerger, *X-ray Diffraction*, Wiley, New York, and Chapman and Hall, London, 1942.

circular slit allows only one layer line to pass
Rotation Axis
Metal screen
Crystal
Film
Movement of film coupled to rotation of crystal

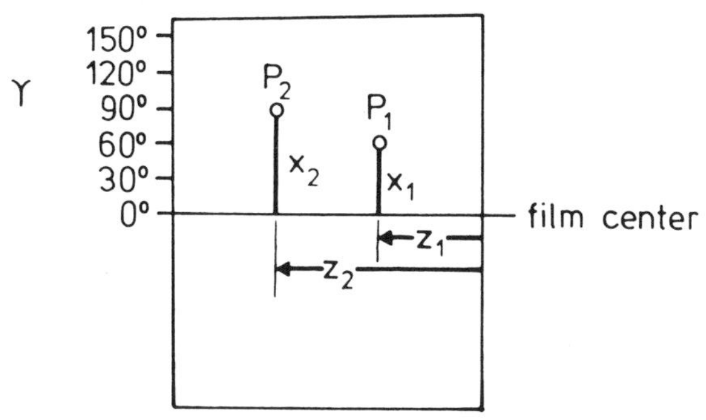

Figure 6.3. Top: schematic operation of a Weissenberg camera. Bottom: film coordinates x and z.

photograph is distorted (Fig. 6.4). On a zero-level photograph, R. L. rows that pass through the R. L. origin appear as straight lines. From the construction shown in Fig. 6.5, left

$$\theta = \Upsilon/2 = \omega - \omega_o$$

or

$$x = 2z - z_o$$

but since $|\mathbf{H}| = d^*$ is proportional to $\sin\theta$, and x is proportional to θ, equispaced points along the R. L. row are not equispaced along the line in the photograph. Noncentral R. L. rows appear as curves

$$\theta = \omega - \sin^{-1}(r^*/d^*) - \omega_o$$

where r^* is the perpendicular distance of the row from the origin (Fig. 6.5, right). Only when d^* is large compared with r^* do these curves become approximately linear. On upper-layer photographs, more complicated kinds of distortion occur.

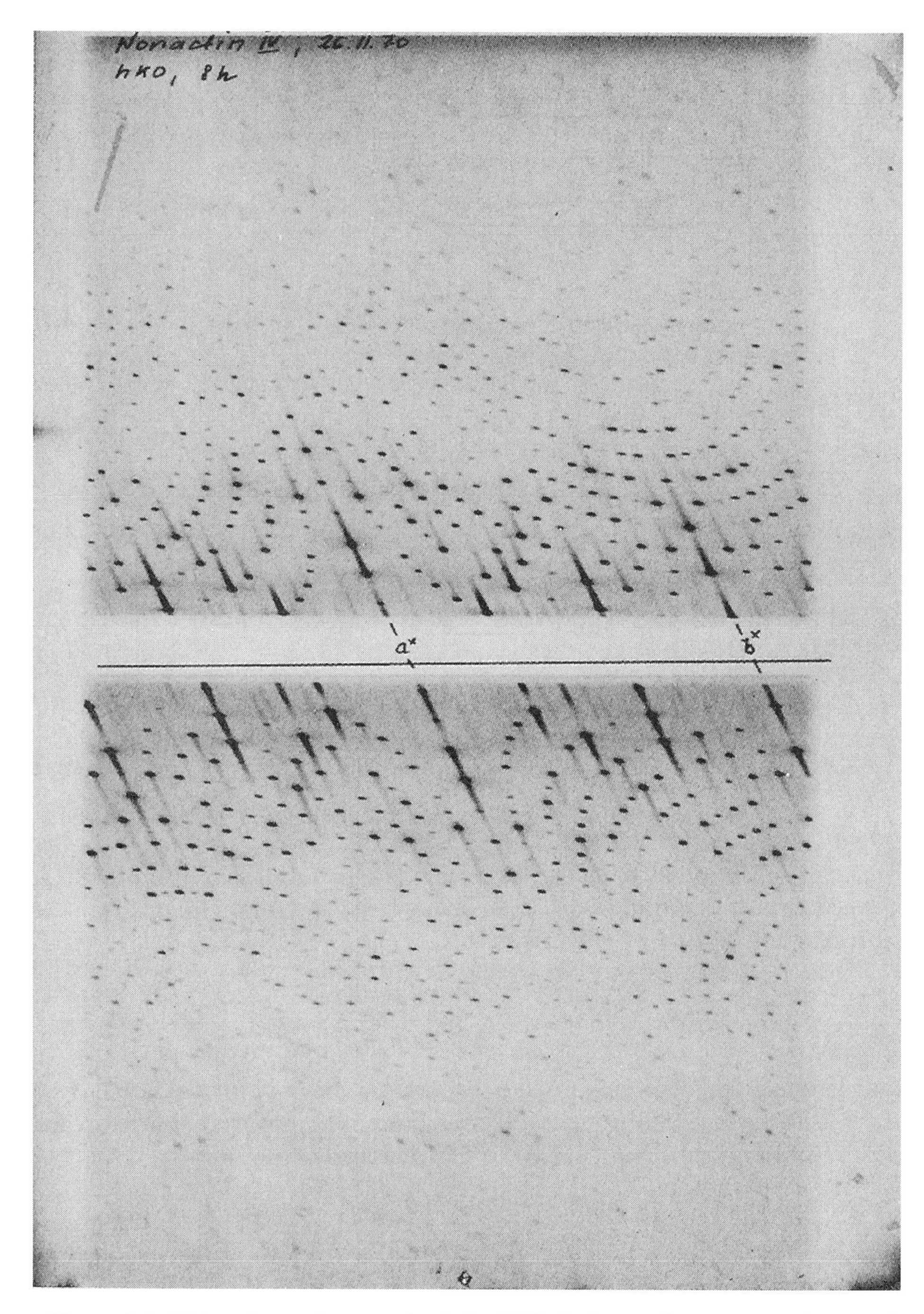

Figure 6.4. Weissenberg photograph of the (*hk*0) R. L. net from a nonactin crystal (Cu $K\alpha$ radiation). Courtesy of Dr. M. Dobler.

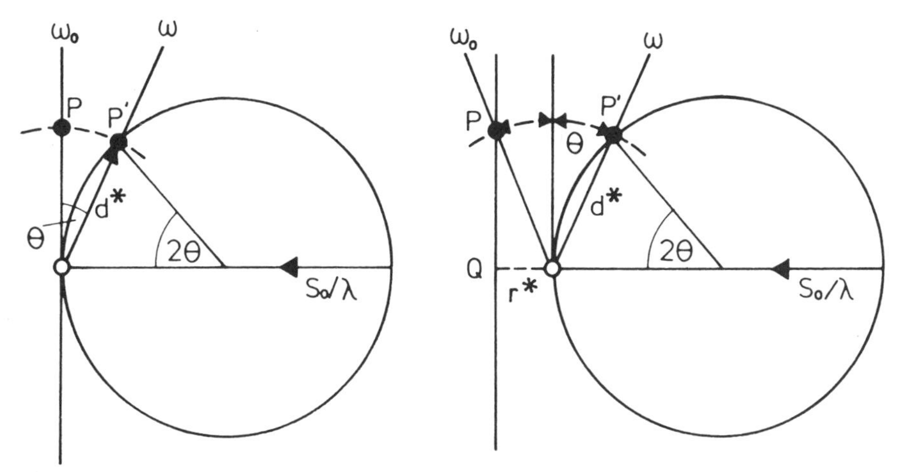

Figure 6.5. Left: for a central R. L. row, $\omega = \omega_0 + \theta$. Right: for a noncentral R. L. row, $\omega = \omega_0 + \theta + \sin^{-1}(r^*/d^*)$. ω_0 corresponds to an arbitrary reference orientation of the R. L.

From a knowledge of the two film-coordinates for the various diffraction spots, the reciprocal lattice net in question can be reconstructed and indices assigned to the spots without much difficulty. The indexing is conveniently accomplished with the aid of special charts. For details, see the excellent book by Buerger.[1]

The relative intensities of the spots can be estimated visually by comparison with an intensity strip, that is, a set of standard spots obtained by exposing a given reflection for different times. The accuracy attainable is about 10%. For many years this was the main method for measuring the intensities of X-ray reflections. For a crystal of any complexity it was quite tedious, involving the processing, indexing, and visual scrutiny of 100 or so separate films. It led to countless crystal structures and to countless headaches. Intensities of diffraction spots on photographic films can now be measured conveniently with microdensitometers.

Precession Photographs

Whereas the R. L. nets recorded on Weissenberg photographs are distorted, the precession camera yields a set of undistorted, scaled photographs of R. L. nets, that can be indexed easily by inspection. In this method the crystal is first oriented with a prominent axis parallel to the incident beam rather than perpendicular to it. A zero-level R. L. plane is then tangent to the Ewald sphere. If the crystal axis is now turned through an angle μ and caused to precess about the incident beam so as to maintain the angle μ with it, then all points on the zero-level R. L. plane within a distance $2 \sin \mu/\lambda$ pass through the Ewald sphere (Fig. 6.6). At

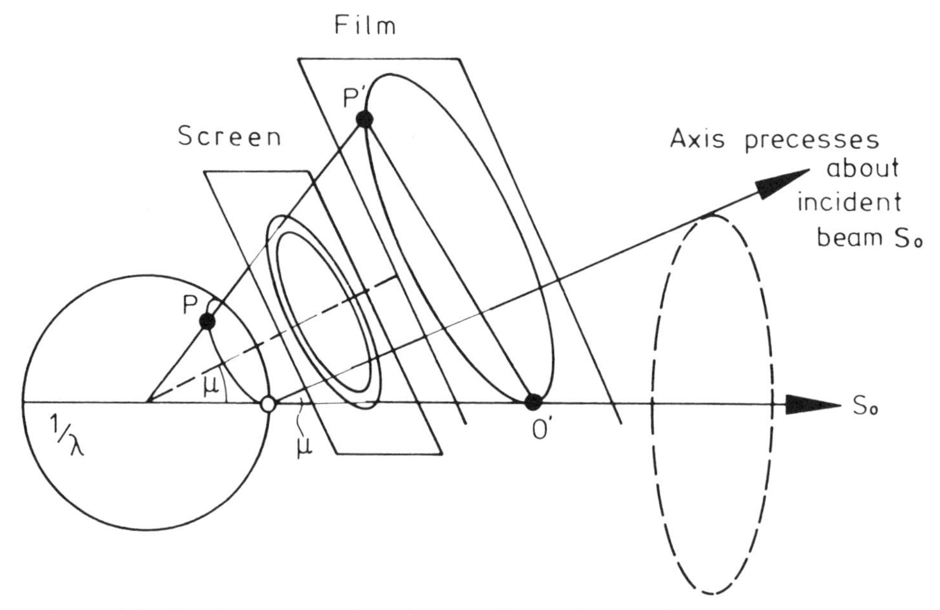

Figure 6.6. Zero-layer precession photograph. As the crystal axis precesses about the incident beam, all points on the zero layer of the R. L. within 2 sin θ/λ pass through the Ewald sphere. The resulting diffracted beams pass through the circular aperture in the screen and can be recorded on a film, whose normal is kept parallel to the crystal axis.

any given point during the precession the plane cuts the Ewald sphere in a circle, and R. L. points lying on this circle will produce reflections. As the crystal axis precesses, the circle rotates round the incident beam. In the precession camera, the film is pivoted about a point in the incident beam and allowed to move so that it is always normal to the crystal axis (i.e., parallel to the R. L. plane) during the precession motion. The reflecting circle on the Ewald sphere is then projected as a circle on the film. Distances O′P′ measured from the film center to any diffraction spot on the film are directly proportional to the length of the corresponding R. L. vector OP. From similar triangles

$$\frac{O'P'}{r} = 2 \sin \theta$$

where r is the crystal-to-film distance. A typical precession photograph is shown in Fig. 6.7. Its orthogonal symmetry portrays the symmetry of the corresponding R. L. net, whose dimensions can be derived easily from measurements on the photograph. For a zero-level photograph, a metal plate containing an annular aperture of appropriate radius has to be

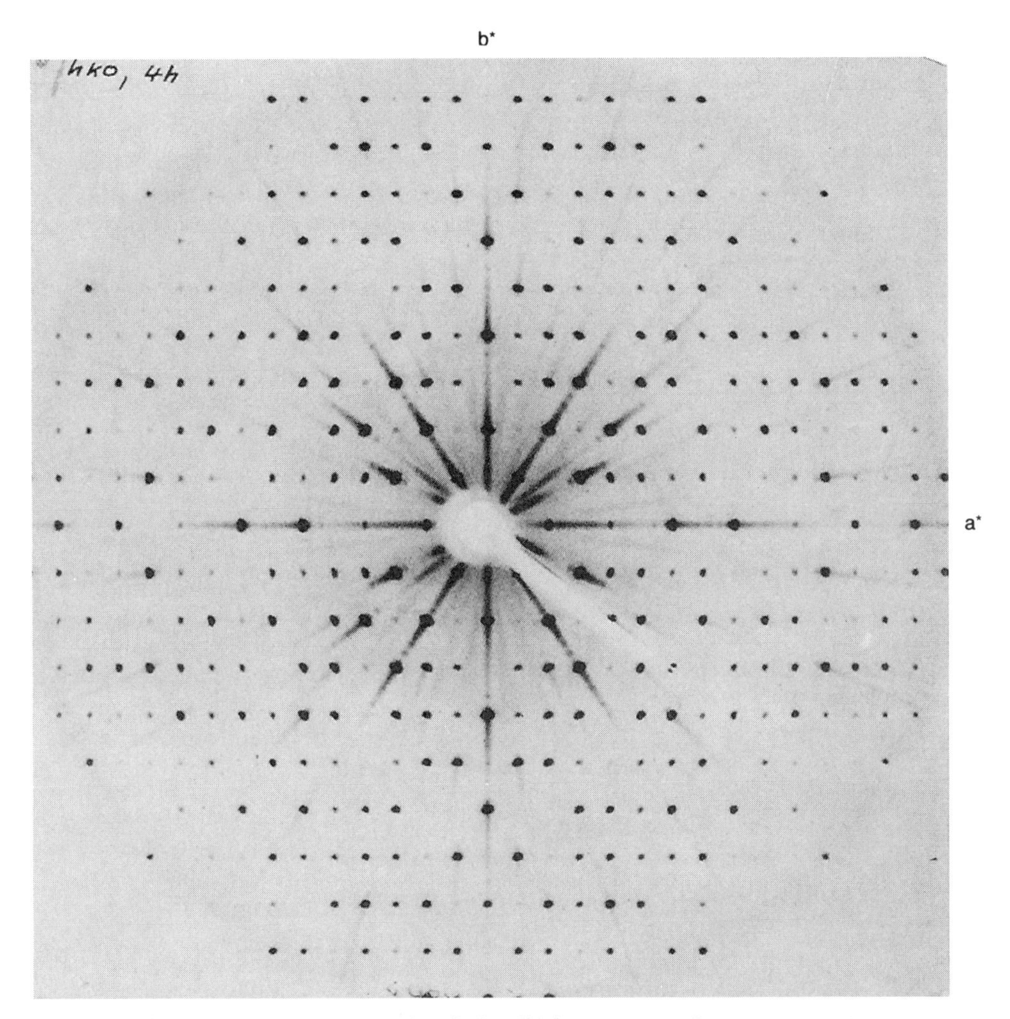

Figure 6.7. Precession photograph of the (*hk*0) R. L. net from a nonactin crystal (Cu $K\alpha$ radiation). Compare with Fig. 6.4. Courtesy of Dr. M. Dobler.

interposed between the crystal and the film to screen out reflections that belong to other levels. Undistorted upper-level photographs can be obtained by appropriate screening and by moving the film to a new position, displaced by rd^* from the zero position (Fig. 6.8). If the crystal has oblique axes, the origin of an upper-level R. L. net will not generally coincide with the center of the photograph. For the first level perpendicular to the **a** axis, for example, $d^* = 1/a$, and a displacement of this magnitude parallel to **a** has nonzero components along the **b*** and **c*** axes in an oblique coordinate system. From Eq. 5.13,

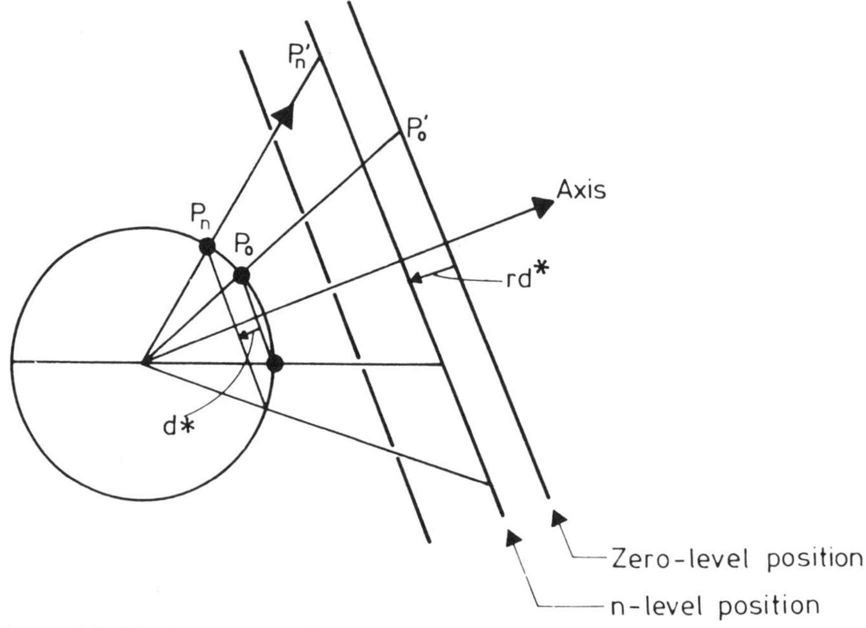

Figure 6.8. Displacements of film and screen required to record an upper-layer precession photograph.

$$\mathbf{a} = (\mathbf{a}\cdot\mathbf{a})\mathbf{a}^* + (\mathbf{a}\cdot\mathbf{b})\mathbf{b}^* + (\mathbf{a}\cdot\mathbf{c})\mathbf{c}^*$$

so

$$\frac{\mathbf{a}}{a^2} = \mathbf{a}^* + \left(\frac{b \cos \gamma}{a}\right) \mathbf{b}^* + \left(\frac{c \cos \beta}{a}\right) \mathbf{c}^*$$

The second and third terms give the offset of the origin.

Single-Crystal Diffractometers

A diffractometer is an instrument for measuring the intensities of diffracted beams individually by counting the number of X-ray photons that arrive at a suitably placed detector. The ancestry of the modern diffractometer goes back to the ionization spectrometer, the instrument that provided the experimental data for the first crystal structure analyses in 1912. This instrument used a simple ionization chamber as detector; radiation entering the chamber produced gaseous ions that were driven by an electric potential to an insulated wire connected to a gold-leaf electroscope. It was a difficult instrument to use, mainly because of the erratic behavior of the primitive X-ray tubes then available, but in the right hands it gave intensity data of a quality quite comparable with that attainable today. As Sir Lawrence Bragg has written: "I think that a main reason why the new world opened up by Laue's discovery was explored to such an extent in Great Britain and not in its country of origin was my father's experience and expertise in making accurate ionization measurements. The examination of the effects by ionization rather than by long photographic exposures was more elastic and adaptable; it lent itself to getting quantitative information, and it was easier to analyze results because the positions both of the crystal and the diffracted beam were measured for each reflection. It paid to resist the temptation of the easy photographic recording, and brave the tricky ionization measurements."[2]

As X-ray analysis spread into other laboratories with less "experience and expertise in making accurate ionization measurements," photographic recording methods, which involved less severe experimental difficulties, gradually came into more general use and ultimately replaced the ionization spectrometer altogether. Another factor was the increase in the number of experimental observations required to solve more complex crystal structures. Reflections had to be measured one by one with the spectrometer, a long and laborious task, whereas many reflections could be recorded simultaneously on a photograph.

However, with the need for greater accuracy in intensity measurements and the development of stabilized X-ray tubes and improved radiation detectors, diffractometers began to come back into use during the 1950s and 1960s. In the earlier models, the orientation of the crystal and detector for each reflection was changed and adjusted by hand, but the advent of computers soon made this unnecessary. Modern diffractometers are usually under the control of a computer, which calculates the required orientation of crystal and counter, drives the circles to the correct settings, and measures background and integrated intensity for each reflection in sequence. Several hundred reflections per day can be measured; and the automated diffractometer can work at night and during weekends and holidays as well. Apart from increased convenience and efficiency, these instruments, when used properly, provide far more accurate intensity measurements than those obtainable by visual estimates of the blackening of the spots on photographic films.

The main disadvantage of the diffractometer is still that (at least with standard instruments) only one reflection is measured at a time, whereas for crystals with large unit cells, several reflections may be produced simultaneously. This disadvantage can be crucial if the diffraction

[2]Sir Lawrence Bragg, *The Development of X-ray Analysis,* D. C. Phillips and H. Lipson, Eds., Bell, London, 1975, p. 32.

pattern has to be recorded as quickly as possible—e.g., if the crystal deteriorates with time or on irradiation.

Most modern diffractometers are of the four-circle type. A schematic drawing of such an instrument is shown in Fig. 6.9. One circle allows the counter to be set at the proper scattering angle (2θ) for each reflection, and the other three circles, ω, χ, and ϕ, bring the crystal into the correct orientation for the reflection to occur. With the 2θ axis vertical and the X-ray beam horizontal, the reflection condition is satisfied when the R. L. vector **H** lies in the equatorial plane at an angle of $90° + \theta$ to the direction of the incident beam—i.e., with its normal bisecting the angle between **s** and $\mathbf{s}_0$ (Fig. 6.10). Starting from any arbitrary orientation of the crystal, this can be accomplished by rotation about the axes χ and ω (comparable with bringing a ship from any arbitrary point on the earth's surface to a particular point on the equator by changing its latitude and longitude). However, the resulting position of the χ circle would sometimes collide with the collimating system or obstruct the diffracted beam. To avoid this, ω can be chosen so that the plane of the χ circle contains the final position of the vector **H**—the so-called bisecting geometry arrangement. The vector **H** is then brought into the correct orientation by rotations about the other two axes, χ and ϕ. Thus, only two of the three crystal-setting circles are actually necessary, but the presence of the third offers a useful, additional degree of freedom.

One design dispenses with the full χ circle but provides for rotation about an axis (K) inclined at 50° to the ω axis. The ϕ circle is mounted

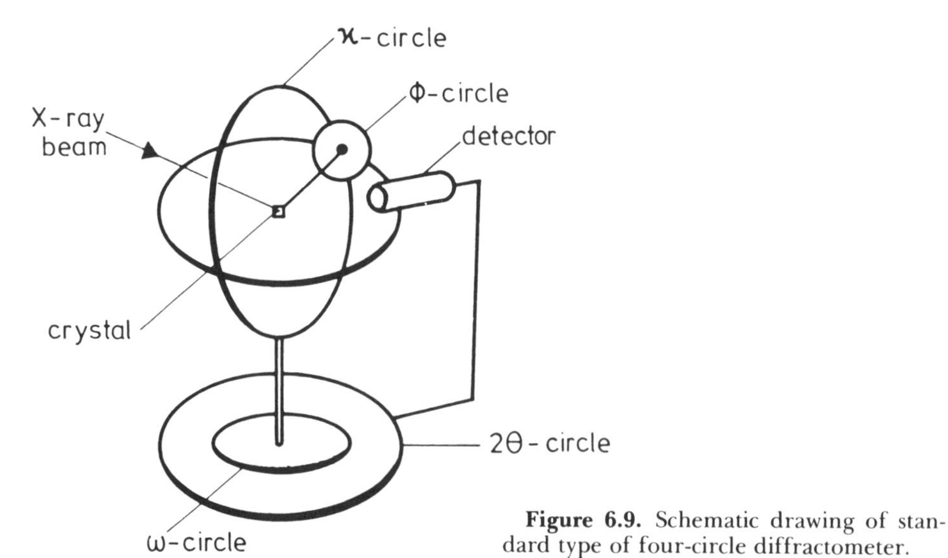

Figure 6.9. Schematic drawing of standard type of four-circle diffractometer.

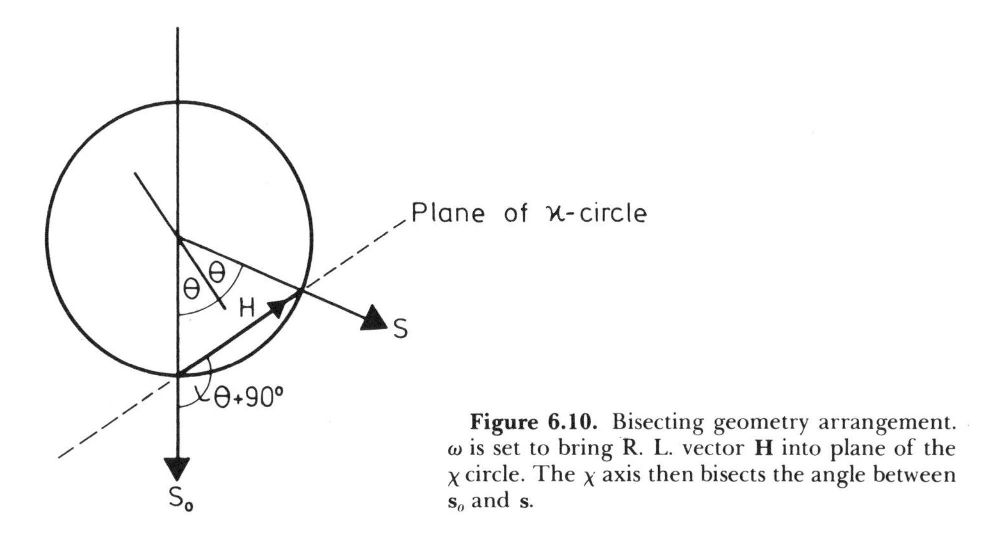

Figure 6.10. Bisecting geometry arrangement. ω is set to bring R. L. vector **H** into plane of the χ circle. The χ axis then bisects the angle between $\mathbf{s}_0$ and **s**.

on a support attached to the K circle; the angle between the ϕ and K axes is also 50°; thus the ϕ- and ω-axes coincide for K = 0 (Fig. 6.11). With this design, a full ω rotation can be carried out without collision with the collimators or obstruction of the diffracted beam. Thus, the R. L. vector **H** can always be brought into the diffracting position by a combination of K and ω rotation (the circles of constant longitude corresponding to rotation about χ are replaced here by circles inclined at 50° to the equator). One advantage of this method is that the crystal may then be rotated about the R. L. vector **H** while this vector remains in diffracting position (azimuthal scan). This is useful for eliminating the possibility of multiple reflection (see section on Double Reflection, Chapter 6) and is brought about by rotation about the ϕ circle with simultaneous adjustment of K and ω.

With both designs, the crystal is attached firmly to a pin on the ϕ-axis and must be brought to the exact center of the instrument. In general, the orientation of the crystal with respect to the diffractometer axes will be unknown at the start, but it can be calculated, together with the unit cell dimensions, once the circle settings for several reflections are determined. The circle settings for all other reflections within the limiting sphere can then be calculated. The calculations are tedious and are best done by a computer that is usually linked to the diffractometer and controls the settings of the circles and other variables that affect the intensity measurements.

During these measurements the intensity of the incident beam must be kept as constant as possible. The total peak intensity I_P of a reflec-

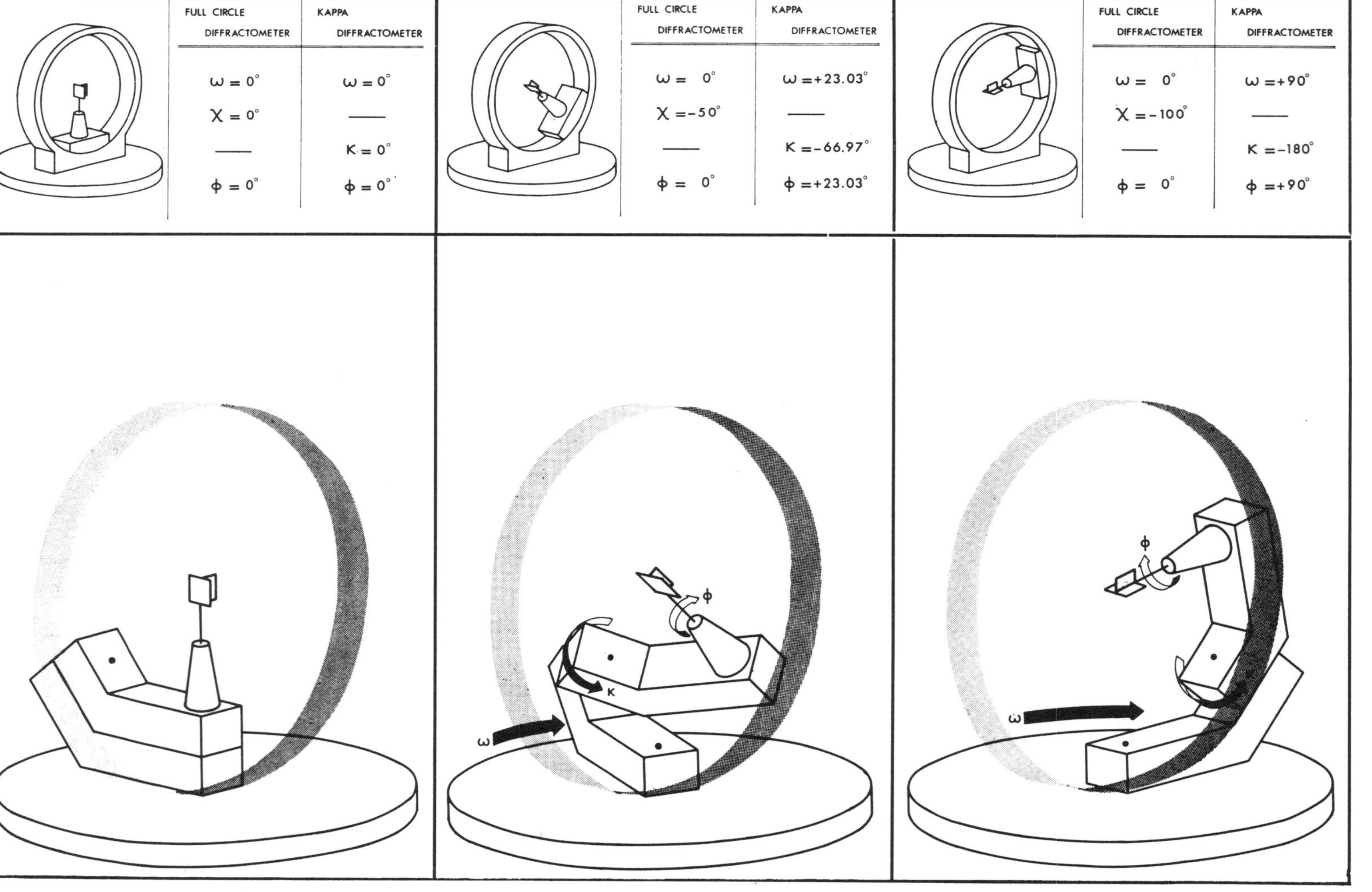

Figure 6.11. Comparison of crystal settings on a conventional and on the Kappa goniometer. Reproduced by permission of ENRAF Nonius, Delft.

tion is measured by summing the individual intensities obtained by moving the R. L. point in small angular steps through the diffracting position (step scan). The scan is usually done by varying ω alone or by varying ω and 2θ together at the same rate, and the scan breadth must include the entire reflection peak. Measurements at the scan limits provide an estimate of the background intensity I_B over the same time interval as that occupied by the scan. The integrated intensity I is then given by the difference

$$I = I_P - I_B$$

The number of counts expected in a given time interval is obviously proportional to the length of the interval. If the average number of counts during such an interval is n, the probability of obtaining k counts is described by the Poisson distribution

$$P(k) = \exp(-n)n^k/k!$$

which converges to the Gaussian distribution with standard deviation $n^{\frac{1}{2}}$ for large n. The standard deviation of I is then

$$\sigma(I) = (I_P + I_B)^{\frac{1}{2}}$$

In a 60-sec scan, under typical conditions, a strong reflection may give a peak count of 10^5 or more and a background count, over the same time, of about 100. The proportional standard deviation $\sigma(I)/I$ is then less than 1%. However, for a peak count of say 120 and a background count of 100, $I = 20$ and $\sigma(I) \sim 15$. For a given crystal and incident beam intensity, the only way to increase the signal-to-noise ratio is to measure over an appreciably longer time interval. Some computer-controlled diffractometers offer the option of a variable scan time that is adjusted according to the peak intensity and background recorded in a preliminary trial scan.

The Integrated Intensity

The total intensity (counts) measured at the detector as the R. L. region is swept at constant angular velocity Ω through the Ewald sphere is known as the integrated intensity. For a small crystal completely bathed in a uniform beam of intensity I_o (expressed in units of energy per unit area per unit time) the integrated intensity is given by

$$I = I_o(r_e)^2(Lp) \cdot \left(\frac{\lambda}{\Omega}\right) \cdot \left(\frac{F}{V}\right)^2 \cdot \lambda^2 v \qquad (6.1)$$

if absorption and extinction are neglected. The quantity $r_e = (e^2/mc^2) = 2.82 \times 10^{-13}$ cm is the "classical radius of an electron;" V is the unit cell volume, and v is the volume of the crystal. The factors L and p, the Lorentz and polarization factors (discussed in the next sections), are dimensionless quantities that depend on the experimental arrangement, as do I_o, λ, and Ω, whereas v depends on the crystal specimen. The quantity

$$P = \frac{I}{I_o}\,\Omega = (r_e)^2(Lp)\left(\frac{F}{V}\right)^2 \lambda^3\, v = Qv \tag{6.2}$$

with the dimension of area is called the integrated reflecting power of the crystal for the reflection in question; Q is the integrated reflecting power per unit volume.

The only factors in the above expressions that vary from one reflection to another are L, p, and F^2. The first two are easily evaluated (see next sections), and the third is the quantity whose value we wish to measure for each reflection. Since the remaining factors are constant from one reflection to another for a given experimental arrangement, they can simply be incorporated in a single proportionality factor that is essentially a scale factor:

$$I_{\mathrm{H}} = K(Lp)F^2_{\mathrm{H}}$$

or

$$F^2_{\mathrm{H}} = K^{-1}(Lp)^{-1}I_{\mathrm{H}} \tag{6.3}$$

Although it can be very useful to have F^2 on an absolute scale, most crystallographers are content to measure the integrated intensities, and hence F^2, on an arbitrary, relative scale. Conversion to an approximate absolute scale is then made with the help of intensity statistics at the start of the structure analysis. Once the structure is solved, the scale factor can be included as an adjustable parameter in the least-squares refinement.

The expression for the integrated intensity (Eq. 6.1) is strictly valid only for an infinitesimally small crystal block. One might have expected I to be proportional to v^2 rather than to v since the scattering amplitude is proportional to the number of unit cells in the block and hence to v. The intensity would then appear to be proportional to v^2. It is true that if we have two blocks of equal cross-sectional area in the incident beam but of different thickness, say in proportion 2:1, the thicker block will produce a scattering amplitude that is twice as great and hence an intensity that is four times as great. However, this holds only at the exact reflection angle where every unit cell is in phase. On the other hand, the width of the diffraction peak given by the thicker block is only half of that given by the smaller, and the net result is that the integrated intensity is twice as great. A similar argument shows that the integrated intensity of a small crystal block is independent of the shape of the block.

For a larger crystal, composed of many small blocks that scatter independently of one another (mosaic blocks), Eq. 6.1 is valid only when attenuation of the incident beam by absorption and scattering is negligible. Otherwise, appropriate corrections have to be applied to avoid or at least to reduce the systematic errors that would be present in the derived F^2 values.

Polarization Factor

In Chapter 3 we noted that the displacement amplitude of a free electron that is subject to an alternating electric field $\mathbf{E}_0 \exp(i\omega t)$ is $\mathbf{x} = -\mathbf{E}_0 e/m\omega^2$ (Eq. 3.15). The amplitude E of the scattered wave is proportional to $\mathbf{x}$. According to electromagnetic theory, the proportionality factor is $(e\omega^2/c^2r)\sin\delta$ so that

$$E = (e^2E_0/mc^2r)\sin\delta$$

where δ is the angle between the electric vector $\mathbf{E}_0$ of the incident wave and the direction in which scattering is observed. The physical constants in this expression do not concern us for the moment since they are independent of angle; we therefore write the corresponding scattered intensity simply as

$$I = KI_0 \sin^2\delta$$

This is zero along the direction of the electric vector $\mathbf{E}_0$ of the incident wave.

Consider now the two extreme situations where the incident beam is polarized with its electric vector (a) perpendicular and (b) parallel to the plane of $\mathbf{s}_0$ and $\mathbf{s}$ (Fig. 6.12). The scattered intensities are:

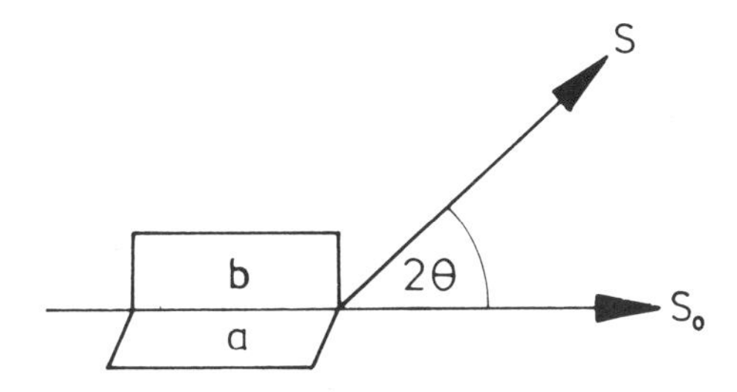

Figure 6.12. The incident beam $\mathbf{s}_0$ can be regarded as being a superposition of beams polarized (a) perpendicular and (b) parallel to the plane of $\mathbf{s}_0$ and $\mathbf{s}$.

$$\text{Case a:} \qquad I_{\perp} = K\, I_o$$

$$\text{Case b:} \qquad I_{\parallel} = K\, I_o \cos^2 2\theta$$

If the incident beam is unpolarized, it can be regarded as being composed of two components of equal intensity, polarized at right angles to each other. The scattered intensity for an unpolarized incident beam is then obtained by averaging:

$$I = K\, I_o\, (1 + \cos^2 2\theta)/2 \tag{6.4}$$

The factor $(1 + \cos^2 2\theta)/2$ thus corresponds to a dependence of the scattered intensity on the scattering angle; it is known as the polarization factor p.

If the incident beam was produced by prior reflection from a crystal (monochromator), the two components (a) and (b) will no longer be of equal intensity. When the original X-ray beam, the monochromated beam, and the scattered beam all lie in the same plane, the polarization factor p takes the form

$$\frac{1 + \cos^2 2\theta \cos^2 2\theta_o}{1 + \cos^2 2\theta_o}$$

where $2\theta_o$ is the reflection angle of the monochromator crystal. This is the situation that usually applies when diffracted intensity is measured in the equatorial plane of a four-circle diffractometer.

Lorentz Factor

For a perfectly parallel, monochromatic, incident beam the diffraction condition is satisfied whenever the Ewald sphere of reflection cuts a reciprocal lattice (R. L.) point. In practice, this exact geometric condition must be modified in several ways. Actual incident beams are neither perfectly parallel nor perfectly monochromatic, so the geometric sphere of reflection must be replaced by a suitably weighted bundle of spheres with slightly different centers and radii (Fig. 6.13). Moreover, for actual, finite, nonideal crystals the R. L. "point" must be replaced by a small region of reciprocal space; the distribution of scattering power within this region will depend on the size and mutual orientation of the ideal regions or domains within the crystal—i.e., on the degree of perfection of the crystal, and this may well be somewhat different from one crystal specimen to another. For a nearly parallel, monochromatic, incident beam and for a given angular setting of the crystal, only a portion of such a region surrounding a R. L. point will intersect the sphere of

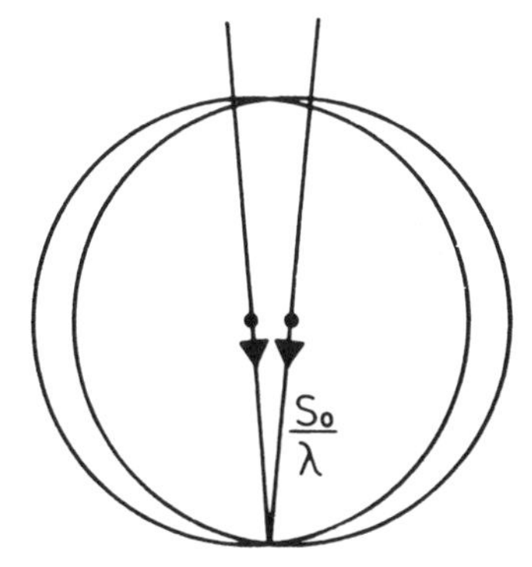

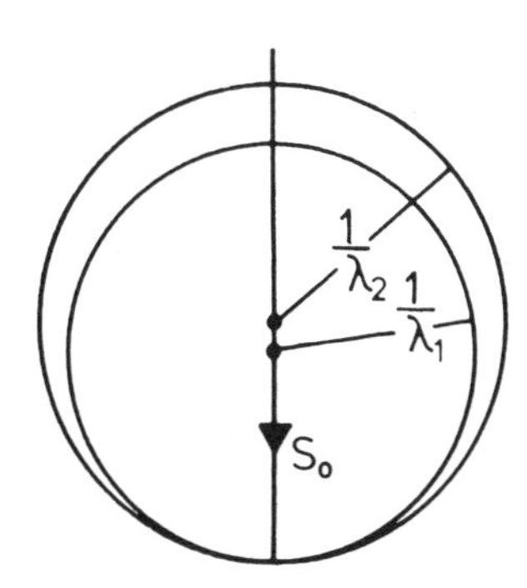

Figure 6.13. Incident radiation not perfectly parallel (left) and not perfectly monochromatic (right).

reflection (Fig. 6.14). To collect the full reflection intensity, the angular setting must be varied to sweep the entire region through the surface of the sphere. Also, if diffracted intensity is not to be lost, the aperture of the radiation detector must be large enough to receive the full range of scattering angle—no problem when photographic methods are used.

The time required for the reflecting region of reciprocal space to sweep through the surface of the sphere of reflection depends not only on the angular velocity Ω but also on the R. L. vector **H** and the wavelength λ. For the special case of an equatorial reflection, where the axis of rotation is perpendicular to the plane containing $\mathbf{s}_o$ and **s**, the terminus of the R. L. vector **H** moves with linear velocity $\Omega|H|$ normal to **H**, and the component of velocity in the radial direction **s** (Fig. 6.15) is

$$\Omega|H| \cos \theta = 2\Omega \frac{\sin \theta \cos \theta}{\lambda} = \frac{\Omega}{\lambda} \sin 2\theta$$

The time required for the reflecting region to pass through the sphere of reflection is obviously inversely proportioned to this. Hence, at con-

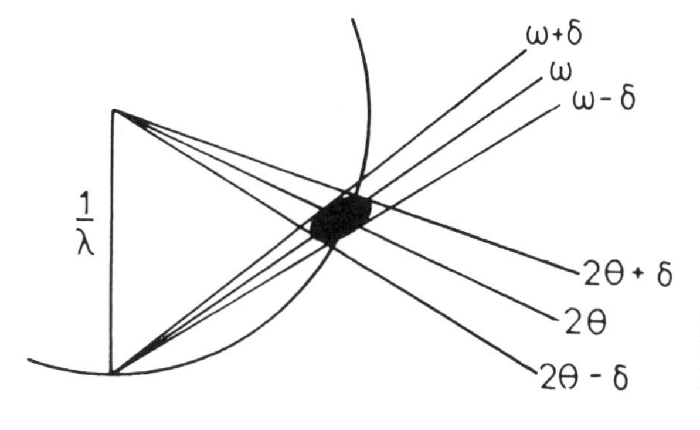

Figure 6.14. Variation of angular settings required to sweep a R. L. point (volume) through the Ewald sphere.

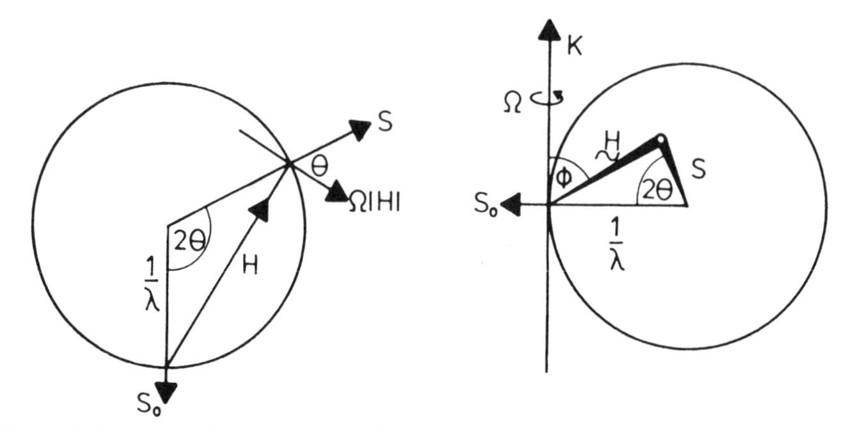

Figure 6.15. Derivation of Lorentz factor for equatorial (left) and general reflection (right).

stant angular velocity Ω, the reflected intensity will be proportional to $\lambda/(\Omega \sin 2\theta)$. This factor, or rather its trigonometric part, is known as the Lorentz or angular velocity factor.

In the general case, the R. L. vector **H** makes some angle ϕ with the rotation axis. The more complicated expression required is best derived by vector analysis.[3] Let **k** be a unit vector in the direction of the rotation axis, so that $\mathbf{k} \cdot \mathbf{H} = |H| \cos \phi$. The terminus of **H** moves with linear velocity $\Omega\mathbf{k} \times \mathbf{H}$, and we require the component of this along **s**, which is:

$$\begin{aligned} \Omega(\mathbf{k} \times \mathbf{H}) \cdot \mathbf{s} &= \Omega\mathbf{k} \cdot (\mathbf{H} \times \mathbf{s}) \\ &= \Omega\mathbf{k} \cdot (\mathbf{H} \times (\mathbf{s}_o + \lambda\mathbf{H})) \\ &= \Omega\mathbf{k} \cdot (\mathbf{H} \times \mathbf{s}_o) \\ &= \Omega(\mathbf{s}_o \times \mathbf{k}) \cdot \mathbf{H} \end{aligned}$$

the component of **H** along the normal to the unit vectors **k** and $\mathbf{s}_o$. Now, the component of **H** along **k** is $|H| \cos \phi$, and the component along $\mathbf{s}_o$ is $|H| \sin \theta$, so that the required component is

$$|H| (1 - \sin^2 \theta - \cos^2 \phi)^{\frac{1}{2}} = |H| (\cos^2 \theta - \cos^2 \phi)^{\frac{1}{2}}$$

and the linear velocity in this direction is

$$2\,\Omega \sin \theta\, (\cos^2 \theta - \cos^2 \phi)^{\frac{1}{2}}/\lambda$$

[3]A. J. C. Wilson, *Elements of X-ray Crystallography,* Addison-Wesley, Reading, Mass., 1970, pp. 120–121.

The Lorentz factor for the general case is the reciprocal of this quantity; the relevant trigonometric part is

$$\frac{1}{2 \sin \theta \, (\cos^2 \theta - \cos^2 \phi)^{\frac{1}{2}}}$$

which reduces to $1/\sin 2\theta$ when $\phi = 90°$.

The Lorentz and polarization factors are conveniently combined in a single trigonometric expression. Thus, for an equatorial reflection

$$Lp = \frac{1 + \cos^2 2\theta}{2 \sin 2\theta}. \tag{6.5}$$

Absorption

As the incident and diffracted beams pass through a crystal their intensities are reduced by absorption. The energy absorbed reappears partly as thermal energy and partly as radiation at a lower frequency—incoherent scattering—but it does not contribute to the coherent scattering. For a given reflection, the diffracted intensity produced by each volume element dv is reduced by the factor $\exp[-\mu(p+q)]$, where μ is the *linear absorption coefficient* and p and q are the lengths of the paths of the incident and reflected beams in the crystal (Fig. 6.16). The intensity diffracted by the crystal as a whole is then reduced by the factor

$$A = \frac{1}{v} \int \exp\,[-\mu(p+q)]\, dv \tag{6.6}$$

assuming the incident beam to have uniform intensity cross-section. The factor A is known as the transmission factor or sometimes as the absorption factor (although some authors prefer to reserve the name absorption factor for the reciprocal of A). Note that $I^A = AI$ where I^A is the actual intensity produced from the crystal, and I is the intensity in the absence of absorption.

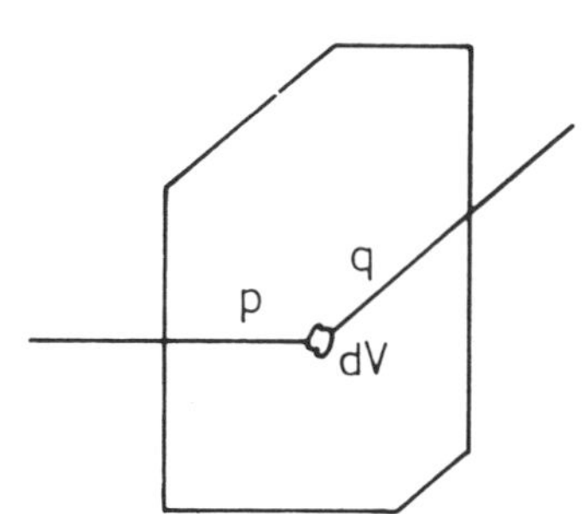

Figure 6.16. Paths of incident and diffracted beams for diffraction by small volume element dV.

For a crystal that contains only one kind of atom, the linear absorption coefficient μ is related to the mass absorption coefficient μ_m of the element in question: $\mu = \mu_m\rho$ where ρ is the density of the crystal. The value of μ_m is practically independent of the state of aggregation or combination of the element, but it depends on the wavelength of the radiation. Values of μ_m for different elements and for different radiations are tabulated in Volume III of *International Tables for X-ray Crystallography*. In general, for a given radiation, μ_m increases with increasing atomic number of the element, and for a given element, μ_m increases with increasing wavelength. However, there are sharp discontinuities, known as absorption edges (see Chapter 3), arising from the sudden onset or disappearance of specific excitation processes. The K absorption edge of an element occurs at a wavelength where the quantized energy of the incident radiation coincides with the ionization energy of a K electron. On the long wavelength side, the energy is insufficient to excite the ionization process; on the short wavelength side the transition probability is at a maximum, corresponding to high absorption of the incident beam. A large mass absorption coefficient for an element is always accompanied by a large imaginary contribution to the atomic form factor—i.e., by large anomalous scattering for the atom in question.

For a crystal that contains several kinds of atoms, the linear absorption coefficient is given by

$$\mu = \rho \sum c_i \, \mu_{mi} \tag{6.7}$$

where c_i is the mass fraction of element i. Thus μ can be calculated from a knowledge of the composition and density of the crystal.

The transmission factor A varies from reflection to reflection since it depends on the directions of the incident and diffracted beams as well as on the size and shape of the crystal specimen. When these factors are known, A can be calculated in principle although in practice the integrals must be evaluated by approximate numerical methods. Moreover, when the product μt is large, the value of A is very sensitive to the shape of the crystal, which may be difficult to measure accurately if it is irregular.

Instead of correcting the observed intensities for the effect of absorption, it is generally preferable to reduce absorption effects as far as possible by avoiding the use of X-radiation for which μ is excessively high and by working with very small crystals. But not too small! The intensity diffracted by a crystal is proportional to its volume and should be of the order of $l^3 \exp(-\mu l)$ for a crystal of linear dimension l. It is therefore maximized when l is approximately $3/\mu$. This is the "optimum size" as far as intensity is concerned, whereas the optimum size for avoiding absorption errors is zero. Some value between these limits could be chosen as a reasonable compromise.

When absorption effects cannot be avoided (e.g., anomalous scattering experiments), it may be advisable to try to grind the crystal to a spherical shape. Even for a sphere, the integral (Eq. 6.6) cannot be evaluated in analytical form, but it has been tabulated as a function of μR and θ. Fig. 6.17 shows that the transmission factor has an approximately linear dependence on $\sin^2 \theta$; the intensities of reflections at low $\sin \theta$ are reduced by absorption more than those at high $\sin \theta$—just the reverse of the effect of the temperature factor. In fact, the absorption error in this case could be partially absorbed in the isotropic temperature factor, leading to artificially low values of vibrational parameters (it might even make some of them negative, corresponding to an apparent "sharpening" of the electron density peaks.) Even for nonspherical crystals, neglect of absorption errors affects the apparent vibrational parameters of the atoms much more than it affects the apparent positions. This occurs because the absorption error varies rather smoothly in reciprocal space and does not undergo very rapid oscillations like the structure amplitude. The effect on the electron-density distribution will

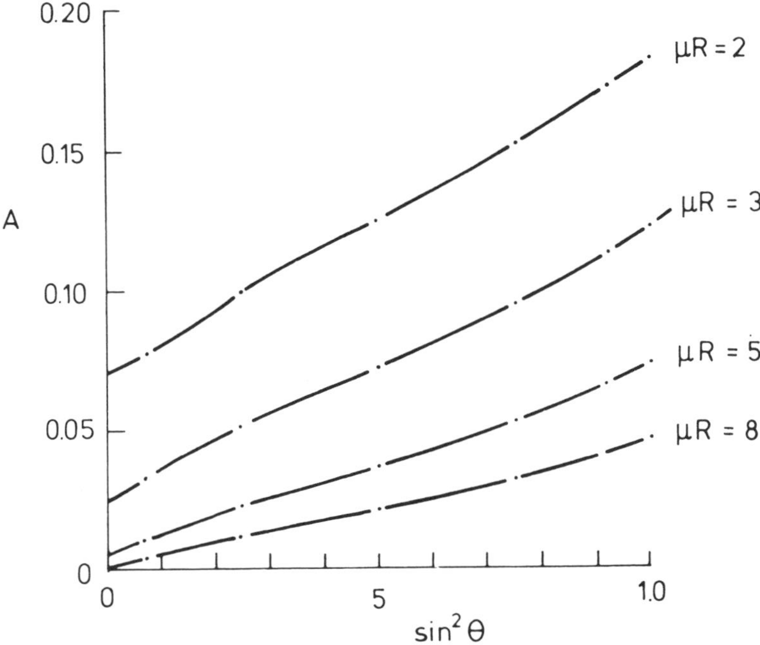

Figure 6.17. Variation of transmission factor A for sphere against $\sin^2\theta$ for different varlues of μR.

be to convolute the individual electron-density peaks with a spurious shape function, one that will generally sharpen the peaks rather than smear them out in real space. Of course, for nonspherical crystals the effect may be strongly anisotropic. In studies where physical significance is to be ascribed to deviations of the individual electron-density peaks from spherical symmetry, accurate corrections must be applied for absorption, or better still, the experiments should be carried out so that absorption errors become negligible.

Extinction

During the latter refinement stages of a crystal structure analysis, it is often possible to detect what seems to be a systematic error in the observed $|F_{\mathbf{H}}|$ values: for very strong reflections there is a tendency for $|F_{\mathbf{H}}|_o$ to be systematically smaller than $|F_{\mathbf{H}}|_c$. This kind of error, if present, is recognized most clearly by examining not the $|F_{\mathbf{H}}|$ values themselves but the corresponding intensities $(I_{\mathbf{H}})_o$ and $(I_{\mathbf{H}})_c$; the latter are calculated by applying *Lp* corrections in reverse to the $|F_{\mathbf{H}}|_c$ values. A scatter plot of I_o/I_c versus I_c for a representative number of reflections will then reveal that the ratio is scattered around unity for weak and medium intensities but that it becomes increasingly smaller than unity for strong reflections. Typical plots of this kind can often be fitted, approximately, to an exponential function:

$$(I_o/I_c) = \exp(-gI_c)$$

where the appropriate value of g is best found by taking logarithms and estimating the slope of the correlation line.

This kind of systematic error is called *extinction* (not to be confused with space-group extinctions), and its influence on the $|F_{\mathbf{H}}|_o$ values can easily be removed by multiplying the observed intensities by $\exp(gI_c)$ and recalculating a set of corrected $|F_{\mathbf{H}}|_o$ values. This is not cheating, it is only making an empirical correction for a discrepancy that, whatever its origin, has been recognized as being systematic. The introduction of an extinction correction must obviously improve the R factor, but if the extinction error is not too severe, it has little effect on the positional and thermal parameters derived from the least-squares analysis. However, the $|F_{\mathbf{H}}|_o$ values used in difference syntheses designed to measure deformation densities (Chapter 8) must be corrected for extinction. This is because the strong intensities, those most affected by extinction, occur mainly in the low $\sin\theta/\lambda$ region, where the information about the valence electron distribution is concentrated.

Extinction is primarily the result of inadequacy of the scattering theory outlined in Chapter 1. This theory—the kinematic theory—tacitly assumes that the incident beam is not affected by its passage through the crystal. Obviously, however, quite apart from absorption processes (previous section), the incident beam is weakened as it travels through the crystal and is scattered at successive volume elements. Eq. 6.1 and 6.2, which state that the integrated intensity and reflecting power of a reflection are proportional to the crystal volume, are valid only for crystals so small that the attenuation of the incident beam can be neglected. Otherwise, for a large enough crystal, these equations would lead to $I > I_0$. In the more complete dynamical theory, due to Ewald,[4] the mutual interactions between the electric fields of the incident and scattered waves are taken into account, but the equations that result are too complex to be solved except for a few special cases.[5]

In 1914, Darwin showed that when attenuation of the incident beam was taken into account for a strong reflection (he used rocksalt, NaCl, as an example), the incident beam should be completely extinguished after it had penetrated the crystal to only a very small depth (total reflection).[6] This would make the reflected intensity independent of the normal photoelectric absorption coefficient μ. However, absorption was already known to influence the reflected intensity.[7] In addition, the width of the diffraction peak would be much narrower than that observed. Finally, since the peak width should be inversely proportional to the effective penetration depth, and this in turn should be inversely proportional to $|F_{\mathrm{H}}|$, the integrated intensity (for total reflection) would have to be proportional to $|F_{\mathrm{H}}|$ rather than to $|F_{\mathrm{H}}|^2$, whereas the latter dependence was known to be a better approximation for most actual crystals. As a way out of this dilemma, Darwin proposed that most crystals are imperfect—i.e., that the reflecting planes are not perfectly

[4]P. P. Ewald, *Ann. Phys. Leipzig* **54,** 519 (1917). For a more recent account see P. P. Ewald, *Acta Crystallogr.* **A25,** 103 (1969).

[5]Although the dynamical theory was introduced only a few years after the discovery of X-ray diffraction, its full importance was not appreciated until much later, when the art of growing perfect or nearly perfect crystals had been developed. With the recognition that such crystals show certain phenomena (Pendellösung fringes, Borrmann absorption) that cannot be explained in terms of the simple kinematic theory, there has been a marked resurgence of interest in the dynamical theory. A good account of recent developments is given in L. V. Azároff, R. Kaplow, N. Kato, R. J. Weiss, A. J. C. Wilson, and R. A. Young, *X-Ray Diffraction,* McGraw-Hill, New York, 1974, Chapters 4 and 5. However, the kinematic theory, with a few minor modifications to allow for extinction effects, is still the main basis for crystal structure analysis.

[6]C. G. Darwin, *Phil. Mag.* **27,** 315 (1914).

[7]W. H. Bragg, *Proc. Roy. Soc., London* **A89,** 430 (1914).

parallel. Because of the divergence of the incident beam, only a small part of it, a narrow ray, would be totally reflected by the surface layers; the remainder would be propagated further until, over a certain angular range, other rays would encounter suitably oriented sets of planes at some depth in the crystal. This event would obviously involve the absorption coefficient μ, it would broaden the diffraction peak, and, as Darwin showed, it would also lead to an approximately F^2 dependence of the integrated intensity.

In his 1914 paper, Darwin described the nonparallelism of the reflecting planes as being a result of internal twisting in the crystal, but he later[8] assumed (mainly to simplify the mathematical treatment) that the imperfect crystal could be regarded as being built from small blocks, each one perfect but separated by discontinuities in the periodicity and slightly disoriented with respect to one another. Since there is no exact phase relationship between waves scattered by different blocks, the diffracted intensity can be taken as the sum of the intensities from the individual blocks that are oriented to diffract at that angle—in other words, the blocks can be regarded as being optically independent. A crystal with these hypothetical properties is known as a mosaic crystal. Microscopic or even normal visual examination shows that some crystals are indeed built from bundles of needlelike fibers or of thin plates, slightly disoriented with respect to one another. Such crystals typically yield poor diffraction diagrams with markedly broad maxima extending over an angular range of several degrees. For most imperfect crystals, however, although the existence of optically independent regions seems well established, there is no direct evidence that these regions are necessarily segregated into physically distinguishable blocks.

The concept of a mosaic crystal leads to a distinction between *primary* and *secondary* extinction. Primary extinction refers to the attenuation of the incident beam as it passes through a single block. If $|F_H|$ is large, more-or-less total reflection of the correctly oriented part of the incident beam will occur at the surface layers of the block. Since the shielded lower layers cannot contribute their proper share to the diffraction, the net intensity of the radiation reflected by the block is reduced below the expected kinematic value. This reduction is known as primary extinction, and it depends on the size of the blocks. The larger the blocks, the more severe will be the primary extinction. If the crystal blocks are so small that primary extinction becomes negligible, then the crystal may be

[8]C. G. Darwin, *Phil. Mag.* **43**, 800 (1922).

described as ideally imperfect. It is also possible that a particular ray of the incident beam passes through one block, loses some intensity to the diffracted beam, and then encounters a second block that happens to lie exactly parallel to the first one. The second block is thus shielded by the first and hence contributes less than its proper share to the diffraction. This kind of intensity loss is known as secondary extinction, and it depends on the thickness of the crystal, on the mean thickness of the blocks, and on their alignment (mosaic spread). The thicker the crystal and the thicker and better aligned the blocks are, the more severe the secondary extinction should be. In general, both primary and secondary extinction would be expected to contribute to the overall extinction of large, imperfect crystals.

The mosaic crystal is only a concept, a model introduced for mathematical simplicity, and should not be taken too literally. Nevertheless, what about the numerical values of the parameters that describe the model? How small would the mosaic blocks have to be for the crystal to be regarded as ideally imperfect—i.e., for primary extinction to become negligible? The answers depend on the value of Q_H, the integrated reflecting power per unit volume (Eq. 6.2), on the wavelength, and on the scattering angle. One result of the Darwin theory[8] is that for reflection from a perfect crystal the diffracted intensity should be proportional to the crystal thickness up to a certain critical thickness where saturation of the diffracted intensity sets in. The critical thickness t_o can be estimated to be of the order of

$$t_o \approx \frac{V}{r_e \lambda |F_H|} = \left(\frac{\lambda}{Q_H \sin 2\theta}\right)^{\frac{1}{2}} \tag{6.8}$$

or about 10^{-3} to 10^{-4} cm for a strong reflection ($F/V \sim 10^{24}$ cm^{-3}, $r_e = 2.8 \times 10^{-13}$ cm, $\lambda \sim 10^{-8}$ cm). Thus, in the mosaic model, primary extinction should be negligible when the mean thickness t of the mosaic blocks is smaller than about 10^{-4} cm.

Suppose we have blocks of this thickness. How large would the mosaic spread have to be for secondary extinction to become negligible as well? We assume the crystal thickness (T) to be 0.2 mm, fairly typical for a crystal used for accurate diffractometer measurements, so that the incident ray will pass through T/t blocks. The width of the diffraction peak from an individual block is λ/t in order of magnitude (Chapter 1). Let M be the width of the composite diffraction peak. Then we have room for Mt/λ peaks from individual blocks without overlapping. The condition for no overlapping is then

$$\frac{T}{t} < \frac{Mt}{\lambda}$$

or

$$M > \frac{\lambda T}{t^2}$$

With $T = 2 \times 10^{-2}$ cm, $\lambda = 10^{-8}$ cm; $t = 10^{-4}$ cm, we obtain $M > 2 \times 10^{-2}$ or about 1°. For appreciable secondary extinction for a crystal of this size, the block size would have to be much greater than 10^{-4} cm or the mosaic spread much smaller than 1°.

During recent years, interest in the extinction problem has been reawakened by the necessity for reliable corrections in accurate electron-density studies (Chapter 8). Several attempts have been made to estimate the dependence of the extinction factor $y = I/I_k$ (where I is the integrated intensity, and I_k is its true kinematic value) on various real or assumed parameters, such as wavelength, scattering vector, crystal size and shape, mean size of mosaic blocks, extent of misalignment of mosaic blocks, and so forth. Once the theoretical dependence of y on these can be established, the unknown parameters can be derived, say from a least-squares fit to the observed I values, and the extinction factor y can be evaluated for each reflection. Thus I_k (and hence $\boldsymbol{F}^2$) could be derived from the observed I values. One should remember that adequate extinction corrections can be based on a physically unreasonable extinction model or even on no model at all.

In 1967, Zachariasen[9] proposed a general theory leading to "a universal formula for the integrated intensity of the incident beam, valid over the entire range from the perfect to the ideal mosaic crystal." Zachariasen starts off from a pair of coupled differential equations (Darwin energy transfer equations) for the mutual transfer of intensity between incident and diffracted rays. If absorption is neglected, these equations reduce to

$$\begin{aligned} \frac{\partial I_o}{\partial S_o} &= -\sigma I_o + \sigma I \\ \frac{\partial I}{\partial S} &= \sigma I_o - \sigma I \end{aligned} \tag{6.9}$$

where σ is the diffracting power per volume element, and S_o and S are distances traversed by the incident and diffracted rays within the crystal. These equations are to be solved subject to the boundary conditions that $I_o = \mathcal{I}_o$ when $S_o = 0$ and $I = 0$ when $S = 0$, but to make any progress at all

[9] W. H. Zachariasen, *Acta Crystallogr.* **23,** 558 (1967).

a cascade of further definitions and approximations has to be introduced. In his treatment of secondary extinction, Zachariasen distinguishes two types of imperfect crystals: type I, where the extinction depends mainly on g, a quantity inversely proportional to the mosaic spread, and type II, where the extinction depends mainly on $\bar{t}$, the mean radius of the mosaic blocks. The Zachariasen theory has been modified in a number of ways,[10,11] and adapted for use in least-squares calculations from which $\bar{t}$ and g may be calculated,[11] but it has also come in for serious criticisms, which have been summarized by Lawrence.[12]

One basic problem is that the distinction between primary and secondary extinction is rather arbitrary with reference to real, imperfect crystals. The limiting cases of ideally perfect and imperfect crystals represent extremes that are rarely met; the typical crystals encountered in X-ray analysis are somewhere in between, and their imperfections cannot really be described in terms of a simple two-parameter model. Thus, the physical significance of derived quantities such as $\bar{t}$ and g remains questionable.

A distinction between primary and secondary extinction is really equivalent to a distinction between coherent and incoherent scattering, and this is artificial, as Kato has pointed out.[13] Every interaction between incident and diffracted waves is, in principle, coherent (for a static arrangement of scattering matter), and the apparent incoherence is merely a consequence of averaging over a statistical ensemble of distortions of various kinds. Kato has shown that when lattice distortions are present, the dynamical theory leads to a description of primary and secondary extinction in terms of two distances. One is the coherence length or primary extinction distance Λ, of the same order of magnitude as the "critical thickness" of a mosaic block (Eq. 6.8). The other is the correlation length τ of the lattice phase factors. If $\tau \gg \Lambda$, there is a coherent interaction (interference) between the primary and diffracted waves, equivalent to primary extinction. If $\tau \ll \Lambda$, the intensity fields satisfy a set of energy-transfer equations similar to Eq. 6.9 but where the physical meaning of the coupling constants is different. The theory is independent of the mosaic model, although this can be treated as a special case.

One way of reducing extinction effects is to cool the crystal rapidly, e.g., by immersion in liquid air. Some crystals shatter completely when

[10] M. J. Cooper and K. D. Rouse, *Acta Crystallogr.* **A26,** 213 (1970).
[11] P. J. Becker and P. Coppens, *Acta Crystallogr.* **A30,** 129, 148 (1974).
[12] J. L. Lawrence, *Acta Crystallogr.* **A33,** 232 (1977).
[13] N. Kato, *Acta Crystallogr.* **A32,** 453, 458 (1976).

subjected to such drastic treatment. In those that survive, the rapid cooling presumably produces internal strains that decrease the size of optically coherent regions. Recent experiments show that for certain molecular crystals, application of a temperature gradient of the order of 20° per cm during the intensity measurement leads to a marked reduction in extinction.[14] The integrated intensity of extinction-affected reflections increases, in some cases up to the value expected by the kinematic theory but never beyond it. The increase is mainly in peak intensity, sometimes accompanied by a detectable broadening of the peak. The effect is reversible—on removing the temperature gradient, the intensity instantly reverts to its original value. Application of a larger and larger gradient ultimately leads to a sudden increase in the peak widths; this is irreversible.

More than 40 years ago, Sakisaka and Sumoto observed analogous intensity changes from a quartz plate showing large primary extinction by applying a temperature gradient of 167° per cm in a direction normal to the incident beam.[15] They attributed the marked increase in diffracted intensity to deterioration of crystal perfection by unequal thermal expansion of the crystal along the temperature gradient. The molecular crystals used in the newer experiments[14] are far from ideal—some are even disordered—but they have much larger thermal expansion coefficients than quartz. Presumably, these crystals contain fairly large optically coherent regions, and the strains set up by even small temperature gradients are sufficient to impair the optical perfection of such regions. The diminution in crystal perfection must occur without any segregation into physically discrete blocks since this could hardly occur reversibly. Such a breakdown presumably occurs only with the larger temperature gradients that produce irreversible peak broadening.

Double Reflection

Suppose that a crystal is oriented with respect to the incident beam so that two reciprocal lattice points happen to intersect the Ewald sphere simultaneously (Fig. 6.18):

$$\mathbf{H}_1 = \frac{\mathbf{s}_1 - \mathbf{s}_0}{\lambda}$$

$$\mathbf{H}_2 = \frac{\mathbf{s}_2 - \mathbf{s}_0}{\lambda}$$

[14]P. Seiler and J. D. Dunitz, *Acta Crystallogr.* **A34,** 329 (1978).
[15]Y. Sakisaka and I. Sumoto, *Proc. Math. Phys. Soc. Japan* **13,** 211 (1931).

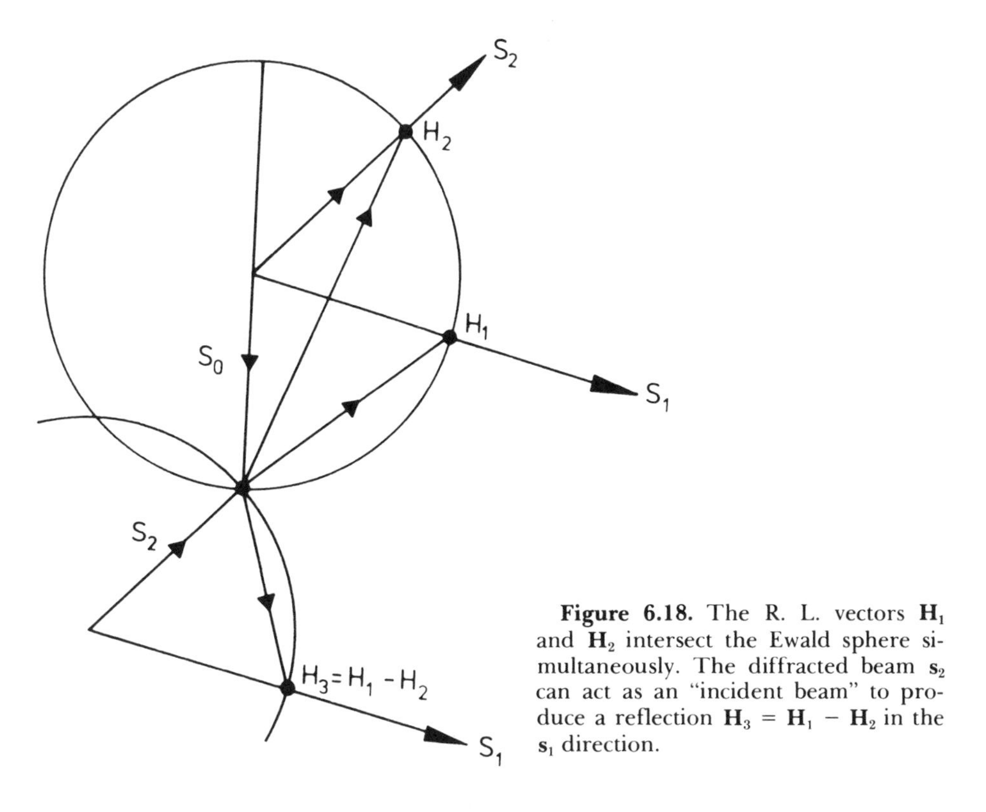

Figure 6.18. The R. L. vectors $\mathbf{H}_1$ and $\mathbf{H}_2$ intersect the Ewald sphere simultaneously. The diffracted beam $\mathbf{s}_2$ can act as an "incident beam" to produce a reflection $\mathbf{H}_3 = \mathbf{H}_1 - \mathbf{H}_2$ in the $\mathbf{s}_1$ direction.

where $\mathbf{s}_1$ and $\mathbf{s}_2$ are unit vectors in the directions of the two diffracted beams, $\mathbf{s}_0$ being along the incident beam. We then have

$$\mathbf{H}_1 - \mathbf{H}_2 = \frac{\mathbf{s}_1 - \mathbf{s}_2}{\lambda}$$

which states that the diffracted beam from $\mathbf{H}_2$ can act as an "incident beam" for the $\mathbf{H}_1 - \mathbf{H}_2$ reflection. In other words, the beam diffracted in the $\mathbf{s}_1$ direction will be partly the geniune $\mathbf{H}_1$ reflection but partly the superimposed $\mathbf{H}_1 - \mathbf{H}_2$ reflection produced by double reflection. Similarly, the reflection $\mathbf{H}_2 - \mathbf{H}_1$ produced by double reflection of $\mathbf{s}_1$ may be superimposed on the $\mathbf{H}_2$ reflection. This is known as the Renninger effect after its discoverer.

In general, the intensity of the doubly reflected beam is quite small compared with the intensity of the singly reflected beams since it depends on the product of the reflecting powers of the two reflections involved. When the double reflection is superimposed on a reasonably strong normal reflection, it merely increases the intensity of the latter by a small, usually insignificant fraction. However, if the normal reflection is weak, its apparent intensity may be appreciably enhanced. In particular, double reflection can lead to apparent exceptions from systematic space-group absences. Fortunately, the diffraction peaks produced by double reflection typically have a different shape from those produced by normal, single reflection. This difference is usually fairly obvious when the intensities are recorded photographically, but it may be lost in diffractometer measurements unless special precautions are taken.

The chance of double reflection is greatest when the plane defined by the incident and diffracted beams coincides with a densely populated reciprocal lattice net—i.e., when this plane is normal to a prominent zone axis of the crystal. In diffractometer measurements, a suspected double reflection can easily be checked by rotating the crystal about the reciprocal lattice vector **H**. The intensity of a normal reflection should be relatively insensitive to the rotation angle (with allowance for differences in absorption and possibly extinction), but the intensity produced by a double reflection should vanish except for a narrow range of rotation angle where the special condition happens to be satisfied.

PART TWO

MOLECULAR STRUCTURE

7. Crystal Structure Analysis and Chemistry

"But who has won? . . . "
"*Everybody* has won, and all must have prizes."

Growth of Information on Molecular Structures

It is difficult to imagine what chemistry would be like today if X-ray crystallography had never been discovered.[1] From spectroscopy (Raman, infrared, and especially microwave) we would have precise information about the shapes and dimensions of perhaps a few hundred simple or highly symmetrical molecules. From gas-phase electron diffraction (which might have been discovered independently of diffraction by crystals) we would have similar information about a few other relatively simple and volatile molecules. Any conclusions about the atomic arrangements in more complex molecules would still have to be based on the ingenuity and imagination of chemists prepared to make heroic extrapolations from limited data on small molecules. We would know nothing about the atomic arrangements in ionic crystals, minerals, metals, and alloys, and next to nothing about the conformations of complex organic molecules, including the biologically important polymers—proteins, polysacharides, and polynucleotides.

The first X-ray analyses were of alkali halide crystals, which were immediately shown to be built from an alternating pattern of cations and anions. Today we take the idea of ionic compounds so much for granted that it may be hard to imagine the extent to which traditional chemical

[1]The events that lead to an important discovery often seem to contain a large element of chance. X-rays were discovered in 1895, and although countless experiments were made with them during the following years, no one had apparently thought of diffraction by crystals until 1912. In an autobiographical memoir, Laue describes how the idea struck him during a conversation with Ewald, who was then working on problems connected with the passage of light through a crystalline medium. Laue wondered what would happen if the wavelength of the light were shorter than the interparticle distances, and he persuaded Friedrich and Knipping to make the experiment. At the time, most people (including Sommerfeld, it is said) would have regarded the experiment as a waste of time. If the idea had not come to Laue in 1912 or if he had not pursued it further, how long would we have had to wait for the discovery?

concepts were upset by these early results. As late as 1927 the idea of ionic compounds was still strongly opposed by conservative chemists: "Prof. W. L. Bragg asserts that in sodium chloride there appear to be no molecules represented by NaCl. The equality in number of sodium and chlorine atoms is arrived at by a chess-board pattern of these atoms; it is a result of geometry and not of a pairing-off of these atoms.... Chemistry is neither chess nor geometry, whatever X-ray physics may be.... It were time that chemists took charge of chemistry once more and protected neophytes against the worship of false gods; at least taught them to ask for something more than chess-board evidence."[2]

Once crystal structure data began to accumulate, all sorts of trends and regularities became apparent, leading to new kinds of systematic descriptions (basic structural types for elements and simple ionic compounds, tables of standard metallic, ionic, covalent, and nonbonded atomic radii, and so forth). A new branch of chemistry—structural chemistry—gradually developed, and one of its main proponents became a leading figure in the new theoretical chemistry that was simultaneously developing and also in the new molecular biology that was to arise a few years later. The full title of Linus Pauling's book,[3] on the series of Baker Lectures delivered in 1937–1938, is "The Nature of the Chemical Bond and the Structure of Molecules and Crystals. An Introduction to Modern Structural Chemistry."

Today we have a vast and rapidly increasing amount of structural information. Several thousand crystal structures of all degrees of complexity are solved each year, mostly by methods that require only minimal human intervention at the various stages of the analysis and often by persons who do not consider themselves crystallographers. What used to be a major research project for an entire laboratory can now be accomplished by a graduate student in partial fulfilment of his Ph.D. requirements. Indeed, structures are now being solved so rapidly that the rate-determining step in completing the work is often not data collection or computing time, as it formerly was, but writing the results for publication. Time to ponder the significance of the results is a luxury that few present-day crystallographers can afford. One might have hoped that the increased ease and rapidity of crystal structure analysis would have left more time for thinking, but the contrary seems to be true. The

[2]H. E. Armstrong, *Nature (London)* **120,** 478 (1927). To Armstrong is also attributed the comment, "Curved arrows never hit the mark," in response to Robinson's introduction of curved arrows to indicate displacements of electron pairs in chemical reactions.

[3]L. Pauling, *The Nature of the Chemical Bond,* Cornell University Press, Ithaca, N.Y. 1st ed., 1939; 2nd ed., 1940; 3rd ed., 1960.

structural facts may be trying to tell us something, but we have no time to listen. Another long-overdue paper is waiting to be written.

Even if the paper is written and published, will it ever be read? No one can possibly read all the thousands of papers on crystal structure that do finally get published, much less digest their contents. As might be expected, there are many duplications; some are intentional (e.g., when an analysis is repeated as a check, or to obtain improved accuracy), but most are simply the result of unawareness. Sometimes the same structure is analyzed almost simultaneously in different laboratories; if this becomes known before publication, every effort should be made to combine the results in a joint paper.

Thanks to the efforts of the Cambridge Crystallographic Data Centre,[4] it is now a fairly simple matter to locate the literature references to published work on organic and organometallic compounds. A series of reference books[5] published by the Centre contains bibliographic information on all such structures published since 1935 (name of compound, elementary formula, names of authors, literature reference). The entries are arranged in 86 chemical classes, with extensive cross-references, and each volume contains cumulative indexes (formula, metal, author). The Data Centre also maintains a series of computer-based files (bibliographic, structural data, chemical connectivity) that are available on magnetic tape through the Centre and a number of affiliated Centres.[6] Programs are available for searching and retrieving file entries, for geometrical calculations, and for plotting molecular diagrams directly from filed atomic coordinates. With almost 20,000 organic and organometallic crystal structures published so far, and the number doubling every four or five years (see Fig. 7.1 for the development during the period 1960–1972), these services are almost essential for speedy and reliable location and retrieval of structural information on organic molecules. Structural inorganic chemistry seems to have lagged in this respect, and efforts should be made to close the gap as soon as possible. However, a biblio-

[4]O. Kennard, D. Watson, F. Allen, W. Motherwell, W. Town, and J. Rodgers, *Chem. Britain* **11**, 213 (1975).

[5]*Molecular Structures and Dimensions. Bibliography,* O. Kennard, D. G. Watson, et al., eds., Vols. 1 and 2, 1935–69; Vol. 3, 1969–71; Vol. 4, 1971–72; Vol. 5, 1972–73; Vol. 6, 1973–74; Vol. 7, 1974–75; Vol. 8, 1975–76. Published for the Crystallographic Data Centre Cambridge and the International Union of Crystallography by Bohn, Scheltema, and Holkema, Utrecht.

[6]At time of writing (December 1976), Affiliated Data Centres have been established in France, Germany, Israel, Italy, Japan, New Zealand, Scandinavia, Switzerland, and the United States. For full details, apply to Cambridge Crystallographic Data Centre, University Chemical Laboratory, Lensfield Road, Cambridge, England.

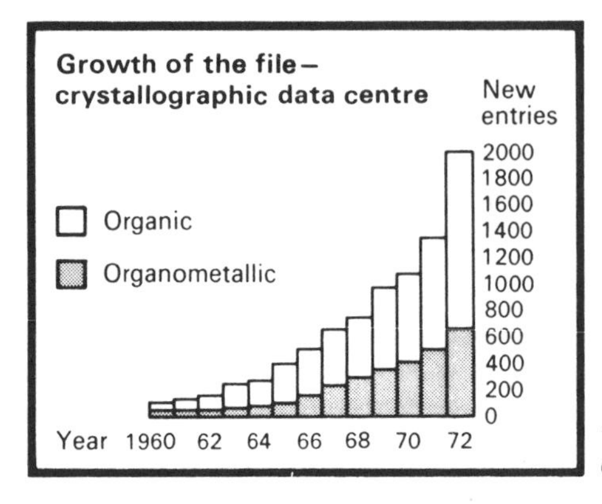

Figure 7.1. Structures included in Cambridge Data File according to publication year, 1960–1972. From Kennard et al., *Chem. Britain* **11**, 213 (1975).

graphic index of bond types in inorganic crystal structures (*BIDICS*) is available and can serve as a guide to the literature.[7]

The remainder of this chapter explores some of the ways in which crystal structure studies have influenced chemistry. First, two main aspects must be considered. One is the analytical aspect, the determination of molecular structure in the classical chemical sense—the derivation of molecular constitution and configuration as expressed in a structural formula or graph. The second aspect is concerned with the metrical details of molecular structure, with questions of conformation, including interatomic distances and angles. This type of information is not available, except very indirectly, from classical chemical methods. Apart from X-ray and neutron crystallography, it can be provided by gas-phase electron diffraction or by spectroscopic methods, mainly microwave and nuclear magnetic resonance (nmr). Microwave spectroscopy is limited, however, to small molecules that possess a permanent dipole moment, and nmr spectroscopy depends to a large extent on the interplay between known features of the molecular structure and unknown features that are to be determined. For simple molecules in the gas phase, electron diffraction is roughly comparable to microwave spectroscopy in terms of power and accuracy, although there is, of course, no restriction to molecules with a permanent dipole moment. In addition to information about equilibrium molecular parameters, information about vibrational amplitudes can also be derived. The main limitation of these other methods is that with increasing molecular complexity, conclusions become more and more indefinite.

[7] *BIDICS (A Bond Index to the Determination of Inorganic Crystal Structures),* I. D. Brown, Ed. Available from Institute for Materials Research, McMaster University, Hamilton, Ontario.

Molecular Structural Formulas

Most of the early X-ray analyses of organic crystals were concerned with compounds of known structural formula—the object was to obtain information about the geometric arrangement of the atoms. Thus, for example, the β-isomers of hexachlorocyclohexane and hexabromocyclohexane were studied in 1926 in an attempt to settle the question of the shape of the cyclohexane ring.[8]

In 1890 Sachse pointed out that two puckered "strain-free" forms of cyclohexane were possible;[9] his "symmetrical" and "unsymmetrical" forms correspond exactly to the later chair and boat forms. Sachse saw that there are two types of positions for substituents in the chair form, which are interconvertible by inversion of the chair, and he also recognized that the boat form is flexible, in contrast to the chair form, which is rigid for constant CCC angle. These ideas were at variance with the then current dogma, based on Baeyer's edict[10] that the cycloparaffin rings possess planar carbon skeletons. Sachse's views met with initial hostility, then with neglect, but they were revived a generation later by Mohr in 1918.[11] Only a few years earlier, the structure of diamond had emerged[12] as one of the first fruits of X-ray analysis, so that Mohr was now in a position to interpret the carbon skeletons in aliphatic and alicyclic molecules as portions cut out of the diamond structure. He also saw that rapid inversion of the chair form might explain the experimental failure to isolate more than a single isomer of a monosubstituted cyclohexane derivative, and he noted that *cis* and *trans* fused decalins (decahydronaphthalenes) would not be so readily interconvertible and should be isolable. The experimental verification soon followed, so that by the mid-1920s the theory of puckered, strain-free rings was more or less accepted, although it was not known for sure whether cyclohexane existed in the symmetrical chair form or in the unsymmetrical flexible form or as a mixture of both forms.

Hendricks and Billeke showed that crystals of hexachlorocyclohexane and hexabromocyclohexane are cubic, space group $Pa3(T_h^6)$, with four molecules per unit cell;[8] the required site symmetry of the molecule is then $\bar{3}(S_6)$, which is compatible with the planar form and with the symmetrical chair form but not with the unsymmetrical boat form, for

[8] S. B. Hendricks and C. Billeke, *J. Am. Chem. Soc.* **48,** 3007 (1926).
[9] H. Sachse, *Ber. Deutsch. Chem. Ges.* **23,** 1363 (1890).
[10] A. Baeyer, *Ber. Deutsch. Chem. Ges.* **18,** 2268 (1885).
[11] E. Mohr, *J. Prakt. Chem.* **206,** 315 (1918).
[12] W. H. Bragg and W. L. Bragg, *Proc. Roy. Soc., London* **33,** 277 (1913); *Nature (London)* **21,** 557 (1913).

example. Since the planar form could be excluded by the chemical evidence, the conclusion would seem to be clear cut. It is therefore puzzling to read the authors' account of their results: "It is of interest to note that Mohr's theory of strainless rings . . . is not compatible with our conclusions. . . . His three-dimensional formulas have a center of symmetry in one case, but not a plane of symmetry. Four carbon atoms of a given cyclohexane ring are coplanar, the 1,4 carbon atoms being equidistantly placed above and below this plane. Such a representation would not be tenable on the basis of our conclusions."[13]

Two years later, in a second paper, Dickinson and Billecke were able to determine the parameters of the halogen atoms and to derive rough values for the positions of the carbon atoms "in agreement with a cyclohexane ring of tetrahedral carbon atoms."[14] The chair form, with substituents in equatorial positions, was established. The second paper makes no reference either to Mohr or to Sachse.

During this period, other regularities in the structures of simple organic molecules were established. It was recognized, for example, that the CH_2 groups of long paraffin chains tend to adopt a planar zigzag arrangement, the C—C distance and CCC angle being approximately the same as in diamond. It was found that conjugated systems tend to be coplanar, that double bonds are generally about 0.2 Å shorter than single bonds, and that triple bonds are even shorter and associated with colinear arrangements of atoms, X—C≡C—X. The structures of aromatic hydrocarbons were investigated so thoroughly by Robertson and his co-workers that there was very little left in this field for other investigators. Robertson showed, for example, that the planar hexagons of carbon atoms in condensed polycyclic hydrocarbons are not quite regular, thus posing a variety of problems to the theoretical chemists of the time who were not slow to take up the challenge.[15]

Only a few analyses during this early period were concerned with

[13]Looking back after more than 50 years, one can only surmise that Hendricks and Billeke had not looked at Mohr's or Sachse's original papers (referred to in footnotes) but had read the background material in the 1924 *Annual Reports of The Chemical Society* (also mentioned in a footnote). There Ingold reviewed the growing evidence for the Sachse-Mohr theory and illustrated his arguments with diagrams that could have been misinterpreted as indicating puckered cyclohexane rings with unequal C–C distances and CCC angles. With equal distances and angles the chair form has, of course, $\bar{3}m$ (D_{3d}) symmetry and is thus completely compatible with the crystallographic evidence.

[14]R. G. Dickinson and C. Billeke, *J. Am. Chem. Soc.* **50,** 764 (1928).

[15]For an extensive discussion of these early contributions to structural organic chemistry, see J. M. Robertson, *Organic Crystals and Molecules,* Cornell University Press, Ithaca, N.Y., 1953.

molecules of unknown structural formula, and most of these were limited to the determination of unit-cell dimensions and space group. This kind of information could sometimes indicate that a proposed molecular formula was wrong, but it was insufficient to establish the correct formula conclusively. One remarkable exception is the 1923 analysis of hexamethylenetetramine, $C_6H_{12}N_4$. The crystals are cubic, space group $I\bar{4}3m(T_d^3)$, with two molecules in the unit cell, and crystallographic symmetry considerations require that the four nitrogen atoms occupy vertices of a regular tetrahedron and that the six carbons occupy vertices of a regular octahedron.[16] One must also admit that a number of incorrect deductions about molecular symmetry were drawn in those early days from erroneous interpretations of space-group evidence. For example, a structure for pentaerythritol, $C(CH_2OH)_4$, was proposed in 1923 with the central carbon atom at the apex of a tetragonal pyramid.[17]

Only after heavy-atom methods (including isomorphous replacement) were developed and applied to the phthalocyanines[18] could X-ray analysis be used to elucidate the atomic arrangements in complex molecules of unknown or only partially known structure with any real hope of success. By the end of the 1940s there was an impressive list of complex natural product molecules whose structures had been determined in this way—cholesterol,[19] calciferol,[20] penicillin,[21] strychnine,[22,23] to mention the more important ones. The analysis of penicillin was begun when almost nothing was known about its molecular structure and was pursued in close collaboration with the chemical investigations that were being carried out simultaneously. The two strychnine analyses, carried

[16]R. G. Dickinson and A. L. Raymond, *J. Am. Chem. Soc.* **45,** 22 (1923).

[17]The correct structure, in which the molecule has $\bar{4}(S_4)$ crystallographic symmetry, was established later by F. J. Llewellyn, E. G. Cox, and T. H. Goodwin (*J. Chem. Soc.* **1937,** 883). An account of the earlier controversy regarding the molecular symmetry is given in *Strukturbericht,* Vol. 1, pp. 643 ff.

[18]J. M. Robertson, *J. Chem. Soc.* **1935,** 615; **1936,** 1195. J. M. Robertson and I. Woodward, *ibid.* **1937,** 219; **1940,** 36.

[19]C. H. Carlisle and D. Crowfoot, *Proc. Roy. Soc., London* **A184,** 64 (1945).

[20]D. Crowfoot and J. D. Dunitz, *Nature (London)* **162,** 608 (1948); for the complete structure analysis see D. C. Hodgkin, B. M. Rimmer, J. D. Dunitz, and K. N. Trueblood, *J. Chem. Soc.* **1963,** 4945.

[21]D. Crowfoot, C. W. Bunn, B. W. Rogers-Low, and A. Turner-Jones, *The Chemistry of Penicillin,* H. T. Clarke, J. R. Johnson and Sir R. Robinson, Eds., Princeton University Press, Princeton, N.J., 1949, p. 310.

[22]C. Bokhoven, J. C. Schoone, and J. M. Bijvoet, *Proc. Koninkl. Neder. Akad. Wetenschap.* **50,** 825 (1947); **51,** 990 (1948); **52,** 120 (1949); *Acta Crystallogr.* **4,** 275 (1951).

[23]J. H. Robertson and C. A. Beevers, *Nature (London)* **165,** 690 (1950); *Acta Crystallogr.* **4,** 270 (1951).

out independently and on different derivatives, fully confirmed the molecular structure that had only recently been proposed for this important alkaloid[24,25] after many years of intensive chemical studies.

This encroachment into traditional preserves of organic chemistry proceeded at an ever increasing pace during the following decade, as better facilities for recording the experimental data and handling the calculations became available. The list of complex natural products—steroids, terpenes, alkaloids, carbohydrates—and of complex organometallic compounds, whose structures were determined exclusively by X-ray analysis, sometimes on a few milligrams of material, became very long. We may single out the structure analyses of vitamin[26] and coenzyme[27] B12, where more than 100 nonhydrogen atoms had to be located to establish all details of constitution, configuration, and conformation. These were compounds whose structures contained features not previously encountered in natural product chemistry—the corrinoid ligand system and the cobalt–carbon bond. Up to this time, the discovery and structural elucidation of new types of atomic groupings had been virtually a prerogative of chemists. With the B12 analyses a new situation had arisen: "To a large extent, knowledge of the specific chemistry of new structural types was formerly a by-product of the constitutional analysis by chemical degradation; in the present case, the subsequent acquisition of such knowledge remains as a task for natural product chemistry."[28]

With the advent of direct methods, the widespread availability of electronic computers, and the development of computer-controlled automatic diffractometers during the 1960s, X-ray crystallography gradually ousted degradative organic chemistry as the main source of information on the structure of natural products. The number of complete structure analyses (organic and organometallic only) mentioned in the bibliographic compilation[5] is 8 for 1945, 18 for 1950, 34 for 1955, 118 for 1960, 416 for 1965, and approximately 1000 for 1969; it is now running at the rate of several thousand per year. This has profoundly affected the strategy of chemical investigations. Some chemists even began to talk about "the crisis in organic chemistry"; presumably they meant that the

[24]R. Robinson, *Nature (London)* **159,** 263; **160,** 18 (1947).

[25]R. B. Woodward, W. J. Brehm, and A. L. Nelson, *J. Am. Chem. Soc.* **69,** 2250 (1947).

[26]D. C. Hodgkin, J. Kamper, J. Lindsey, M. Mackay, J. Pickworth, J. H. Robertson, C. B. Shoemaker, J. G. White, R. J. Prosen, and K. N. Trueblood, *Proc. Roy. Soc., London* **A242,** 228 (1957).

[27]P. G. Lenhert and D. C. Hodgkin, *Nature (London)* **192,** 937 (1961); for the complete structure analysis see P. G. Lenhert, *Proc. Roy. Soc., London* **A303,** 45 (1968).

[28]A. Eschenmoser, *Pure Appl. Chem.* **7,** 297 (1963).

traditional degradative methods, which had contributed so much to the development of organic chemistry in the past, were in danger of becoming obsolete. Organic chemistry not only survived the crisis but became stronger as a result of it. In Barton's words:[29]

> I became convinced that the solution of structural problems in organic chemistry is in most cases much more quickly done by X-ray crystallography than it is by organic chemistry. This represented a complete change in the activities of organic chemists, because always in the past we had spent half our time on degradative and half on synthetic work. But in the early 1960's everybody realized that the degradative work was no longer going to be needed. We were not going to discover new reactions, new arrangements, new chemical phenomena, by chemical degradation. We would have to discover them instead by synthesis. This has not been to the disadvantage of organic chemistry at all. In the last 10 years we have seen a turning of the talent that used to be spent on degradative work to synthetic work, and organic synthesis is advancing very rapidly. We are now discovering new reactions at a rate which was never achieved before, even in the earliest days of organic chemistry.

To appreciate the significance of Barton's words, one has to recall that 19th century structure analysis was *entirely* a matter of chemistry. Structural formulas for thousands of compounds were derived by an intricate chain of logical reasoning based on simple experimental facts. Substances were weighed, dissolved, allowed to react, and the products were isolated, purified, and subjected to elemental analysis. No other possibilities were available. The structure assigned to a compound was essentially a kind of summary of the reactions that the compound could undergo. As a simple example, the formula $CH_3 \cdot CO \cdot OH$ for acetic acid could be regarded as a concise expression of the facts: (a) that one H atom behaves differently to the other three, (b) that the two oxygen atoms behave differently, i.e., that one can be replaced by another atom or grouping more easily than the other, (c) that the two carbon atoms behave differently, one being detachable as carbon dioxide, the other as methane or a methyl derivative.

It is a remarkable triumph of this kind of logic that the structural formulas derived by deductive reasoning from chemical phenomena correspond so closely with the actual atomic arrangements as finally revealed by physical methods, and it would be a pity if this kind of reasoning has really become obsolete by progress in crystal structure analysis. It would still stand as one of the most impressive monuments of the human intellect.

[29]Sir Derek Barton, *Chem. Britain* **9**, 149 (1973).

Nowadays, when new natural products of an unusual type are isolated, their structures are often determined by X-ray analysis before much is known about the chemical reactions they undergo. Just as structure could be inferred from the chemical behavior of a compound, so certain general aspects of the chemical behavior can be inferred from the structure by simply applying the rules in reverse. This may have a limited kind of practical utility, but the process involves little intellectual challenge and is essentially uninteresting. Besides, in contrast to the degradative methods, it provides no insight into the specific chemical behavior of the compounds.

It can also be claimed that every structure analysis of a synthetic product tells us *something* about chemical reactivity—viz., that a given sequence of reactions carried out under specific conditions led to a particular product or products with this or that structure. This information can be important, of course, especially when there is some question about the identity of the products, but more often it merely confirms what is already known from other evidence.

Occasionally, however, the confrontation of chemical experimentation with crystal structure studies leads to an apparent contradiction that can produce new and unexpected insights. An old canon of classical chemistry stated that the synthesis of a natural product molecule by a sequence of steps directed toward the proposed structure could be regarded as a more or less incontravertible *proof of structure.* Indeed, until crystal structure analysis and other spectroscopic methods appeared, synthesis was virtually the only proof of structure available to the chemist. In nearly all cases, structures established by synthesis were in full agreement with those found by physical methods. If not, something had to be wrong, either with the X-ray analysis—crystallographers have been known to make mistakes—or with the chemistry.

An unexpected contradiction arose in the case of patchouli alcohol, a sesquiterpenoid compound whose structure (I) had been derived from the results of a long series of chemical experiments lasting more than 50 years and apparently confirmed in 1961 by total synthesis.[30] There was no reason whatsoever to doubt the correctness of the structure, and our involvement with patchouli alcohol had quite a different motivation. We wanted to determine the Cr—O—C bond angles in a chromic acid ester of a tertiary alcohol—information that we hoped might throw light on the details of a proposed mechanism for the oxidation of primary

[30] G. Büchi, R. Erickson, and N. Wakabayashi, *J. Am. Chem. Soc.* **83,** 927 (1961); G. Büchi and W. McLeod, *ibid.* **84,** 3205 (1962).

and secondary alcohols by chromic acid.[31] The patchouli alcohol diester was the only diester that could be obtained as a reasonably stable crystal—most of the others tended to explode violently at the slightest provocation.

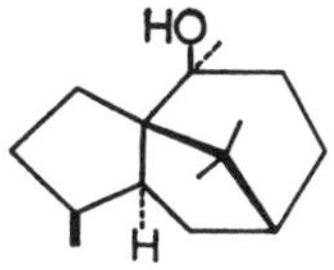

(I) 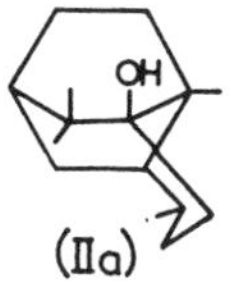 (IIa) 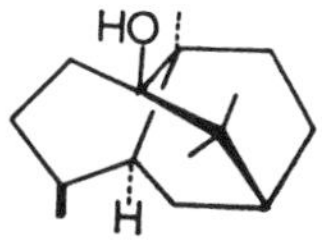 (IIb)

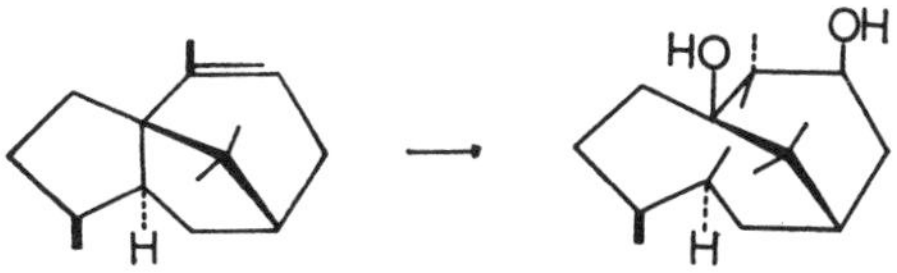

(III) → (IV)

We were quite perplexed when X-ray analysis of the chromate diester led to a different molecular skeleton (II) from the one we expected. Had the alcohol rearranged during esterification? Hardly, but even this remote possibility had to be considered. When hydrolysis of the ester yielded back authentic patchouli alcohol, we were convinced that the structure derived from degradation studies and confirmed by synthesis was wrong. In other words, the synthesis had led to a different compound from the one at which it was aimed.

The only possible explanation[32] is that an unprecedented rearrangement had occurred in one of the first degradative steps—the pyrolysis of patchouli alcohol acetate to give α-patchoulene (III)—and that the reverse rearrangement had occurred in one of the latter steps of the synthesis—the oxidation of α-patchoulene with peracetic acid (III→IV). Since neither rearrangement had been recognized, patchouli alcohol and α-patchoulene were assumed to have identical carbon skeletons whereas they are actually different. It seems likely that these unprecedented and hence unexpected rearrangements would have remained forever unrecognized had it not been for this almost accidental X-ray analysis. So much for planned research.

[31] J. Roček, F. H. Westheimer, A. Eschenmoser, L. Moldoványi, and J. Schreiber, *Helv. Chim. Acta* **45,** 2554 (1962).

[32] M. Dobler, J. D. Dunitz, B. Gubler, H. P. Weber, G. Büchi, and J. Padilla O., *Proc. Chem. Soc.* **1963,** 383.

Conformational Analysis

Of all possible spatial arrangements of atoms in a molecule of given constitution and configuration, those that correspond to potential energy minima are known as conformations. The definition, like the concept itself, is imprecise—the term *constitution* implies that the molecule is describable in terms of a unique classical structural formula, the term *configuration* overlaps with *conformation,* and how are we to know whether a proposed spatial arrangement corresponds to a potential energy minimum or not? However, we do know that the physical and chemical properties of many classes of compounds can be correlated with spatial factors that are not defined by specification of constitution and configuration alone, and we need a concept that encompasses these factors. Generally, conformations are arrangements that arise by rotation about bonds, and they may be described by specification of relevant torsion angles. Moreover, we can be reasonably confident that any particular arrangement of atoms observed in a molecular crystal cannot be far from an equilibrium structure of the isolated molecule.

X-ray analysis thus provides information about the preferred conformations of molecules although it has nothing to say about the energy differences between conformations or the energy barriers that separate them. This information has to be obtained by other methods. Energy differences can be derived, in principle, from measurements of equilibrium concentrations of the relevant conformational isomers, and energy barriers can be obtained from measurements of interconversion rates.[33]

Some of the most dramatic examples of the influence of conformational factors occur in cyclic molecules. The physical and chemical properties of the *n*-alkanes, $CH_3—(CH_2)_n—CH_3$, show a regular dependence on chain length, with a typical zigzag alternation between even and odd values of n. On the other hand, the corresponding properties of the cycloalkanes $(CH_2)_n$ do not vary so regularly with ring size but often show pronounced maxima or minima around $n = 8–10$,[34] a behavior that must be related to the spatial arrangement of the atoms in these cyclic molecules. Whereas the structures of the small rings ($n \leq 6$) are fixed or severely constrained by geometric factors (see Chapter 9), the larger rings possess torsional degrees of freedom and can exist in many possible conformations. X-ray analyses of crystalline medium-ring

[33]For a useful account of recent progress in these areas, see *Internal Rotation in Molecules,* W. J. Orville-Thomas, Ed., Wiley, New York, 1974.

[34]For examples, see J. Sicher in *Progress in Stereochemistry,* W. Klyne and P. B. D. de la Mare, Eds., Vol. 3, Butterworths, London, 1962, p. 262.

(n = 8–12) compounds helped to establish the stable conformations of these and related molecules.[35] Together with thermochemical data and vibrational frequencies for a few standard molecules, these results provided an experimental basis for the evolution of molecular mechanics calculations from a rather crude, unreliable tool into a refined instrument for conformational analysis.[36]

An example of how conformational factors can elucidate otherwise puzzling aspects of chemical reactivity is the acetolysis of cyclodecyltoluene-*p*-sulfonate,[37] a reaction with several features that call for explanation. First, at 50°C the reaction proceeds about 500 times faster than the acetolysis of the corresponding cyclohexyl derivative under similar conditions. Second, the carbonium ion formed in the first step eliminates a proton, but the cycloolefin thus formed is not the thermodynamically more stable *cis* isomer but mainly the *trans* isomer in the ratio of about 4:1 (the equilibrium mixture is about 30:1 in favor of *cis*). Third, experiments with ^{14}C and ^{2}H labeled compounds show that the reaction is accompanied by hydride shifts coupled with rearrangement of the initially labeled cyclodecyl cation; the distribution of label in the *trans*-cyclodecene product (Fig. 7.2) requires that transannular 1,5- or 1,6-shifts, which are something of a rarity, must be invoked in addition to the normal 1,2-shifts.

We should like to be able to understand this behavior in terms of specific structural features of low-energy conformations of cyclodecane and the cyclodecylcarbonium ion. X-ray analysis showed that of the many possible conformations of cyclodecane, one occurred in a wide variety of crystalline derivatives and hence must be somewhat more stable than the others.[35] This conformation (Fig. 7.3) has some obviously unfavorable features; two triangles of hydrogen atoms, above and below the center of the ring, are squeezed tightly together to give nonbonded H···H contacts that are about 0.5 Å shorter than the sum of the van der Waals' radii,[38] and some of the CCC angles are widened to nearly 120°

[35]For reviews see J. D. Dunitz, *Perspectives in Structural Chemistry*, J. D. Dunitz and J. A. Ibers, Eds., Vol. 2, Wiley, New York, 1968, p. 1; *Pure Appl. Chem.* **25,** 495 (1971).

[36]For recent reviews see C. Altona and D. H. Faber, *Topics Curr. Chem.* **45,** 1 (1974); J. D. Dunitz and H. B. Bürgi, *International Review of Science, Physical Chemistry,* Series 2, Vol. 11, J. M. Robertson, Ed., Butterworths, London, 1975, p. 81; O. Ermer, *Structure Bonding* **27,** 163 (1976); N. L. Allinger, *Adv. Phys. Org. Chem.* **13,** 2 (1976).

[37]V. Prelog, W. Küng, and T. Tomljenović, *Helv. Chim. Acta* **45,** 1352 (1962); V. Prelog, *Pure Appl. Chem.* **6,** 545 (1963).

[38]According to a neutron-diffraction analysis [O. Ermer, J. D. Dunitz, and I. Bernal, *Acta Crystallogr.* **B29,** 2278 (1973)] the shortest transannular H···H distances in cyclodecane-1,6-*trans*-diol are 1.91–1.98 Å, between type III methylene groups (see Fig. 7.3).

$\bigcirc\overset{\star}{C}H-O-SO_2-C_6H_4-CH_3$ label in starting material

Distribution of label in product

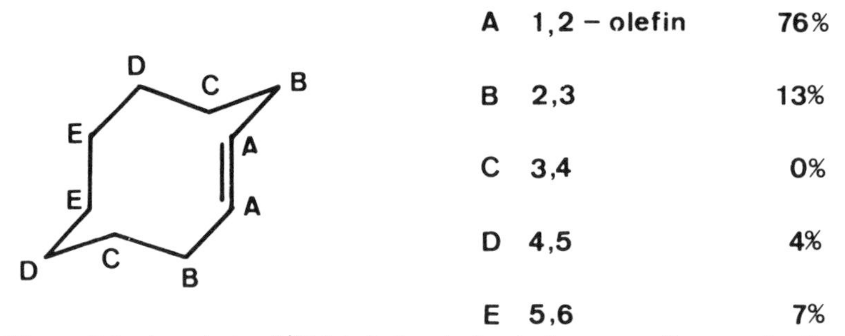

A	1,2 – olefin	76%
B	2,3	13%
C	3,4	0%
D	4,5	4%
E	5,6	7%

Figure 7.2. Acetolysis of ^{2}H-labeled cyclodecyltoluene-*p*-sulfonate: distribution of label in starting material and product [results of V. Prelog, W. Küng, and T. Tomljenović, *Helv. Chim. Acta* **45,** 1352 (1962)]. The *trans*-cyclodecene conformation shown is merely for illustration; it is not one of the low-energy conformations.

instead of the normal 112.5° found in unstrained polymethylene chains. However, from thermochemical studies[39] the molecule was known to have a strain energy of about 12 kcal/mol, so some unfavorable features were to be expected. At the time, one just had to assume that the alternative, unobserved conformations would be even more strained—an assumption that seems justified by the results of most force-field calculations done in the intervening years. Such calculations confirm that the observed conformation is indeed slightly more stable than its nearest competitors but only by a few kcal/mol.[40]

Let us look at this stable conformation in more detail. It has virtual $2/m$ symmetry, with three kinds of nonequivalent carbon atoms and six kinds of hydrogens. To make a carbonium ion, we have to convert one tetrahedral carbon atom into a trigonal one, which provides an opportunity to reduce the intraannular H···H repulsions. The worst interactions occur between pairs of hydrogen atoms attached to opposite type III carbons, so we can assume that the most stable cyclodecyl cation is the

[39]J. Coops, H. van Kamp, W. A. Lambregets, J. Visser, and H. Decker, *Rec. Trav. Chim.* **79,** 1226 (1960).
[40]See O. Ermer, *Structure Bonding* **27,** 163 (1976) for details.

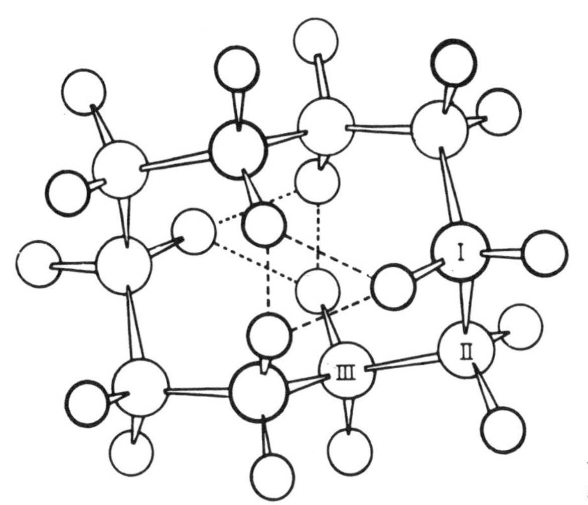

Figure 7.3. Energetically favored conformation of cyclodecane.

one with its trigonal carbon atom at a type III position (Fig. 7.4). By removing one hydrogen from its intraannular position in this way we relieve about a third of the total strain energy of cyclodecane, roughly 4 kcal/mol. The resulting relative stabilization of the carbonium ion transition state or intermediate is roughly the right order of magnitude to account for the rate increase relative to the cyclohexyl case where this strain-relief stabilization is not present.

The next step is the elimination of a proton to give the cycloolefin. For a favorable transition state the empty p orbital of the cation should be parallel to the C—H bond to be cleaved, and this condition is nearly satisfied for the elimination leading to the *trans* isomer, the angle between the p orbital and the relevant C—H bond being only about 5°. For the alternative elimination leading to the *cis* isomer, this angle is

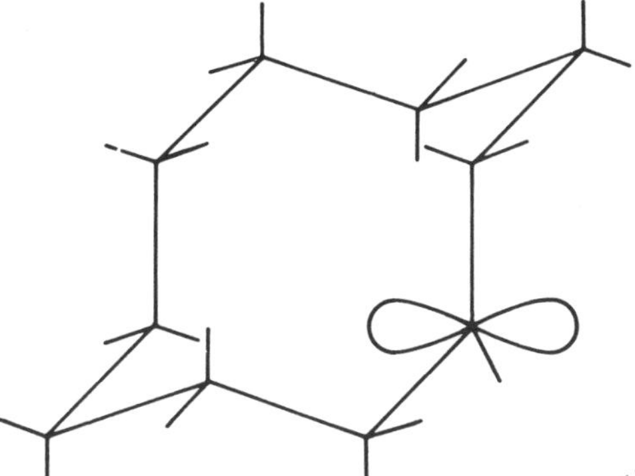

Figure 7.4. Probable stable conformation of cyclodecyl cation.

about 30°, so an appreciable conformational change would have to occur to achieve the required stereoelectronic conditions. Such a change would require energy, leading to a higher transition state. Thus, the observed conformation, plus a few reasonable assumptions, can also explain why the less stable *trans* isomer is preferentially formed in a kinetically controlled elimination step.

As for the unusual 1,5- or 1,6-transannular hydride shifts that have to be postulated to account for the distribution of label in the product (Fig. 7.2), it is clear that in the assumed conformation of the cyclodecyl cation, the empty *p* orbital points nearly towards the intraannular hydrogen atom at position 5. The leap across the ring is about as safe as it could possibly be.

In a second example from medium-ring chemistry, it is a cyclooctane derivative that shows nonconformist behavior. The Beckmann or Schmidt rearrangements of symmetrical alicyclic diketoximes generally yield mixtures of the two possible isomeric cyclodiamides.[41] However, the

HO−N=C ... C=N−OH ⟶ O=C ... C=O / H−N ... N−H + O=C ... N−H / H−N ... C=O

eight-membered ring is anomalous in this respect because 1,5-cyclooctanedione yields exclusively the mirror-symmetric isomer.[42,43] Now cyclooctane is the only cycloalkane that has a stable conformation with a mirror plane as its sole symmetry element. This conformation **1** shows severe transannular H···H repulsion between the inward pointing hy-

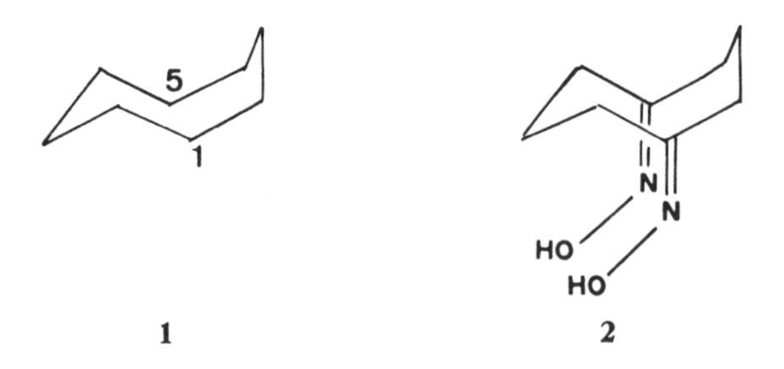

[41] M. Rothe, *Chem. Ber.* **95,** 183 (1962).

[42] G. I. Glover and H. Rapoport, *J. Am. Chem. Soc.* **86,** 3397 (1964); G. I. Glover, R. B. Smith, and H. Rapoport, *ibid.* **87,** 2003 (1965).

[43] F. K. Winkler, P. Seiler, J. P. Chesick, and J. D. Dunitz, *Helv. Chim. Acta* **59,** 1417 (1976).

drogen atoms of the methylene groups marked 1 and 5. This energetically unfavorable interaction is erased if the two tetrahedral carbons are replaced by trigonal ones, but it is preserved if the trigonal atoms are introduced at other positions in the ring. Conformation **2** should thus be particularly favorable as far as ring strain is concerned, and it is also more favorable than the one with reversed oxime configurations, which would lead to unduly short contacts between the O atoms and the pseudo-axial H atoms at positions 2 and 4. Thus there are good grounds for expecting that the mirror symmetry of the stable conformation of cyclooctane should be preserved in its 1,5-dione dioxime and hence also in the rearrangement product. A crystal structure analysis of cyclooctane-1,5-dione dioxime[43] shows that the molecules indeed have the mirror-symmetric conformation **2**. In fact, the asymmetric unit of the crystal consists of three independent molecules of dioxime linked by a system of O—H···N hydrogen bonds into distinct trimeric units with approximate $C_{3h}(\bar{6})$ symmetry (Fig. 7.5). The horizontal mirror plane corresponds to the approximate mirror plane of the individual molecules, which are then related to one another by the approximate threefold rotation axis.

These two examples show how molecular conformations observed in crystals can sometimes provide insight into otherwise puzzling details of chemical reactions in solution. We are often asked: how do you know that the conformation you observe in the solid state is the same as that occurring in solution? The answer is that we do *not* know unless our crystal structure results are backed up by other physical measurements (infrared or Raman spectra) that might show the presence of only one conformation under a variety of conditions. In a crystal, the molecules

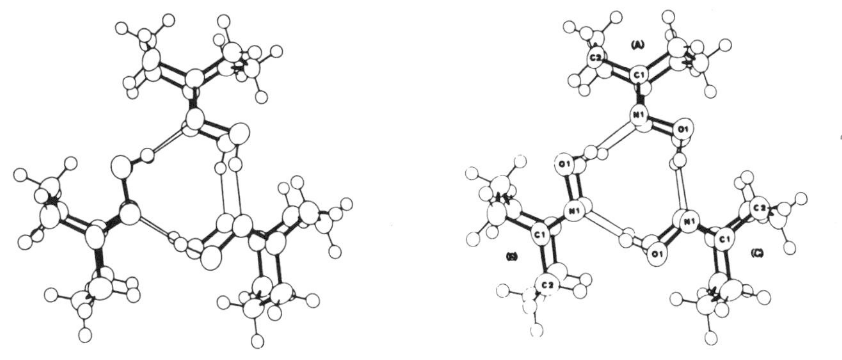

Figure 7.5. Stereoscopic view of the trimeric unit in crystals of cyclooctane-1,5-dione dioxime. From Winkler, Seiler, Chesick, and Dunitz, *Helv. Chim. Acta* **59,** 1417 (1976).

are all more or less locked in a single conformation (or rarely in more than one), whereas a solution may contain an equilibrium mixture of several low-energy conformations. However, if we find essentially the same conformation of a molecule, or a part of one, in half a dozen different crystalline or molecular environments, we can be confident that this conformation does correspond to a fairly definite energy minimum and should hence occur, though perhaps not exclusively, in solution.

One compound that appears to be a complex equilibrium in solution, with the major component different from the solid-state conformation, is caprylolactam (1-azacyclononan-2-one). From infrared and molecular polarization evidence, it was inferred[44] that this molecule occurs in the crystal as a *trans* amide and in solution as an equilibrium mixture of *cis* and *trans* amides in the ratio of ca. 4:1. From subsequent infrared studies, it was suggested that the *trans* amide group in the solid-state conformation is markedly nonplanar,[45] as later confirmed by crystal structure analysis.[46] The relative stabilization of the transoid form in the crystal can be attributed to the more favorable hydrogen-bonding pattern that can be attained. In solution, the multiple bands for the N—H stretching vibration[45,47] indicate that more than one *cis* conformation must be present and possibly more than one transoid conformation as well. The availability of several almost equienergetic conformations seems to be a common feature of medium-ring molecules, as shown by molecular mechanics calculations and as suggested by the frequent occurrence of disordered crystal structures.

It must be emphasized that chemical reactivity depends on many factors besides purely conformational ones; an exclusive preoccupation with the geometric aspects of molecular structure can sometimes prevent us from recognizing other factors that might be important or even decisive in influencing the rate of a reaction. Nevertheless, there is no doubt that for atomic groupings to react with each other, they must approach each other. Approach directions and distances observed in crystals may thus serve as useful guides for postulating the kinds of approach that can occur in solution. Moreover, for understanding solid-state reactions, this sort of information is crucial.

[44]R. Huisgen, H. Brade, H. Walz, and I. Glogger, *Chem. Ber.* **90,** 1437 (1957).
[45]H. E. Hallam and C. M. Jones, *J. Mol. Struct.* **1,** 413 (1967).
[46]F. K. Winkler and J. D. Dunitz, *Acta Crystallogr.* **B31,** 276 (1975).
[47]C. Y. S. Chen and C. A. Swenson, *J. Phys. Chem.* **73,** 2999 (1969); J. Smoliková, M. Havel, S. Vašičkova, A. Vitek, M. Svoboda, and K. Bláha, *Collect. Czech. Chem. Commun.* **39,** 293 (1974).

Solid-State Organic Chemistry

Reactions in the solid state often show quite different behavior from reactions in solution. Of course, the study of chemical reactions that occur in solids is somewhat limited—it is not possible to mix reagents or vary concentrations, as in the liquid and vapor phases. Indeed, solids are typically inert, and the changes that do occur are often regarded as more of a nuisance than anything else. Normally, energy has to be provided, thermally or photochemically, to make solid-state reactions proceed at a measurable rate; another possibility is reaction between the molecules of the crystal and small molecules that can diffuse into the crystal from its surroundings. All these types of reactions can occur at crystal surfaces, voids, or inclusions as well as in the ordered bulk of the crystal, but it is with the latter that we are mainly concerned here.[48]

The most characteristic feature of the crystalline state is its three-dimensional regularity. The molecules in a given crystal are usually frozen in a definite conformation, and interactions with neighboring molecules are fixed or at least severely restricted by the crystal packing. This is in fundamental contrast to gases or fluids, where molecules are free to adopt different conformations and where intermolecular encounters occur more or less at random. This disparity provides a general basis for understanding the more striking differences in chemical behavior between solids and liquids—e.g., (1) compounds that show similar chemical reactivity in solution show no resemblance as far as their solid state reactivity is concerned; (2) a given compound often reacts differently in the solid state from in the melt (or in solution); even when reaction occurs in both phases, the solution may yield a mixture of products, and the solid-state reaction may yield a single product; (3) polymorphic crystal forms of the same compound often show great differences in chemical reactivity. Whatever other factors may be implicated, it is clear that solid-state reactivity is largely controlled by details of molecular conformation and packing in the particular crystal structure that is involved. This is therefore an area where crystal structure studies are essential, and the basic idea behind such studies was stated several years ago in the introduction to the pioneering series of investigations of

[48]For some excellent recent reviews of various aspects of organic solid-state chemistry, see: G. M. J. Schmidt, *Pure Appl. Chem.* **27,** 647 (1971); M. D. Cohen and B. E. Green, *Chem. Britain* **9,** 490 (1973); J. M. Thomas, *Phil Trans. Roy. Soc., London* **A277,** 251 (1974); M. D. Cohen, *Angew. Chem., Intern. Ed.* **14,** 386 (1975); I. C. Paul and D. Y. Curtin, *Science* **187,** 19 (1975).

solid-state photochemical reactions by Schmidt and his collaborators:[49]

> The following papers are mainly concerned with the thesis that the course of certain types of solid-state reactions are determined by the geometry of the reactant lattice, or in Hertel's phrase, by "topochemical" factors. This thesis may be formulated as the postulate that *reaction in the solid state occurs with a minimum amount of atomic or molecular movement*. . . . It follows that for each reaction type there should exist an upper limit for such distances beyond which reaction can no longer occur. Furthermore, bimolecular reactions are expected to take place between nearest neighbours, which in turn suggests that the molecular structure of the product might be a function of the geometric relation in the crystal lattice of reactant molecules.[50]

One of the first and still one of the most extensive studies of this kind was of the photochemical dimerization of cinnamic acid derivatives to give cyclobutane derivatives. In solution, *trans*-cinnamic acids are either photoinactive, or they yield mixtures of the various possible stereoisomers. Irradiation of crystalline derivatives leads, however, to stereospecific products, which depend on the type of crystal packing. In the α-type structures, nearest-neighbor double bonds are 3.6–4.1 Å apart and related by inversion centers. These crystal modifications yield the centrosymmetric dimer, α-truxillic acid, on irradiation (Fig. 7.6). The β-type structure, with nearest-neighbor double bonds related by a crystal translation of 3.6–4.1 Å, yields the head-to-head mirror-symmetric β-truxinic acid, while the γ-type, in which nearest-neighbor double bonds are separated by more than 4.1 Å is stable to light.[51,52] The classification into α, β, and γ types can be made from a knowledge of cell dimensions. In some cases, e.g., *o*-ethoxy-*trans*-cinnamic acid, all three modifications of the same compound can be obtained by recrystallization from different solvents.

These results have stimulated an enormous amount of work on organic solid-state reactions that is impossible to summarize here.[48] There are obvious applications to synthesis, in particular to asymmetric synthesis from nonchiral reactants in chiral crystals;[53,54] optical yields as high as 70% have been obtained.

There are also some obvious difficulties. One of the main problems in

[49]These papers have now been collected and reprinted in G. M. J. Schmidt et al., *Solid State Photochemistry,* D. Ginsburg, Ed., Verlag Chemie, Weinheim, New York, 1976.
[50]M. D. Cohen and G. M. J. Schmidt, *J. Chem. Soc.* **1964,** 1996.
[51]M. D. Cohen, G. M. J. Schmidt, and F. I. Sonntag, *J. Chem. Soc.* **1964,** 2000.
[52]G. M. J. Schmidt, *J. Chem. Soc.* **1964,** 2014.
[53]K. Penzien and G. M. J. Schmidt, *Angew. Chem., Intern. Ed.* **8,** 608 (1969).
[54]B. S. Green, M. Lahav, and G. M. J. Schmidt, *Mol. Cryst. Liquid Cryst.* **29,** 187 (1975).

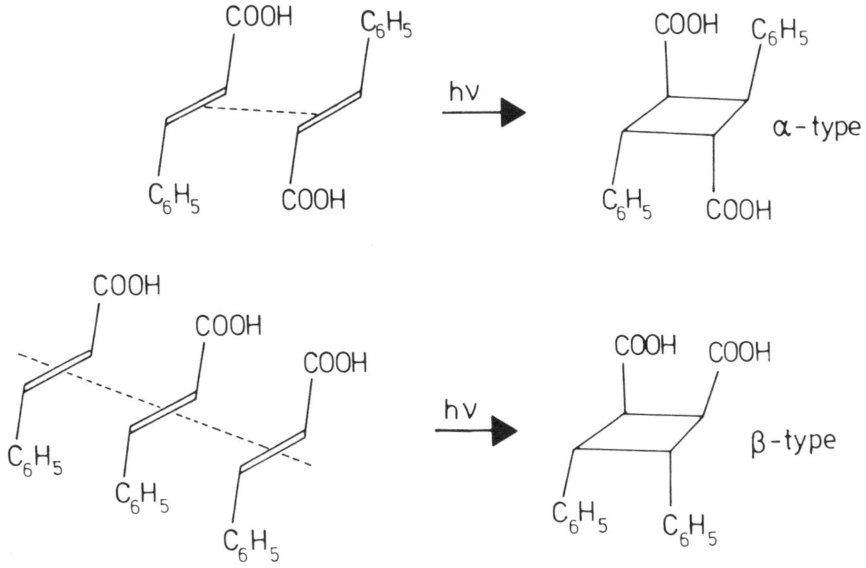

Figure 7.6. Lattice-controlled photodimerization of *trans*-cinnamic acid to stereospecific products.

controlled synthesis is the difficulty of achieving the desired type of crystal structure in any given case, for the factors that control crystal packing are not yet well understood. A given crystal structure corresponds to a free energy minimum, which can usually (i.e., ignoring disordered structures) be identified with a potential energy minimum, a balance between attractive and repulsive forces. There is no shortage of concepts for apportioning the potential energy into component parts—hydrogen bonds, donor–acceptor interactions, electrostatic interactions, polarizabilities, steric repulsions, van der Waals' attractions, and so on, but we have to concede that except for a few simple cases any predictions about how molecules are going to pack must still be regarded as speculative. Moreover, enthalpy differences associated with polymorphic phase transformations often amount to only a few hundred cal/mol or even less, showing that quite different modes of molecular packing correspond to virtually equienergetic minima. Polymorphism appears to be a widespread phenomenon, and one might conjecture that almost all compounds are polymorphic if the right conditions for crystallizing the different forms could be found.

As the factors that control molecular packing become better understood, important progress in synthetic solid-state chemistry can be expected. One such factor is the tendency of polarizable atoms in different molecules to come as close together as possible, and this has been exploited for the stereospecific synthesis of the tricyclic diketone shown below from a suitable halogen-substituted 1,5-diphenylpenta-1,4-diene-

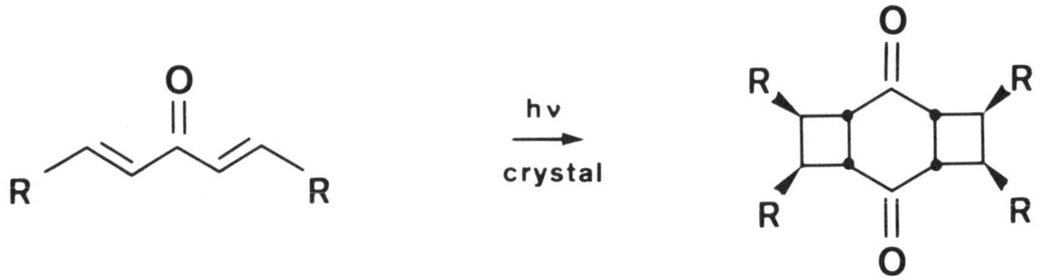

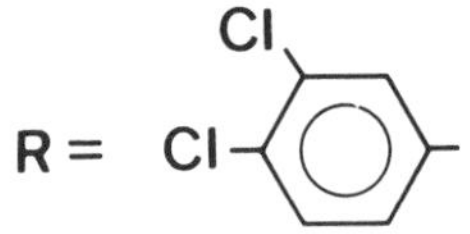

hv
crystal

R =

3-one in a single photochemical step.[55] Four carbon-carbon bonds for the outlay of a single photon!

The topochemical postulate mentioned above—that reaction in the solid state occurs with a minimum of atomic and molecular movement—has been outstandingly successful in creating order from chaos for many solid-state reactions, but it has its limitations. In solid-state photodimerizations, some molecules (e.g., *cis*-cinnamic acid derivatives) photoisomerize before the dimerization step. In other cases, molecules may have to undergo rotation at their respective sites before they find a suitable reaction partner. The photolysis of azobisisobutyronitrile provides an example; here one of the radicals formed in a first dissociation step has to rotate about its C—C≡N axis before abstracting a hydrogen atom from the other radical (Fig. 7.7).[56]

The role of crystal defects is also far more important than was first realized. The real solid has surfaces, grain boundaries, dislocations, substitutional impurities, occluded solvent, and possibly other defects. Molecules at or close to these defects have a higher potential energy than those in the bulk crystal and may be more reactive than those in the perfectly ordered regions. Reactions at defects may lead to side reactions, but they may also yield the major product if the bulk crystal is unreactive and excitation energy transfer is rapid. Moreover, once any

[55]B. S. Green and G. M. J. Schmidt, *Tetrahedron Lett.* **1970,** 4249.

[56]A. B. Jaffe, K. J. Skinner, and J. M. McBride, *J. Am. Chem. Soc.* **94,** 8510 (1972); A. B. Jaffe, D. S. Malament, E. P. Slisz, and J. M. McBride, *J. Am. Chem. Soc.* **94,** 8515 (1972).

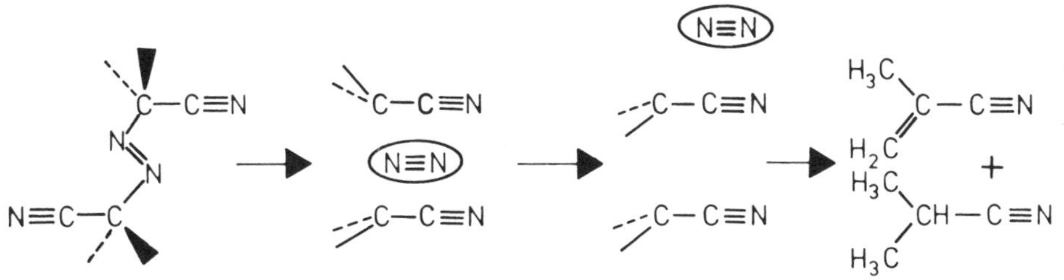

Figure 7.7. Disproportionation of radicals formed by solid-state photolysis of azobis-isobutylnitrile.

solid-state reaction has proceeded to the extent that the concentration of product becomes appreciable, the regular three-dimensional arrangement of the initial reactant must be more or less severely disrupted.

The crystallization process itself can be regarded as a simple chemical reaction. At one extreme, only weak van der Waals intermolecular interactions are involved, but intermediate degrees of intermolecular interaction are to be seen in hydrogen bonding, which can often be regarded as representing an incipient stage of a proton-transfer reaction. Donor–acceptor interactions can also be regarded as incipient chemical reactions;[57] they are especially important in molecular compounds where two or more different molecules crystallize in a single ordered phase. Molecular compounds can result from localized or delocalized donor–acceptor interactions,[58] from hydrogen bonding, or simply by enclosure of small molecules in cavities formed by a host structure (clathrate compounds).

Reaction Intermediates

X-ray analysis of reaction intermediates can, in principle, yield important information about the nature of the molecular transformations that are involved, but only a few such studies have been carried out, presumably because of the difficulties of isolating and crystallizing such compounds.

The Wittig reaction is often used in synthetic organic chemistry for replacing carbonyl groups by carbon—carbon double bonds:

$$(C_6H_5)_3P{=}CH_2 + O{=}C\begin{smallmatrix}R_1\\R_2\end{smallmatrix} \rightarrow (C_6H_5)_3P{=}O + CH_2{=}C\begin{smallmatrix}R_1\\R_2\end{smallmatrix}$$

[57]This viewpoint is clearly enunciated with many examples by H. Bent, *Chem. Rev.* **68,** 587 (1968).

[58]For a review of the chemistry, spectroscopy, and structure of π-molecular compounds see F. H. Herbstein, *Perspectives in Structural Chemistry,* J. D. Dunitz and J. A. Ibers, Eds., Vol. 4, Wiley, New York, 1971, p. 169.

The reaction between hexafluoroacetone and the cumulene $(C_6H_5)_3P{=}C{=}P(C_6H_5)_3$, yielding $(C_6H_5)_3P{=}C{=}C(CF_3)_2$ and $(C_6H_5)_3P{=}O$ as final products, can be stopped at an intermediate stage to give an adduct, whose four-membered ring structure

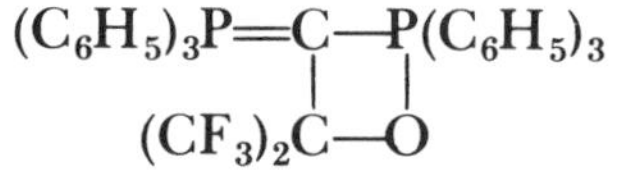

$$
\begin{array}{l}
(C_6H_5)_3P{=}C{-}P(C_6H_5)_3 \\
\qquad\qquad\;\, | \quad\; | \\
\qquad\; (CF_3)_2C{-}O
\end{array}
$$

was proposed from chemical and spectroscopic evidence.[59] This structure was confirmed by X-ray analysis,[60] which showed, however, that the formation of the four-membered ring is incomplete. The P—O distance of 2.01 Å is about 0.30 Å longer than a normal P—O single bond, and the P—C bond distance in the ring (formally a single bond) is 1.76 Å, equal to the exocyclic P=C bond distance. The C—C and C—O ring distances are more or less normal. This indicates that in the intermediate, the P—O bond is only partially formed (one can write structures with formal charges on the P and O atoms with no bond between them).

Another remarkable example was provided several years ago by Mills and Robinson.[61] Reaction of *tert*-butylacetylene with cobalt octacarbonyl gives a compound, $Co_2(CO)_4(HC_2\,Bu^t)_3$, that decomposes to 1,2,4-tri-*tert*-butylbenzene on treatment with bromine. With one mol of unsubstituted acetylene and two mol of *tert*-butylbenzene the final product is *o*-di-*tert*-butylbenzene. The crystalline intermediate isolated from the latter reaction contains molecules with the structure shown in Fig. 7.8. The three alkyne groups are linked into a six-carbon chain with the two terminal atoms, carrying the *tert*-butyl substituents about 3.2 Å apart; the trimer is ready to cyclize but held in suspended animation, as it were, by complexation to the two cobalt atoms. Cleave the Co—Co bond, and the preformed organic intermediate must collapse into the aromatic product. The X-ray analysis gives us a glimpse of an unstable intermediate whose structure could not possibly have been inferred by any other method, but even here, although the path leading from the intermediate to the 1,2-disubstituted benzene product can readily be envisaged, the path leading from the separated alkyne molecules to the cobalt complex can only be guessed at.

[59]G. H. Birum and C. N. Mathews, *Chem. Commun.* **1967,** 123.
[60]G. Chioccola and J. J. Daly, *J. Chem. Soc., A* **1968,** 568.
[61]O. S. Mills and G. Robinson, *Proc. Chem. Soc.* **1964,** 187.

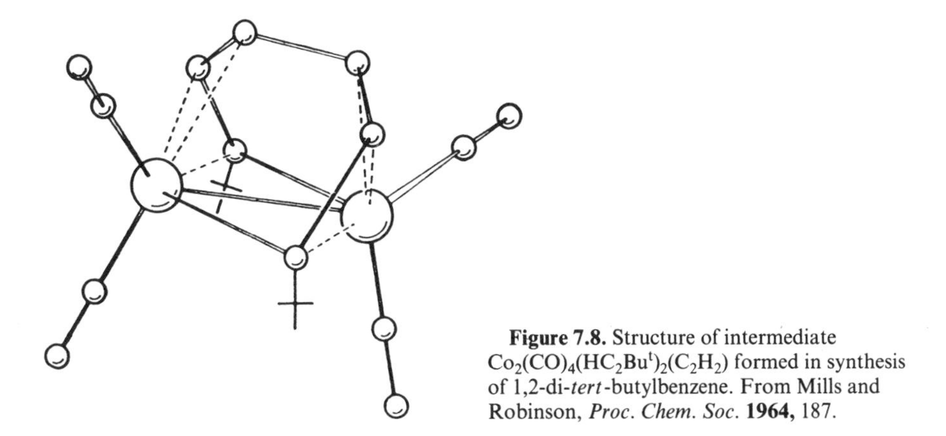

Figure 7.8. Structure of intermediate $Co_2(CO)_4(HC_2Bu^t)_2(C_2H_2)$ formed in synthesis of 1,2-di-*tert*-butylbenzene. From Mills and Robinson, *Proc. Chem. Soc.* **1964,** 187.

Molecular Potential Energy Surfaces

Any discussion of chemical reactivity is impossible without a knowledge of the structures of educt and product molecules. What is transformed into what? This information, though essential, is not nearly enough to allow the intricate details of structural transformations to be derived. Traditional methods of studying reaction mechanisms involve the measurement of reaction rates and the influence of temperature, concentration, solvent, and subtle molecular variations on these rates. Though these methods provide some degree of conceptual classification, they still leave the structural details of the transformation processes in the dark. The powerful structural probe furnished by X-ray analysis does not seem to offer much help in this respect because it is limited to the study of systems at equilibrium, and what we are after is the structural description of how molecular systems pass from one equilibrium state to another.

The concept of the potential energy surface of a reacting system needs to be introduced here since it provides a common language in which both static and dynamic properties of molecules can be discussed. It is assumed (Born-Oppenheimer approximation)[62] that the potential energy of a polyatomic system depends only on the relative spatial arrangement of the atomic nuclei. For N such particles, we require $3N-6$

[62]The basic idea behind the Born-Oppenheimer approximation is the separation of nuclear and electronic motion in the Schroedinger equation; the ground-state energy of the system is calculated with the nuclei assumed to be stationary, and the calculation is repeated for a range of other stationary arrangements. The motions of the nuclei are then supposed to follow classical mechanical trajectories, subject to the calculated energy surface.

coordinates to define their relative arrangement[63] provided $N > 2$ (for $N = 2$, we still need one coordinate, the internuclear distance). Every point in this multidimensional space represents some arrangement of the N particles and can be associated with a potential energy (we pass over the difficulty of actually calculating this quantity).

Equilibrium arrangements of atoms—i.e., stable molecules—correspond to the energy minima, and molecular transformations correspond to the passage of a representative point from one energy minimum to another. Actually, we deal with very large numbers of molecules, each of which takes a somewhat different trajectory on the surface, depending on the exact details of its random encounters with other molecules; nevertheless, if the molecules do not have too much kinetic energy, the paths will tend to run along the energy valleys that connect the minima over the lowest pass (transition state) between them. We can refer to these valleys as minimum energy paths or reaction paths. Knowledge of these paths would tell us how molecules must approach one another for a given reaction to proceed and how changes in interatomic distances or angles are coupled as bonds are broken and new ones formed. However, it is very difficult to obtain this information, either from experiments or from theoretical calculations.

Chemical reactions are often visualized in terms of a reaction coordinate/energy diagram of the type shown in Fig. 7.9 for a simple one-step reaction involving no intermediates. The reaction coordinate, extending along the abscissa, is to be thought of as a curve in a multidimensional space. The educt and product molecules are stable species, i.e., they correspond to energy minima with respect to all deformation coordinates, but species corresponding to intermediate points on the reaction coordinate are unstable, i.e., they would spontaneously collapse into educt or product unless acted on by external forces. The species corresponding to the transition state at the top of the energy profile would appear to be in unstable equilibrium, but transition-state theory says that all particles arriving there from the left will pass to the right, and vice-versa—like a tennis-ball passing over the net.[64]

In transition-state theory, the species corresponding to the energy maximum is considered to be in "virtual equilibrium" with the stable species, leading to the result that the number of molecules that reach

[63]We need three coordinates per particle ($3N$) less three to define the position of the system as a whole and three to define its orientation, except for a two-particle system, whose orientation is defined by only two coordinates. See Chapter 9 for further discussion.

[64]A clear account of transition-state theory is given by K. J. Laidler, *Theories of Chemical Reaction Rates,* McGraw-Hill, New York, 1969.

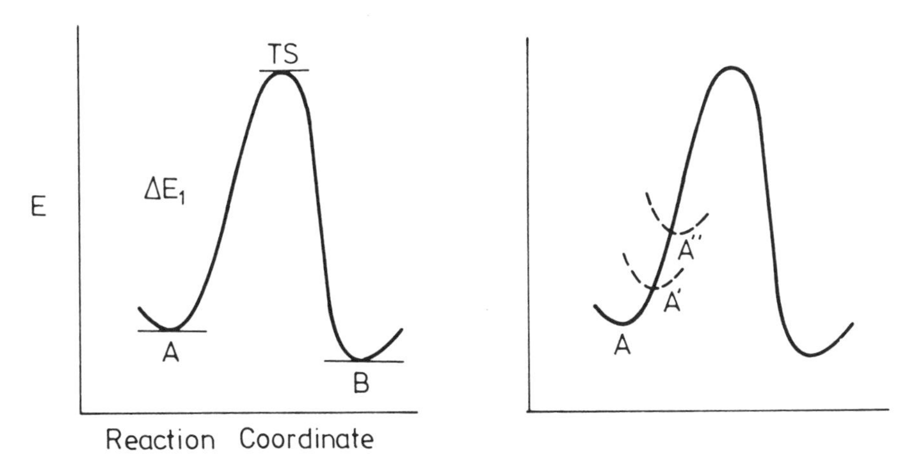

Figure 7.9. Left: reaction coordinate/energy diagram for a reaction involving no intermediates. Right: displacement of equilibrium structure A along reaction coordinate.

the transition state, say from the left, is proportional to exp $(-\Delta E_1/RT)$ in accordance with the Boltzmann distribution. From kinetic data it is then possible to infer the height of the energy barrier. However, as far as any particular molecule is concerned, the voyage over the barrier begins, takes its course, and ends in a very short time, on the order of $(kT/h) \sim 10^{-13}$ sec at $T = 300°K$; thus, it is hardly possible to make direct observations about the sequence of geometric changes that occur along the reaction coordinate. For qualitative descriptions of the overall process, we have had to rely mainly on mechanistic interpretations of the results of chemical kinetic studies. Quantum mechanical calculations of potential surfaces for simple systems are being carried out,[65] but there are theoretical and practical difficulties that rapidly become insurmountable with increasing complexity of the system. For more complicated systems, calculations at various levels of approximation can still be made, but a judicious blend of theoretical insight and chemical intuition is needed to simplify the reacting system to its essentials.[66] One soon reaches the point where one does not really know how reliable the results of the calculations are.

On the experimental side, we can determine the structures of stable

[65]For a recent review see R. F. W. Bader and R. A. Gangi, "Ab Initio Calculation of Potential Energy Surfaces," in *Theoretical Chemistry*, Vol. 2, The Chemical Society, London, 1975.

[66]For some recent examples, see M. Simonetta and A. Gavezzoti, *Struc. Bonding* **27,** 1 (1976); S. Scheiner, W. N. Lipscomb, and D. A. Kleier, *J. Am. Chem. Soc.* **98,** 4770 (1976); M. J. S. Dewar, G. J. Fonken, S. Kirschner, and D. E. Minter, *ibid.* **97,** 6750 (1975) and the review by G. Klopman, *Ann. Rev. Chem. Soc.* **72B,** 47 (1975).

molecular species by diffraction methods, we can assess the relative depths of the corresponding energy minima from equilibria or from thermochemical measurements, and we can estimate the heights of energy barriers from kinetic studies. For a few very simple or highly symmetric molecules, the behavior of the potential energy hypersurface near the equilibrium structure can be derived from analysis of vibrational spectra. All these pieces of information are essential, but they are inadequate even for a qualitative description of a reaction path.

Imagine that by making suitable changes in the species A (Fig. 7.9), we could move the potential energy minimum a little along the reaction coordinate. For example, for a *cis-trans* isomerization of a planar group, we might try to destabilize the planar *trans* amide group by building it into a small or medium ring:

CH_3–X=Y–CH_3 (trans) CH_3–X=Y–CH_3 (cis)

X=Y (bridged by a ring, trans)   X=Y (bridged by a ring, cis)

If we could carry out such changes in a systematic manner, we would obtain a series of molecules with structures resembling that of the initial species A but showing deformations corresponding to the displacement along the reaction coordinate, which, in our example, would be mainly the change in torsion angle about the X—Y bond but possibly involving changes in other structural parameters as well. Of course, the molecules thus obtained would not be exactly the same as the intermediate A → B species because the structural changes would not necessarily be restricted to those involving the reaction coordinate. However, we might hope that by observing the structural parameters of a sequence of increasingly deformed molecules, we could describe the main features of the reaction path for the isomerization in question.

Cis-Trans Isomerization of Amides

This approach has been applied to the *cis-trans* isomerization of the secondary amide group.

$$\mathrm{O{=}C(R_1)\text{—}N(R_2)(H)} \leftrightarrow \mathrm{O{=}C(R_1)\text{—}N(H)(R_2)}$$

Open-chain *N*-alkyl amides occur predominantly in the planar *trans* form, as shown by nmr and infrared studies of dilute solutions.[67] The amount of *cis* form depends on the substituents; it does not exceed 5% (except for *N*-alkyl formamides, where it may be 10–20%) and is essentially zero when R_1 and R_2 are bulky substituents. For *N*-methylacetamide, the activation energy for the *cis* → *trans* isomerization is about 19 kcal/mol, with the reverse process about 2.5 kcal/mol more costly,[68] and similar values are reported for other amides.[67] In small-ring lactams the amide group is necessarily *cis;* it can adopt a transoid conformation only when the ring is large enough to accommodate a torsion angle of approximately 180° about one of its bonds. This occurs at the nine-membered ring, caprylolactam, which exists in solution as an equilibrium mixture of several forms, mainly *cis,* but in the crystal exclusively as a transoid amide that shows marked deviations from planarity.

Before discussing the details of these deviations, a few words are necessary about the description of out-of-plane deformations. The $3N-6$ coordinates needed to specify the relative positions of N atoms may be chosen such that $2N-3$ lie in a given plane; the remaining $N-3$ then describe displacements from this plane, and they may be chosen in many different ways. It is convenient to distinguish two basically different kinds of out-of-plane deformations: those that involve twisting about bonds, and those that involve out-of-plane bending at the atoms concerned. These are illustrated for the simple case of ethylene (C_2H_4, $N=6$) in Fig. 7.10; the S_1 deformation is a pure twist, and S_2 and S_3 represent the symmetric and antisymmetric out-of-plane bending deformations. An alternative choice can be made by taking linear combinations of S_2 and S_3 to produce the set shown in the lower part of Fig. 7.10, where bendings are now localized at the individual carbon atoms. We also have a choice of coordinates to describe the magnitudes of these bending deformations; here we follow Warshel, Levitt, and Lifson[69] in choosing the coordinates χ, the dihedral angles between planes containing the central bond and the substituents at each carbon atom.[70]

The amide group ($N = 6$) also has three out-of-plane deformations, which may again be described in terms of two out-of-plane bending

[67]W. E. Stewart and T. H. Siddall, *Chem. Rev.* **70,** 517 (1970).

[68]T. Drakenberg and S. Forsén, *Chem. Commun.* **1971,** 1404.

[69]A. Warshel, M. Levitt, and S. Lifson, *J. Mol. Spectrosc.* **33,** 84 (1970).

[70]Various ways of choosing out-of-plane deformation coordinates and some relationships between them are described by H. B. Bürgi and E. Shefter, *Tetrahedron* **31,** 2976 (1975).

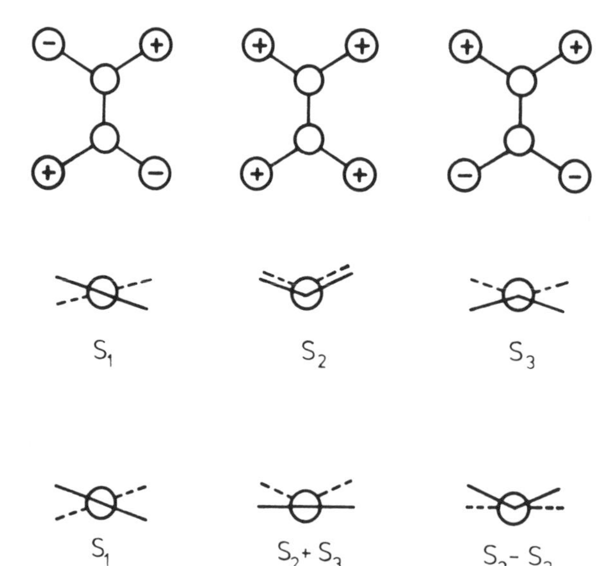

Figure 7.10. The three out-of-plane deformations of a planar X_2Y_4 molecule. The S_1 deformation represents a pure twist about the central bond; the S_2 and S_3 deformations represent symmetric and antisymmetric out-of-plane bendings. The linear combinations $S_2 + S_3$ and $S_2 - S_3$ represent bendings that are localized at either end of the molecule.

coordinates χ_N and χ_C, and a twisting coordinate τ. These quantities can be expressed in terms of the four torsion angles about the C′—N bond,[71] $\omega_1 = \omega(C_\alpha C'NC_\alpha)$, the main-chain or ring torsion angle, $\omega_2 = \omega(OC'NH)$, $\omega_3 = \omega(OC'NC_\alpha)$, $\omega_4 = \omega(C_\alpha C'NH)$, which are related by the condition $(\omega_1 + \omega_2) - (\omega_3 + \omega_4) = 0$ (modulo 2π) (Fig. 7.11):

$$\begin{aligned} \chi_C &= \omega_1 - \omega_3 + \pi = -\omega_2 + \omega_4 + \pi \text{ (modulo } 2\pi) \\ \chi_N &= \omega_2 - \omega_3 + \pi = -\omega_1 + \omega_4 + \pi \text{ (modulo } 2\pi) \qquad (7.1) \\ \tau &= (\omega_1 + \omega_2)/2 \end{aligned}$$

with the side condition $|\omega_1 - \omega_2| < \pi$, which states that a case with, say, $\omega_1 = 170°$, $\omega_2 = -175°$, should be treated as if $\omega_1 = 170°$, $\omega_2 = 185°$. Alternatively, the twisting coordinate can be redefined as $\tau' = \omega_1 + \omega_2$, which puts χ_N, χ_C, and τ' on the same scale but dispenses with the distinction between cisoid and transoid conformations. In general, values of χ_N and τ obtained from X-ray analysis are usually affected by uncertainty in the position of the amide hydrogen and are hence much less accurate than χ_C values.

The conformation found in the caprylolactam crystal[46] is shown in Fig. 7.12, and it is remarkably similar to that calculated for the corresponding cycloolefin, *trans*-cyclononene, with a carbon-carbon double bond in

[71] F. K. Winkler and J. D. Dunitz, *J. Mol. Biol.* **59**, 169 (1971).

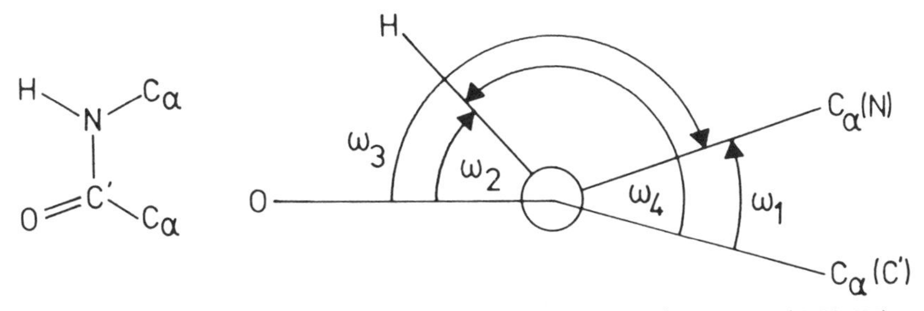

Figure 7.11. The four torsion angles, $\omega_1 = \omega(C_\alpha C'NC_\alpha)$, $\omega_2 = \omega(OC'NH)$, $\omega_3 = \omega(OC'NC_\alpha)$, and $\omega_4 = \omega(C_\alpha C'NH)$ are not independent since, for example, $\omega_1 - \omega_3 = \omega_4 - \omega_2$.

place of the amide group.[72] This similarity underlines the importance of ring-closure conditions (which are very similar in the two cases) in determining the general pattern of ring torsion angles. In particular, the important torsion angle ω_1(C(8)—N—C(1)—C(2)) is 148.4°, compared with 150.5° for the corresponding calculated angle in the cycloolefin. The other three torsion angles around the amide link are ω_2(H—N—C—O) = 177°, ω_3(C—N—C—O) = −25.8°, ω_4(H—N—C—C) = −9°, leading to χ_N = 23.1° (estimated standard deviation of 2.2°), χ_C = −5.8(0.5)° and τ = 162.9(1.1)°. From Eq. 7.1

$$\omega_1 = \tau + (\chi_C - \chi_N)/2$$

so that all three deformations are working cooperatively to change ω_1 away from 180°—i.e., to bring the ring members C(2) and C(8) closer than they would be for a *trans*-planar conformation of the amide group.

Starting from a coplanar group, it is always cheaper in energy to achieve a given torsion angle by a combination of out-of-plane bending distortion and twisting distortion than by a pure twisting distortion.[73,74] The actual amount of each kind of distortion will depend on local features of the potential energy surface. In our case, the torsion angle ω_1 equals $(\tau' + \chi_C - \chi_N)/2$, and it must be fixed within narrow limits by the ring-closure conditions since it is virtually the same in the lactam and the cycloolefin, where the double bond is certainly much more resistant to out-of-plane deformation than the amide link. If we assume that the potential function for out-of-plane deformation of the amide bond is

[72] O. Ermer and S. Lifson, *J. Am. Chem. Soc.* **95,** 4121 (1973).

[73] See Ref. 35, pp. 57–60, for a discussion of this point in connection with the nonplanar double bond of *trans*-cyclodecene.

[74] A recent X-ray analysis of a *trans*-cyclooctene derivative containing no heavy atoms shows that the large deviation of the C—C=C—C torsion angle (137.7°) from 180° results from approximately equal amounts of pure twisting and out-of-plane bending (O. Ermer, *Angew. Chem., Intern. Ed.* **13,** 604 (1974)).

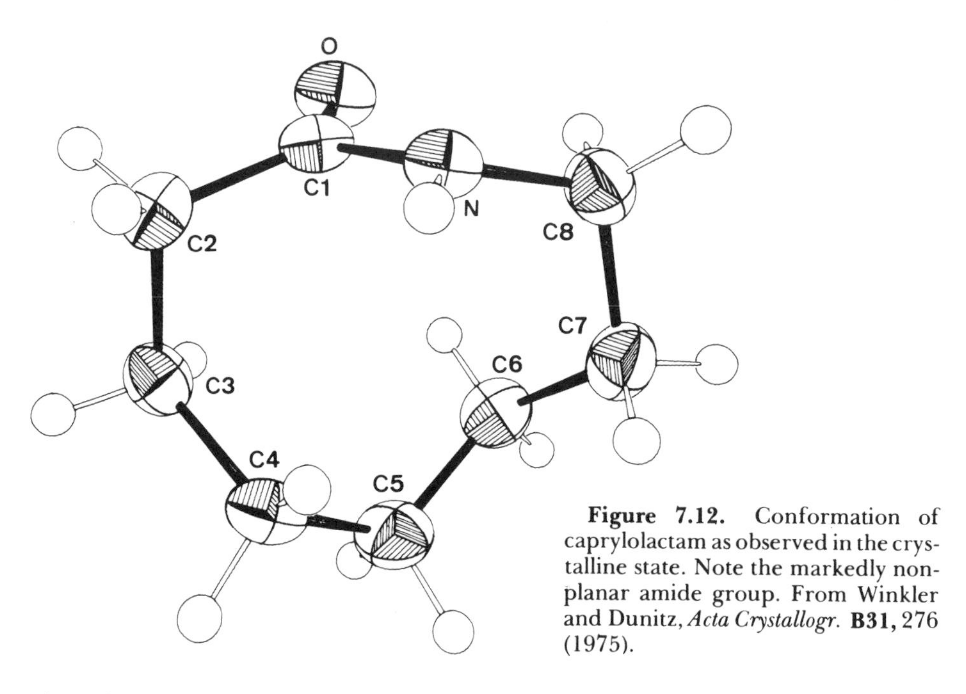

Figure 7.12. Conformation of caprylolactam as observed in the crystalline state. Note the markedly nonplanar amide group. From Winkler and Dunitz, *Acta Crystallogr.* **B31,** 276 (1975).

given by a sum of quadratic terms in τ', χ_N, and χ_C with no cross terms, i.e.,

$$2V = c_1\tau'^2 + c_2\chi_N^2 + c_3\chi_C^2 \tag{7.2}$$

then minimization of V with respect to τ', χ_N, χ_C under the constraint ω_1 = constant leads to the relationships $c_1\tau' = c_2\chi_N = c_3\chi_C$. In other words, the relative contributions of twisting and out-of-plane bending distortions should be inversely proportional to the quadratic potential constants in Eq. 7.2. With $\tau' = -34.2°(325.8°)$, $\chi_N = 23.1°$, $\chi_C = -5.8°$ we obtain $c_1:c_2:c_3 = 1:1.5:6$ approximately. The carbonyl end of the amide group thus seems to have a much greater resistance to out-of-plane bending than the nitrogen end. The same conclusion can be drawn from observations of other cyclic and open-chain amides[75,76] and from results of vibrational analyses.[75]

Since out-of-plane bending at the carbon atom (χ_C) appears to play a relatively minor role in the out-of-plane deformations of the amide group, we may ignore it entirely in a first approximation to the description of the corresponding energy function. Various lines of evidence suggest that the energy profile along the χ_N coordinate is rather flat

[75] J. D. Dunitz and F. K. Winkler, *Acta Crystallogr.* **B31,** 251 (1975).
[76] S. E. Ealick and D. van der Helm, *Acta Crystallogr.* **B33,** 76 (1977).

around $\chi_N = 0$. Coplanarity of the atoms of the amide group ensures optimal interaction between trigonally hybridized carbon and nitrogen atoms, but this is achieved at the energy cost required to bring the three bonds at N into a common plane (roughly 5–6 kcal/mol, by analogy with the inversion barrier of ammonia). A slight degree of pyramidalization at N can thus be said to correspond to worse π orbital overlap but to better hybridization, and the superposition of these effects should produce a rather flat potential along the χ_N coordinate or even a double minimum with a small barrier at $\chi_N = 0$. The experimental evidence that might allow us to decide between these two possibilities is equivocal. Costain and Dowland's microwave analysis of formamide[77] supported the double-minimum picture ($\chi_N \sim 19°$, barrier $\sim$ 1.1 kcal/mol), but a more recent study[78] gives a very flat single-minimum quartic potential. Quantum-mechanical calculations for *N*-methylacetamide[79,80] and for formamide, acetamide, and *N*-ethylacetamide[80] are also indecisive; the CNDO/2 approximation gives double minima with small barriers, while the INDO approximation gives flat single minima.[81] According to fairly elaborate ab initio calculations for formamide,[82] the sign of the small energy difference between the planar and puckered structures is changed by altering the basis set. In any case, the microwave evidence and the quantum-mechanical calculations refer to molecules in the gas phase and may not be directly relevant to amides in condensed phases; here, it may be argued, the increase in C′—N double bond character due to N—H···O hydrogen bonding should increase the resistance to out-of-plane bending at nitrogen. Bond lengths and bond angles observed for amides in condensed phases differ significantly from the gas-phase values, and the pattern of changes generally supports the idea that hydrogen bonding (incipient protonation at carbonyl oxygen) is a major factor.[75]

The simplest expressions for the τ', χ_N energy surface that allow for

[77]C. C. Costain and J. M. Dowling, *J. Chem. Phys.* **32,** 158 (1960).

[78]E. Hirota, R. Sugisaki, C. J. Nielsen, and G. O. Sørensen, *J. Mol. Spectrosc.* **49,** 251 (1974).

[79]G. N. Ramachandran, A. V. Lakshminarayanan, and A. S. Kolaskar, *Biochem. Biophys. Acta* **303,** 8 (1973); G. N. Ramachandran and A. S. Kolaskar, *idem.,* 385.

[80]A. S. Koleskar, A. V. Lakshminarayanan, K. P. Sarathy, and V. Sasisekharan, *Biopolymers* **14,** 1081 (1975).

[81]The results of these calculations (Refs. 79, 80) are described in terms of a different coordinate system from the one used here. The other coordinate system has its axes along χ_N and ω_1 (our notation). For the analogous deformations of ethylene, τ and χ are mutually orthogonal symmetry coordinates, whereas ω and χ are nonorthogonal coordinates.

[82]D. H. Christensen, R. N. Kortzeborn, B. Bak, and J. J. Led, *J. Chem. Phys.* **53,** 3912 (1970).

the features mentioned above contain at least four potential constants. Out of many possibilities, we choose one that has its minimum at the planar structure, is sinusoidal along the τ' axis, and reduces to a harmonic potential along the χ_N axis:

$$V(\tau', \chi_N) = \frac{V_o}{2}(1 - \cos \tau') + p_N \chi_N^2 + q_N(1 - \cos \tau')[\exp(-\alpha \chi_N^2) - 1]$$

The values of V_o, p_N, q_N, and α can be estimated by fitting the potential to four experimental data or assumptions:

(a) $V(180°, \chi_N)$ has a minimum at $\chi_N = 60°$ (pyramidal grouping with tetrahedral bond angles as in an amine).

(b) $V(180°, 0) - V(180°, 60°) = 6$ kcal/mol (inversion barrier of ammonia).

(c) The energy barrier to *cis-trans* isomerization is 20 kcal/mol.

(d) The quadratic constant p_N is set at 6 kcal/mol radians^2, an estimate based on a survey of results of various vibrational analyses.

With all energy quantities in kcal/mol and angles in degrees, we obtain $V_o = 26.0$, $p_N = 1.83 \times 10^{-3}$, $q_N = 9.075$, $\alpha = 3.28 \times 10^{-4}$.

The function is shown in Fig. 7.13. It should not be taken too seriously for it is oversimplified in several ways. A potential function that purports to be applicable to amides in general can hardly avoid being oversimplified. Ours has periodicity 2π along τ' (π along τ) and hence does not distinguish between *cis* and *trans* amides. Because of the mirror symmetry across the lines $\tau' = 0, 180°$ and $\chi_N = 0$, deformations corresponding to points in the $(+, +)$ and $(+, -)$ quadrants are considered to be equivalent, but they are actually distinct (one combination leads to eclipsing of the nitrogen lone pair with the C═O bond, the other to an antiplanar orientation). Interactions among substituents, which would vary from one amide to another have been ignored. Nevertheless, Fig. 7.13 does express essential features of the τ', χ_N potential surface. The points marked correspond to deformations actually observed in the highly strained nonplanar amides, caprylolactam,[46] and compounds I–IV.[76]

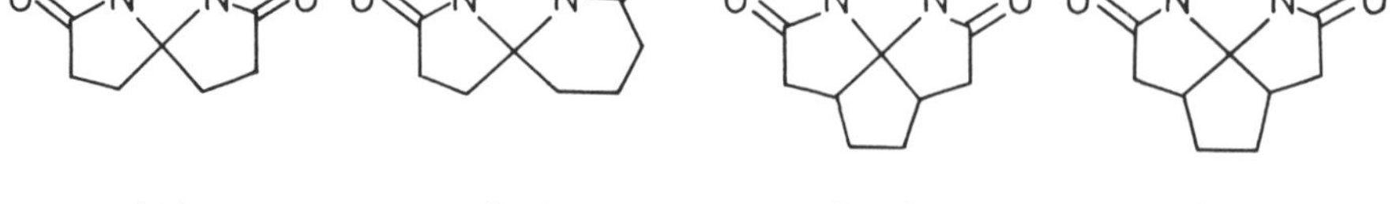

(I) (II) (III) (IV)

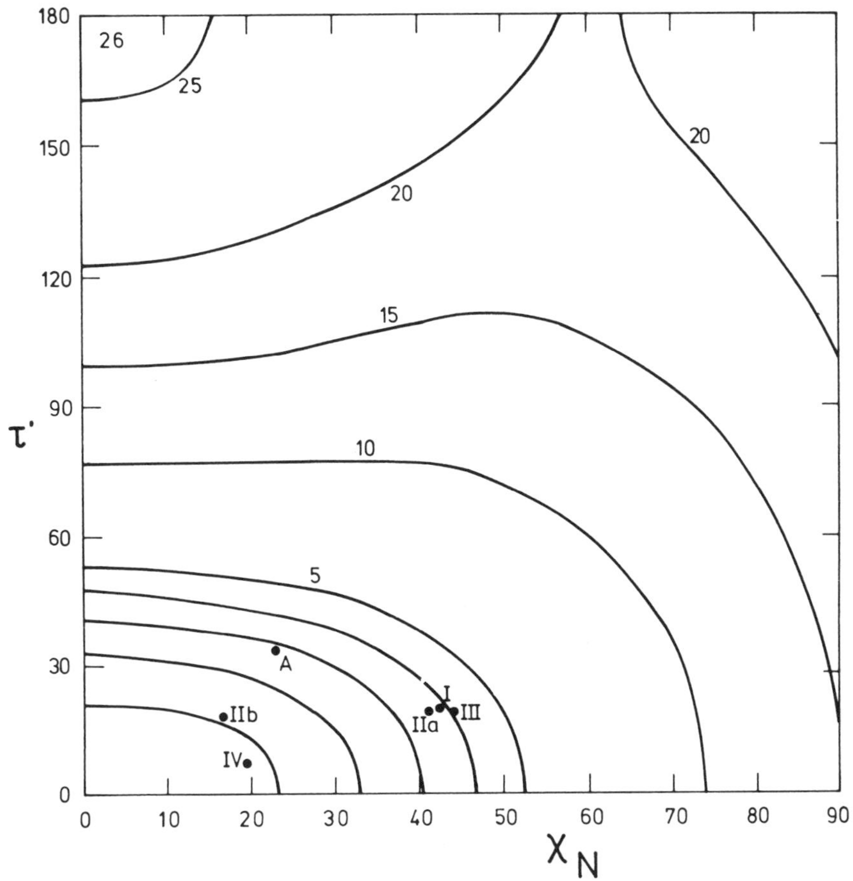

Figure 7.13. Approximate form of $V(\tau',\chi_N)$ energy surface for amides. The points marked correspond to deformations observed in caprylolactam (A) and compounds I–IV.

The caprylolactam point is somewhat displaced from the minimum of the energy profile along the line $\tau' + \chi_N$ = constant, but the energy difference with respect to this minimum is only about 0.1 kcal/mol. For the other four points, involving fused ring systems, the constraints have not been investigated. On the whole, the observed deformations seem quite compatible with the approximate potential function and support the idea that τ' and χ_N are strongly coupled along the energy valley leading from planar amide to the transition state for *cis-trans* isomerization.

It is sometimes assumed that the energy increment associated with out-of-plane deformation of the amide group is described by a potential of the form:

$$V(\omega) = \frac{V_o}{2}(1 - \cos 2\,\omega) \qquad (7.3)$$

and ab initio SCF LCAO calculations on formamide[82,83] seem to support the approximate validity of this potential with $V_o \sim 20$ kcal/mol. As long as χ_C and χ_N are zero, there is no difficulty in applying this formula since the four torsion angles around the amide bond are either equal or supplementary. In general, however, this condition does not apply. For example, for caprylolactam the four torsion angles are $\omega_1 = 148°$, $\omega_2 = 177°$, $\omega_3 = -26°$, $\omega_4 = -9°$, and it is not clear which value should be used to estimate the energy increment. With the main chain torsion angle ω_1 we would obtain 5.5 kcal/mol, but with ω_2 we would obtain only 0.1 kcal/mol, for example. From Fig. 7.13 the approximate strain energy of the amide group of caprylolactam is about 2.9 kcal/mol, so it is severely overestimated in one case and underestimated in the other. The free energy of activation for interconversion of *cis* and *trans* isomers of caprylolactam has been estimated as 17 kcal/mol from ^{13}C nmr studies,[84] about 3 kcal/mol smaller than the value for a strain-free amide. This difference can be taken as a rough experimental estimate of the strain energy of caprylolactam, and it is in excellent agreement with the value estimated from Fig. 7.13.

What about the structurally similar enamine grouping?

Should this be planar, like the amide group, because the nitrogen lone pair can conjugate with the C═C double bond, or should it be non-planar, with a pyramidal N atom, because the conjugation with C═C is so much weaker than with C═O that the energy stabilization attributable to conjugation is insufficient to outweigh the energy cost of bringing the three bonds at N into a common plane? The recent structure analysis of

[83]M. Perricaudet and A. Pullman, *Int. J. Peptide Protein Res.* **5**, 99 (1973). The calculations were performed with all bond angles assumed equal to 120° (i.e., $\chi_C = 0$, $\chi_N = 0$) and constant bond lengths. The effect of varying χ_N as the twist angle changes was also neglected in the calculations described in Ref. 82.

[84]K. L. Williamson and J. D. Roberts, *J. Am. Chem. Soc.* **98**, 5082 (1976). These authors use $\omega_3 = 26°$ to calculate the distortion energy from Eq. 7.3, obtaining the value 3.8 kcal/mol, but they give no reasons for selecting this particular torsion angle out of the four at their disposal.

(V) 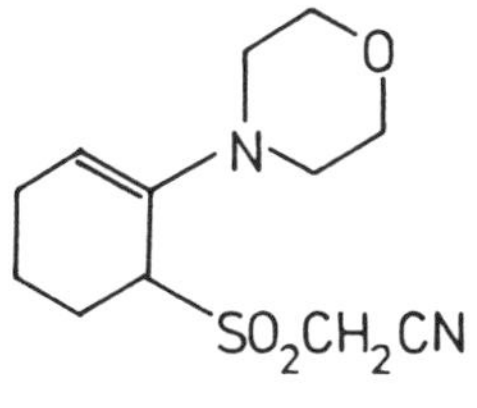 (VI)

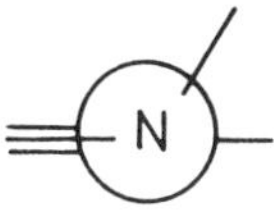

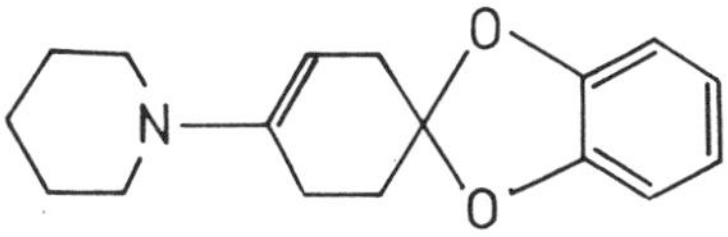

(VII) 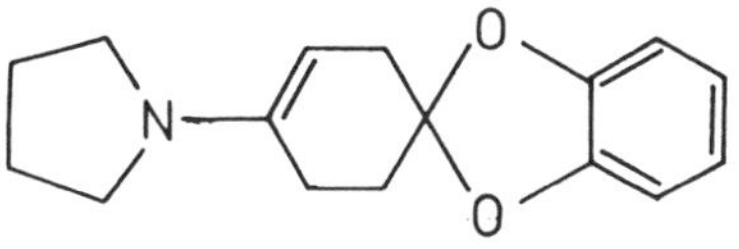 (VIII)

3-cyanomethylsulfonyl-2-morpholinocyclohexene (V)[85] shows a distinctly pyramidal nitrogen atom; the N—C(sp^2) bond length of 1.42 Å is about 0.1 Å longer than the typical value in an amide group, and one of the N—C morpholine bonds virtually eclipses the C═C bond,[86] the other being inclined at a steep angle to the N—C═C plane (VI). The same features occur in the piperidine enamine VII.[87] On the other hand, analyses of the analogous VIII and of other pyrrolidene enamines show N atoms that are palpably nonplanar but distinctly less pyramidal than in the morpholine and piperidine enamines VI and VII.[88] Thus, although the nitrogen atom of the enamine grouping seems to be essentially pyramidal rather than planar, the variability in the degree of pyramidality suggests that the energy profile along the enamine nitrogen inversion path is rather flat.

From Crystal Structure Data to Chemical Reaction Paths

In contrast to the *cis-trans* isomerization discussed in the previous section, most chemical reactions involve the breaking of some bonds and

[85]M. P. Sammes, R. L. Harlow, and S. H. Simonsen, *J. Chem. Soc., Perkin II* **1976,** 1126.

[86]In aliphatic aldehydes and carboxylic acids, the C_α—C_β bond is nearly always in *syn*-planar orientation to the C═O bond (J. D. Dunitz and P. Strickler, in *Structural Chemistry and Molecular Biology,* A. Rich and N. Davidson, Eds., Freeman, San Francisco, 1968, p. 595). In terms of the bent-bond description of double and triple bonds (L. Pauling, in *Theoretical Organic Chemistry,* IUPAC Kekulé Symposium, Butterworths, London, 1959), the *syn*-planar, eclipsed conformation C_β—C_α—C═O corresponds to the usual staggered disposition of bonds about the central C_α—C bond. The same kind of description would apply here.

[87]R. Hobi, Doctoral Dissertation, E.T.H., Zürich 1977; K. L. Brown and R. Hobi, unpublished results.

[88]C. Kratky, K. L. Brown, R. Hobi, and L. Damm, unpublished results.

Table 7.1. Characteristic distances for various types of bonds found in organic molecules

Bond	Type	Value (Å)	Bond	Type	Value (Å)
C—C	C—CH$_2$—CH$_2$—C	1.54	C—N	—>C—NH$_2$	1.47
	C—CH$_2$—CH=C	1.51		RCO—NH·R	1.33
	C—CH$_2$—C≡C	1.46	C=N		1.28-1.38
	C=CH—CH=C	1.46	C—F	—>C—F	1.38
	C=CH—C≡C	1.44		=CH—F	1.33
	C≡C—C≡C	1.38	C—Cl	—>C—Cl	1.77
	C—CH$_2$—C(ar)	1.51		=CH—Cl	1.71
	C(ar)—C(ar)	1.51		≡C—Cl	1.64
C=C	C(ar)=C(ar)	1.39	C—Br	—>C—Br	1.94
	C—CH=CH—C	1.33		=CH—Br	1.87
	>C=C=C<	1.31		≡C—Br	1.80
	>C=C=C=C<	1.28	C—I	—>C—I	2.14
C≡C		1.20		=CH—I	2.07
C—O	R—O—R	1.43	C—H		1.06-1.10
	RCO—OR	1.31-1.36	N—H		1.01
	RCOO—R	1.43-1.45	O—H		0.96
	C(ar)—OR	1.36	S—S	—S—S—	2.05
C=O	>C=O	1.19-1.22	O—O	—O—O—	1.48
	=C=O	1.16	N—N	>N—N<	1.45
	—COO$^-$	1.26	N=N	—N=N—	1.25
C—S	R—S—R	1.82	N—O	>N—O—	1.36
C—S	>C=S	1.71		—NO$_2$	1.22
C—P	—>C—P	1.84			
C—Si		1.87			

the formation of others. So far we have more or less taken it for granted that, given the structure of a molecular crystal, we can always tell a bond from a nonbond. Bond distances are much shorter than nonbonded ones; the various types of covalent bond are associated with characteristic distances (Table 7.1),[89] and contact distances between pairs of atoms not bonded to each other and not forced together by other bonds are typically much larger than these. In fact, nonbonded or van der Waals radii (Table 7.2) are generally about 0.7–1.0 Å larger than corresponding covalent radii, so that there is a gap of 1.5–2.0 Å between typical bonded and nonbonded interatomic separations.

Particular arrangements of bond lengths, bond angles, and torsion angles within a given molecule may force pairs of nonbonded atoms into closer contact than that given by the appropriate nonbonded radii. For example, the 1,3-distance between carbon atoms is about 2.55 Å in a paraffin chain, about 2.40 Å in benzene, and about 2.20 Å in a cyclo-

[89]Based largely on the list of selected bond lengths in *Interatomic Distances, Supplement*, L. E. Sutton, Ed., Special Publication No. 18, The Chemical Society, London, 1965.

Table 7.2. Van der Waals radii of atoms (in Å), after Pauling[3]

		H	1.2		
N	1.5	O	1.4	F	1.35
P	1.9	S	1.85	Cl	1.80
As	2.0	Se	2.0	Br	1.95
Sb	2.2	Te	2.2	I	2.15

Radius of methyl group (CH_3), 2.0 Å. Half-thickness of aromatic molecule, 1.85 Å.

butane ring. These intramolecular distances are much shorter than typical nonbonded contact distances between carbon atoms (3.4–3.6 Å), but they are much longer than typical bonded distances (1.2–1.5 Å) and hence can easily be distinguished from the latter. In the absence of such constraints, nonbonded contacts that are appreciably shorter than normal have to be attributed to specific attractive interactions between the atoms concerned. Hydrogen bonds, between pairs of electronegative atoms separated by a proton, constitute the best known class of such interactions.

During the past few years, however, several examples of interatomic distances that are intermediate between the traditional bonding and nonbonding ranges have been uncovered. In the donor-acceptor molecular compound formed by trimethylamine and iodine, $(CH_3)_3N\cdots I—I$, the I—I distance is 2.83 Å, about 0.15 Å longer than in gaseous iodine, the N··· I—I grouping is linear, and the N··· I distance is 2.3 Å, about 0.25 Å longer than the value expected for a covalent bond (ca. 0.70 + 1.33 = 2.03 Å) but much less than the sum of the van der Waals radii (1.5 + 2.15 = 3.65 Å).[90] Anomalously long bonds (or anomalously short nonbonded contacts) also occur in many other complexes of strong electron donors (trimethylamine, pyridine, 4-picoline, hexamethylenetetramine, 1,4-dioxane, 1,4-dithiane, sulfur, 1,4-diselane, etc.), with electron acceptors such as the halogens (especially I_2), diiodoacetylene, iodoform, and so forth.[91,92]

Long bonds also occur in organic molecules. A remarkable example is the pair of 9,9′ carbon—carbon bonds in the photoisomer of bi(anthracene-9,10-dimethylene) (IX), which are 1.77 Å long.[93] This is 0.23 Å longer than a standard C—C bond and would correspond to a bond

[90] K. O. Strømme, *Acta Chem. Scand.* **13,** 268 (1959).
[91] O. Hassel and C. Rømming, *Quart. Rev.* **16,** 1 (1962).
[92] H. A. Bent, *Chem. Rev.* **68,** 587 (1968).
[93] M. Ehrenberg, *Acta Crystallogr.* **20,** 182 (1966).

number of about ½, according to the relationship $\Delta d = -0.71 \log n$ proposed by Pauling,[94] leaving some doubt whether the compound should be formulated as IX or as X, (or various equivalent forms with no bonds between the 9 and 9′ positions). It has been noted that long bonds (1.61–1.64 Å) tend to occur in molecules where a σ bond is nearly parallel to four adjacent π orbitals (XI); the weakening of the σ bond is attributed to through-bond coupling.[95]

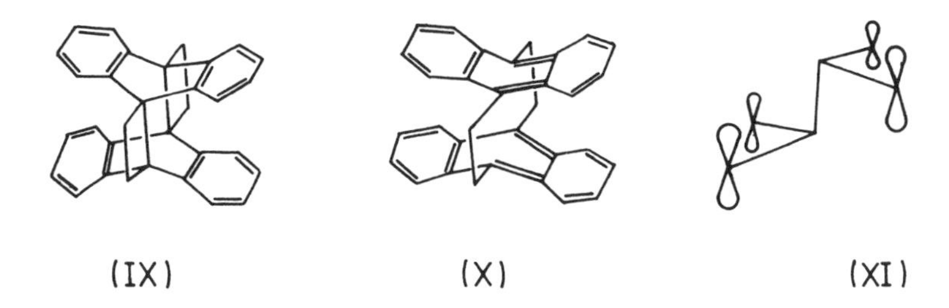
(IX) (X) (XI)

Another even longer "bond" occurs in XII, where the central C—C distances are 1.78 Å and 1.84 Å in the two independent molecules present in the crystal.[96] The possibility that the crystal is composed of a random mixture of isomers, some with the tetracyclic structure XII and some with the bridged 10-annulene structure XIII, has been completely eliminated by a careful analysis of the vibration ellipsoids. If a C—C distance of 1.84 Å can correspond to a long "bond," what about the distance of 1.89 Å[97] between the nonbonded carbon atoms in bicyclo[1.1.1]pentane XIV or about the transannular N···C distance of 1.99 Å[98] in the eight-membered ring of the alkaloid clivorine XV?

The above examples suggest that the distance criterion between bonds

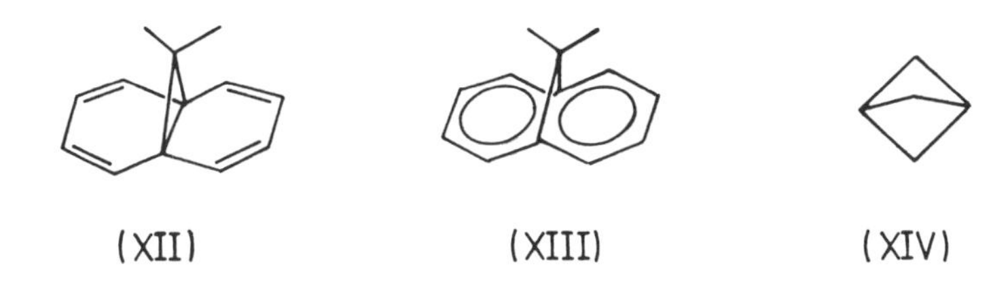
(XII) (XIII) (XIV)

[94] L. Pauling, *J. Am. Chem. Soc.* **69,** 542 (1947).

[95] D. A. Dougherty, W. D. Hounshell, H. B. Schlegel, R. A. Bell, and K. Mislow, *Tetrahedron Lett.* **1976,** (39), 3479.

[96] R. Bianchi, G. Morosi, A. Mugnoli, and M. Simonetta, *Acta Crystallogr.* **B29,** 1196 (1973).

[97] A. Padwa, E. Shefter, and E. Alexander, *J. Am. Chem. Soc.* **90,** 3717 (1968).

[98] K. B. Birnbaum, *Acta Crystallogr.* **B28,** 2825 (1972).

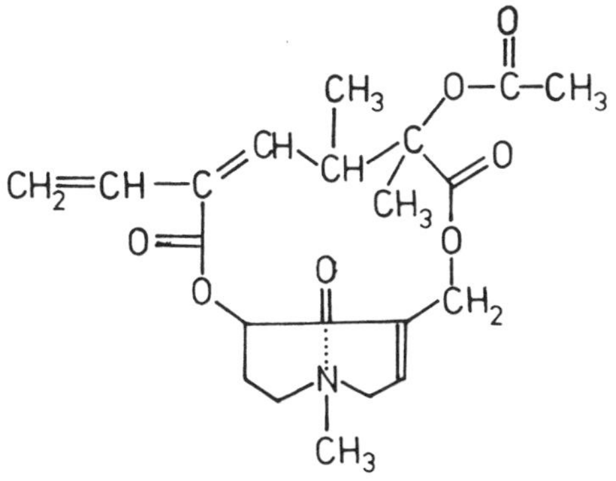

Clivorine
(XV)

and nonbonded contacts may not be so sharp after all. Indeed, the structural evidence supports the view that for certain classes of molecules there is a more or less continuous transition from nonbonding to covalent bonding. In the following sections, we discuss some of this evidence and its implications.

Linear Triatomic Species

Structure analyses of many crystals containing different counter-ions show that the I···I···I grouping is linear or nearly so (bond angle 175°–180°) and that the I··· I distances lie between the standard values—i.e., 2.67 Å for covalent bonding and 4.30 Å for nonbonded contact (Table 7.3). If one distance is plotted against the other, the scatter plot shown in Fig. 7.14a is obtained; there is an obvious correlation between the two distances. Very similar experimental correlations are obtained for the S—S···S grouping in thiathiophthenes (XVI, Fig. 7.14b), the O—H···O grouping in various hydrogen-bonded structures (Fig. 7.14c), and other linear or nearly linear triatomic systems.[99]

(XVI)

[99]H. B. Bürgi, *Angew. Chem.* **87,** 461 (1975); *Angew. Chem. Intern. Ed.* **14,** 460 (1975). The experimental data on which Fig. 7.14 is based are described in this article.

Table 7.3. Interatomic distances (Å) in polyiodide anions[a]

Compound	$d_1(I_1—I_2)$	$d_2(I_2—I_3)$	$\sigma(d)$
I_2 (g)	2.67	4.30[b]	—
I_2 (crystal)	2.70	3.54	—
$[(C_2H_5)_4N]I_7$	2.735	3.435	0.003
NH_4I_3	2.791	3.113	0.004
$[(C_2H_5)_4N]I_5$	2.81	3.17	0.015
Cs_2I_8	2.84	3.00	0.03
CsI_3	2.842	3.038	0.002
$[(C_5H_5)_2Fe]I_3$	2.85	2.97	0.02
	2.89	2.97	0.02
$[(C_2H_5)_4N]I_3$	2.892	2.981	0.004
	2.912	2.961	0.004
$[(C_2H_5)_4N]I_7$	2.904	2.904	0.003
$[(C_6H_5)_4As]I_3$	2.919	2.919	0.001
$[(C_5H_5)_2Fe]I_3$	2.93	2.93	0.02
$[(C_2H_5)_4N]I_3$	2.928	2.928	0.003
	2.943	2.943	0.003

[a]Data from Bürgi, *Angew. Chem., Intern. Ed.* **14,** 460 (1975). This reference contains the original literature sources.
[b]Van der Waals diameter.

In his 1968 review of donor-acceptor interactions[92] Bent suggested that "certain kinds of attractive intermolecular interactions may be viewed as incipient valence shell expansions and often as the first stage of bimolecular nucleophilic displacement reactions." In particular, he drew attention to the correlation between the two I···I distances observed in the linear I_3^- anion and noted that "the hyperboloid curve may be presumed to show, approximately, the changes that occur in the distances between nearest neighbors in the linear exchange reaction $I^1 + I^2I^3 = I^1I^2 + I^3$." Similarly, the hyperboloid curves shown in Fig. 7.14b,c may be presumed to depict the corresponding changes in interatomic distance for the thiathiophthene valence isomerization reaction (XVI) and for proton transfer between a pair of oxygen atoms (OH···O↔O···HO).

There is an obvious resemblance between the curves of Fig. 7.14 and the minimum energy path on the calculated potential energy surface for the linear H_3 system—i.e., for the simplest of all exchange reactions, $H_2 + H \leftrightarrow H + H_2$, where the nuclei are constrained to lie in a line. Many quantum mechanical calculations have been performed for this system;[65] the surface shown in Fig. 7.15 is based on the extensive ab

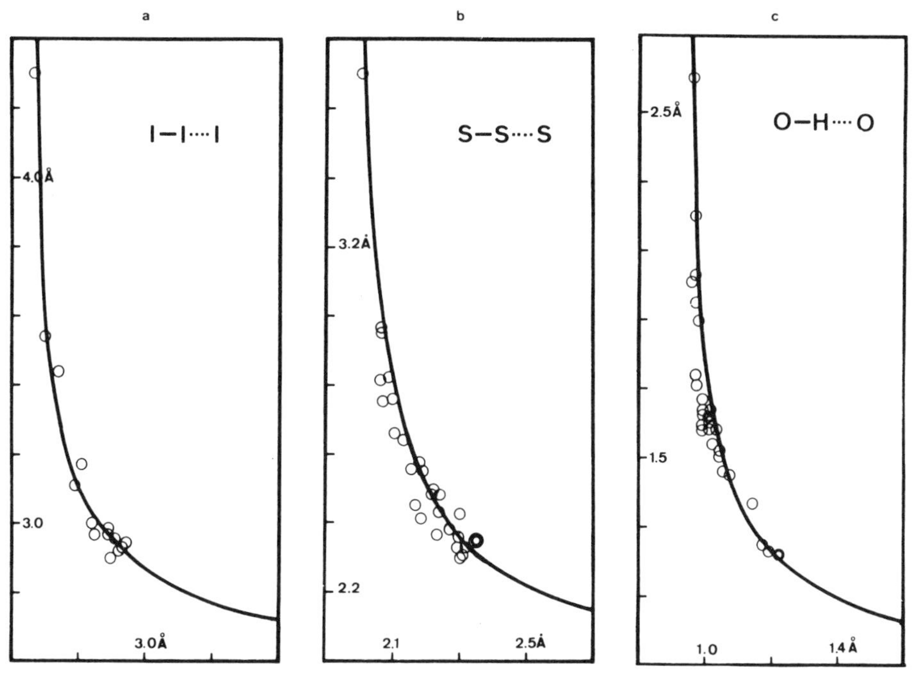

Figure 7.14. Correlation plots of interatomic distances in linear triatomic systems. (a) Triiodide anions, (b) thiathiophthenes, and (c) O—H··O hydrogen bonds. From Bürgi, *Angew. Chem., Intern. Ed.* **14,** 460 (1975).

initio calculations performed by Liu.[100] The profiles along the top and right represent the potential energy of an H_2 molecule (equilibrium distance 1.401 bohrs = 0.741 Å) as function of internuclear distance, with the third H atom so far off that its interaction with the H_2 molecule is negligible. As this atom begins to approach H_2, the potential energy of the system increases, but the energy increase can be minimized by adjusting the H_2 internuclear distance, say d_1, for a given value of d_2, leading to the minimum energy path depicted by the dashed curve. The symmetrical arrangement, $d_1 = d_2 = 0.930$ Å, corresponds to a saddle point or transition state, which is calculated to be 9.8 kcal/mol above the energy of isolated H_2 + H. At the transition state, the reaction coordinate is tangent to the minimum energy path and corresponds to the antisymmetric stretching mode of the activated complex. This mode is associated with an imaginary frequency since the second derivative of the energy along the path is negative at the saddle point.

[100]B. Liu, *J. Chem. Phys.* **58,** 1925 (1973).

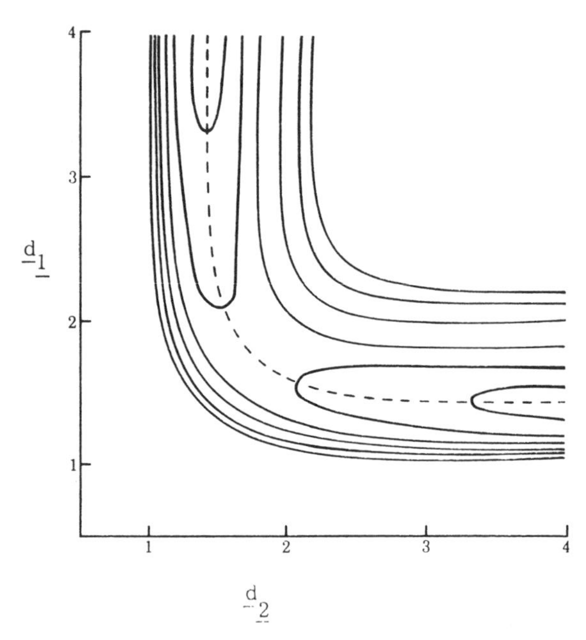

Figure 7.15. Potential energy surface for linear H_3 system, drawn from data calculated by Liu [*J. Chem. Phys.* **58**, 1925 (1973)]. Distances are in bohr units (1 bohr = 0.529 Å), and contour lines are drawn at energy intervals of 0.01 hartrees (1 hartree = 27.21 eV).

The lengthening of 0.19 Å of the H···H distance at the transition state is not too different from the corresponding lengthenings found in the symmetrical I···I···I, S···S···S, and O···H···O structures, which all lie in the range 0.25–0.30 Å. Despite the enormous chemical disparities among these systems, the general similarity between the minimum energy path of Fig. 7.15 and the correlation curves of Fig. 7.14 might encourage us to believe that the latter do represent analogous minimum energy paths for the reactions in question. However, even if this interpretation is correct, we should still refrain from referring to the symmetrical points on these paths as transition states because we do not know if they correspond to saddle points or to dips in the energy surface.

The correlation curves in Fig. 7.14 are not just smooth curves drawn to fit the scatter plots as well as possible; they have a special analytical form derived from a simple intuitive model of chemical bonding. In 1947 Pauling[94] found that he could make sense out of the variable coordination numbers and interatomic distances observed in metals by relating the distances to what he called the bond number n. The length of a fractional bond $d(n)$ was related to that of a single bond by:

$$d(n) - d(1) = \Delta d = -c \log n \tag{7.4}$$

The constant c was assigned the value 0.60 Å (decadic logarithms), but it can be expected to vary somewhat, depending on the nature of the atoms involved. In discussing the linear triatomics we adopt the Pauling relationship with the additional assumption that the sum of the two bond numbers equals unity for all related pairs of distances—i.e., we assume conservation of bond number along the minimum energy path. This leads immediately to the relationship:

$$10^{-\Delta d_1/c} + 10^{-\Delta d_2/c} = 1 \qquad (7.5)$$

where Δd_1 and Δd_2 are related bond distance increments. The constant c can be evaluated from $c = \Delta d'/\log 2 = 3.32\ \Delta d'$, where $\Delta d'$ is the bond distance increment in the symmetrical structure ($n = \frac{1}{2}$). For each scatter plot we then have to estimate two distances: $d(1)$, the single-bond length, to which the increments Δd are referred, and $\Delta d' = d(\frac{1}{2}) - d(1)$, in order to evaluate c. The curves of Fig. 7.14 are based on the estimates (in Å):[99]

$d(1) = 2.67$, $\Delta d' = 0.26$ ($c = 0.85$) for I···I···I

$d(1) = 2.02$, $\Delta d' = 0.31$ ($c = 1.03$) for S···S···S

$d(1) = 0.96$, $\Delta d' = 0.26$ ($c = 0.85$) for O···H···O

and they fit the experimental scatter plots as well as can be desired. With $d(1) = 0.741$ Å and $d(\frac{1}{2}) = 0.930$ Å ($c = 0.62$ Å), the sum of the bond numbers for the H_3 system lies between 1.00 and 1.02 for all points along Liu's calculated minimum energy path (Fig. 7.15), so that bond number conservation appears to hold very well here too. The assumption of bond order conservation was first introduced by Johnston to infer properties of reaction paths for some gas-phase reactions;[101] as he remarked, "the assumption cannot be purely correct, but it cannot be totally wrong either, and it provides a definite starting point for the discussion."

A survey of experimental results for other symmetrical three-center four-electron systems X···Y···X (Table 7.4) found in crystals indicates that $\Delta d'$ varies only between narrow limits (0.21–0.26 Å) for such systems. There is just a hint that the more stable trihalide ions ICl_2^- and IBr_2^- are associated with slightly lower $\Delta d'$ values than the less stable ones; ICl_2^- and IBr_2^- occur only as symmetrical or nearly symmetrical structures, whereas the I_3^- species is often markedly asymmetrical in crystal environments (Table 7.3). Similarly, the smaller value of $\Delta d'$ for HF_2^- compared with O···H···O may reflect the greater stability of the former, which is calculated[102] to be 44 kcal/mol lower in energy than

[101] H. S. Johnston, *Adv. Chem. Phys.* **3,** 131 (1960).
[102] F. Keil and R. Ahlrichs, *J. Am. Chem. Soc.* **98,** 4787 (1976).

Table 7.4. Values of $d(1)$ and $d(\frac{1}{2})$ for various linear triatomic systems[a]

XYX	$d(1)$ (Å)	$d(\frac{1}{2})$ (Å)	$\Delta d'$ (Å)	c (Å)
O···H···O	0.957 (H_2O, g)	1.22[c]	0.26	0.86
$(F\cdots H\cdots F)^-$	0.917 (HF, g)	1.13[d]	0.21	0.70
$(Br\cdots Br\cdots Br)^-$	2.284 (Br_2, g)	2.54[e]	0.26	0.86
$(I\cdots I\cdots I)^-$	2.667 (I_2, g)	2.93[f]	0.26	0.86
$(Cl\cdots Br\cdots Cl)^-$	2.138 (BrCl, g)	2.40[g]	0.26	0.86
$(Cl\cdots I\cdots Cl)^-$	2.321 (ICl, g)	2.55[h]	0.23	0.76
$(Br\cdots I\cdots Br)^-$	2.485 (IBr, g)[b]	2.71[i]	0.23	0.76

[a]Except for IBr, $d(1)$ values are taken from *Interatomic Distances, Supplement,* L. E. Sutton, Ed., Special Publication No. 18, The Chemical Society, London, 1965.
[b]T. S. Jaseja, *J. Mol. Spectrosc.* **5,** 445 (1960).
[c]Averaged value from data listed by J. C. Speakman in *Molecular Structure by Diffraction Methods,* G. A. Sim and L. E. Sutton, Eds., Vol. 3, The Chemical Society, London, 1975, p. 89.
[d]KHF_2; J. A. Ibers, *J. Chem. Phys.* **40,** 402 (1964).
[e]From data given by G. L. Breneman and R. D. Willett, *Acta Crystallogr.* **B25,** 1073 (1969).
[f]See Table 7.3.
[g]$[(C_2H_5)_4N](BrCl_2)$; W. Gabes and K. Olie, *Cryst. Struct. Commun.* **3,** 753 (1974).
[h]$KICl_2$ and $KICl_2\cdot H_2O$; S. Soled and G. B. Carpenter, *Acta Crystallogr.* **B29,** 2104 (1973).
[i]$KIBr_2\cdot H_2O$; S. Soled and G. B. Carpenter, *Acta Crystallogr.* **B29,** 2556 (1973).

HF + F^-. Ab initio calculations[103] for HF_2^- and $(HO\cdots H\cdots OH)^-$ yield $\Delta d'$ values of 0.21 Å and 0.25 Å, respectively, in full agreement with experimental values. For H_3^-, an experimentally unobserved species, $\Delta d'$ is calculated[102] to be 0.34 Å (cf. 0.19 Å for neutral H_3, also unobserved).

Apart from the O···H···O and I_3^- systems, which are frequently observed as asymmetric arrangements in crystals (Fig. 7.14a, c), structural information about asymmetric arrangements of the other linear triatomics listed in Table 7.4 is sparse. However, the examples that are known seem to obey the bond number conservation rule quite well. For instance, for the Br_3^- system with $d(1) = 2.284$ Å, $\Delta d' = 0.26$ Å ($c = 0.86$ Å), the observed distances in $CsBr_3$[104] ($d_1 = 2.44$ Å, $d_2 = 2.70$ Å) lead to $n_1 = 0.68$, $n_2 = 0.32$, and the observed distances in PBr_7[105] ($d_1 = 2.39$ Å, $d_2 = 2.91$ Å) lead to $n_1 = 0.82$, $n_2 = 0.17$.

To summarize, in all known linear triatomics XYX (three-center four-electron systems) the two independent interatomic distances are correlated by relationships of the form defined by Eq. 7.5, based on the assumption of bond number conservation. The parameter $c = 3.32\ \Delta d'$

[103]A. Støgård, A. Strick, J. Almlöf, and B. Roos, *Chem. Phys.* **8,** 405 (1975).
[104]G. L. Breneman and R. D. Willett, *Acta Crystallogr.* **B25,** 1073 (1969).
[105]*Ibid.* **23,** 467 (1967).

does not vary much from one system to another (among those that are stable enough to occur in crystal structures). There are, however, indications that the lower the energy of the symmetric X···Y···X species is with respect to XY + X, the lower is the value of c. For each system the correlation curve may be presumed to run close to the minimum energy path on the corresponding X···Y···X potential energy surface. We thus have the possibility of mapping reaction paths from crystal structure data, the reactions here being either exchange reactions (XY + X $\leftrightarrow$ X + YX) or dissociations (XYX $\rightarrow$ XY + X), depending on the energy variation along the path.

The Bond Number Relationship

The bond number n that appears in Eq. 7.4 is clearly a useful quantity for interpreting the observed structures of linear triatomic systems (and of more complex ones too, as we shall see). It is not synonymous with the "bond order" of molecular orbital theory,[106] although it may be related to it in some way,[99] and seems to have no rigorous basis in terms of current theoretical models of chemical binding. In his original definition of bond number,[94] Pauling merely stated that "n is the number of shared electron pairs involved in the bond. The logarithmic relation [our Eq. 7.4] is, of course, to be expected in consequence of the exponential character of interatomic forces."

Eq. 7.4 can be derived intuitively by considering the properties of the Morse function

$$V(R-R_0) = D\{1 - \exp[-b(R-R_0)]\}^2 \tag{7.6}$$

which is known to be a good approximation to the potential energy curves of diatomic molecules; D is the binding energy, R_0 is the equilibrium distance, and b is the Morse constant. The energy of a polyatomic molecule is a function of many variables, but the Morse curve can still be regarded as a reasonable approximation to the energy change as a given bond is stretched or contracted. For a given bond, it is simpler to rewrite the Morse function as

$$\begin{aligned} V(r) &= D[1 - \exp(-br)]^2 \\ &= D[1 - 2\exp(-br) + \exp(-2br)] \end{aligned}$$

with the origin of r at the equilibrium distance R_0. If the bond number n

[106]See, for example, E. Heilbronner and H. Bock, *Das HMO-Modell und seine Anwendung*, Vol. 1, Verlag Chemie, Weinheim, 1968, p. 138.

is now identified with a factor that multiplies the attractive term by n, leaving the repulsion terms unaltered, we obtain a modified function (Fig. 7.16)

$$V_n(r) = D[1 - 2n \exp(-br) + \exp(-2br)] \tag{7.7}$$

for a fractional bond. The first derivative is

$$\frac{\partial V_n}{\partial r} = 2bD \exp(-br) [n - \exp(-br)]$$

which equals zero when

$$r = -(1/b)\ln n = -(2.3/b) \log n$$

This is the same as Eq. 7.4 with $c = 2.3/b$.

The Morse constant b for a diatomic molecule can be evaluated from a knowledge of the fundamental frequency and dissociation energy.[107] Like c (or $\Delta d'$) it does not vary over a wide range from one molecule to another; however, values of $2.3/b$ are about 50% larger than the c values to which they should correspond in cases where a comparison can be made—e.g.,

	b	$2.3/b$	c (Table 7.4)
Br_2	1.97 Å^{-1}	1.18 Å	0.86 Å
I_2	1.86	1.24	0.86

Moreover, evaluation of V_n at the new minimum gives a binding energy of Dn^2 for a fractional bond, whereas an approximately linear relationship would seem to be required. Although the analogy with Morse curves is enlightening, it should not be pressed too far.

Ligand Exchange at Tetrahedral Cadmium

The explicit program of deriving chemical reaction coordinates from crystal structure data was first announced in an important paper by Bürgi.[108] He showed how structural data pertaining to five-coordinated cadmium complexes in various crystal environments could be interpreted to give a detailed description of the structural changes occurring in the course of the nucleophilic displacement reaction at tetrahedrally coordinated cadmium, with a trigonal bipyramidal transition state or intermediate (XVII). The complexes in question all have three equa-

[107]See, for example, E. A. Moelwyn-Hughes, *Physical Chemistry*, 2nd ed., Pergamon Press, Oxford, 1961, p. 417.

[108]H. B. Bürgi, *Inorg. Chem.* **12**, 2321 (1973).

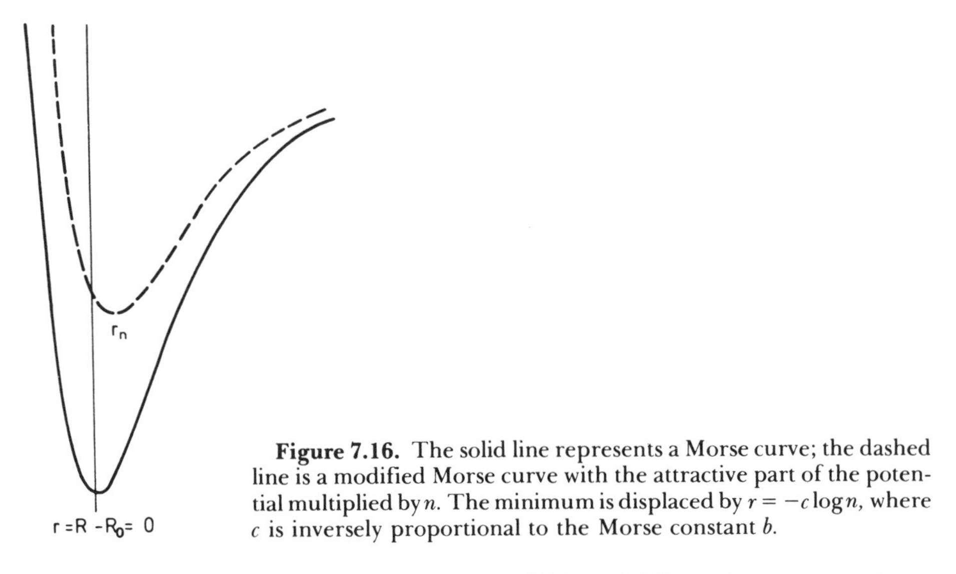

Figure 7.16. The solid line represents a Morse curve; the dashed line is a modified Morse curve with the attractive part of the potential multiplied by n. The minimum is displaced by $r = -c \log n$, where c is inversely proportional to the Morse constant b.

torial sulfur ligands, but the fourth and fifth axial ligands are sometimes iodine, sometimes sulfur, and sometimes oxygen.

S S
X ···x··· Cd ···y··· Y
S

(XVII)

The axial cadmium–ligand distances x and y were therefore brought to a common basis by subtracting the appropriate sums of covalent radii to obtain distance increments, Δx and Δy. The correlation plot (Fig. 7.17) shows the same general behavior as the curves in Fig. 7.14 for linear triatomic systems; the curve represents the analytical function (Eq. 7.5) with c = 1.05 Å ($\Delta x = \Delta y = \Delta d'$ = 0.32 Å). A more interesting correlation is obtained by plotting Δx and Δy against Δz, the displacement of the Cd atom from the plane of the three equatorial sulfur atoms. This is shown in Fig. 7.18, where each structure is represented by *two* sample points (Δx, Δz) and (Δy, $-\Delta z$) with opposite signs of Δz. The extreme point for $\Delta x = 0$ ($\Delta y = \infty$) corresponds to the geometry of an ideal CdS_4 tetrahedron with the fifth ligand Y at an infinite distance. From Fig. 7.18 it is clear that as Δy decreases and Δx increases, the CdS_3 fragment flattens until for $\Delta x = \Delta y$ = 0.32 Å, we have a trigonal bipyramidal struc-

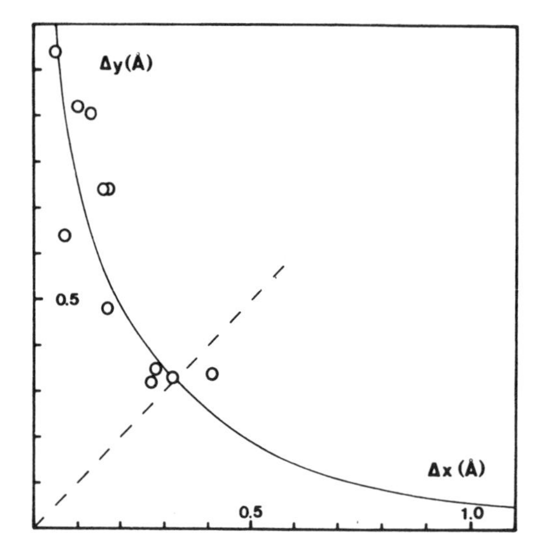

Figure 7.17. Correlation of the axial distance increments Δx and Δy in X—CdS_3—Y complexes. From Bürgi, *Inorg. Chem.* **12,** 2321 (1973). Reprinted by permission.

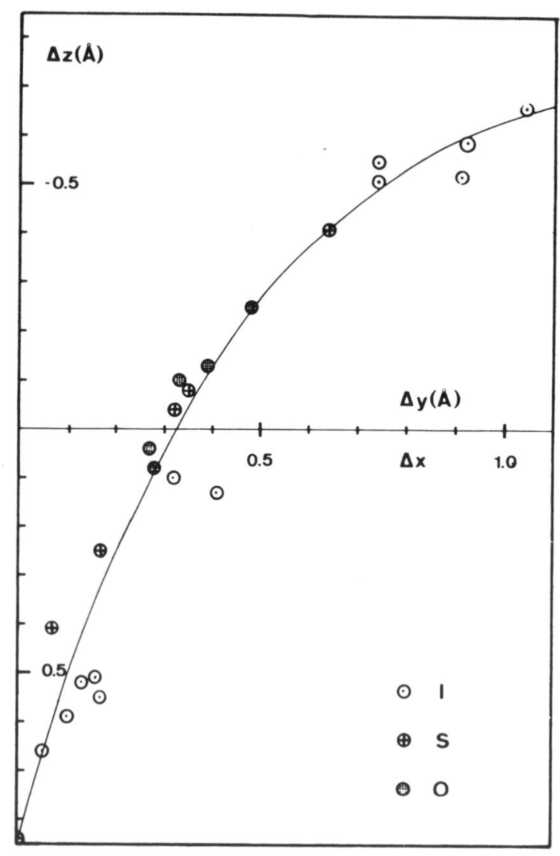

Figure 7.18. Correlation of Δx and Δy against Δz (displacement of Cd from the plane of equatorial ligands) in X—CdS_3—Y complexes. From Bürgi, *Inorg. Chem.* **12,** 2321 (1973). Reprinted by permission.

ture ($\Delta z = 0$). These changes have an obvious interpretation in terms of the ligand-exchange reaction

$$XCdS_3 + Y \leftrightarrow [X\cdots CdS_3\cdots Y] \leftrightarrow X + CdS_3Y$$

and are very reminiscent of the kind of process that is believed to occur in S_N2 type nucleophilic substitution reactions. In other words, we can take the view that the atomic arrangements observed in different crystal environments represent a sequence of points frozen along the path of this reaction, which would appear to be otherwise inaccessible to experimental study.

The smooth curve in Fig. 7.18 is based on a slight extension of the bond-number conservation assumption. Thus far, we have used only the relationship between bond number and bond distance increment, $\Delta d = -c \log n$, but now we need a relationship between bond number and the out-of-plane displacement Δz of the Cd atom. Bürgi assumed that

$$\begin{aligned} n_x &= (-3 \cos \theta + 1)/2 \\ n_y &= (3 \cos \theta + 1)/2 \end{aligned} \tag{7.8}$$

where θ is the angle XCdS. For the tetrahedral species CdS_4, $\cos \theta = -1/3$, leading to $n_x = 1$, $n_y = 0$; for the trigonal bipyramidal arrangement, $\cos \theta = 0$, leading to $n_x = n_y = 1/2$. The assumptions thus satisfy the bond number requirements for the initial, final, and symmetric intermediate stages of the reaction. We then have

$$\begin{aligned} \Delta x &= -c \log n_x = -c \log[(-3 \cos \theta + 1)/2] \\ \Delta y &= -c \log n_y = -c \log[(3 \cos \theta + 1)/2] \end{aligned} \tag{7.9}$$

Since the equatorial Cd—S bond distances do not differ by more than 2–3% from their mean value, 2.52 Å (the same as in cadmium sulfide), we can express Eqs. 7.9 in terms of $\Delta z = -2.52 \cos \theta$:

$$\begin{aligned} \Delta x &= -c \log[(\Delta z + 0.84)/1.68] \\ \Delta y &= -c \log[(-\Delta z + 0.84)/1.68] \end{aligned} \tag{7.10}$$

With $c = 1.05$ Å, this is the curve drawn in Fig. 7.18.

The same experimental data can be looked at from a slightly different point of view—as a description of the breakdown of the symmetrical trigonal bipyramidal structure. For this purpose, it is convenient to express the distortions in terms of symmetry displacement coordinates of an idealized structure with D_{3h} symmetry.[109] The symmetry coordinates of interest are:

$$\begin{aligned} S_2 &= (\Delta x + \Delta y)/\sqrt{2} \\ S_3 &= (-\Delta x + \Delta y)/\sqrt{2} \\ S_4 &= (\Delta\alpha_{1x} - \Delta\alpha_{1y} + \Delta\alpha_{2x} - \Delta\alpha_{2y} + \Delta\alpha_{3x} - \Delta\alpha_{3y})/\sqrt{6} \end{aligned} \tag{7.11}$$

where the $\Delta\alpha$ values are deviations (from 90°) of the angles between axial ligands (X, Y) and equatorial ligands (1, 2, 3). These coordinates are illustrated in Fig. 7.19.

Fig. 7.20, left, shows that the decay of the trigonal bipyramidal structure begins with a simultaneous deformation along the antisymmetric stretch coordinate S_3 and the antisymmetric out-of-plane bending coordinate S_4, which belong to the same irreducible representation (A_2'') of the D_{3h} point group. If the curve really corresponds to a minimum energy path, then for small displacements at least it should coincide with the direction of the major axis of the potential energy ellipsoid in the S_3, S_4 space, and we have a basis for estimating the sign and relative magnitude of the cross term in the quadratic form

$$2V = F_{33}S_3^2 + F_{44}S_4^2 + 2F_{34}S_3S_4$$

from the initial slope of the curve. The relationship is

$$\tan 2\gamma = 2F_{34}/(F_{44} - F_{33})$$

where γ is the angle between the initial slope and the S_4 axis.

The symmetry coordinates S_2 and S_3 belong to different irreducible representations, A_1' and A_2'', respectively. Fig. 7.20, right, shows that at the start $S_2 \sim 0$—i.e., there is no deformation along the symmetric stretch coordinate. As ligand X moves closer to the cadmium atom, ligand Y moves away by the same amount. The symmetrical deformation only comes into play as the reaction proceeds, and its contribution tends to cancel the contribution of S_3 on one ligand and to reinforce it for the other. For very large displacements the reaction path tends to approach the line of unit slope ($S_2 = S_3$) asymptotically. In this region an advance along the reaction path leaves the Cd—X distance unchanged and only increases the Cd—Y distance.

[109]Symmetry displacement coordinates are linear combinations of the internal displacement coordinates that transform according to the irreducible representations of the point group in question, and methods of deriving such coordinates are described in most books on applied group theory. Normal vibration coordinates constitute a special set of symmetry coordinates.

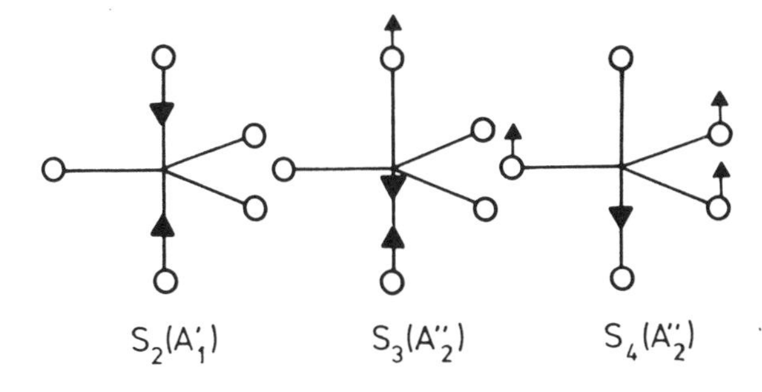

Figure 7.19. Symmetry displacement coordinates (and their irreducible representations) for trigonal bipyramidal (D_{3h}) structure: S_2, symmetric axial stretch; S_3, antisymmetric axial stretch; S_4, out-of-plane bend.

The value of 0.32 Å for the axial bond increment $\Delta d'$ in the symmetrical trigonal bipyramidal structure lies at the upper limit of the values found for symmetrical linear triatomics (Table 7.4). Not much experimental information is on hand for evaluating $\Delta d'$ increments for other comparable systems—it is uncommon for a tetrahedral molecule XMY_3 and a trigonal bipyramidal one XMY_3X both to correspond to observable species. The best studied system is probably PCl_4^+/PCl_5. The tetrahedral cation occurs in crystalline phosphorus pentachloride $(PCl_4)(PCl_6)$[110] and in several other crystal structures, but the reported P—Cl distances range from 1.90 to 1.98 Å; the most reliable value is perhaps 1.944 Å, the mean distance in $(PCl_4)_2Ti_2Cl_{10}$ and $(PCl_4)Ti_2Cl_9$,[111] corrected for thermal motion. In the neutral molecule, the axial P—Cl distance is 2.124 Å,[112] leading to $\Delta d' = 0.18$ Å. From a survey of experimental data and results of quantum mechanical calculations, Bürgi has estimated $\Delta d'$ for some other systems.[113] He obtains relatively low values for PH_4^+/PH_5 (0.07-0.14 Å), PF_4^+/PF_5 (0.12 Å), AsF_4^+/AsF_5 (0.13 Å), SiH_4/SiH_5^- (0.11-0.14 Å), SiF_4/SiF_5^- (0.13 Å), $R_3NSiH_3/(R_3N)_2SiH_3$ (0.18 Å), where the pentavalent species has either been observed or is expected to be not prohibitively high in energy—e.g., SiH_5^- is calculated to be stable with respect to H^- and SiH_4.[102] Relatively large values are obtained for CH_4/CH_5^- (0.65 Å), $CH_3Cl/CH_3Cl_2^-$ (0.61 Å), $CH_3F/CH_3F_2^-$ (0.39 Å), where the pentavalent species is calculated to have a high energy.[102] There is at least a rough correlation between $\Delta d'$ and the relative energy of the pentavalent species with respect to the tetrahedral one. Bürgi has noted[113] that $\Delta d'$ tends to become smaller as the electronegativity difference between the axial ligand X and the central atom M increases.

[110] D. Clark, H. M. Powell, and A. F. Wells, *J. Chem. Soc.* **1942,** 642; H. Preiss, *Z. Anorg. Chem.* **51,** 380 (1971).
[111] T. J. Kistenmacher and G. D. Stucky, *Inorg. Chem.* **10,** 122 (1971).
[112] W. J. Adams and L. S. Bartell, *J. Mol. Struct.* **8,** 23 (1971).
[113] H. B. Bürgi, to be published.

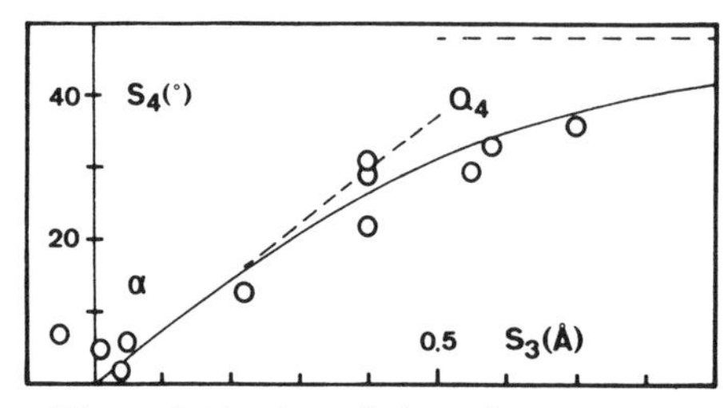

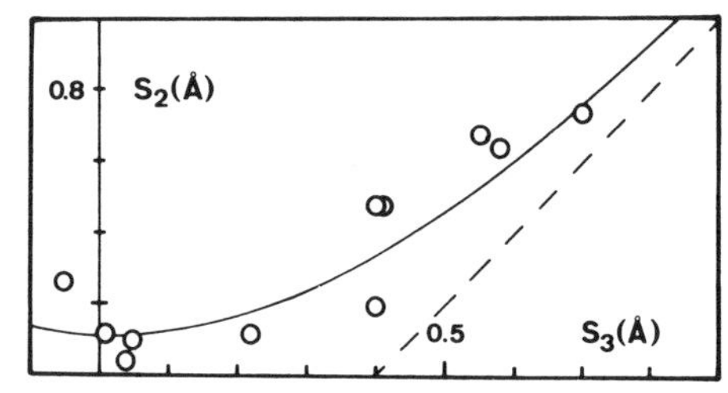

Figure 7.20. Correlation of symmetry displacements for $XCdS_3Y$ structures. Left: $S_3(A_2'')$ against $S_4(A_2'')$; right: $S_3(A_2'')$ against $S_2(A_1')$. From Bürgi, *Inorg. Chem.* **12,** 2321 (1973). Reprinted by permission.

Again this electronegativity difference runs roughly parallel to the stability of the pentavalent species.

Tetrahedral Molecules and Ions

Molecules that are expected to show exact tetrahedral symmetry as isolated particles often show significant deviations from this symmetry in a crystal environment. The distortions are sometimes simply dismissed as minor blemishes in an otherwise harmonious edifice, although several authors have looked for and found various kinds of systematic behavior among them.[114–119] Others have tried to analyze the deformations in terms of crystal packing forces.[120] Here we are more interested in the nature of the deformations than in their origin. We take the view, outlined in the previous sections, that a sequence of distorted structures, arranged in the right order, maps out a minimum energy path in the appropriate parameter space.

Strictly speaking, the symmetry of a molecule in a crystal is its site symmetry, the symmetry of its environment. Thus, unless the molecule occupies a special position of the space group, it has no symmetry. Nevertheless, molecules in general positions often appear to be approximately symmetrical.[121] A tetrahedral MX_4 molecule can be distorted in many ways—there are nine (i.e., $3\times5 - 6$) independent distortion

[114] W. S. McDonald and D. W. J. Cruickshank, *Acta Crystallogr.* **22,** 37 (1967).
[115] W. H. Baur, *Trans. Am. Crystallogr. Assoc.* **6,** 129 (1970).
[116] W. H. Baur, *Acta Crystallogr.* **B30,** 1195 (1974).
[117] I. D. Brown and R. D. Shannon, *Acta Crystallogr.* **A29,** 266 (1973).
[118] S. J. Louisnathan and G. V. Gibbs, *Mater. Res. Bull.* **7,** 1281 (1972).
[119] G. A. Lager and G. V. Gibbs, *Am. Mineral.* **58,** 756 (1973).
[120] J. A. McGinnety, *Acta Crystallogr.* **B28,** 2845 (1972).
[121] For a mathematician, an object (e.g., a molecule) either possesses a certain symmetry or it does not—the concept of approximate symmetry is meaningless. Nevertheless, we shall continue to use it.

modes—and the deformed molecule could have the actual or approximate symmetry of some subgroup of T_d.[122] Many such molecules in crystal environments are distorted from T_d symmetry in one of two ways—one preserving approximate C_{3v} symmetry, the other approximate C_{2v} symmetry. An analysis in terms of symmetry coordinates shows that the two distortion modes are closely related; indeed, the C_{2v} type distortion can be regarded as a combination of two equal C_{3v} type distortions, as indicated in Fig. 7.21, and vice-versa. We limit our discussion to the C_{3v} distortions, which have more obvious implications for reaction path studies (S_N1 reaction); they are also simpler to analyze since they can be described in terms of only three parameters whereas the C_{2v} distortions need four. The results are remarkable:[123] the C_{3v} distortion paths for a variety of MX_4 and MX_3Y species (excluding transition metal complexes) are virtually identical.

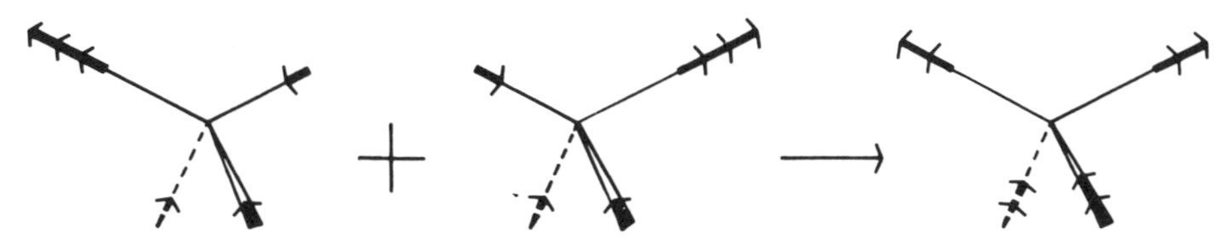

Figure 7.21. The combination of two equal C_{3v} type distortions leads to a C_{2v} type distortion.

The three independent structural parameters for a MX_4 molecule with C_{3v} symmetry are conveniently chosen as r_2, r_1 and θ, where r_2 is the length of the unique M—X bond, r_1 is the length of the three symmetry-related bonds, and θ is the angle between the threefold axis and one of these bonds (Fig. 7.22). Each molecule observed can be represented by a pair of corresponding sample points (r_1, θ) and (r_2, θ), as illustrated in Fig. 7.23 for SO_4 groups in different environments.[124] As the axial bond r_2 becomes longer, the equatorial bonds r_1 shorten and θ decreases. The two sets of points lie close to two smooth curves that intersect at the tetrahedral angle, where $\delta r_2/\delta\theta \sim -3\delta r_1/\delta\theta$. It is only for small distortions that the mean bond distance remains constant. The r_1 curve can be extrapolated nicely to $\theta = 90°$, where the sample point refers to the

[122]The subgroups in question are $T_d(m3)$, $D_{2d}(\bar{4}2m)$, $C_{3v}(3m)$, $D_2(222)$, $C_{2v}(mm2)$, $C_2(2)$, $C_s(m)$. The remaining subgroups $T(23)$, $S_4(\bar{4})$, $C_3(3)$ are not possible point groups of an AX_4 molecule where X is a monoatomic ligand.

[123]P. Murray-Rust, H. B. Bürgi, and J. D. Dunitz, *J. Am. Chem. Soc.* **97,** 921 (1975).

[124]Most of the experimental data for the correlations discussed in this section were compiled from the literature by P. Murray-Rust. For a compilation of data for PO_4 tetrahedra, see Ref. 116.

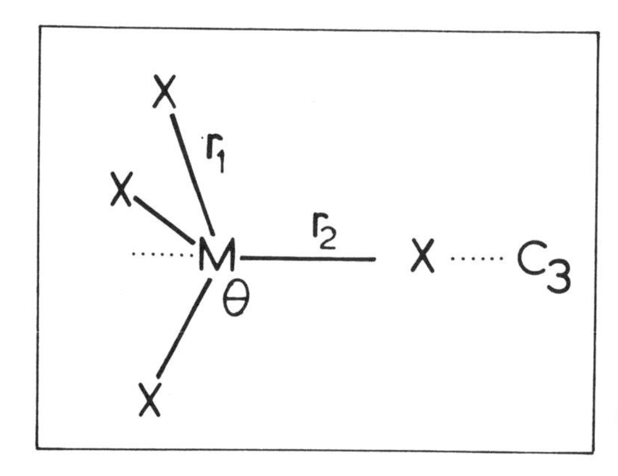

Figure 7.22. Definition of structural parameters r_1, r_2, θ for MX_4 or MX_3Y molecules with C_{3v} symmetry. From Murray-Rust, Bürgi, and Dunitz, *J. Am. Chem. Soc.* **97**, 921 (1975). Reprinted by permission.

planar SO_3 molecule in the gas phase. The two curves taken together can be regarded as mapping the path of a hypothetical reaction, $SO_4^{2-} \rightarrow SO_3 + O^{2-}$, heterolytic removal of a ligand from the sulfate ion to give neutral SO_3 plus a doubly negatively charged oxide ion.

Results for YSO_3 species (thiosulfates, sulfonates, fluorosulfonates, and amidosulfonates) yield a very similar r_1, θ dependence (Fig. 7.23) to that found for the sulfates; the data are insufficient to define the r_2, θ curves for individual Ys. The fact that the sample points for sulfites (where Y is an electron pair) lie on the same curve can be taken as a structural expression of the analogy between the oxidation–reduction process $SO_3 + 2e \leftrightarrow SO_3^{2-}$ and the Lewis acid–base reaction $SO_3 + :Y \leftrightarrow YSO_3$. The pyramidality of the SO_3 unit clearly depends on the electronegativity of Y; ligands of high electronegativity produce a more flattened pyramid, those of low electronegativity a more pointed one, in agreement with Gillespie's ideas based on the VSEPR (valence shell electron pair repulsion) model.[125]

Essentially the same correlation plots are obtained for PO_4^{3-} and $AlCl_4^-$ tetrahedra (Fig. 7.24) as for sulfates. Likewise, the YPO_3 and $YSnCl_3$ (Fig. 7.25) plots are similar to the YSO_3 plot and show the same kind of dependence on the electronegativity of Y. The $YSnCl_3$ points, in particular, are drawn from a very heterogeneous range of examples (e.g., $SnCl_4$, π-$(C_5H_5)Fe(CO)_2SnCl_3$, $SnCl_3^-$, etc.) in which the formal oxidation state of the metal atom ranges from +4 to +2. Data for YBF_3, $YAlBr_3$, $YSiF_3$, $YGeBr_3$, $YSn(CH_3)_3$, YSF_3, and $YClO_3$ show the same trend but are too meager or inaccurate for definite conclusions to be drawn. The

[125]R. J. Gillespie, *Molecular Geometry*, Van Nostrand-Reinhold, London and New York, 1972, Chap. III.

tetrahedral species BF_4^- and ClO_4^- were also examined, but as very weak bases they usually show only minor deviations from T_d symmetry; besides, the structural data for these species are often inaccurate because of disorder or high thermal motion in the crystal. The anionic groupings BO_4, AlO_4, and SiO_4 usually occur in highly condensed networks with extensive sharing of oxygen atoms between tetrahedra.

The most surprising result perhaps is that apart from differences in the absolute values of r_1 and r_2, all the curves appear to be the same despite differences in the central atom (S, P, Al, Sn, etc.) and in the ligands. Once again we might ask whether this unexpected invariance might not be just another manifestation of bond number conservation. In fact, we can try to derive an analytical expression for the r_1, r_2, inter-

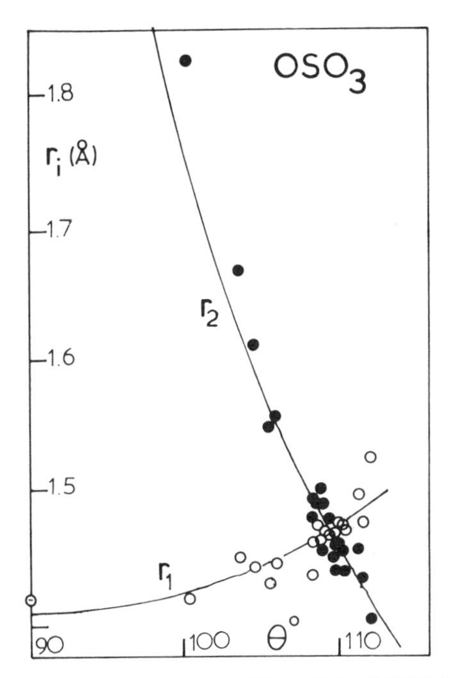

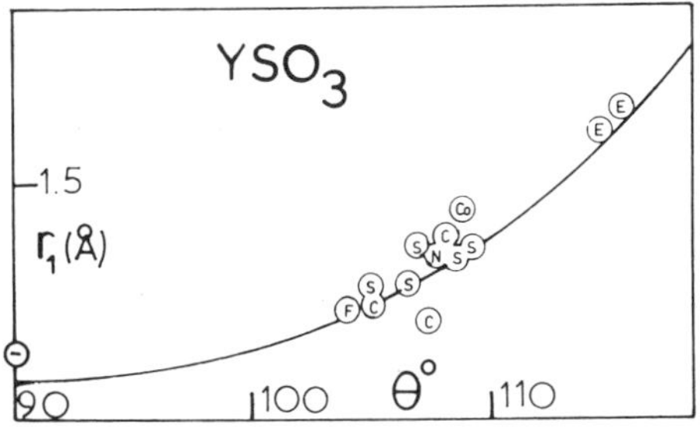

Figure 7.23. Correlation plots for OSO_3 and YSO_3 tetrahedra with approximate C_{3v} symmetry observed in various crystal structures. Black circles show r_2,θ correlations, open ones show r_1,θ. The letters in the open circles define the element Y in YSO_3 (E = electron pair, Θ = planar SO_3). From Murray-Rust, Bürgi, and Dunitz, *J. Am. Chem. Soc.* **97**, 921 (1975). Reprinted by permission.

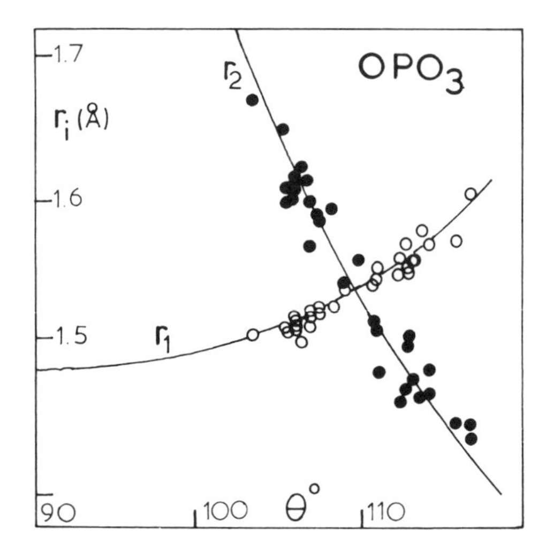

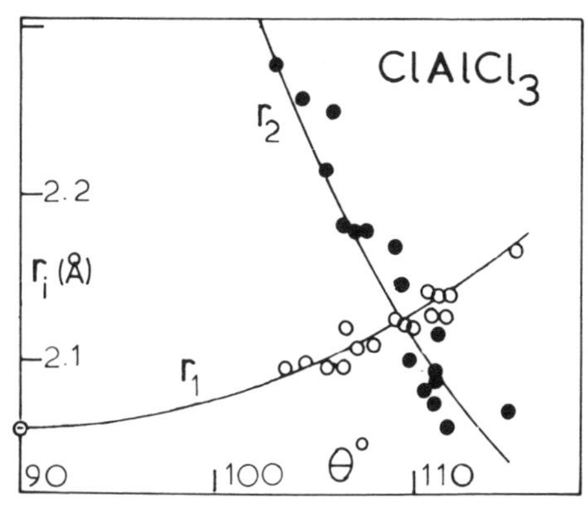

Figure 7.24. Correlation plots for OPO_3 and $ClAlCl_3$ tetrahedra with approximate C_{3v} symmetry. From Murray-Rust, Bürgi, and Dunitz, *J. Am. Chem. Soc.* **97**, 921 (1975). Reprinted by permission.

dependence by assuming, as before:

(a) $\Delta r_i = -c \log n_i$, where n_i is the bond number

(b) $n_2 + 3n_1 = 4$ (bond number conservation)

(c) $n_2 = -3 \cos \theta$

This at least gives the "correct" values of n_2 for the regular tetrahedron ($\cos \theta = -\frac{1}{3}$, $n_2 = 1$) and for the planar MX_3 unit ($\cos \theta = 0$, $n_2 = 0$). The functions

$$\Delta r_2 = -c \log(-3 \cos \theta) \qquad \Delta r_1 = -c \log\left(\tfrac{4}{3} + \cos \theta\right) \qquad (7.12)$$

are shown in Fig. 7.26a. Although the Δr_2 curve has the right general shape, the Δr_1 curve clearly differs from the experimental correlations; in particular, it shows quite the wrong behavior at $\theta = 90°$. However, instead of abandoning the notion of bond number conservation, we can try (c′) $n_2 = 9 \cos^2 \theta$

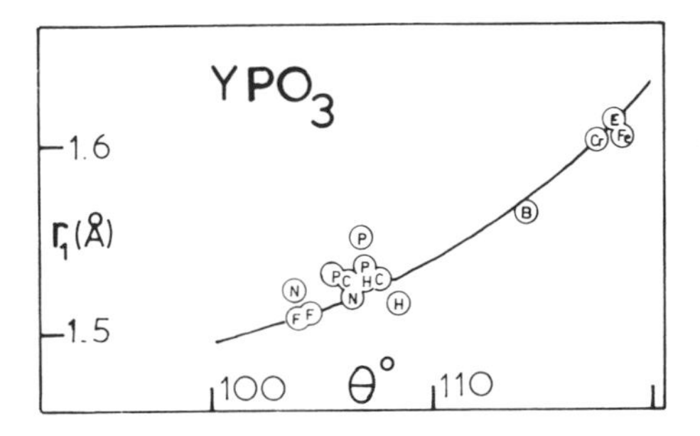

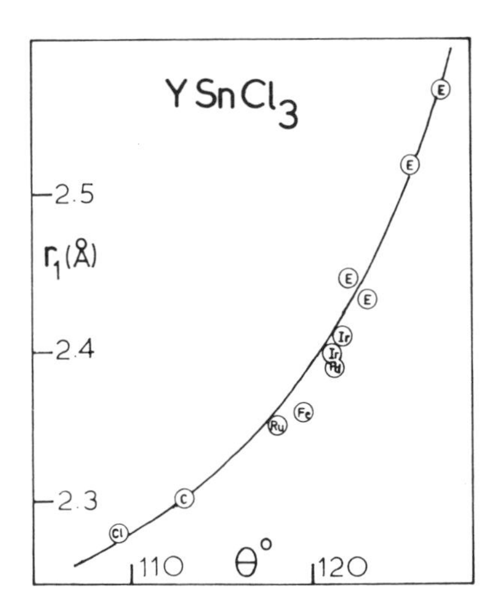

Figure 7.25. Correlation plots for YPO_3 and $YSnCl_3$ tetrahedra. Letters in the circles denote the element Y. From Murray-Rust, Bürgi, and Dunitz, *J. Am. Chem. Soc.* **97,** 921 (1975). Reprinted by permission.

which gives the same values as assumption (c) for the two special values of θ. The functions

$$\Delta r_2 = -c \log (9 \cos^2 \theta) \qquad \Delta r_1 = -c \log (\tfrac{4}{3} - 3 \cos^2 \theta) \qquad (7.13)$$

are shown in Fig. 7.26b. Now both curves have the right general shape. When c is determined by least-squares fitting of these functions to the individual data sets, we obtain the results shown below.

Bond	r_o (tetrahedral)	c
Al—Cl	2.125 Å	0.49 Å
Sn—Cl	2.227	0.48
S—O	1.472	0.51
P—O	1.534	0.47

Within experimental error the c values are equal for the four systems for which extensive experimental data are available. This is shown dramati-

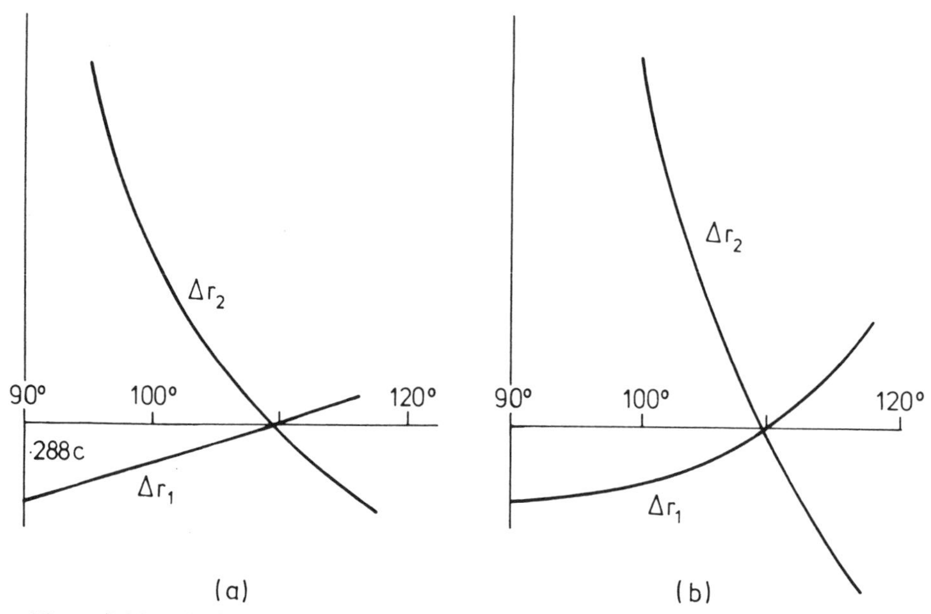

Figure 7.26. The functions (a) $\Delta r_2 = -c \log(-3 \cos \theta)$; $\Delta r_1 = -c \log(4/3+\cos \theta)$ and (b) $\Delta r_2 = -c \log(9 \cos^2 \theta)$; $\Delta r_1 = -c \log(4/3-\cos^2 \theta)$ derived by assuming bond-number conservation for tetrahedral molecules with C_{3v} symmetry.

cally in Fig. 7.27, where all the sample points (about 200) are assembled on one diagram. For this purpose the observed bond distances r_1, r_2 have been replaced by bond distance differences Δr_1, Δr_2 (between distances observed and those of the corresponding T_d species).

It is worth stressing that the extrapolation to infinte Δr_2 for $\theta = 90°$ is a property of the model, not of the experimental data, which are, as they stand, just as compatible with a finite intersection at a Δr_2 value of 1 Å or more. More extensive data are needed for very deformed molecules with θ less than 100° in order to define the experimental curve better in this region. Eqs. 7.13 also imply that Δr_1 would become infinite for $\theta = 131.81°$ ($\cos \theta = -2/3$).

The observed deformations can be interpreted as mapping the path for the S_N1 type reaction because they correspond to heterolytic weakening and ultimately fission of the axial bond to give a planar MX_3 species (e.g., the Lewis acids SO_3 and $AlCl_3$). In view of the similarity between the deformation patterns of different tetrahedral species it would seem that this path is not very sensitive to the nature of the central atom or of the ligand. As mentioned earlier, the r_1,θ curves for YMX_3 species can also be interpreted in terms of removal of an electron pair from a tricoordinated pyramidal MX_3 unit—e.g., $SO_3^{2-} \rightarrow SO_3 + 2e^-$. The sample

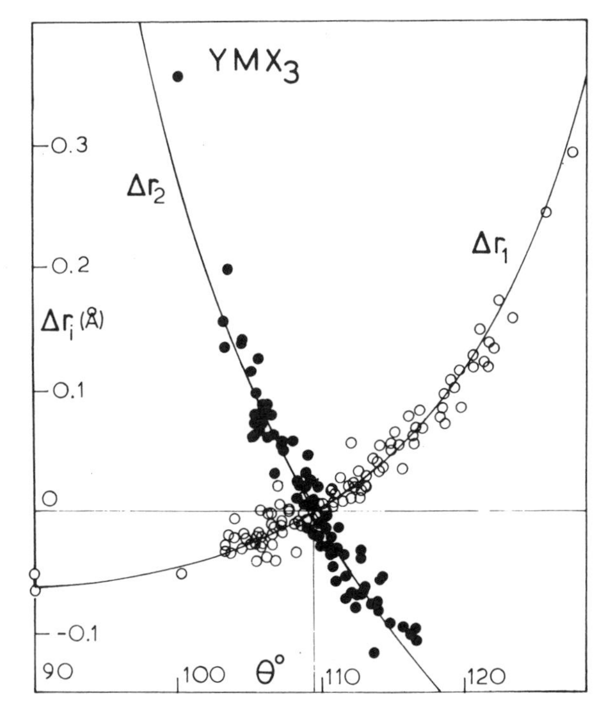

Figure 7.27. Correlation plots for all experimental data shown in Figs. 7.23, 7.24, and 7.25 referred to a common origin. The smooth curves are drawn for $c = 0.50$ Å. From Murray-Rust, Bürgi, and Dunitz, *J. Am. Chem. Soc.* **97,** 921 (1975). Reprinted by permission.

points that map this oxidation process lie on the same curve as those mapping the S_N1 dissociation of MX_4. The position of any YMX_3 species on the curve can thus be taken as a measure of the electron-donating power of Y with respect to X.

How far can it be assumed that these curves describe the minimum energy path for the S_N1 dissociation of carbon compounds? The data for deformations of carbon tetrahedra are too sparse or inaccurate to answer this question. However, the relevant parameters from an electron-diffraction analysis of *tert*-butyl chloride[126] lie very close to our curves ($\theta = 107.3°$, $\Delta r_2 = 0.057$ Å, $\Delta r_1 = -0.014$ Å compared with values of 0.050 and -0.017 Å expected for this angle); this agreement is encouraging.

An alternative although closely related analysis of the experimental data for distorted MX_4 molecules can be given in terms of symmetry coordinates.[127] The four M—X distances r_1, r_2, r_3, r_4 transform in the point group T_d as $A_1 + T_2$, the six angles θ_{12}, θ_{13}, θ_{14}, θ_{23}, θ_{24}, θ_{34} as $A_1 + E + T_2$, with a functional dependence of the A_1 angular coordinate on

[126] R. L. Hilderbrandt and J. D. Wieser, *J. Chem. Phys.* **55,** 4648 (1971).
[127] P. Murray-Rust, H. B. Bürgi, and J. D. Dunitz, *Acta Crystallogr.* **B34,** 1787, 1793 (1978).

the five others. (For infinitesimal angular distortions, positive and negative deviations from 109° 28′ cancel, so the A_1 angular displacement coordinate is zero.) For an arbitrary distortion of a reference molecule (with M—X distance r') from T_d symmetry, distortion components that preserve C_{3v} symmetry can be extracted:

$$\begin{aligned}
D_1(A_1) &= \frac{1}{2}\,(r_1+r_2+r_3+r_4-4r') \\
D_{3a}(T_2) &= \frac{1}{\sqrt{12}}\,(3r_1-r_2-r_3-r_4) \\
D_{4a}(T_2) &= \frac{1}{\sqrt{6}}\,(\theta_{12}+\theta_{13}+\theta_{14}-\theta_{23}-\theta_{24}-\theta_{34}) \\
D_5(A_1) &= \frac{1}{\sqrt{6}}\,(\theta_{12}+\theta_{13}+\theta_{14}+\theta_{23}+\theta_{24}+\theta_{34}-656.83^\circ)
\end{aligned} \tag{7.14}$$

The displacements along the other six symmetry coordinates are:

$$\begin{aligned}
D_{2a}(E) &= \frac{1}{\sqrt{12}}\,(2\theta_{12}-\theta_{13}-\theta_{14}-\theta_{23}-\theta_{24}+2\theta_{34}) \\
D_{2b}(E) &= \frac{1}{2}\,(\theta_{13}-\theta_{14}-\theta_{23}+\theta_{24}) \\
D_{3b}(T_2) &= \frac{1}{\sqrt{6}}\,(2r_2-r_3-r_4) \\
D_{3c}(T_2) &= \frac{1}{\sqrt{2}}\,(r_3-r_4) \\
D_{4b}(T_2) &= \frac{1}{\sqrt{12}}\,(2\theta_{12}-\theta_{13}-\theta_{14}+\theta_{23}+\theta_{24}-2\theta_{34}) \\
D_{4c}(T_2) &= \frac{1}{2}\,(\theta_{13}-\theta_{14}+\theta_{23}-\theta_{24})
\end{aligned} \tag{7.15}$$

These vanish if the distorted molecule retains exact C_{3v} symmetry, and they are small if it has approximate C_{3v} symmetry ($r_2 \sim r_3 \sim r_4$; $\theta_{12} \sim \theta_{13} \sim \theta_{14}$; $\theta_{23} \sim \theta_{24} \sim \theta_{34}$).

When the experimental data for the various types of MX_4 molecules are expressed in this way, it becomes clear that the distortion process begins with a simultaneous displacement along the D_{3a} and D_{4a} coordinates, which both transform as T_2. For PO_4 tetrahedra, the ratio of displacements D_{3a}/D_{4a} is nearly constant at about −0.010 Å/deg, and approximately the same ratio is obtained for the other systems for which extensive experimental data are available. As long as the totally symmetric D_1 displacement is negligible, the mean bond length is obviously constant. It has been mentioned (in the discussion of Fig. 7.23) that this condition holds only for small distortions from T_d symmetry.

A similar analysis can be made for the distortion path that preserves C_{2v} symmetry. The nonvanishing symmetry coordinates here include one that transforms as E as well as A_1 and T_2 components. However, if the T_2 components are plotted against one another, their ratio is the same as for the C_{3v} distortion.

Structural Correlation Principle

We have shown that it is possible to derive information about reaction paths by searching for certain kinds of correlations among the structural parameters for a group of related molecules or crystals. The idea is to select some structural fragment or subunit that occurs in a wide variety of environments and try to correlate changes that occur in one structural parameter with changes that occur in others.[128] If correlations between these parameters can be found, the various copies of the fragment can be ordered in a sequence that can be interpreted in terms of a gradual deformation of the fragment, just as a series of "still" pictures, arranged in the right order, can be viewed cinematically. Although the actual equilibrium arrangements of atoms observed in the various copies depend on a complicated interplay of intra- and intermolecular forces that are rarely understood in detail, it is a useful working hypothesis to assume that the sequence of changes in the structure of the fragment will occur along a potential energy valley in the relevant parameter space. Each sample point (since observed in an equilibrium structure) corresponds exactly to an energy minimum under the perturbation operative in that structure, but we cannot assess the effect of individual perturbations in particular structures. The scatter of sample points may be caused partly by experimental inaccuracies in the data, but some scatter must also result from the different perturbations that act on our fragment in its different environments. Similar perturbations, e.g., solvent effects, can have drastic effects on chemical reactivity.

[128]In this connection it is only correlations between geometrically independent structural parameters that are of relevance. For cyclohexane rings in the chair conformation, for example, the excellent correlation between the mean CCC bond angle θ and the mean CCCC torsion angle ω that would doubtless be obtained from structural data expresses no more than the relation $\cos \omega = \cos \theta/(1 + \cos \theta)$, which can more easily be derived from geometry.

However, if the irregularities in the distribution of sample points are not too large, the effect of these perturbations can be averaged to give a more or less well-defined smooth curve that should approximately follow a minimum energy path in the parameter space. In other words, the subsystem is expected to deform along the path of least resistance. Whether such a path leads to unstable species that revert to the initial subsystem or to stable products depends on the way the energy varies along the path. This variation will depend on the nature of the atoms in the fragment and also on its environment (cf. solvent effects).

We can summarize these arguments in terms of what could be called the principle of structural correlation: if a correlation can be found between two or more independent parameters describing the structure of a given structural fragment in various environments, then the correlation function maps a minimum energy path in the corresponding parameter space.

A problem arises in defining exactly what is meant by "the corresponding parameter space." In our discussion so far we have ignored the perturbations that are effective in producing the observed deformations in different crystal structures. For example, the deformations of linear trihalide ions were described in terms of a reaction path for the displacement,

$$X^- + X_2 \leftrightarrow [X_3^-] \leftrightarrow X_2 + X^-$$

but each observed deformation actually occurs as a response to an interaction between the X_3^- fragment and a particular environment containing cations. Symmetrical X_3^- fragments tend to occur when the interaction with cations is weak or itself symmetrical, whereas large antisymmetric deformations tend to occur when one end of the fragment interacts with a cation much more strongly than the other end. It is therefore more correct to reformulate the reaction as:

$$X_3^- + A^+ \leftrightarrow X_2 + XA$$

and similarly for some of the other examples discussed. However, because the environments in which our fragments occur are so heterogeneous, it is virtually impossible to define a parameter space that encompasses the structural parameters of the fragment *and* its environment. We have to restrict ourselves to the changes in the structure of the fragment itself and map the deformation path in "the corresponding parameter space." This deformation path is obtained by projecting sample points in a higher dimensional space on the subspace of the fragment, and it may be said to describe the response of the fragment to external forces. For the YMX_3 fragments described in the previous section, the deformation of the MX_3 unit depends essentially on the nature of atom Y; here the perturbation is a part of the molecule itself, but we can still conceptually isolate the MX_3 part and map its response to the changing force exercised by Y (as in Fig. 7.25, for example). The path is virtually the same as that of the MX_3 unit in an MX_4 molecule that is subject to an external force.

It is an open question how closely the deformation path derived from structural correlations follows a minimum energy path in the potential energy surface of the isolated fragment. We would expect a close correspondence for energy valleys with very steep sides, a less close one for relatively shallow valleys and large external perturbations, but too little reliable information is available for detailed comparisons.

If the smooth curve through the experimental sample points is accepted as a minimum energy path, then by hypothesis the energy must increase along directions normal to this curve. Unfortunately, we do not know how sharply the energy rises normal to the curve, but even worse we know nothing about the energy variation along the curve—i.e., about where the minima and maxima occur.[129] In the five-coordinated Cd example, the structural data are just as compatible with a chemical reaction that involves decomposition of the trigonal bipyramidal species $XCdS_3Y$ to give the tetrahedral products $XCdS_3$ or CdS_3Y as with the ligand exchange reaction in terms of which the data were discussed. We know the path, but we do not know where the journey begins or where it ends. To fill out the picture we require a background drawn from other sources. The search for correlations among structural data can be carried out as an exercise in itself, but the patterns that emerge will undoubtedly become more significant when set against the facts of experimental chemistry, their interpretation in mechanistic terms, and the results of quantum-mechanical studies.

[129]This may go a little too far. The fact that a sample point can be observed at all means that the corresponding energy cannot be very high compared with neighboring points on the surface. It is thus possible to draw guarded inferences from the distribution of a large enough number of sample points. For example, if the sample points show a large scatter from the curve in certain regions and a small scatter in others, we might infer that in the regions of large scatter, the energy increase normal to the path is relatively small. On the other hand, if we find two regions of the parameter space with many sample points, separated by a region containing none, we might infer that the energy in the sparsely populated region is markedly higher than in the densely populated regions. However, the distributions in question could also have arisen merely by an unfortunate choice of structures. We have to depend largely on published data to ensure a reasonable sample size, and it is unlikely that the number of structures of a given type for which data are available will follow the Boltzmann distribution.

It is remarkable that the structural correlation approach leads to minimum energy paths that can be expressed as simple analytical functions derived from the assumption of bond number conservation. To a first approximation, minimum energy paths for reactions of low activation energy (the only ones we are concerned with) can be treated as paths of constant energy. If such reactions involve bond breaking and bond formation, the energy cost in stretching the bond to be broken has to be compensated by a comparable energy gain from partial formation of a new bond. The minimum energy path is then the path of constant binding energy, and it would seem that this invariance can be expressed as an invariance of Pauling bond numbers. Although these quantities lack theoretical basis, their usefulness in deriving analytical functions to fit the observed correlations cannot be denied.

Nucleophilic Addition to Carbonyl Groups

In the previous sections we discussed reaction paths for two basic processes involving a tetrahedral center: (1) addition of a nucleophile to give a five-coordinated trigonal bipyramidal structure; (2) removal of a ligand with its electron pair to give a three-coordinated trigonal planar structure. With carbon as the central atom, addition generally leads to high energy species; the trigonal bipyramidal structures formed are almost always transition states or transient intermediates that are not directly observable. On the other hand, the tetrahedral/trigonal interconversion may be energetically uphill or downhill, depending on the nature of the ligands and on many other factors. One of the most important processes of this type is nucleophilic addition to a carbonyl group. The resulting tetrahedral species may be a stable product, but it is often an intermediate or a transition state; it then reacts further, generally by elimination (i.e., reverse of addition) to produce a new trigonal center. The typical sequence of steps may be described as:

(1) Approach of nucleophile to carbonyl group
(2) Passage through a first transition state
(3) Formation of tetrahedral intermediate
(4) Passage through a second transition state
(5) Formation of product

Various proton-transfer processes occur during these steps (acid or base catalysis). Such sequences of reactions (XVIII) occur in the formation and hydrolysis of peptides, carboxylic acid derivatives, acetals and hemiacetals, ketals and hemiketals, and so forth, and have been the subject of many investigations. Here we discuss features of the reaction paths for addition/elimination processes involving carbonyl centers, as revealed by structural correlations.

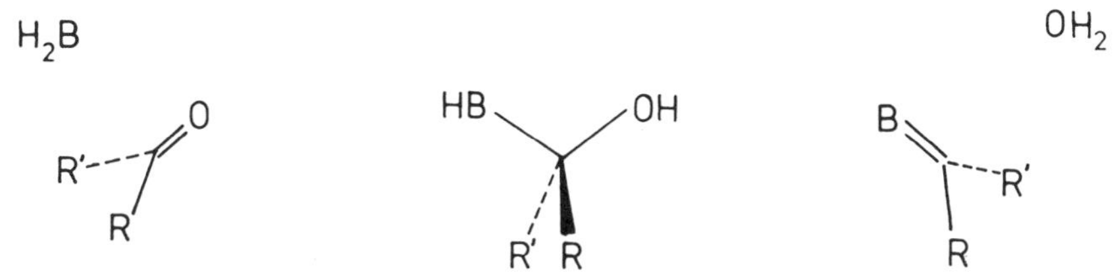

(XVIII)

The most complete information available concerns molecules containing amino and carbonyl groups. Evidence for strong transannular interactions between such groups in medium rings was obtained many years ago by spectroscopic methods,[130] and more recently several examples of unusually short N···C=O distances have been uncovered by crystal structure analysis. We have already mentioned the natural product clivorine (XV), where the N···C=O distance of 1.99 Å seems much too

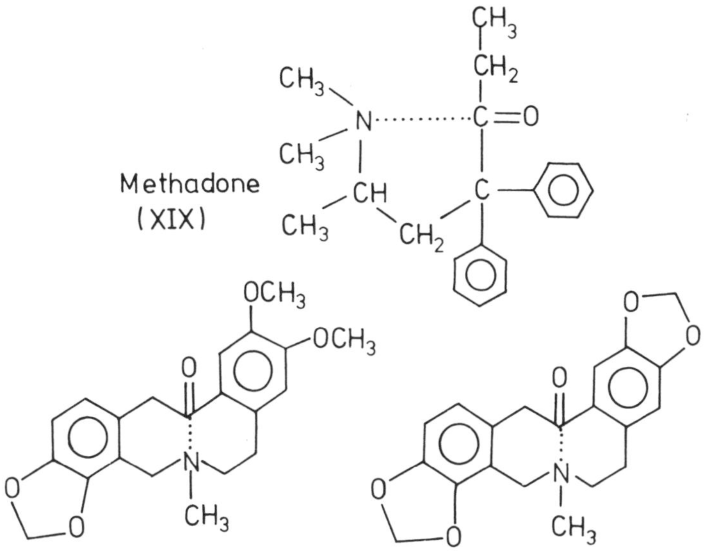

Methadone (XIX)

Cryptopine (XX)

Protopine (XXI)

[130]For a review, see N. J. Leonard, *Rec. Chem. Progr.* **17**, 243 (1956).

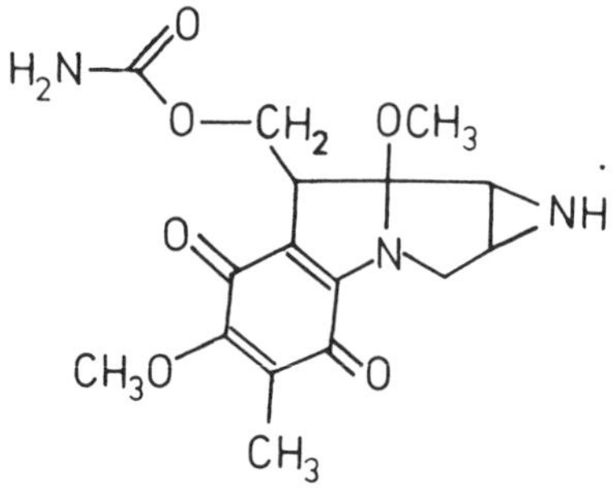

Mitomycin A (XXIII)

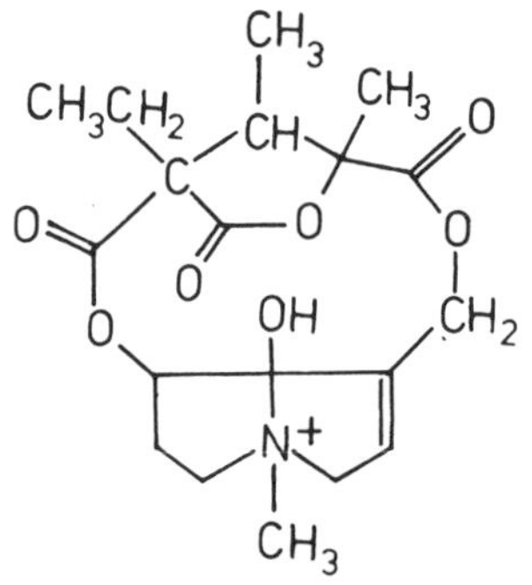

Retusamine (XXII)

long for a bond but much too short for a nonbond. In this molecule and in others showing short N···C=O distances the carbonyl grouping RR′C=O deviates from its usual coplanar geometry. An initial survey,[131] based on six molecules (XV, XIX, XX, XXI, XXII, XXIII) for which structural data were available (Table 7.5), showed that in all cases the N, C, and O atoms lay in an approximate local mirror plane (with respect to the R, R′ substituents, carbon atoms in every case) but that the carbonyl carbon was displaced from the plane of its three substituents *towards* the approaching nitrogen atom (Fig. 7.28). The observed out-of-plane displacements Δ correlated nicely with the N···C distances d_1, as seen in Fig. 7.29, where the smooth curve represents the analytical function

$$d_1 = -1.701 \log \Delta + 0.867 \text{ Å} \tag{7.16}$$

with Δ expressed in Å. A maximum value of Δ is obtained when d_1 equals the standard C—N single bond distance of 1.479 Å.[132] Substitution leads to $\Delta_{max} = 0.437$ Å, and Eq. 7.16 can be rewritten as:

$$d_1 = -1.701 \log n + 1.479 \text{ Å}$$

with $n = \Delta/\Delta_{max}$. Alternatively, we could assume $n = (\Delta/\Delta_{max})^2$ (which gave better results for the tetrahedral molecules discussed earlier), leading to

$$d_1 = -0.85 \log n + 1.479 \text{ Å}$$

[131] H. B. Bürgi, J. D. Dunitz, and E. Shefter, *J. Am. Chem. Soc.* **95,** 5065 (1973).

[132] According to more recent results by G. I. Birnbaum [*J. Am. Chem. Soc.* **96,** 6165 (1974)] a more appropriate value for the C—N single bond distance inthis type of molecule would be about 0.02 Å longer. However, the change would not affect any of the subsequent conclusions.

Table 7.5. Distances and angles found in molecules containing interacting amino and carbonyl group[a,b]

Molecule	d_1 (Å)	d_2 (Å)	Δ (Å)	∠ N···C=O
A, methadone (XIX)[c]	2.910	1.214	0.064	105.0°
B, cryptopine (XX)[d]	2.581	1.209	0.102	102.2
C, protopine (XXI)[e]	2.555	1.218	0.115	101.6
D, clivorine (XV)[f]	1.993	1.258	0.213	110.2
E, retusamine (XXII)[g]	1.64	1.38	0.36	110.9
F, *N*-brosylmitomycin A (XXIII)[h]	1.49	1.37	0.42	113.7
Senkirkine (XXIV)[i]	2.292	1.213	0.115	109.3
XXV[j]	2.76	1.203	0.016	107.6
XXVI[k]	2.457	1.217	0.098	111.1
Monoclinic XXVII[l]	2.76	1.211	0.023	112.5
Orthorhombic XXVII[l]	2.69	1.216	0.054	114.5
XXVIII[m]	2.606	1.215	0.061	102.2
XXIX[m]	2.594	1.202	0.062	98.6
XXX[m]	2.557	1.218	0.088	104.4

[a]From Bürgi, Dunitz, and Shefter, *J. Am. Chem. Soc.* **95,** 5065 (1973).
[b]Distances d_1, d_2, and Δ are defined in Fig. 7.28. The six molecules used in the initial survey are labeled A–F.
[c]H. B. Bürgi, J. D. Dunitz, and E. Shefter, *Cryst. Struct. Commun.* **2,** 667 (1973); E. Bye, *Acta Chem. Scand.* **B28,** 5 (1974).
[d]S. R. Hall and F. R. Ahmed, *Acta Crystallogr.* **B24,** 346 (1968).
[e]S. R. Hall and F. R. Ahmed, *Acta Crystallogr.* **B24,** 337 (1968).
[f]K. B. Birnbaum, *Acta Crystallogr.* **B28,** 2825 (1972).
[g]J. A. Wunderlich, *Acta Crystallogr.* **B23,** 846 (1967).
[h]A. Tulinsky and J. H. van der Hende, *J. Am. Chem. Soc.* **89,** 2905 (1967).
[i]G. I. Birnbaum, *J. Am. Chem. Soc.* **96,** 6165 (1974).
[j]M. Kaftory and J. D. Dunitz, *Acta Crystallogr.* **B31,** 2912 (1975).
[k]M. Kaftory and J. D. Dunitz, *Acta Crystallogr.* **B31,** 2914 (1975).
[l]M. Kaftory and J. D. Dunitz, *Acta Crystallogr.* **B32,** 1 (1976).
[m]W. B. Schweizer, Doctoral Dissertation 5948, E.T.H., Zürich, 1977.

If we assume that the sum of the C—O and C—N bond numbers stays constant at all stages of this reaction, we obtain for the C—O distances

$$d_2 = -c \log(2-n) + 1.426 \text{ Å}$$

where the additive constant is the standard C—O single bond length.

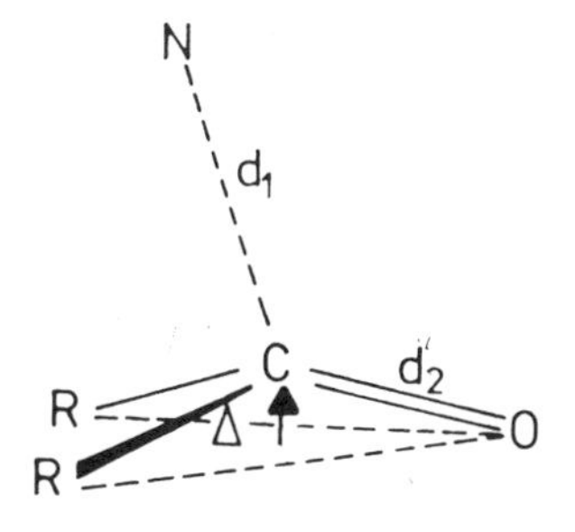

Figure 7.28. When the carbon atom of the carbonyl grouping RR′C=O interacts with a nearby nucleophile, it is typically displaced from the plane RR′O toward the nucleophile.

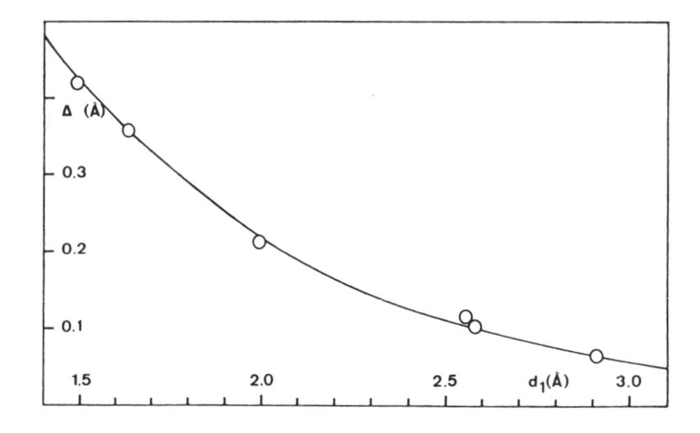

Figure 7.29. Correlation plot of Δ vs. d_1 (smooth curve obtained from Eq. 7.16). From Bürgi, Dunitz, and Shefter, *J. Am. Chem. Soc.* **95,** 5065 (1973). Reprinted by permission.

The observed C—O distances tend to cluster around 1.21–1.22 Å, except for the extreme cases with Δ close to Δ_{max}, suggesting that the dependence of n on (Δ/Δ_{max}) is nonlinear. The second power dependence, with $c \sim 0.71$ Å, reproduces the observed trend fairly well.

When the data from these six crystal structures are projected on to the NCO plane, they give a vivid picture of the geometric changes that occur during the addition reaction (Fig. 7.30). As the nucleophile approaches the carbonyl C atom, the plane containing C and the two alkyl substituents bends away, and the C—O distance increases. The experimental points leave no doubt that the approach of the nucleophile is along a direction at approximately 109° to the C—O bond and not perpendicular to it, as might have been expected. The arrows show the probable orientation of the lone-pair orbital of the approaching nitrogen atom—we cannot observe this, but it may be assumed to lie close to the local threefold axis of the tertiary amino group involved in all cases except F (Table 7.5). There is obviously a preferred orientation for nucleophilic attack of an amine on a carbonyl group, with the lone pair

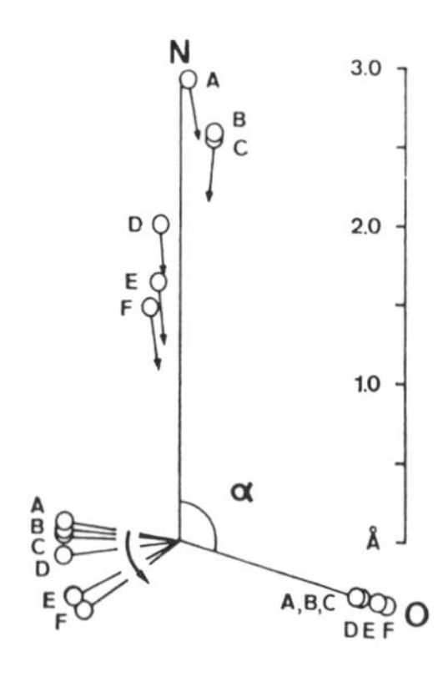

Figure 7.30. Reaction coordinate projected on the NCO plane showing nitrogen (top: arrows indicate the estimated direction of lone pair), carbonyl oxygen (bottom right), and bisector of RCR′ (bottom left). From Bürgi, Dunitz, and Shefter, *J. Am. Chem. Soc.* **95,** 5065 (1973). Reprinted by permission.

roughly coincident with the N···C direction, i.e., at about 109° to the C—O bond.

There is no lack of explanations for the preference of the nucleophile to approach the C atom from the rear of the C═O bond rather than perpendicular to the RR′C═O plane. It can be explained by the charge distribution, by the bent-bond model, or by MO models at various levels of sophistication. In terms of the bent-bond model (Fig. 7.31), the addition process is formally equivalent to an S_N2 displacement at unsaturated carbon. A simple MO model (with hydride ion as nucleophile) is illustrated in Fig. 7.32. Here the highest occupied MO involves the 1*s* orbital of H^- in an in-phase combination with the π^* orbital of the carbonyl group and in an out-of-phase combination with the π orbital. The oblique approach to the carbon atom clearly leads to more favorable net overlap than the perpendicular one. This qualitative conclusion is confirmed by SCF-LCGO calculations for various nuclear arrangements of the model system $CH_2{=}O + H^- \rightarrow CH_3O^-$, corresponding to nucleophilic addition of hydride ion to formaldehyde to yield methanolate anion;[133,134] the calculated reaction path for this reaction shows striking similarities to the one derived from structural correlations for amine addition.

The additional structural information on N···C═O interactions that has become available in the last few years (Table 7.5) confirms the general conclusions of the initial survey, but it also shows that some qualifications are necessary. Depending on the substituents, amino groups vary in their nucleophilicity. One measure of the relative nucleophilicity of a tertiary amino group is the sum of the bond angles at N, which is about 330° for a typical, pyramidal, trialkyl amine and 360° for a planar amine. For most of the examples in Table 7.5 the angle sum at N is in the range 330°–340°, but the amino group in XXV, with its aryl substituent, has angle sum 355.9° and is evidently a poor nucleophile, in keeping with the general chemical experience that anilines are much less nucleophilic than aliphatic amines. Other structural expressions of the delocalization of the N lone pair in XXV are seen in the relatively short N—C_{ar} distance of 1.39 Å, comparable with that found in other aniline derivatives, and in the orientation of the benzene ring plane, with its normal roughly parallel to the N lone-pair direction. According to Eq. 7.16, the N···C═O distance of 2.76 Å in XXV should correspond to a Δ of 0.077 Å, about five times larger than that observed.

[133] H. B. Bürgi, J. M. Lehn, and G. Wipff, *J. Am. Chem. Soc.* **96,** 1956 (1974).
[134] H. B. Bürgi, J. D. Dunitz, J. M. Lehn, and G. Wipff, *Tetrahedron* **30,** 1563 (1974).

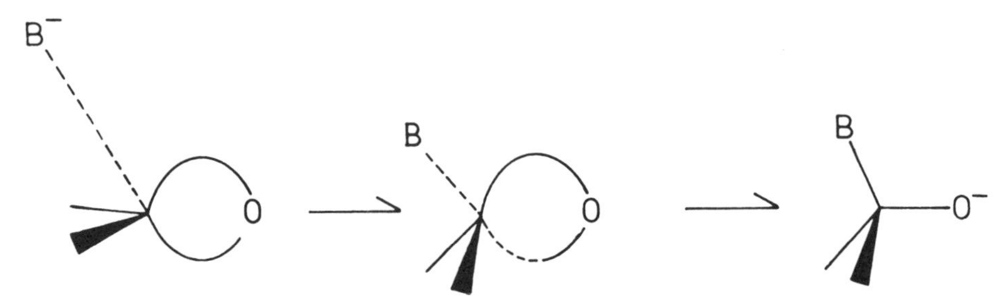

Figure 7.31. Bent-bond model of nucleophilic addition to a carbonyl group.

The anilinic N evidently induces a much lower degree of pyramidalization at the carbonyl than a typical amino group.

O
N
CH_3

(XXV)

One objection to the conclusions drawn from the initial survey is that it is very difficult to exclude the possibility that the "preferred" N···C═O approach angle of 100°–110° might be imposed by geometric constraints in the molecules or by other factors that have nothing to do with the

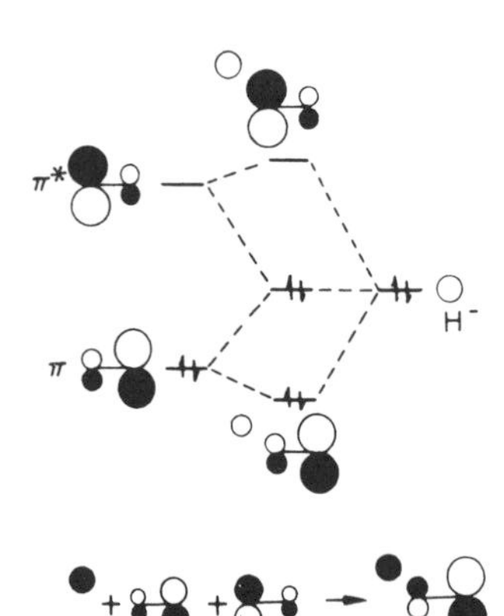

Figure 7.32. MO model of interactions between 1s orbital of H^- and the π(C—O) and π*(C—O) orbitals. The highest occupied MO is shown below the orbital interaction scheme. From Bürgi, Dunitz, Lehn, and Wipff, *Tetrahedron* **30,** 1563 (1974).

incipient nucleophilic addition process. Apart from methadone (XIX), the N···C=O approaches examined in the initial survey represent cases of transannular interaction in eight- or ten-membered rings, where geometric constraints may obviously be relevant. The last five analyses included in Table 7.5 were all carried out to test this possibility.

In XXVII, with "natural" bond angles (e.g., those of a Dreiding model of the molecule) the N···C=O angle would be approximately 90°. In

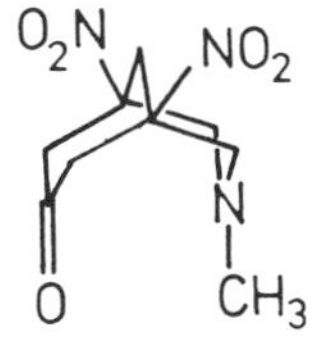

(XXVII)

both crystal modifications it is found[135] that the molecule is slightly disorted from the ideal geometry to increase the N···C=O angle to about 110° (Fig. 7.33).

An even more striking example is provided by the analysis of the 1,8-disubstituted naphthalene derivatives XXVIII–XXX.[136] For an undistorted naphthalene skeleton (Fig. 7.34a), the N···C=O angle would be close to 90° and the nitrogen lone pair would not be directed towards the carbonyl C atom. In most 1,8-disubstituted naphthalenes, both substituents are splayed outward, as in Fig. 7.34b, which would lead to a still worse misalignment of the nitrogen lone pair. In molecules XXVIII–XXX the exocyclic C—C bond is splayed outward, but the C—N bond is

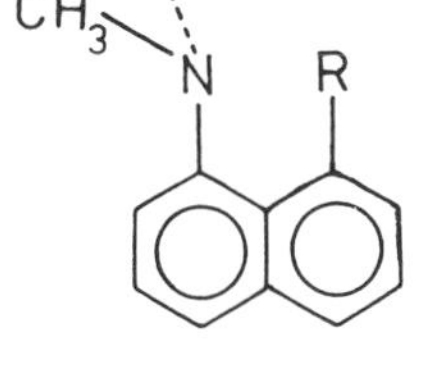

(XXVIII) R = COOH
(XXIX) R = $COOCH_3$
(XXX) R = $CO \cdot CH_3$

[135] M. Kaftory and J. D. Dunitz, *Acta Crystallogr.* **B32,** 1 (1976).
[136] W. B. Schweizer, Doctoral Dissertation No. 5948, E. T. H., Zürich, 1977; W. B. Schweizer, G. Procter, M. Kaftory, and J. D. Dunitz, *Helv. Chim. Acta* **61,** 2783 (1978).

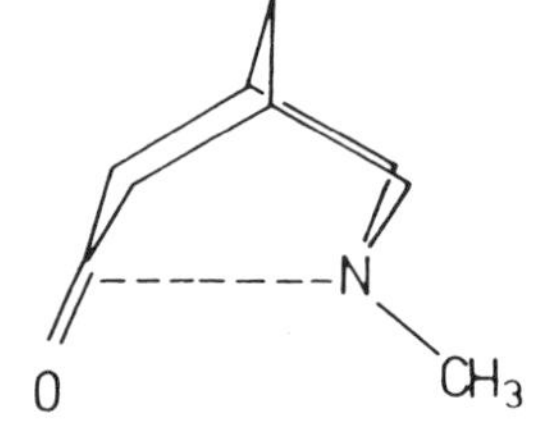

Figure 7.33. Schematic of 3-azabicyclo[3.3.1]nonan-7-one skeleton. Left: idealized molecule. Right: distorted, as observed by crystal structure analysis of XXVII.

splayed inward; this gives an N···C═O angle of about 100° and tends to orient the lone pair more towards the N···C direction (Fig. 7.34c). Thus, the usual pattern of bond angles is distorted to bring the amino nitrogen into a more favorable position for attack than is possible in the undistorted molecule. The crystal structure of the amino acid XXVIII is particularly interesting because it contains two independent molecules in the asymmetric unit. One molecule shows the pattern of bond-angle distortions mentioned above, but the other has both exocyclic bonds splayed outward, with the carboxyl group nearly in the naphthalene plane rather than perpendicular to it. The explanation is that the crystal is actually a 1:1 molecular compound of the amino acid, which shows the N···C═O interaction, and the corresponding zwitterion, which does not. Interatomic distances and angles for both molecules are shown in Fig. 7.35. Another case where amino acid and zwitterion occur in the same crystal structure is anthranilic acid.[137]

Table 7.5 reveals that very small changes in some of the molecules or in their environments may be associated with large changes in the geometry of the N···C═O fragment. The most striking comparison is per-

(XXIV) Senkirkine

haps the one between the closely similar molecules, clivorine (XV) and senkirkine (XXIV), with a difference of about 0.30 Å in the N···C═O

[137]C. J. Brown, *Proc. Roy. Soc., London* **A302,** 185 (1968).

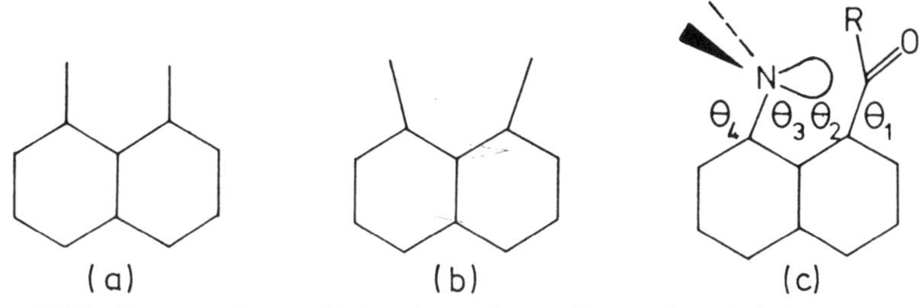

Figure 7.34. Pattern of exocyclic bonds: (a) for undistorted naphthalene skeleton, (b) as observed in typical 1,8-disubstituted naphthalenes, and (c) as observed in molecules XXVIII–XXX, where the bond angles θ_1, θ_2, θ_3, θ_4 are (degrees):

	θ_1	θ_2	θ_3	θ_4
XXVIII	116	124	117	123
XXIX	118	122	117	124
XXX	117	123	116	123

distance. Analogous structural changes in XXVI may be produced by solvent effects, for the position of the carbonyl absorption maximum

O
N
CH_3
(XXVI)

of this compound is strongly solvent dependent: 1666 cm^{-1} (C_6H_{12}), 1656 cm^{-1} (CCl_4), 1641 cm^{-1} (CH_3CN), 1631 cm^{-1} (CH_2Cl_2), 1613 cm^{-1} ($CHCl_3$).[138] Birnbaum[139] found that in compounds of this type there is an approximately linear correlation between carbonyl absorption frequency and N···C=O distance; according to his correlation, the N···C distance in XXVI would cover a range of roughly 0.5 Å, from about 2.60 Å in cyclohexane solution to about 2.20 Å in chloroform. Of course, we do not know what the distances actually are in solution, but one might well suspect that they do differ appreciably from one solvent to another. The potential energy variation along the reaction path for this type of molecule is presumably rather flat, and the position of the energy minimum may then be susceptible to small external perturbations. Moreover, shallow minima are likely to be strongly anharmonic, leading possibly to

[138]N. J. Leonard, private communication (1960).
[139]G. I. Birnbaum, *J. Am. Chem. Soc.* **96,** 6165 (1974).

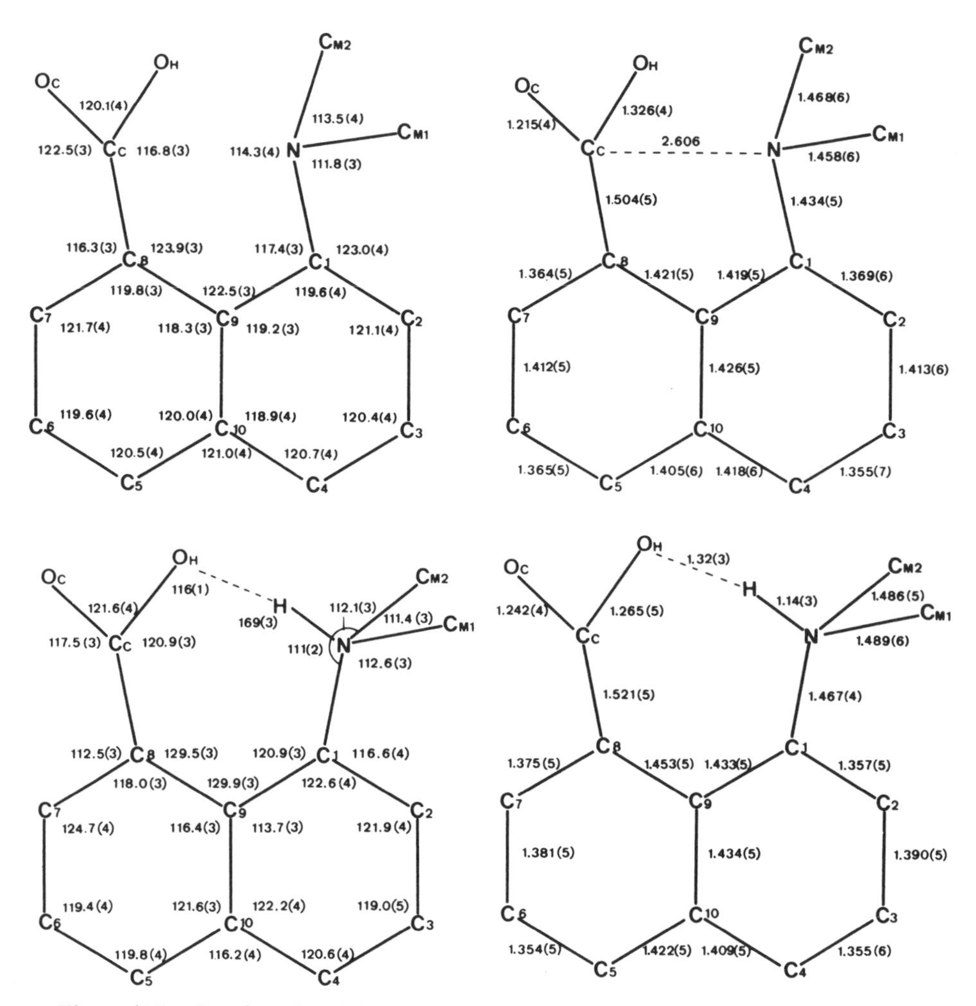

Figure 7.35. Bond angles (left) and distances (right) for 8-dimethylamino-1-naphthoic acid (XXVIII) and its zwitterion, which occur in the same crystal structure.

"observed" atomic positions (averages over the molecular vibrations) that might differ appreciably from equilibrium positions. Crystal structure studies at different temperatures may throw light on this problem.[140]

Crystal structure data for O···C═O interactions (oxygen as nucleophile) have not been neglected in the search for structural correlations. The picture that emerges from such a survey[141] is not as clear as in the

[140]Recent results indicate that the anharmonicity is not so pronounced after all. A low-temperature (98 K) analysis of XXVI has yielded a molecular geometry that is virtually identical to that obtained at room temperature. In particular there is no detectable change in the N···C distance [E. Bye and J. D. Dunitz, *Acta Crystallogr.* **B34,** 3245 (1978)].

[141]H. B. Bürgi, J. D. Dunitz, and E. Shefter, *Acta Crystallogr.* **B30,** 1517 (1974).

N···C=O case, mainly because O···C=O interactions are generally weaker and even more sensitive to perturbations. The many examples in the literature form a heterogeneous collection; intramolecular examples are found in dicarboxylic acids, diesters, nitroaldehydes and ketones, and so forth, while intermolecular ones involve mostly interactions between pairs of carbonyl groups and include several well known cases—e.g., triketoindane—that have been discussed by various authors in terms of dipole-dipole or donor-acceptor interactions.[142] Although observed N···C distances cover the whole range from about 1.5 Å (covalent bonding) to about 3 Å, there is a pronounced gap in the distribution of O···C distances between about 1.5 and 2.6 Å (the shortest intermolecular O···C distances observed are about 2.77 Å). The gap may be the result of an unfortunate choice of examples (we know of no 1,5-transannular O···C=O interactions in eight-membered rings), but it may also occur because the O···C=O interaction energy is not large enough to overcome the nonbonded repulsion energy to give equilibrium structures in the intermediate region.

Fig. 7.36 shows O···C=O distances (on the long-distance side of the gap) plotted against O···C=O angles and out-of-plane displacements Δ. Although it is not nearly as informative as Fig. 7.29, it does suggest certain similarities with N···C=O interactions. The angle of approach α is not as constant as in the N···C=O case, but there is a clear preference for the range 90°–110°. Similarly, the tendency for the out-of-plane displacement Δ to increase with decreasing O···C distance can be discerned in spite of considerable scatter. The scatter may arise from several causes besides experimental inaccuracies in the data. As already mentioned, the examples constitute a mixed bag, and quite different kinds of perturbation may be operative from case to case, with energies of the same magnitude as those involved in the interaction we try to distinguish. In some intermolecular examples the carbonyl C atom interacts with *two* nucleophilic O atoms, one on either side of the carbonyl plane, so that the net displacement Δ is the resultant of two displacements in opposite directions. Nevertheless, the data show clearly that for distances greater than 2.6 Å an averaged O···C=O displacement curve would lie well below the corresponding N···C=O curve. Fig. 7.36 suggests that for a given nucleophile···carbon distance, the displacement Δ with oxygen as nucleophile is only about a third of the displacement with nitrogen as nucleophile, and this is confirmed by comparative studies of 1,8-disub-

[142]W. Bolton, *Acta Crystallogr.* **18,** 5 (1965); W. Fedeli and J. D. Dunitz, *Helv. Chim. Acta* **51,** 445 (1968).

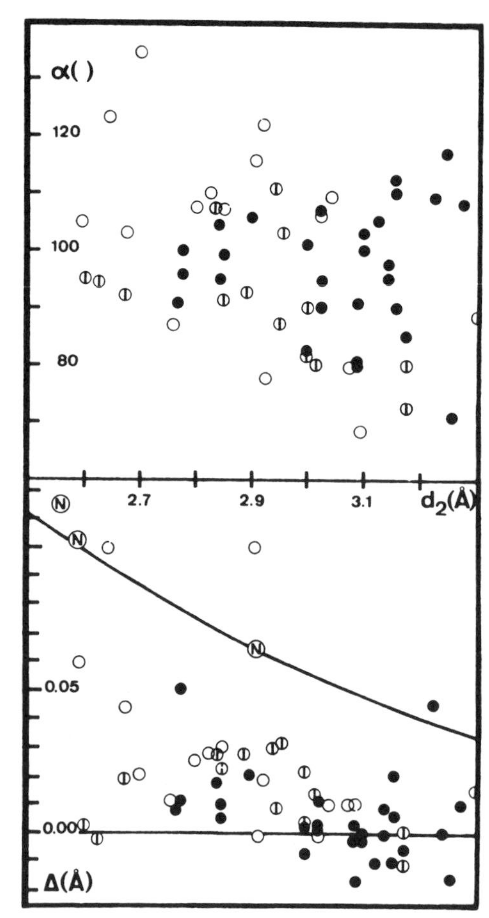

Figure 7.36. Scatter plots of O···C=O distance against O···C=O angle (above) and against out-of-plane displacement Δ. Intramolecular interactions are designated by marked circles ⦶ for dicarboxylic acids and by open circles for other types of molecules; intermolecular interactions are designated by filled circles. A portion of the curve Ⓝ relating N···C=O distance and Δ is also shown for comparison. From Bürgi, Dunitz, and Shefter, *Acta Crystallogr.* **B30,** 1517 (1974).

stituted naphthalene derivatives with methoxyl groups replacing the dimethylamino groups of XXVIII and XXX.[136]

It seems reasonable to assume that for these long-distance interactions, pyramidalization of the carbonyl group (considered in isolation) will increase the potential energy, whereas the nucleophile–carbon interaction will lower it. Since Δ (N···C=O) is about three times larger than Δ (O···C=O) in the range 2.5–3.0 Å, the corresponding energy cost should be about ten times greater, assuming a quadratic dependence on Δ.[143] This suggests that the N···C=O interaction energy is at least one order of magnitude greater than the O···C=O interaction energy at the

[143]The pyramidalization energy can be estimated from the out-of-plane bending force constant of a carbonyl group to be about 200 Δ^2 kcal/mol Å^2, equal to 0.5 kcal/mol for $\Delta = 0.05$ Å.

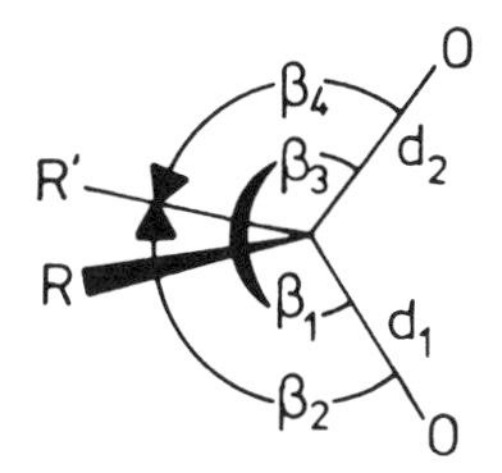

Figure 7.37. Definition of d_1, d_2 and β_1, β_2, β_3, β_4 for a tetrahedral carbon center with two oxygen substituents.

same distance, which provides semiquantitative expression to the common chemical experience that nitrogen is a "better" nucleophile than oxygen.

At the other side of the gap there are extensive structural data for tetrahedral subunits with two geminal C—O bonds (diols, ketals, hemiketals, acetals, hemiacetals). The two C—O distances are usually slightly different, but any correlation that may be present between the structural parameters is obscured by scatter arising from various possible sources (differences in oxygen substituents, hydrogen bonding, variation in O—C—O—R torsion angles, etc.). However, there does seem to be a significant correlation between the quantities $\Delta d = d_2 - d_1$ and $\Delta\beta = \beta_1 + \beta_2 - \beta_3 - \beta_4$ (Fig. 7.37), which correspond to a pair of antisymmetric stretching and bending displacement coordinates of a tetrahedron with C_{2v} symmetry. The scatter plot of Δd against $\Delta\beta$ (Fig. 7.38) shows a clear tendency for large positive values of Δd to be associated with large positive values of $\Delta\beta$. This means that as one C—O bond is stretched and the other is shortened, a coupling between the antisymmetric stretching and bending coordinates comes into play to bring the shorter bond towards the RCR′ plane and the longer bond away from that plane; it implies that in the decomposition of a tetrahedral intermediate RR′CXY the leaving group X does not depart along the initial direction of the C—X bond. Instead, the direction of departure of X is continually adjusted to maintain an XCY angle of about 110°. The same kind of angle constancy was found in the preferred direction of approach of the nucleophile for N···C=O interactions, and the present result merely confirms that with oxygen as nucleophile the reverse decomposition proceeds along the same reaction path.

Structural data for tetrahedral carbon atoms RCXYZ with three electronegative substituents (orthoesters and amide acetals) are very sparse and inaccurate. However, the available data suggest[141] that here too there is a coupling between antisymmetric bending and stretching coordinates that might be relevant to the decomposition mode.

Among the many factors that might influence the relative lengths of the geminal C=O bonds, one that seems important is the conformation

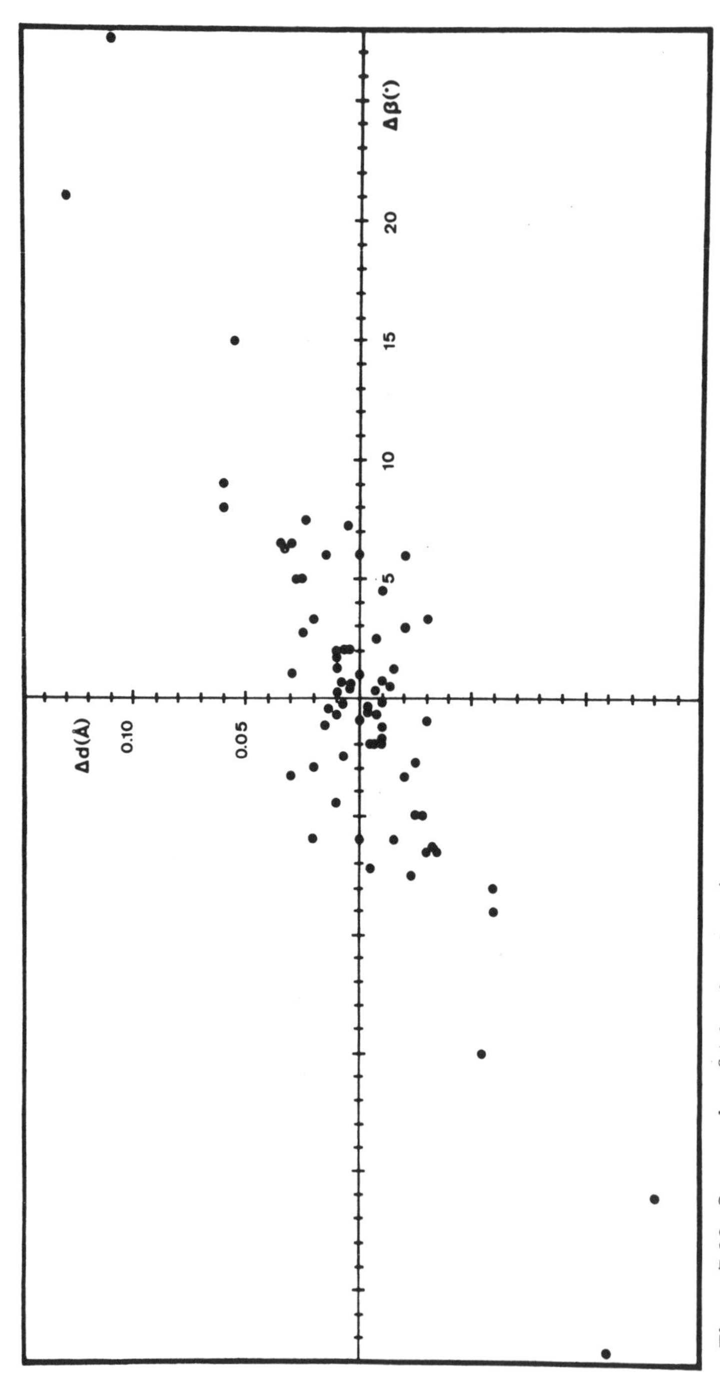

Figure 7.38. Scatter plot of $\Delta d = d_2 - d_1$ against $\Delta\beta = \beta_1 + \beta_2 - \beta_3 - \beta_4$ (see Fig. 7.36 for definitions of symbols) for tetrahedral carbon centers that carry two oxygen substituents. From Bürgi, Dunitz, and Shefter, *Acta Crystallogr.* **B30,** 1517 (1974).

of the RO—C—O′R′ fragment. Qualitative MO perturbation theory[144] suggests that for a X—C—Y system where both X and Y are electronegative substitutents, the C—X bond should become longer (and weaker) when it is antiperiplanar (*app*) to one of the lone-pair orbitals on Y.[145,146,147] Interaction of the occupied lone-pair orbital n_Y with the unoccupied σ^*(C—X) orbital gives a modified orbital $n_Y + \lambda\sigma^*$(C—X) that is antibonding with respect to C—X and bonding with respect to C—Y. This interaction, which is maximal for a coplanar arrangement of the two orbitals, lowers the energy of the occupied orbital and is net stabilizing. SCF-LCGO ab initio calculations for methanediol, HO—CH_2—O′H′, show that when the O—C bond is *app* to a lone pair on O′, it is slightly lengthened (by about 0.02 Å) and weakened, whereas the C—O′ bond is slightly shortened (by about 0.01 Å) and strengthened.[148,149] Similar changes occur when either of the OH groups is replaced by NH_2. Corresponding calculations for methanetriol $CH(OH)_3$ and methanetetrol $C(OH)_4$ show that when a C—O bond is *app* to two or three lone-pair orbitals, the bond-lengthening effect is cumulative.[149] However, it is superimposed on another effect that depends on the number of electronegative substitutents: as this number increases, the average C—O distance decreases—by ca. 0.03 Å for one additional OH group and by 0.05 Å for two. This trend can be seen in the observed C—F distances in CH_3F(1.385 Å), CH_2F_2(1.358 Å), CHF_3(1.332 Å), and CF_4(1.317 Å)

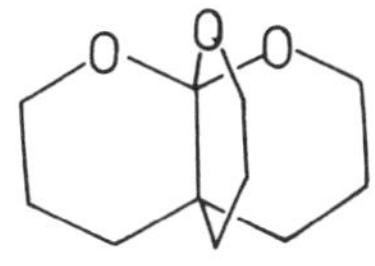

(XXXI)

and in the stable orthoester XXXI, where the mean length of the inner C—O bonds is about 0.035 Å shorter than the mean of the three outer C—O bonds.[150]

The importance of conformational effects, of the kind outlined above,

[144] R. Hoffmann, *Accounts Chem. Res.* **4,** 1 (1971).

[145] E. A. C. Lucken, *J. Chem. Soc.* **1959,** 2954.

[146] C. Romers, C. Altona, H. R. Buys, and E. Havinga, *Topics Stereochem.* **4,** 39 (1969).

[147] S. David, O. Eisenstein, W. J. Hehre, L. Salem, and R. Hoffmann, *J. Am. Chem. Soc.* **95,** 3806 (1973).

[148] G. A. Jeffrey, J. A. Pople, and L. Radom, *Carbohyd. Res.* **25,** 117 (1972).

[149] J. M. Lehn, G. Wipff, and H. B. Bürgi, *Helv. Chim. Acta* **57,** 493 (1974).

[150] S. H. Banyard and J. D. Dunitz, *Acta Crystallogr.* **B32,** 318 (1976).

has been recognized for many years in carbohydrate chemistry, especially in connection with the "anomeric effect"—i.e., the tendency for the exocyclic C—O bond at the anomeric carbon to adopt the axial rather than the equatorial position in pyranoses.[151] More recently, extensive studies by Deslongchamps and co-workers show that conformational effects have a decisive role in the hydrolysis of esters and amides, where the nature of the product depends on the orientation of lone-pair orbitals in the tetrahedral intermediate hemi-orthoester or hemi-orthoamide.[152] In particular, cleavage of a C—O or C—N bond can occur only if the bond in question is oriented *app* to lone-pair orbitals on the other two heteroatoms (O or N).

In conclusion, we mention two cases where the structural correlation approach has provided some insight into actual chemical phenomena involving nucleophilic addition to carbonyl groups. The first example concerns the rapid equilibration of the keto acid XXXII and the isomeric lactone alcohol XXXIV.[153] Infrared and nmr spectra show that in solution the two isomers are present in approximately equal concentration. The overall activation enthalpy for the isomerization appears to be quite low, about 8–13 kcal/mol, depending on the experimental conditions; in addition to ring closure/opening, a proton transfer step is

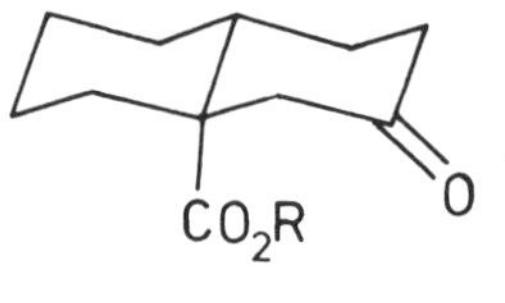

CO_2R O | OR C O O

(XXXII) R=H (XXXIV) R=H

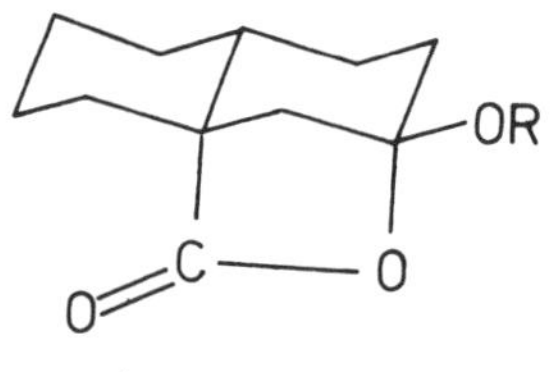

(XXXIII) R=CH_3 (XXXV) R=CH_3

involved, and this appears to be rate limiting in aprotic solvents. Crystallization from several solvents yields the same crystal modification, which turns out to contain exclusively molecules in the closed form (XXXIV). A crystal structure analysis shows that at the geminally substituted tetrahedral center, the ring C—O bond is long (1.51 Å compared with about 1.44 Å for the corresponding bond in a normal γ-lactone), and the exocyclic C—O bond is short (1.38 Å compared with ca. 1.43 Å in an alcohol). There is also a marked asymmetry in the bond

[151] J. T. Edwards, *Chem. Ind. (London)* 1102 (1955).
[152] P. Deslongchamps, *Tetrahedron* **31**, 2463 (1975).
[153] D. J. Chadwick and J. D. Dunitz, *J. Chem. Soc., Perkin II*, in press.

angles, in the sense expected from our previous correlations (Fig. 7.38). The pattern of bond length and bond angle differences at the tetrahedral center (Δd = 0.13 Å, $\Delta\beta$ = 19.5°) thus indicates an appreciable displacement along the reaction path leading from XXXIV to XXXII. In keeping with this, a lone-pair orbital of the hydroxyl oxygen is *app* to the long C—O bond, and a hydrogen bond is formed from the hydroxyl group of one molecule to the carbonyl oxygen of another (incipient proton transfer). This latter feature is, of course, absent in the crystal structure of the methyl ether (XXXV). However, the *app* relationship of the methoxy lone pair to the ring C—O bond is retained, and so is the pattern of bond distances and angles at the tetrahedral center, although it is not quite as extreme (Δd = 0.11 Å, $\Delta\beta$ = 16.9°). In contrast to the keto acid ↔ lactone alcohol isomerization, the rate of the equilibration XXXIII ↔ XXXV is immeasurably slow, so that both species can be isolated separately. Crystal structure analysis of the keto ester XXXIII shows that in this molecule the ester group is oriented so that the methoxyl O is ideally placed for attack on the keto C atom with an O···C=O angle of 104°. This structure shows a relatively short O···C distance of 2.77 Å, and the keto C atom is displaced by 0.036 Å from the plane of its three bonded neighbors towards the O atom. Thus the structures on both sides of the transition state for the isomerization XXXIII ↔ XXXV show features that indicate displacement along the corresponding reaction path.

Our second example concerns hydride reduction of unsymmetrically substituted succinimide derivatives (XXXVI) to give lactam alcohols. For bulky substitutents R_1 and R_2, the reaction shows a remarkable regio-

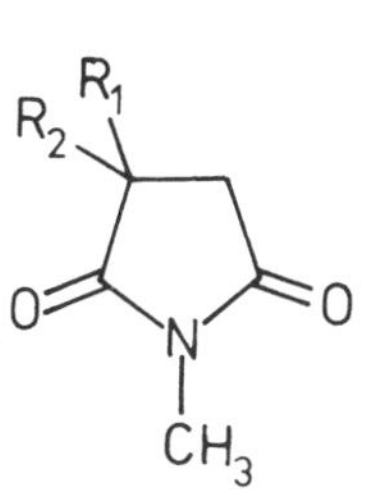

(XXXVI) 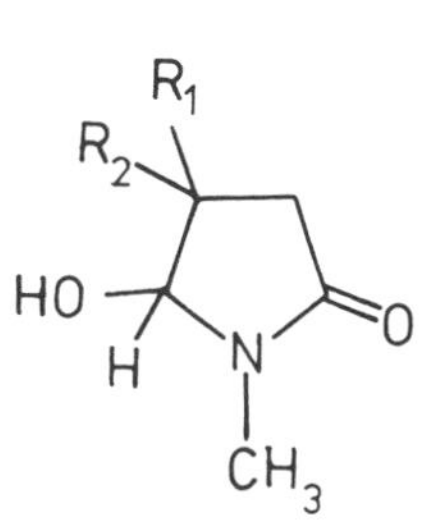 (XXXVII)

specificity, the carbonyl group attacked being exclusively the one on the more hindered side of the molecule (XXXVII).[154] Similar behavior has been reported for the analogous hydride reduction of cyclic anhydrides

[154] J. B. P. A. Wijnberg, H. E. Schoemaker, and W. N. Speckamp, *Tetrahedron* **34,** 179 (1978).

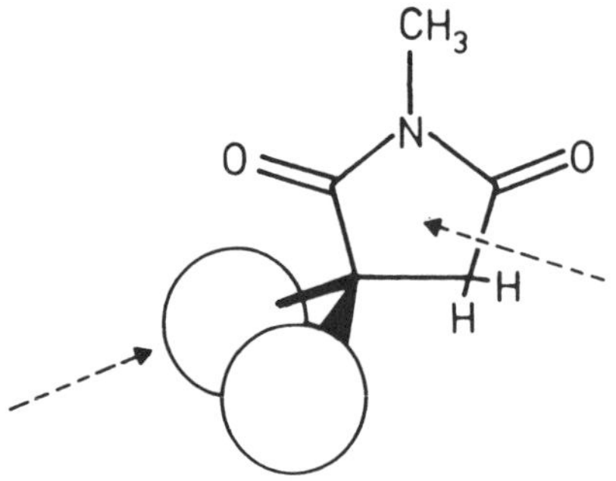

Figure 7.39. The rear approach to the carbonyl group adjacent to the substituents is less hindered than that to the other carbonyl group.

to yield lactones.[155] The observed regiospecificity is completely consistent with our finding that nucleophiles attack preferentially from the rear of the carbonyl group because for such an attack the approach to the carbonyl group adjacent to the substitutents is less hindered than the approach to the other carbonyl group (Fig. 7.39). The possibility that the two carbonyl groups have intrinsically different electronic character in the disubstituted imide itself can be eliminated from the results of a crystal structure analysis of the 3-phenyl-3-benzyl derivative (which shows 100% regiospecificity in the hydride reduction); within experimental error, the two carbonyl groups are structurally equivalent as far as the ring conformation is concerned.[156]

Other possible stereochemical implications of a nonperpendicular approach of nucleophiles to carbonyl groups have been discussed by Baldwin[157] and by Anh and Eisenstein.[158]

Other Structural Correlations

Earlier we discussed the reaction path for removal of an apical ligand from a pentacoordinated trigonal bipyramidal species to yield a tetracoordinated tetrahedral species and an isolated ligand (see section on Ligand Exchange at Tetrahedral Cadmium). Another mechanistically important reaction of trigonal bipyramidal molecules is ligand reorganization—i.e., interconversion of apical and equatorial ligands. From nmr experiments it is known that such reorganization processes can occur with low activation energies, and several reaction paths have been proposed.[159] In the Berry-Smith process,[160] the trigonal pyramid is imagined

[155] D. M. Bailey and R. E. Johnson, *J. Org. Chem.* **35,** 3574 (1970).
[156] R. B. Rosenfield and J. D. Dunitz, *Helv. Chim. Acta* **61,** 2176 (1978).
[157] J. Baldwin, *J. Chem. Soc., Chem. Commun.* **1976,** 738.
[158] N. T. Anh and O. Eisenstein, *Nouv. J. Chim.* **1,** 61 (1977).
[159] For a review see R. R. Holmes, *Accounts Chem. Res.* **5,** 296 (1972).
[160] R. S. Berry, *J. Chem. Phys.* **32,** 933 (1960).

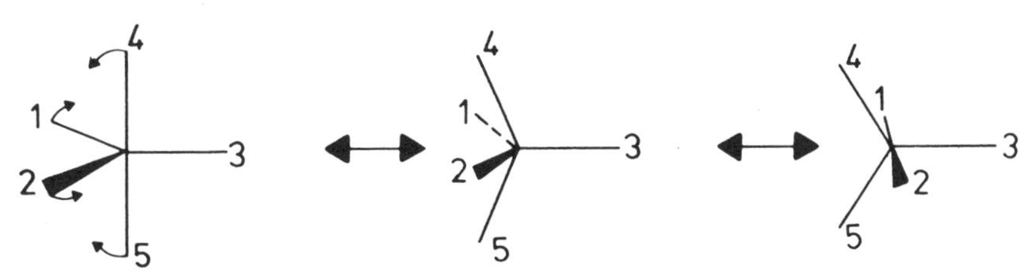

Figure 7.40. Berry-Smith process for interconversion of apical and equatorial ligands in a trigonal bipyramidal molecule.

to deform to a low-lying transition state or intermediate with square pyramidal geometry with retention of C_{2v} symmetry for all points on the reaction path (Fig. 7.40).[161] Pentacoordinated species have been observed in many crystal structures, sometimes as nearly perfect trigonal bipyramids, sometimes as square pyramids, but usually as more-or-less distorted versions of these ideal forms. Although the observed structures differ with respect to the central atom and the nature of the ligands, they may be arranged in a sequence that vividly illustrates the smooth transition along the Berry-Smith path from D_{3h} to C_{4v} symmetry (Fig. 7.41). Muetterties and Guggenberger described the deformation process in terms of the dihedral angles between adjacent faces of the coordination polyhedra,[162] but an alternative description can be given in terms of bond angles at the central atom. The latter description shows that two distinct paths, differing with respect to the angle between axial and equatorial ligands in the square pyramidal structure, can be discerned.[99]

Structural data for chlorocuprate complexes, with coordination number 4, 5, or 6 have also been analyzed.[163] The observed $CuCl_n$ structures can be described as more-or-less distorted versions of four idealized arrangements—square planar (D_{4h}), tetrahedral (T_d), trigonal bipyramidal (D_{3h}), and octahedral (O_h)—and can be considered as frozen-in intermediates along the interconnecting reaction paths. Smooth curves were obtained for three such paths: (a) the D_{4h}/T_d interconversion of $CuCl_4^{2-}$ (b) dissociation of $CuCl_5^{3-}$ to $CuCl_4^{2-}$, and (c) dissociation of $CuCl_6^{4-}$ to $CuCl_4^{2-}$.

It is quite a jump from these inorganic examples to the conformation of the five-membered ring. As discussed in Chapter 9, the out-of-plane

[161]The Berry-Smith process is often described as a pseudorotation. The term pseudorotation implies not only motion along a cyclic coordinate but also one along an essentially *isoenergetic* path. To talk of pseudorotation between different "isomers" thus seems to involve a contradiction of terms.

[162]E. L. Muetterties and L. J. Guggenberger, *J. Am. Chem. Soc.* **96,** 1748 (1974).

[163]P. Murray-Rust and J. Murray-Rust, *Acta Crystallogr.* **A31,** S64 (1975).

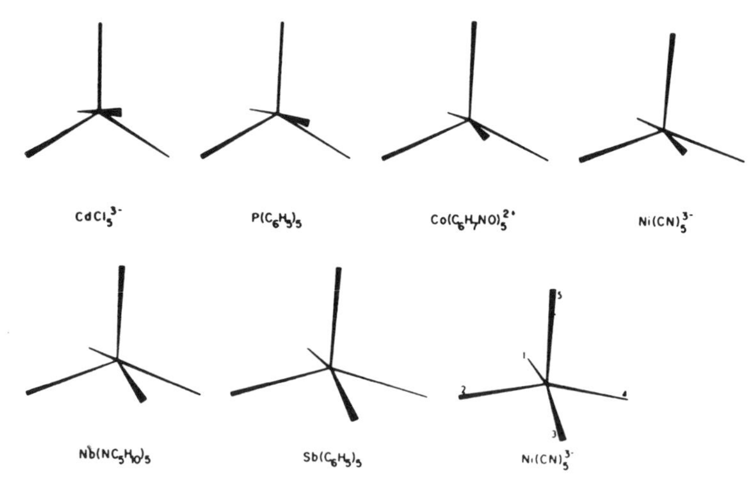

Figure 7.41. Transition from D_{3h} to C_{4v} symmetry as illustrated by a sequence of observed structures of MX_5 molecules. From Muetterties and Guggenberger, *J. Am. Chem. Soc.* **96,** 1748 (1974). Reprinted by permission.

deformations of a pentagon can be described by two parameters: one is a puckering amplitude that can be expressed in terms of ω_0, the maximum torsion angle achievable in a pseudorotational circuit, and the other is a phase angle ϕ' that adopts the values 0°, 36°, 72°... for symmetrical (C_2) twist forms and 18°, 54°, 90°... for envelope forms. Altona, Geise, and Romers examined the conformation of ring D in steroids[164] and found that ω_0 has a relatively constant value (about 47°) whereas ϕ' ranged from about −20° to +10°, with a slight preference for small positive values[165] (Fig. 7.42). These findings imply that the corresponding two-dimensional potential energy surface has a somewhat arc-shaped trough near the minimum at $\omega_0 \sim 47°$, $\phi' \sim 0°$. The potential energy clearly increases much more sharply along ω_0 than along ϕ'. Indeed, were it not for the ring fusion, which constrains the torsion angle about the 13–14 bond to a large positive value (+40° to +50°), the arc-shaped trough would be drawn out into a complete circle (see Chapter 9).

A similar analysis[166] of out-of-plane deformations of furanose rings in β-purine and β-pyrimidine nucleosides and nucleotides shows two ranges of preferred conformations, one with $\phi' \sim 2°$–15°, the other with $\phi' \sim 140°$–175° (the reference conformations with $\phi' = 0°$ and $\phi' = 180°$ are defined in Fig. 7.43). These ranges correspond to the so-called C_3'-*endo* and C_2'-*endo* groups of conformations ($\phi' = 18°$ and $\phi' = 162°$ represent

[164]C. Altona, H. J. Geise, and C. Romers, *Tetrahedron* **24,** 13 (1968).
[165]Our phase angle ϕ' is half of the phase angle Δ defined in Ref. 164.
[166]C. Altona and M. Sundaralingam, *J. Am. Chem. Soc.* **94,** 8205 (1972).

envelope forms with C_3' and C_2', respectively, out-of-plane). The observed ω_0 values lie mostly in a narrow range of about 35°–40°—i.e., the furanose ring is somewhat less puckered than cyclopentane.

The conformational changes that occur in Na^+, K^+, and NH_4^+ complexes of the ionophoric antibiotic compound nonactin (XXXVIII) have

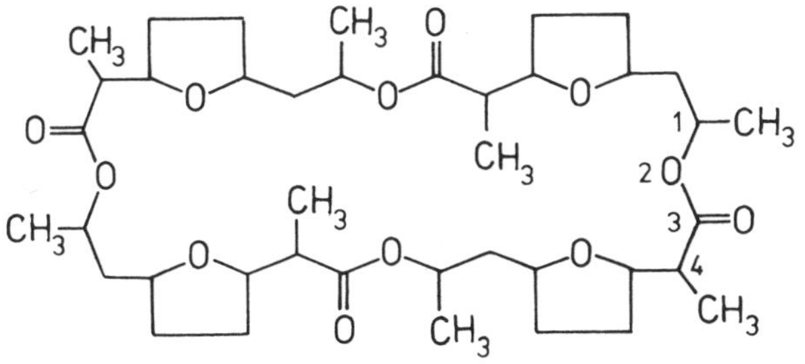

(XXXVIII)

been analyzed by Neupert-Laves and Dobler.[167] In all three complexes the cations are surrounded by the four carbonyl oxygens and by the four ether oxygens of the tetrahydrofuran rings in approximately cubic coordination. The cation—O (ether) distances are about 2.8 Å, but the

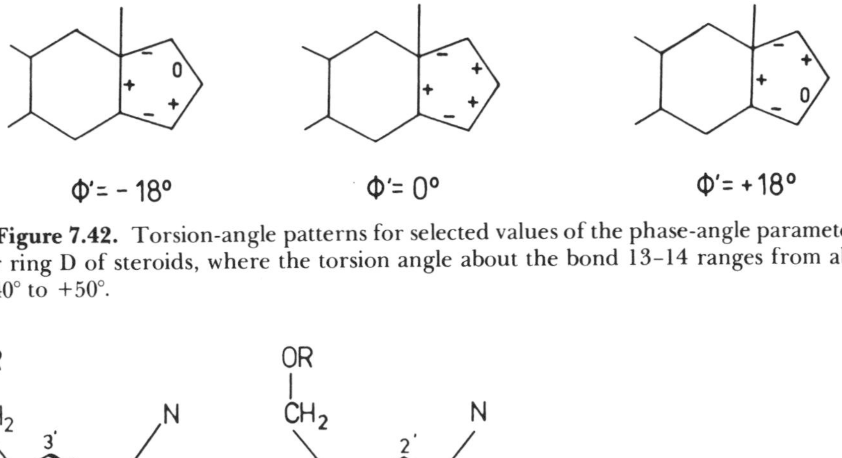

$\Phi' = -18°$ $\Phi' = 0°$ $\Phi' = +18°$

Figure 7.42. Torsion-angle patterns for selected values of the phase-angle parameter ϕ' for ring D of steroids, where the torsion angle about the bond 13–14 ranges from about +40° to +50°.

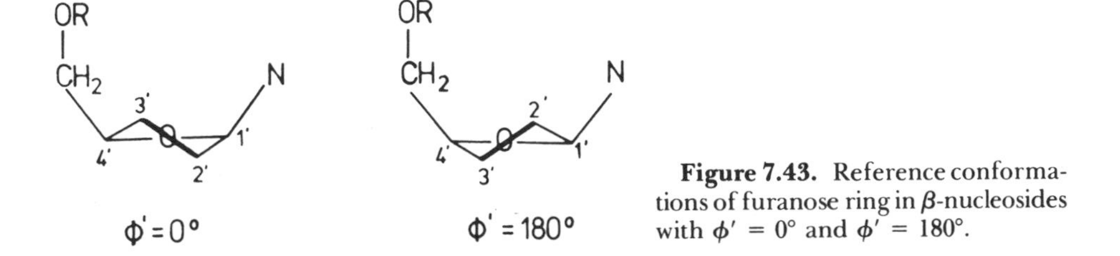

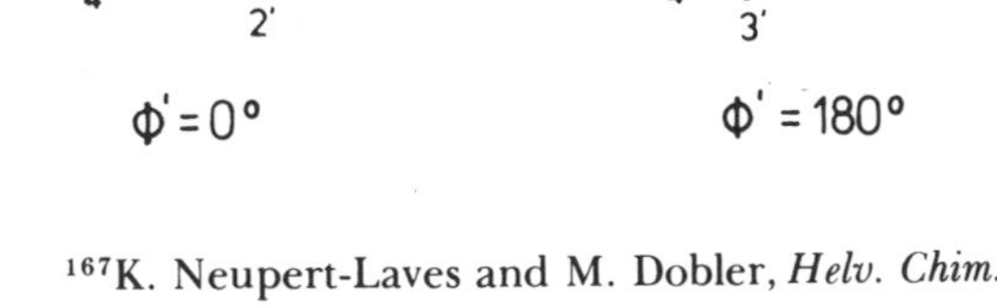

Figure 7.43. Reference conformations of furanose ring in β-nucleosides with $\phi' = 0°$ and $\phi' = 180°$.

[167] K. Neupert-Laves and M. Dobler, *Helv. Chim. Acta* **59,** 614 (1976).

cation—O (carbonyl) distances vary considerably; they are 2.4 Å in the Na^+ complex, 2.8 Å in the K^+, and 3.1 Å in the NH_4^+ complex. The overall contraction or dilation of the cube is thus associated mainly with an inward or outward displacement of the carbonyl oxygens, and this is brought about by gradual rotation of the planes of the ester groups, involving differences of 20°–30° in the torsion angles about the C(1)—O(2) and C(3)—C(4) types of bond (Fig. 7.44). Changes in the remaining 24 torsion angles around the macrocycle do not amount to more than a few degrees. The conformational changes attending ion capture probably follow a similar course. It is interesting that the torsion angle differences with respect to the uncomplexed molecule are smallest for the NH_4^+ complex and largest for the Na^+ complex, thus providing a structural analogy to the ion-selectivity order: $NH_4^+ > K^+ > Na^+$.

Data from several crystal structure analyses of 1,6-methano-[10]-annulene derivatives have been used to map the reaction path for the pericyclic ring-closure reaction (XXXIX).[168] A complete treatment of

(XXXIX)

this reaction would be very difficult because of the many internal structural parameters involved and the complicated relationships between dependent and independent variables. To reduce the problem to manageable proportions, several drastic simplifications are necesary, but allowing for these a fairly well-defined reaction path could be mapped from the available experimental data. At long 1···6 distances the path seems to be determined mainly by bond-angle-strain minimization requirements, but it is influenced more and more by a specific attraction between atoms 1 and 6 as their mutual separation approaches the normal bonded distance of about 1.5 Å. Molecule XII with its remarkable C···C distance of 1.8 Å represents an intermediate point on the reaction path, frozen-in by the perturbations associated with its substituents or by its crystal structure. The energy variation along the path must be very flat to allow this structure to be observed at all. In agreement with this, the activation energy for the valence isomerization XII ↔ XIII in solu-

[168] H. B. Bürgi, E. Shefter, and J. D. Dunitz, *Tetrahedron* **31**, 3089 (1975).

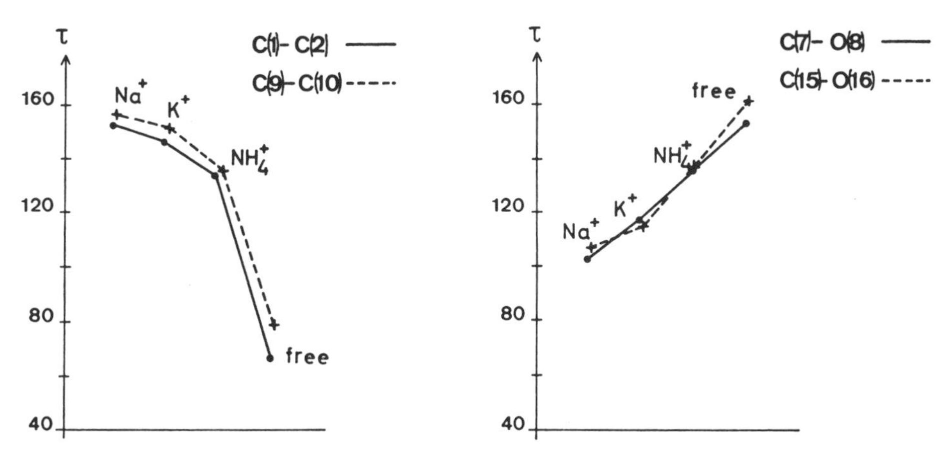

Figure 7.44. Torsion angles around C(7)—O(8), C(9)—C(10) and symmetry equivalent bonds of Na^+, K^+, NH_4^+, and free nonactin. From Neupert-Laves and Dobler, *Helv. Chim. Acta* **59,** 614 (1976).

tion is known from ^{13}C-nmr experiments[169] to be less than 6.6 kcal/mol.

Ermer[170] has found an approximately linear correlation between C—C bond lengthening in saturated and unsaturated (nonconjugated) bicyclic molecules and the net compression of the CCC bond angles involving the bond in question. From this correlation, the coupling constant between C—C bond stretching and CCC bond angle compression could be derived; it corresponds to a repulsive 1,3-C···C interaction. In the methano-[10]-annulenes and related molecules with long C(1)···C(6) distances, the bridge bonds shorten as the central CCC angle is compressed,[168] i.e., the correlation is opposite that found by Ermer. This could be the structural expression of the attractive interaction between C(1) and C(6) that follows from the rules of orbital symmetry conservation.[171]

A study of nonbonded contacts to divalent sulfur (Y-S-Z)[172] reveals that electrophiles tend to approach S in a direction roughly 20° from the perpendicular to the Y-S-Z plane, whereas nucleophiles tend to approach approximately along the extension of one of the bonds to S. Inasmuch as electrophiles should interact preferentially with the highest occupied molecular orbital (HOMO) and nucleophiles should interact

[169]H. Günther, H. Schmickler, W. Bremser, F. A. Straube, and E. Vogel, *Angew. Chem., Intern. Ed.* **12,** 570 (1973).

[170]O. Ermer, *Tetrahedron* **30,** 3103 (1974).

[171]R. B. Woodward and R. Hoffmann, *The Conservation of Orbital Symmetry,* Verlag Chemie, Academic Press, New York, 1970.

[172]R. E. Rosenfield, R. Parthasarathy, and J. D. Dunitz, *J. Am. Chem. Soc.* **99,** 4860 (1977).

preferentially with the lowest unoccupied one (LUMO),[173] this result would suggest that the HOMO is essentially a sulfur *p*-type lone-pair orbital while the LUMO is a σ^* orbital.

The general usefulness of the structural correlation approach has emerged only recently, and it seems likely that many other applications will emerge during the next few years. As long as detailed X-ray determinations of moderately complex crystal structures were major undertakings, the recognition that even small deviations from "normal" bond lengths and angles in molecules are often interrelated in a systematic manner was hardly possible because of the sparsity of accurate data. Such deviations as were found in individual structures were usually dismissed as being caused by experimental error rather than by specific intra- or intermolecular interactions.[174] Now that the bank of existing data becomes augmented by several thousand new crystal structure analyses per year, the analysis of structural correlations seems to be a promising area of research that allows the crystallographer to contribute more directly to problems of reaction mechanism and to the exploration of molecular potential energy surfaces. It is true that the structural correlation principle (see Structural Correlation Principle section earlier in this chapter), which provides the identification of observed correlations with minimum energy paths, lacks any rigorous theoretical foundation and should be regarded merely as a useful working hypothesis. Some of the underlying concepts also need more theoretical backing; in particular, the description of "subunits" needs to be made more precise in structural and energetic terms.

In spite of these deficiencies, the vast literature on crystal and molecular structure and on chemical reactivity now available offers a wide range of possible correlations to be sought. Systematic studies of chemical reactions along these lines have hardly begun, but they could well furnish the material to build a bridge between the "statics" of crystals and the "dynamics" of reacting chemical systems.

[173]See G. Klopman in *Chemical Reactivity and Reaction Paths* G. Klopman, Ed., Wiley, New York, 1974, pp. 83–90.

[174]A. I. Kitaigorodsky, *Molecular Crystals and Molecules,* Academic Press, New York and London, 1973, pp. 186–190.

8. Electron-Density Distributions in Molecules

"Let's hear it," said Humpty Dumpty: "I can explain all the poems that ever were invented—and a good many that haven't been invented just yet."

Electron-Density Difference Maps

In favorable circumstances (good quality crystals and experimental intensity measurements free from serious absorption and extinction errors), the crystallographic R factor (reliability index) may be reduced to a level of 0.03–0.04 or even less at the close of a routine least-squares refinement of atomic positional and thermal parameters. The remaining discrepancies between F_0 and F_c are, of course, attributable partly to experimental errors in the $|F_0|$ values, partly to residual errors in the positional and thermal parameters, and partly to inadequacies in the least-squares model.

There is obviously one serious flaw in the usual least-squares model, and this is the assumption that the electron-density distribution in the crystal can be regarded as a superposition of electron-density peaks corresponding to free, spherically symmetric atoms. This assumption is implied by the use of standard, spherically symmetric atomic scattering factors f_i^0 to represent the scattering of the various atoms, and it is perfectly adequate in the early stages of a crystal structure analysis. However, we would expect that molecule formation is accompanied by a certain redistribution of the electron density of the individual atoms to bring more density into some regions, e.g., between the nuclei, with a corresponding density deficit in other regions. Any such redistribution of electron density should affect the valency shell much more than the inner core, and it should therefore be relatively more important (and more easily detectable) in light atoms than in heavy ones. It should be exceptionally severe for bonded H atoms. This is the reason why X—H bond distances determined by X-ray analysis are characteristically too short—the electron-density peak is displaced from the position of the proton (as determined by neutron diffraction) towards the bonded X atom.

It turns out that the difference density function $D(\mathbf{r})$, the Fourier transform of ($F_0 - F_c$), calculated at the termination of a routine least-squares analysis, often shows systematic features that are quite in keeping, at least qualitatively, with the kind of charge redistribution expected to occur on molecule formation. Such features include, for example, positive density between the nuclei and at positions compatible with lone pairs and often negative density in the immediate vicinity of the nuclei. Thus, the charge redistribution that occurs on passing from a set of isolated atoms to a molecule would seem to be experimentally observable if sufficiently accurate data are available. Some highly interesting results in this direction have been obtained during the last 15 years or so, but before we discuss them in detail, we mention some problems that concern not only the experimental difficulties but also the confrontation between experimental and theoretical results.

Even for the first-row elements C, N, and O, which occur in typical organic molecules, the changes in electron distribution that occur on molecule formation are quite small compared with the total electron density. They can be easily obliterated—by errors in the experimental F values or by general smearing out of the electron density caused by thermal vibration. Thus, meaningful results can be expected only when considerable attention has been paid to eliminating possible sources of error in the measured intensities; in addition, it is advantageous to cool the crystal during the intensity measurements in order to reduce thermal vibration as far as possible. Obviously the presence of second-row and higher elements puts even greater demands on the quality of the data. It is particularly important that the data be free of extinction errors (Chapter 6, section on Extinction), which affect mainly the low-order reflections, because it is these reflections that contain most of the information concerning the valency electrons. Moreover, the difference density in the immediate vicinity of the atomic centers is the difference between two relatively large densities; it is very sensitive to errors in the experimental data and in the model and also to the choice of the scale factor k used to bring the experimentally observed structure amplitudes to an absolute scale. In principle, k is a measurable quantity, but it is usually estimated by adjusting the relative F_0 values to the scale of F_c, which depends, in turn, on the choice of atomic scattering factors used in the structure-factor calculation.

Another difficulty is that the atomic positions derived from a conventional least-squares analysis of X-ray data may not coincide exactly with mean nuclear positions, as determinable, in principle, from neutron diffraction data. The X-ray positions correspond to centroids of electron-density peaks, and for these to coincide with nuclear positions, the individual electron-density peaks should be centrosymmetric and non-overlapping, conditions that are rarely fulfilled exactly. As a result of the asymmetry of the electron-density peaks, apparent X-ray positions may be displaced from the true positions towards bonding or lone-pair electron-density accumulations. In a difference synthesis based on the biased atomic positions, these features would then be somewhat suppressed. As discussed in Chapter 4 the bias due to asymmetry of the electron-density peaks can be reduced by carrying out the least-squares refinement with high-order reflections only or by introducing suitably modified weighting schemes, but it can only be eliminated completely by determining the atomic positions from neutron diffraction data.

Difference syntheses calculated for a model based on neutron-diffraction parameters are often referred to as X–N maps; those based on X-ray data alone are called X–X maps.

Neutron Diffraction

As far as crystal structure analysis is concerned, the interpretation of neutron diffraction data is very similar to that of X-ray data; the main difference is that neutrons are scattered by atomic nuclei, not by the extranuclear electrons. Since atomic nuclei are negligibly small compared with the wavelength of the neutrons commonly used for diffraction experiments (~ 1 Å), neutron scattering amplitudes b_n for atoms at rest correspond essentially to scattering from point atoms and are thus independent of $\sin\theta/\lambda$. Unlike the X-ray scattering amplitude f_n^0, the value of b_n varies irregularly from one element to the next, or, more exactly, from one nuclide to another. Since isotopes of the same element will be distributed at random over each set of equivalent atomic positions, a crystal that is perfectly ordered as far as X-ray diffraction effects are concerned may appear to be disordered (isotopic disorder) when examined by neutron diffraction. By analogy with the X-ray structure factor, the neutron structure factor for the contents of an averaged unit cell may be written

$$F_N(\mathbf{H}) = \sum b_k T_k \exp(2\pi i \mathbf{H}\cdot\mathbf{r}_k)$$

where b_k is actually the mean $\langle b_k \rangle$ of the scattering amplitudes of the isotopes of atom k present, weighted according to their relative abundance, and T_k is the temperature factor of atom k. Some b values (sometimes known as scattering lengths since b has the dimension of length) for representative elements and isotopes commonly occurring in organic compounds are listed in Table 8.1.

Table 8.1. Neutron-scattering data[a] for some elements and isotopes (1 barn = 10^{-28} m^2)

Element or isotope	b (10^{-14} m)	σ_{coh} (barns)	σ_{incoh} (barns)
^{1}H	−0.374	1.8	79.7
^{2}H(D)	0.667	5.6	2.0
C	0.665	5.6	0.0
N	0.940	11.1	0.3
O	0.580	4.23	0.0
F	0.565	3.98	0.0
P	0.51	3.3	0.3
S	0.285	1.02	0.2
Cl	0.958	11.5	3.5
Br	0.679	5.79	0.3

[a]Based on a tabulation given in *Chemical Applications of Thermal Neutron Scattering,* B. T. M. Willis, Ed., Oxford University Press, 1973, Appendix 1.

In addition to the coherent scattering mentioned above, any isotopic disorder present produces an incoherent scattering that is proportional to $\langle b^2 \rangle - \langle b \rangle^2$ (see Chapter 1). The quantity $4\pi\{\langle b^2 \rangle - \langle b \rangle^2\}$ is known as the incoherent scattering cross-section (σ_{incoh}) and is usually expressed in terms of barns (10^{-28} m^2). Values of σ_{incoh} are also given in Table 8.1, together with values of $\sigma_{coh} = 4\pi\langle b \rangle^2$, the coherent scattering cross-section.

Incoherent scattering also occurs when only a single isotopic species with a nonzero nuclear spin is present. The reason is that the general scattering process can be considered in terms of a compound nucleus formed by the target nucleus and the incident neutron. The neutron has a spin of $\frac{1}{2}$. If the target nucleus has spin I, two possible spin combinations, $I + \frac{1}{2}$ and $I - \frac{1}{2}$, with relative abundances $(I + 1)/(2I + 1)$ and $I/(2I + 1)$, can be formed. These have different nuclear energy levels and hence different scattering amplitudes b_+ and b_-. The random occupation of atomic sites by these two species leads to a spin-incoherent scattering in the same way as that produced by a random distribution of isotopes.

In practice, spin incoherent scattering arises mainly from the presence of hydrogen (^{1}H) atoms. For ^{1}H the values of b_+ and b_- have scattering amplitudes of opposite sign with a weighted average close to zero. For this reason, the coherent scattering is quite small, and the incoherent scattering is quite large, so that neutron diffraction measurements of compounds containing ^{1}H are saddled with a high background intensity that can be troublesome. For neutron-diffraction studies of hydrogen-containing compounds, it is therefore advantageous to use perdeuterated samples wherever possible. Deuterium (^2H) has a larger coherent scattering amplitude and gives a much lower incoherent background than hydrogen (^1H).

Neutron diffraction analysis serves to locate hydrogen atoms (preferably as ^2H) much more accurately than is possible by X-ray analysis. It also has the advantages that the atomic positions determined correspond to nuclear positions rather than to centroids of electron-density peaks, and that the derived thermal vibrational parameters correspond more closely to the actual atomic vibrations. In X-ray analysis, deviations of the density peaks from sphericity that are caused by bonding tend to get absorbed in the thermal vibration parameters. As a result, vibrational parameters estimated from X-ray data tend to be somewhat too large.

On the other hand, a neutron-diffraction study is not as easily performed as an X-ray study and is extremely costly both in effort and financial outlay. A further point is that for a neutron-diffraction analysis, a rather large crystal of the compound in question is needed ($>$ 10 mm^3; this is about 1000 times the size required for X-ray analysis).

An Example: *p*-Nitropyridine-*N*-oxide

To illustrate some of these points, it is useful to discuss one example of a deformation density study in some detail; we choose the recent low-temperature, combined X-ray/neutron diffraction studies[1,2] of *p*-nitropyridine-*N*-oxide.[3] The crystals are orthorhombic, space group *Pnma*, $a = 12.53, b = 6.01, c = 7.90$ Å (at room temperature); according to the space-group symmetry, the four molecules in the unit cell have to lie in the crystallographic mirror planes at $y = \pm\frac{1}{4}$. The overall structure is shown in Fig. 8.1. The newer studies include a redetermination of the crystal structure by X-ray diffraction using data measured with Mo$K\alpha$ radiation at room temperature (1162 reflections, $\sin\theta/\lambda$ limit 0.76 Å^{-1}) and at 30 K (2201 reflections, $\sin\theta/\lambda$ limit 1.00 Å^{-1}).[1] In addition to conventional least-squares refinement of both data sets, a separate refinement using only high-order reflections ($\sin\theta/\lambda > 0.75$ Å^{-1}) was done on the low-temperature data. A parallel low-temperature neutron-diffraction analysis (30 K, 451 reflections, $\sin\theta/\lambda$ limit 0.55 Å^{-1})[2] yielded positional and thermal parameters free of bias. For the hydrogen atoms this is especially important; for the C, N, and O atoms the advantage

[1]Y. Wang, R. H. Blessing, F. K. Ross, and P. Coppens, *Acta Crystallogr.* **B32,** 572 (1976).
[2]P. Coppens and M. S. Lehmann, *Acta Crystallogr.* **B32,** 1777 (1976).
[3]For the original crystal structure analysis see E. L. Eichhorn, *Acta Crystallogr.* **9,** 787 (1956).

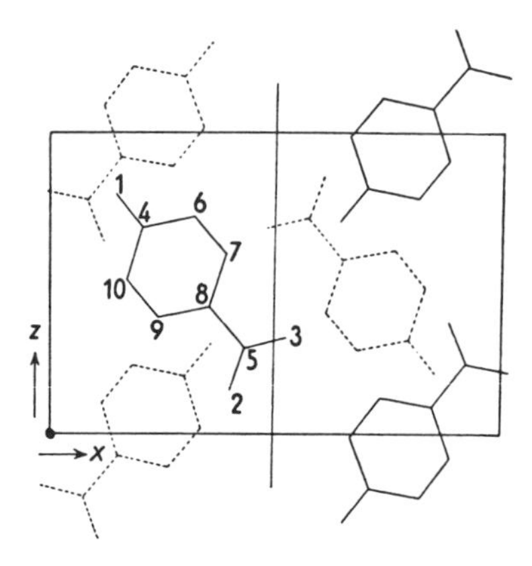

Figure 8.1. Crystal structure of *p*-nitropyridine-*N*-oxide projected on the *xz* plane. The molecules lie in the mirror planes at $y = \frac{1}{4}, \frac{3}{4}$. From Eichhorn, *Acta Crystallogr.* **9,** 787 (1956).

is partially offset by the lower precision in the neutron parameters, whose estimated standard deviations are two to three times larger than those from the X-ray analysis.

The reduction in mean-square vibrational amplitudes obtained by crystal cooling is impressive (Fig. 8.2). At room temperature, typical values for C, N, and O atoms are about 0.04 Å^2; they are largest (up to 0.1 Å^2) for the oxygen atoms in the direction normal to the molecular plane—there is a hint here that the NO_2 group may have a considerable torsional amplitude about the C—N bond. At 30 K, typical mean-square vibrational amplitudes are reduced to about 0.007 Å^2, but they are still somewhat larger (up to 0.015 Å^2) for the oxygens in the direction normal to the molecular plane. The neutron analysis shows that the hydrogen atoms also have much larger vibrational amplitudes (about 0.030 Å^2) normal to the molecular plane than in this plane (about 0.015 Å^2). The analysis of the X-ray vibrational parameters in terms of rigid-body translation and libration yields eigenvalues of the **T** tensor of 0.036, 0.025, and 0.025 Å^2 at room temperature, reduced to 0.006, 0.005, and 0.004 Å^2 at 30 K with a corresponding behavior of the **L** tensor (0.0130, 0.0047, and 0.0046 radians^2 at room temperature; 0.0020, 0.0006, and 0.0006 radians^2 at 30 K). The corrections for apparent bond shortening caused by rigid-body libration are appreciable—up to 0.012 Å—at room temperature but almost negligible (at most 0.002 Å) at 30 K.

Bond distances and angles derived from the neutron and X-ray analyses (30 K, full data, uncorrected for rigid-body libration) are shown in Fig. 8.3. The main differences are in the N—O bond lengths, which are systematically longer, by 0.006–0.010 Å, in the X-ray analysis—as though the oxygen atoms were shifted slightly outwards. However, the results of the high-order X-ray refinement are in almost perfect agreement with those of the neutron analysis.

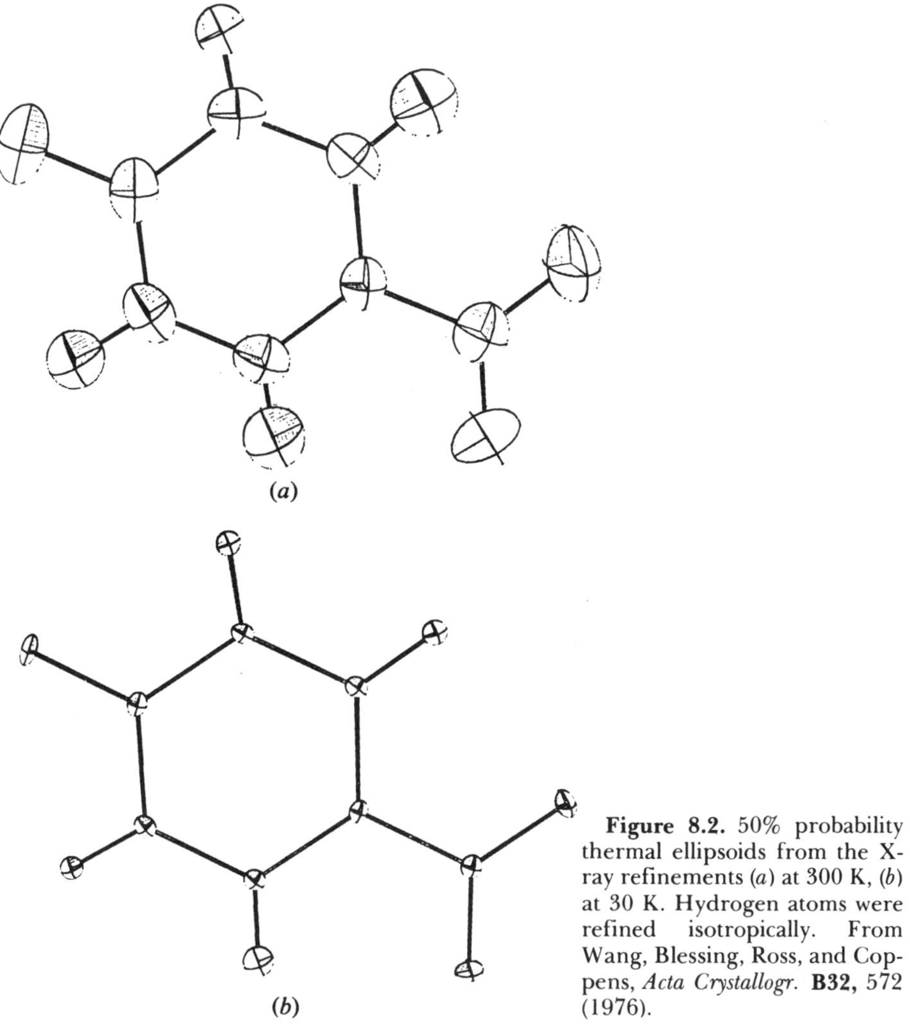

Figure 8.2. 50% probability thermal ellipsoids from the X-ray refinements (*a*) at 300 K, (*b*) at 30 K. Hydrogen atoms were refined isotropically. From Wang, Blessing, Ross, and Coppens, *Acta Crystallogr.* **B32,** 572 (1976).

Deformation density maps were calculated by subtracting neutral, spherical atoms placed at the neutron positions from the observed electron density. Calculations were made with temperature factors derived from high-order refinement of the 30 K X-ray data as well as from the neutron data, and for several $\sin\theta/\lambda$ cutoff values. Figure 8.4 shows the X–N map calculated in the molecular plane with a data cutoff at $\sin\theta/\lambda = 0.75$ Å^{-1}. Although this map obviously contains a certain amount of noise in the form of minor peaks and troughs, devoid of any physical significance, the main features are clearly compatible with the kind of electron-density deformation expected on molecule formation. These include:

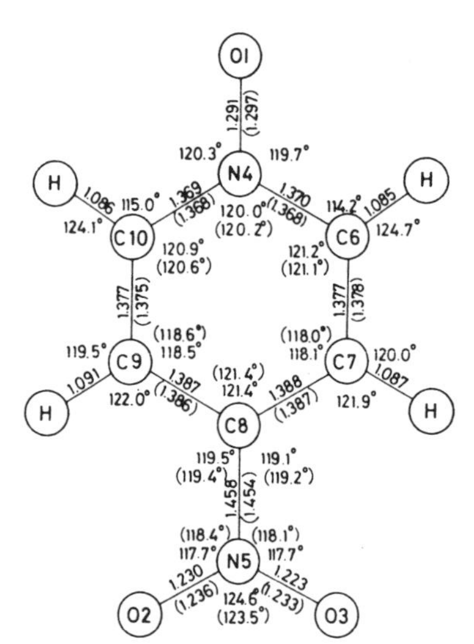

Figure 8.3. Bond lengths and angles from the neutron-diffraction analysis (with X-ray results in parentheses). From Coppens and Lehmann, *Acta Crystallogr.* **B32,** 1777 (1976).

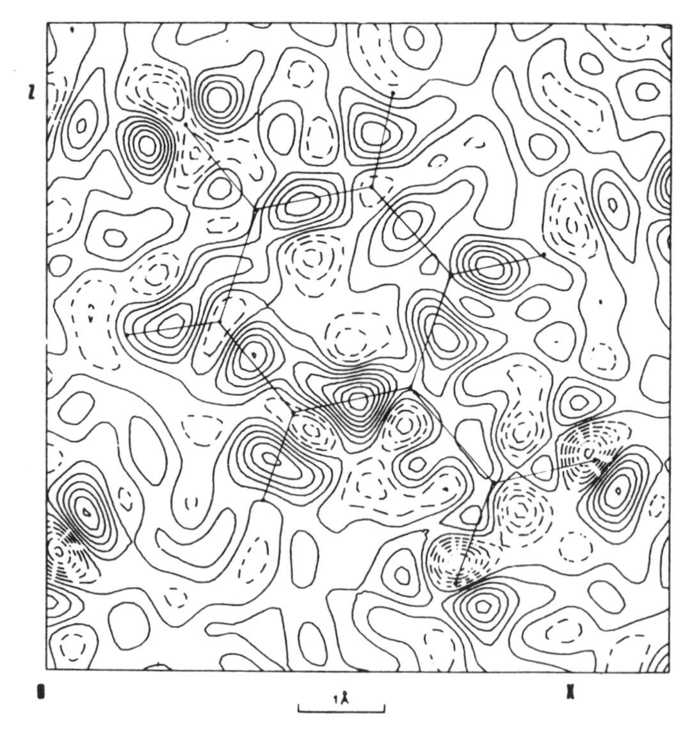

Figure 8.4. X–N map of *p*-nitropyridine-*N*-oxide at 30 K, obtained by subtracting neutral spherical atoms (positions and thermal parameters derived from a neutron-diffraction analysis) from the observed electron density calculated with data cutoff of 0.75 $Å^{-1}$. Contours at intervals of 0.1 e. $Å^{-3}$; negative contour lines broken. From Coppens and Lehmann, *Acta Crystallogr.* **B32,** 1777 (1976).

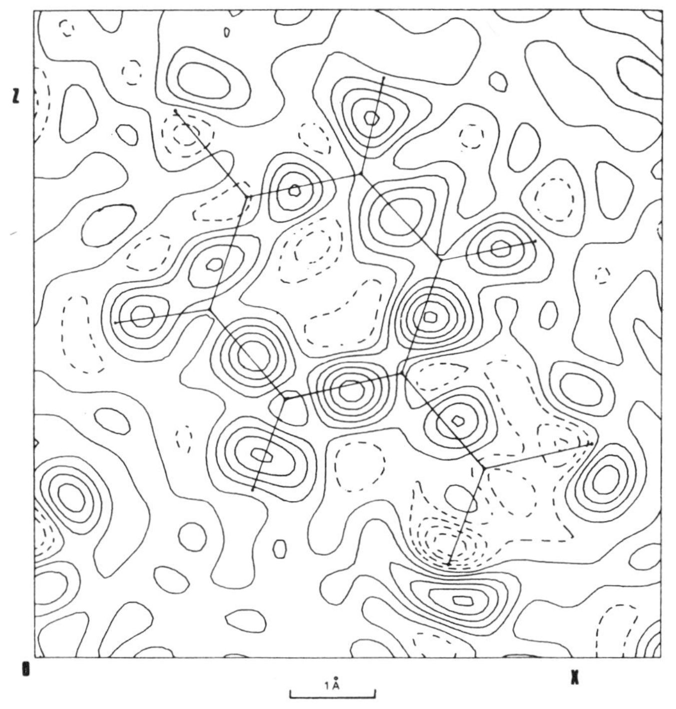

Figure 8.5. X–X map of *p*-nitropyridine-*N*-oxide at 30 K. As Fig. 8.4 but based on X-ray parameters (high-order refinement) for C, N, and O atoms and neutron parameters for H atoms. Data cutoff 0.65 $Å^{-1}$. From Coppens and Lehmann, *Acta Crystallogr.* **B32,** 1777 (1976).

(1) Bonding density of 0.4–0.6 e. $Å^{-3}$ at or near the midpoints of the C—C, C—N and C—H bonds. The density in the terminal N—O bonds is lower, and this is attributed[2] to the subtraction of spherical nitrogen and oxygen atoms with $\frac{5}{4}$ and $\frac{6}{4}$ electrons, respectively, per atomic orbital. If, instead, "prepared" nonspherical atoms with only one electron per bonding orbital were subtracted, the difference density in the N—O bonds would be enhanced. However, such a procedure would involve certain preconceptions about the bonding process.[4]

(2) Density features compatible with lone-pair electrons on the oxygen atoms.

(3) Slight electron deficiencies at or near the atomic positions.

Since Fig. 8.4 shows the deformation density in the molecular plane, any density indicative of π bonding is not apparent there, but the ex-

[4]An earlier suggestion (see Ref. 1) that the low density in the N—O bonds might be attributable to the use of X-ray rather than neutron thermal parameters can be discounted from the results of the neutron diffraction analysis (see Ref. 2).

pected extension normal to the plane can be seen in the appropriate sections.

The X–N maps calculated with cutoff limits of sin θ/λ = 0.65 Å^{-1} and 1.00 Å^{-1} (full data) are similar in general appearance to Fig. 8.4, but the peaks and troughs are, of course, more diffuse in the low-order map and sharper in the high-order one. The lone-pair features are decidedly more prominent in the high-order map, suggesting that lone-pair density peaks are sufficiently concentrated in space to have a nonnegligible contribution to the scattering even at sin θ/λ values greater than 0.75 Å^{-1}. On the other hand, the high-order map contains much more spurious detail than Fig. 8.4, probably because of the relative lack of precision in the high-order intensity measurements.

For comparison with the X–N map of Fig. 8.4, an X–X map (sin θ/λ limit, 0.65 Å^{-1}) based on X-ray parameters for C, N, and O atoms (high-order refinement) and neutron parameters for H atoms is shown in Fig. 8.5. Lone-pair densities in the X–X map are not as clear as in the X–N map, but, on the other hand, the X–N map contains more spurious detail, caused in part by the higher cutoff limit and in part by the lower precision of the neutron positional parameters compared with the X-ray ones. The latter are biased, but they are less affected by random experimental errors. An X–N map can be affected by two sets of experimental errors—those in the neutron-diffraction measurements and those in the X-ray ones, whereas the X–X map is saddled only with the latter.

The quantitative significance of the deformation densities depicted in Figs. 8.4 and 8.5 needs to be assessed carefully, but the main features can be identified, at least qualitatively, with concepts (bonding density, lone-pair electrons, π bonds) that chemists use in discussing the electronic structure of molecules. The results show that at the close of a conventional least-squares refinement, the residual discrepancies between F_o and F_c may depend more on the inadequacies of the spherical atom model than on experimental errors in the data. That is, the F_o values may be more accurate than the F_c values.

Residual Density Features in Organic Molecules

During the past few years similar deformation density studies have been made for many organic crystals. A partial list, based on a tabulation provided in a recent review article,[5] is given in Table 8.2. Although the

[5]P. Coppens, *International Review of Science, Physical Chemistry,* Series 2, Vol. 11, *Chemical Crystallography,* J. M. Robertson, Ed., Butterworths, London and Boston, 1975, p. 21.

objectives in the various analyses are very different from one to another, as are also the techniques used and the quality of the results, certain generalizations are beginning to emerge. The detail of the deformation density is usually greatly improved when the analysis is based on low-temperature data. An example is shown in Fig. 8.6. Several of the analyses reported in Table 8.2 are based on positional and thermal parameters derived from parallel neutron-diffraction studies, and in some of these studies—e.g., glycylglycine (Table 8.2, footnote *s*)—the X–N map is definitely superior to the corresponding X–X map. How-

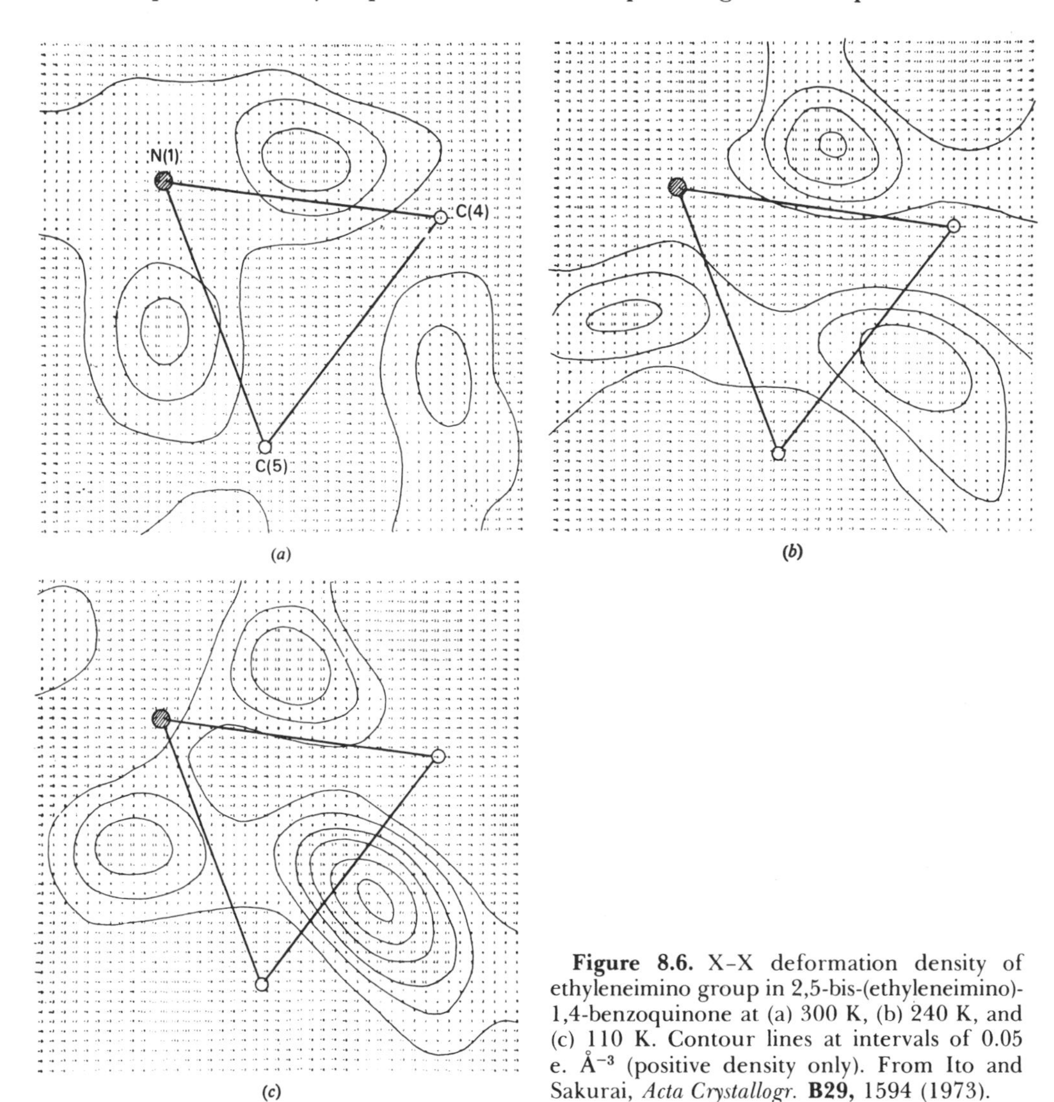

Figure 8.6. X–X deformation density of ethyleneimino group in 2,5-bis-(ethyleneimino)-1,4-benzoquinone at (a) 300 K, (b) 240 K, and (c) 110 K. Contour lines at intervals of 0.05 e. Å^{-3} (positive density only). From Ito and Sakurai, *Acta Crystallogr.* **B29**, 1594 (1973).

Table 8.2. Some charge density studies of organic crystals

Compound	Formula	Details	Footnote
Hydrocarbons			
tribenzocyclododeca-1,5,9-triene-3,7,11-triyne	$C_{24}H_{12}$	X–X	*a*
1,8-bis(phenylethynyl)naphthalene	$C_{26}H_{16}$	X–X	*b*
tetraphenylbutatriene	$C_{28}H_{20}$	X–X	*c*
penta-*m*-phenylene	$C_{30}H_{20}$	X–X	*a*
hexa-*m*-phenylene	$C_{36}H_{24}$	X–X	*a*
Nitrogen-containing			
sym-triazine	$C_3H_3N_3$	X–N	*d*
1,2,3-*cis*-tricyanocyclopropane	$C_6H_3N_3$	X–X	*e*
1,2,3,4-tetracyanocyclobutane (centrosymmetric isomer)	$C_8H_4N_4$	X–X	*f*
tetracyanoethylene	C_6N_4	X–N, X–X	*g*
hexamethylenetetramine	$C_6H_{12}N_4$	X–N	*h*
Oxygen-containing			
oxalic acid dihydrate	$C_2H_2O_4 \cdot 2H_2O$	X–N	*i*
acetylene dicarboxylic acid	$C_4H_2O_4$	X–X	*a,b*
2,5-dimethylhex-3-yne-2,5-diol	$C_6H_{14}O_2$	X–X (90 K)	*j*
cis- and *trans*-2,5-dimethylhex-3-ene-2,5-diol	$C_6H_{16}O_2$	X–X (110 K)	*j*
2,5-dimethylbenzoquinone	$C_8H_8O_2$	X–X	*k*
1,3,5-triacetylbenzene	$C_{12}H_{12}O_3$	X–N	*l*
sucrose	$C_{12}H_{22}O_{11}$	X–N, X–X	*m*
Oxygen- and nitrogen-containing			
ammonium oxalate monohydrate	$C_2O_4(NH_4)_2 \cdot H_2O$	X–N	*n*
α-glycine	$C_2H_5NO_2$	X–N	*o*
cyanuric acid	$C_3H_3N_3O_3$	X–N, X–X (100 K)	*p*
fumaric acid monoamide	$C_4H_5NO_2$	X–X	*q*
uracil	$C_4H_4N_2O_2$	X–X	*r*
α-glycylglycine	$C_4H_8N_2O_3$	X–N, X–X	*s*
p-nitropyridine-*N*-oxide	$C_5H_4N_2O_3$	X–N, X–X (30 K)	*t*
tetracyanoethylene oxide	C_6N_4O	X–N	*u*
2,5-bis(ethyleneimino)-1,4-benzoquinone	$C_{10}H_{10}N_2O_2$	X–X (110, 240, 300 K)	*v*

Sulfur-containing			
2-aminoethylsulfonic acid (taurine)	$C_2H_7NO_3S$	X–X	*w*
thiocytosine	$C_4H_5N_3S$	X–X	*x*
Others			
decaborane	$B_{10}H_{14}$	X–N (113 K)	*y*
benzenechromiumtricarbonyl	$C_9H_6O_3Cr$	X–N (78 K)	*z*

[a]H. Irngartinger, L. Leiserowitz, and G. M. J. Schmidt, *J. Chem. Soc., B* **1970,** 497.
[b]A. E. Jungk and G. M. J. Schmidt, *Chem. Ber.* **105,** 2607, 2623 (1972).
[c]Z. Berkovitch-Yellin and L. Leiserowitz, *J. Am. Chem. Soc.* **97,** 5627 (1975).
[d]P. Coppens, *Science* **158,** 1577 (1967).
[e]A. Hartman and F. L. Hirshfeld, *Acta Crystallogr.* **20,** 80 (1966).
[f]M. Harel and F. L. Hirshfeld, *Abstr. 1st Eur. Cryst. Meet., Bordeaux* (1973).
[g]P. Becker, P. Coppens, and F. K. Ross, *J. Am. Chem. Soc.* **95,** 7604 (1973).
[h]J. A. K. Duckworth, B. T. M. Willis, and G. S. Pawley, *Acta Crystallogr.* **A26,** 263 (1970).
[i]P. Coppens, T. M. Sabine, R. G. Delaplane, and J. A. Ibers, *Acta Crystallogr.* **B25,** 2451 (1969).
[j]A. F. J. Ruysink and A. Vos, *Acta Crystallogr.* **B30,** 1997 (1974); R. B. Helmholdt and A. Vos, *ibid.,* **A33,** 456 (1977).
[k]F. L. Hirshfeld and D. Rabinovich, *Acta Crystallogr.* **23,** 989 (1969).
[l]B. H. O'Connor and F. H. Moore, *Acta Crystallogr.* **B29,** 1903 (1973).
[m]J. C. Hanson, L. C. Sieker, and L. H. Jensen, *Acta Crystallogr.* **B29,** 797 (1973); G. M. Brown and H. A. Levy, *ibid.,* 790.
[n]J. C. Taylor and T. M. Sabine, *Acta Crystallogr.* **B28,** 3340 (1972).
[o]J. Almlöf, A. Kvick, and J. O. Thomas, *J. Chem. Phys.* **59,** 3901 (1973).
[p]G. C. Verschoor and E. Keulen, *Acta Crystallogr.* **B27,** 134 (1971); P. Coppens and A. Vos, *ibid.,* 146.
[q]F. L. Hirshfeld, *Acta Crystallogr.* **B27,** 769 (1971).
[r]R. F. Stewart and L. H. Jensen, *Z. Kristallogr.* **128,** 133 (1969).
[s]J. F. Griffin and P. Coppens, *J. Am. Chem. Soc.* **97,** 3496 (1975).
[t]Y. Wang, R. H. Blessing, F. K. Ross, and P. Coppens, *Acta Crystallogr.* **B32,** 572 (1976); P. Coppens and M. S. Lehmann, *ibid.,* 1777.
[u]D. A. Mathews and G. D. Stucky, *J. Am. Chem. Soc.* **93,** 5954 (1971).
[v]T. Ito and T. Sakurai, *Acta Crystallogr.* **B29,** 1594 (1974).
[w]A. M. O'Connel, *Acta Crystallogr.* **B25,** 1273 (1969).
[x]S. Furberg and L. H. Jensen, *Acta Crystallogr.* **B26,** 1260 (1970).
[y]R. Brill, H. Dietrich, and H. Dierks, *Acta Crystallogr.* **B27,** 2003 (1971).
[z]B. Rees and P. Coppens, *Acta Crystallogr.* **B29,** 2515 (1973).

ever, if sufficiently extensive and accurate high-order X-ray data are available, the bias in the estimated X-ray parameters (except for hydrogen atoms) can be greatly reduced by appropriate refinement techniques. In such cases the advantage to be gained from a parallel neutron-diffraction study seems to be less pronounced.

In nearly all of these studies, distinct electron-density peaks are found at or close to the midpoint of bonds. In unstrained aliphatic molecules the peaks usually have cylindrical symmetry, at least approximately, around the respective internuclear axes, and their heights vary between ca. 0.1 and 0.6 e. $Å^{-3}$, depending on the thermal motion of the atoms involved, as well as on the bond type. In general, C—C bonds are associated with somewhat higher density maxima than C—N, C—O, or N—O bonds. Maxima associated with the bonds in strained three- and four-membered rings are clearly displaced outward from the internuclear axes. This is shown for the ethyleneimine ring in Fig. 8.6, and very similar results were obtained for cyclopropane and ethyleneoxide rings (Table 8.2, footnotes *e, u*). Such pictures provide a beautiful verification of intuitive ideas[6] about the nature of the bonding in this kind of molecule, and we may well feel at this stage that we are "seeing" electrons. Similarly, for the three-center bonds in decaborane the density maxima are clearly displaced inward from the internuclear lines towards the center of the BHB triangles (Fig. 8.7), again in agreement with current views about the nature of the bonding in these "nonclassical" systems.[7] In the X–N map of glycylglycine (Table 8.2, footnote *s*) the bonding density of the amide carbonyl bond seems to be slightly displaced from the bond axis towards the neighboring, positively charged $—NH_3^+$ group of the molecule—a possible indication of a long-range through-space interaction?

Noncylindrical density peaks, elongated normal to the molecular plane, are observed in most unsaturated and aromatic molecules that have been studied. On the other hand, the bonding density associated with triple bonds is again cylindrically symmetric about the bond axis. In cumulenes, successive π bonds along the chain are expected to be mutually orthogonal; this has been observed, as shown by Fig. 8.8, taken from a study of tetraphenylbutatriene.

The situation with respect to lone-pair densities is not as clear, especially because these features are sometimes partially or completely sup-

[6]As first advanced in various formalisms by T. Förster, *Z. Physik. Chem. (B)* **43,** 58 (1939); A. D. Walsh, *Nature (London)* **159,** 712 (1947); and C. A. Coulson and W. E. Moffitt, *Phil. Mag.* **40,** 1 (1949).

[7]R. Hoffmann and W. N. Lipscomb, *J. Chem. Phys.* **37,** 2872 (1962).

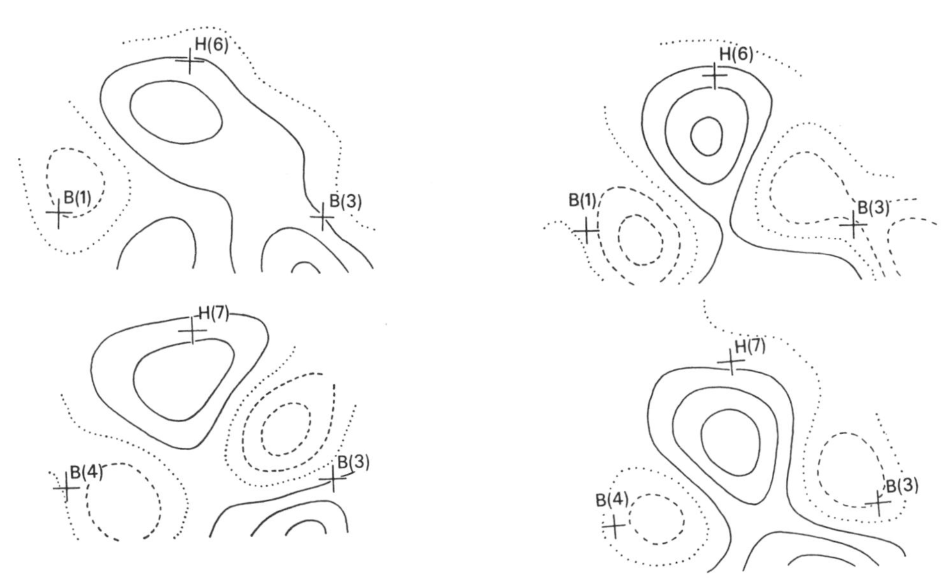

Figure 8.7. Decaborane, X-N difference map in the plane of the B—H—B bridges in two crystallographically independent molecules (left and right). Contour intervals at intervals of 0.1 e. Å^{-3}. From Brill, Dietrich, and Dierks, *Acta Crystallogr.* **B27,** 2003 (1971).

pressed in X-X maps, as a result of bias in the apparent position of terminal atoms. The oxygen lone-pair densities in *p*-nitropyridine-*N*-oxide (Figs. 8.4 and 8.5) appear nearly at 90° to the N—O bond directions; this is particularly evident for the nitroxo group, suggesting that one lone pair is essentially $(2p)^2$ and that the other (not shown in a deformation map) is $(2s)^2$. For other terminal oxygen atoms, two density maxima at about 120° to the bond axis are found (as in cyanuric acid, glycine, and glycylglycine), but sometimes the two peaks coalesce into a broad half-moon shaped feature with its maximum on the extension of the bond. Sometimes only one lone-pair density maximum is found where two would be expected. These differences may be attributed partly to differences in the nonbonded environments of the atoms concerned and partly to differences in thermal motion. For cyano groups the lone-pair density usually appears as a single maximum on the extension of the bond, and similarly for the C=O groups in benzenechromiumhexacarbonyl.[8]

In general, no appreciable density is seen near the midpoints of unsymmetrical hydrogen bonds X—H···Y, in keeping with the essentially electrostatic rather than covalent character of such bonds. However, in view of the tendency (although hardly the rule) for X—H bonds to point

[8]B. Rees and P. Coppens, *Acta Crystallogr.* **B29,** 2516 (1973).

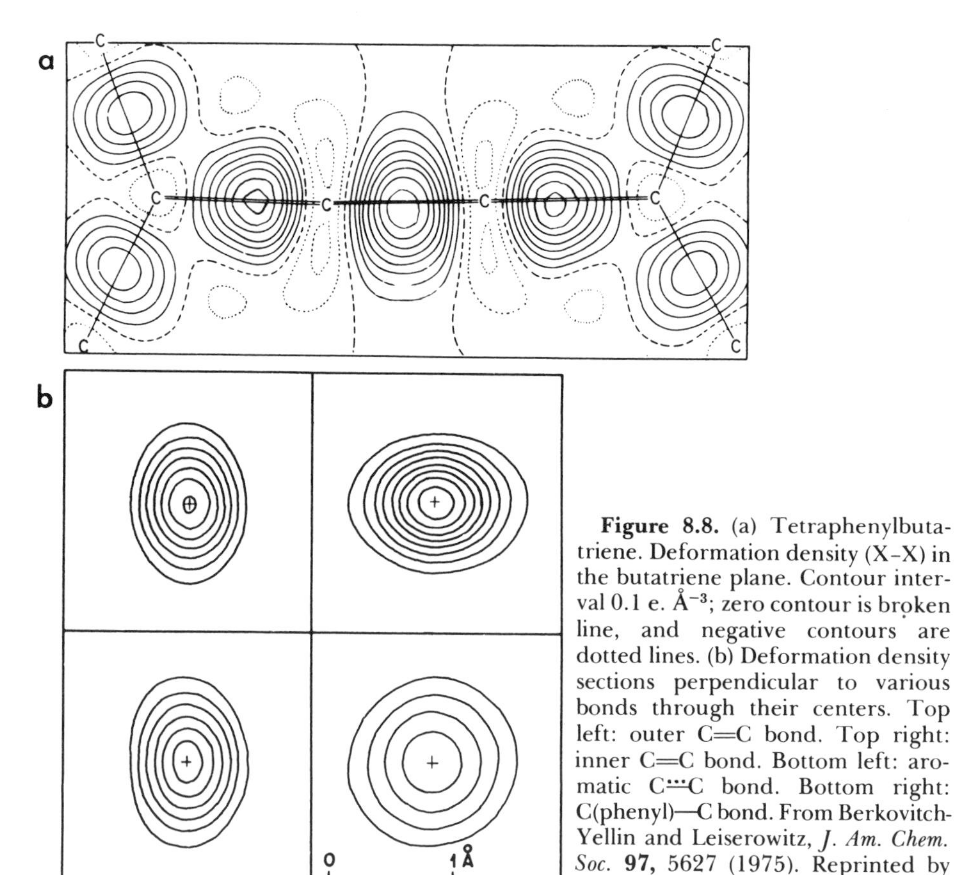

Figure 8.8. (a) Tetraphenylbutatriene. Deformation density (X–X) in the butatriene plane. Contour interval 0.1 e. $Å^{-3}$; zero contour is broken line, and negative contours are dotted lines. (b) Deformation density sections perpendicular to various bonds through their centers. Top left: outer C═C bond. Top right: inner C═C bond. Bottom left: aromatic C⋯C bond. Bottom right: C(phenyl)—C bond. From Berkovitch-Yellin and Leiserowitz, *J. Am. Chem. Soc.* **97**, 5627 (1975). Reprinted by permission.

towards a lone pair of the acceptor atom Y rather than towards Y itself,[9] density on both sides of the H atom should build up in more symmetrical hydrogen bonds. Such a density buildup is discernable in an X–X map of sodium hydrogen bisacetate (at 90 K), where a continuous ridge of electron density about 0.55 e. $Å^{-3}$ high runs along the symmetrical O⋯H⋯O hydrogen bond from one oxygen atom to the other.[10]

Difference Densities by Least-Squares Analysis

In several of the analyses referred to above, the deformation density has been estimated not from an X–X or X–N difference map (as described earlier) but by an alternative method, in which the deformation density is expressed in parametric form and refined by least-squares

[9]L. Leiserowitz and G. M. J. Schmidt, *J. Chem. Soc., A* **1969**, 2372.
[10]E. D. Stevens, M. S. Lehmann, and P. Coppens, *J. Am. Chem. Soc.* **99**, 2829 (1977).

analysis, along with the usual positional and thermal parameters. An example is the very clean representation of the deformation density in tetraphenylbutatriene (Fig. 8.8).

Any function (e.g., the electron-density distribution around an atomic center) can be expressed as a sum of symmetric and antisymmetric functions; the Fourier transforms of the former will be real, and those of the latter will be imaginary. Deformations from the spherical free-atom distributions calculated for atoms at rest can be expressed in a similar way; for atoms at positions of high-site symmetry, the deformation terms are conveniently expressed in terms of spherical harmonics corresponding to monopole, dipole, quadrupole, octupole (*s, p, d, f,* etc.) components with appropriate radial dependencies.[11,12] A closely related approach, designed to handle atoms in low-symmetry environments in organic molecules, has been developed by Hirshfeld[13] and used extensively by him and his group.

In this approach, the nonspherical deformation density around each atom is expressed as a linear combination of localized functions of the type $R_n(r) \cos^n\theta_k$ ($n = 1, 2, 3$) where θ_k is measured from some specified polar axis. Three such functions with $n = 1$ are directed along three mutually perpendicular axes L_i, six functions with $n = 2$ along the face diagonals of the cube so formed, and ten functions with $n = 3$ along these face diagonals and the body diagonals as well. The radial functions $R_n(r)$ are arbitrarily chosen to have the form $r^n \exp(-\alpha r)$ with α fixed for each atom type so as to place the maximum of $R_3(r)$ at a distance ($r = 3/\alpha$) about a third to half of a typical bond distance involving that atom. In addition to these 19 angle-dependent product functions for each atom, three spherically symmetric terms ($n = 0$) with the same kind of radial dependence are invoked, making a total of 22 functions for each atom. Although not strictly necessary, it is convenient to orient the cubic reference frame attached to each atom to match any planes or axes of local symmetry that may be assumed to be present in the molecule. For example, a methylene group bonded to two neighboring atoms could be assumed to have local C_{2v} symmetry, and the reference frame attached to the carbon atom could be oriented accordingly. In this way, the number of independent deformation components can be reduced since some coefficients may vanish and others may be set equal by symmetry.

The Fourier transforms of these deformation functions with variable

[11]B. Dawson, *Proc. Roy. Soc., London* **A298,** 255 (1967).
[12]K. Kurki-Suonio, *Acta Crystallogr.* **A24,** 379 (1968).
[13]F. L. Hirshfeld, *Acta Crystallogr.* **B27,** 769 (1971).

coefficients are added to the free-atom scattering factor before multiplication by the temperature factor. The coefficients are then evaluated, along with the usual position and vibration parameters, by least-squares analysis. Once the coefficients are known, the total deformation density is easily reconstructed, either for the molecule at rest or for the molecule in the crystal, taking vibrational smearing into account. The residual difference map (including the deformation density in the calculation of F_c) can provide a check of whether significant features not included in the deformation density model remain to be accounted for. In practice, the number of coefficients can be reduced by introducing restrictions based on assumed equivalence of deformation terms that belong to different atoms or by imposing rigid-body constraints on the atomic vibration parameters, at the cost of some loss in generality.

The application of the method is not free from difficulties which arise mainly from extensive correlation among the least-squares parameters;

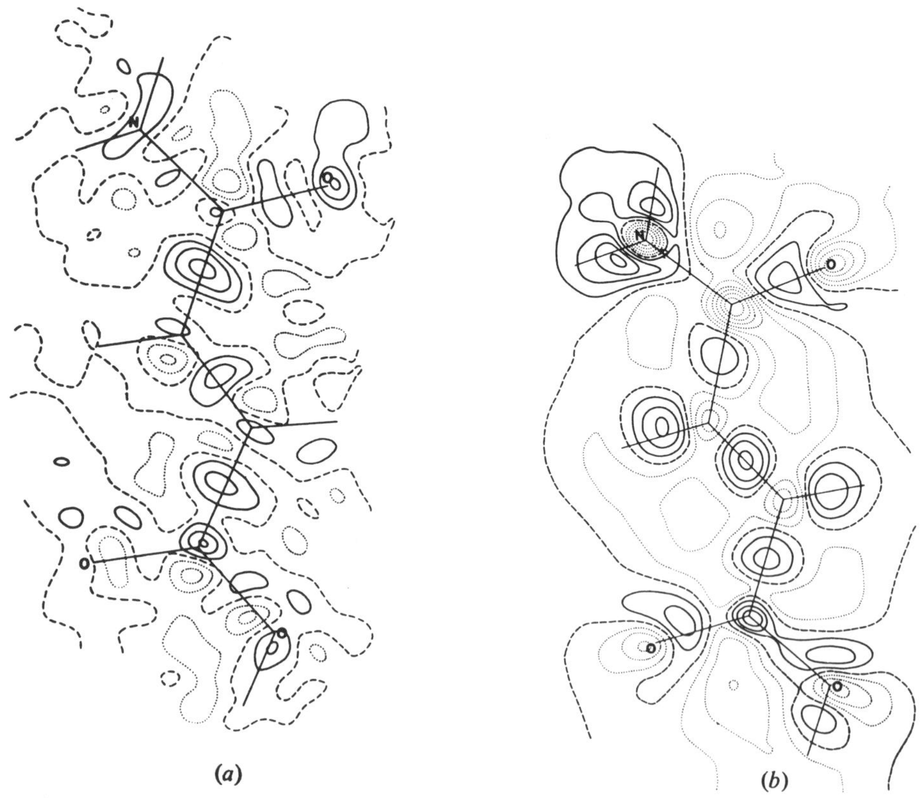

Figure 8.9. Fumaric acid monoamide: X–X difference density in mean molecular plane. (a) Conventional difference map. (b) By least-squares refinement, including vibrational smearing. Contour interval, 0.1 e. Å^{-3}. From Hirshfeld, *Acta Crystallogr.* **B27,** 769 (1971).

moreover, the physical significance of the individual deformation terms is obscure to say the least. Nevertheless, within these limitations, the method can extract maximal information from a given set of data, and the additional clarity with respect to a conventional X–X map, prepared with the same set of data, can be considerable; this is indicated by Fig. 8.9 for the deformation density of fumaric acid monoamide in the molecular plane. Of course, the smooth appearance of the least-squares deformation density is partly a result of including only a limited number of deformation terms, and its regularity is partly imposed by the introduction of various symmetry constraints among the coefficients. However, in this case (and in others) the improvement in overall clarity more than compensates for the inherent limitations of the method.

Hirshfeld[14] claims that the usual bias in vibrational parameters derived from X-ray analysis can be almost completely eliminated by refining these parameters together with charge deformation parameters.

Comparison with Theoretical Deformation Densities

In spite of the exciting possibilities that seem to be present in confronting theoretical charge density calculations with experimental results, very few reliable comparisons are available. The problem is more difficult than it first seems. The usual X-N or X-X maps correspond to dynamic densities, averaged over the lattice and molecular vibrations, whereas charge densities obtained by quantum-mechanical calculations are based on a fixed configuration of the atomic nuclei and are therefore instantaneous or static densities. Quantitative comparison is possible only if the effect of thermal motion is subtracted from the dynamic densities or added to the static ones. Another problem is that reliable ab initio calculations are available only for rather small molecules in the gaseous state, whereas the best experimental results are obtained for more complex molecules that are crystalline at experimentally accessible temperatures. In any case, small molecules are not particularly suitable for experimental deformation density studies because they tend to crystallize with small unit cells that do not yield any reflections in the low sin θ/λ range where information about charge deformation is largely concentrated.

Strictly speaking, the static density calculated for an equilibrium configuration of atomic nuclei is unobservable since any experimental measurement would have to include the effect of zero-point vibration, at least. Moreover, the static densities corresponding to different instan-

[14]F. L. Hirshfeld, *Acta Crystallogr.* **A32**, 239 (1976).

taneous configurations of nuclei that occur during a molecular vibration are slightly different, so that the best possible estimate of a dynamic density, averaged over the different nuclear configurations, would involve an enormous number of calculations. The other extreme is to assume that the static density can simply be convoluted with the vibrations,[15,16] an approximation that should hold less well for the molecular vibrations than for the lattice vibrations. Fortunately, the former are usually much less important than the latter. Calculations indicate that the dynamic density of small, fairly rigid molecules—e.g., acetylene, azide ion, and so forth—can be calculated with tolerable accuracy from a knowledge of the static density and the mean-square vibrational amplitudes, or rigid-body thermal parameters, as obtained from neutron or modified X-ray refinements.

The alternative approach (mentioned earlier) is to derive the static density by least-squares refinement of the coefficients of a limited set of deformation terms. These may be chosen, as described in the previous section, as having prescribed functional forms, or they can be made to correspond approximately to products of appropriate orbitals. The static density in a molecule can be defined as

$$\rho = 2 \sum_{\mu} \sum_{\nu} \sum_{i} C_{i\mu} C_{i\nu}\, \phi_{\mu}\, \phi_{\nu} = \sum_{\mu} \sum_{\nu} P_{\mu\nu}\, \phi_{\mu}\, \phi_{\nu} \tag{8.1}$$

where $C_{i\mu}$ is the coefficient of the basis orbital ϕ_{μ} in the ith molecular orbital, and the sum extends over all occupied molecular orbitals. The quantities

$$P_{\mu\nu} = \sum_{i} C_{i\mu} C_{i\nu} \tag{8.2}$$

are called population parameters. Atom and bond populations can be defined as

$$q_{\mathrm{A}} = \sum_{\mu} P_{\mu\mu}(A) \tag{8.3}$$

$$q_{\mathrm{AB}} = \sum_{\mu(\mathrm{A})} \sum_{\nu(\mathrm{B})} P_{\mu\nu} S_{\mu(\mathrm{A})\nu(\mathrm{B})} \tag{8.4}$$

where S is the overlap integral over the orbitals μ on atom A and ν on atom B. The gross Mulliken population of atom A is obtained by dividing bond populations q_{AB} equally between atoms A and B, and summing over all bonds emanating from A

[15] A. F. J. Ruysink and A. Vos, *Acta Crystallogr.* **A30,** 497 (1974).
[16] E. D. Stevens, J. Rys, and P. Coppens, *Acta Crystallogr.* **A33,** 333 (1977).

$$q'_A = q_A + \tfrac{1}{2} \sum_B q_{AB} \tag{8.5}$$

In principle, atom and bond populations can be obtained from X-ray diffraction data by including appropriate parameters in a least-squares refinement using generalized X-ray scattering factors, which are Fourier transforms of orbital products.[17] However, the results of population refinement depend very strongly on the type and number of orbitals that constitute the basis set.[18] For cyanuric acid, for example, a good description of the experimental density can be obtained with several sets of least-squares parameters that lead to different values for the atom and bond populations.

The theoretical difference density $\delta\rho$ is obtained by subtracting spherical free-atom densities ρ_A from the calculated static density ρ. The main difficulty is that the deformation density in the bonding and lone-pair regions is only a fraction of the total density near the atomic centers, and it is extremely sensitive to the choice of the basis set used to calculate the molecular and free-atom densities. Minimum basis sets (one atomic orbital per electron) are inadequate for this purpose.[19] A study by Cade shows the sensitivity of theoretical difference densities calculated for HF and O_2 to the choice of basis set, especially in the bond region.[20] The difference density in this region is far too small, even when saturated *sp* basis sets are used, and it is satisfactorily reproduced only when *d* orbitals are introduced. For cyanuric acid, the vibration-averaged difference density calculated with a minimal *sp* basis set fails completely to reproduce bonding density accumulations revealed in the experimental deformation map (Table 8.2, footnote *p*) and vastly exaggerates the lone-pair densities; the difference density calculated with the semi-empirical INDO method gives equally poor results.

Of course, experimental deformation maps are also sensitive to assumptions about what is subtracted from the total electron density, but the free-atom densities are accurately known. Thus, there is no ambiguity here if the deformation density is defined as representing the redistribution of the electrons on molecule formation. The effects of basis set truncation in theoretical difference densities are of a different nature and invalidate certain types of calculation.

Despite these difficulties, some theoretical static difference densities

[17] R. F. Stewart, *J. Chem. Phys.* **50,** 2485 (1969).
[18] D. S. Jones, D. Pautler, and P. Coppens, *Acta Crystallogr.* **A28,** 635 (1972).
[19] W. N. Lipscomb, *Trans. Am. Cryst. Assoc.* **8,** 79 (1972).
[20] P. E. Cade, *Trans. Am. Cryst. Assoc.* **8,** 1 (1972).

based on limited basis sets seem to reproduce the qualitative features seen in experimental deformation maps.[21,22,23] The agreement in these cases may be partially attributable to fortuitous cancellation of errors of opposite sign; the inadequacy of the calculation tends to underestimate the difference density, and the neglect of thermal motion tends to overestimate it. In other cases—e.g., dicyanogen—the correct relative heights of bond and lone-pair peaks is obtained only after vibrational smearing of the theoretical difference density;[24] the lone-pair peak, being sharper than the bond peak in the static difference density, is more affected by vibrations. In most of these comparisons, it is the theoretical deformation density calculated for a small molecule (e.g., cyanogen) that is compared with the experimental deformation density of a particular fragment (e.g., a cyano group) of a larger molecule. The agreement obtained here, and also the qualitative agreement between experimental densities for chemically similar groups in different molecular environments, implies at least a limited transferability of the more prominent charge-deformation features from molecule to molecule, which is encouraging for future work in this field. On the other hand, the influence of different crystal and molecular environments on certain deformation features, particularly lone pairs, may be considerable. This question has hardly begun to be studied.

Net Charges on Atoms

Given the experimental electron density in a crystal, it would seem simple to calculate the net charges on the atoms by integrating the electron density over the appropriate regions of space. The integrations could be performed for the total density, the deformation density (subtracting spherical, neutral, free-atom densities), or the valence density (subtracting spherical core densities), with the proviso that termination of series errors would have to be allowed for in the first case. Even if the fine details of these density functions were sensitive to various factors, such as bias in the positional and vibrational parameters, one might hope

[21] E. A. Laws, R. M. Stevens, and W. N. Lipscomb, *J. Chem. Phys.* **56,** 2029 (1972); *J. Am. Chem. Soc.* **94,** 4467 (1972): boron hydrides.

[22] R. M. Stevens, E. Switkes, E. A. Laws, and W. N. Lipscomb, *J. Am. Chem. Soc.* **93,** 2603 (1971): cyclopropane.

[23] J. Almlöf, A. Kvick, and J. O. Thomas, *J. Chem. Phys.* **59,** 3901 (1973): glycine.

[24] P. Becker, P. Coppens, and F. K. Ross, *J. Am. Chem. Soc.* **95,** 7604 (1973); P. Coppens, *Acta Crystallogr.* **B30,** 255 (1974).

that the integrated densities would be much less susceptible. The only real problem is in defining the volume over which the integrations are to be carried out. There is no unique or generally accepted answer to this because of the lack of agreement about what constitutes an atom in a molecule. This uncertainty carries with it a corresponding indeterminacy in the values of the atomic charges. Despite the ubiquity and usefulness of models that involve the concept of atomic charges, we have to admit that the concept itself is indefinite and hard to quantify.

Even for a typical "ionic" crystal such as NaCl, the definition of the atomic charges is not completely unequivocal because the electron density does not fall to zero along the direction of closest approach of the atoms. To carry out the necessary integrations, one must decide where one atom ends and the other begins, and the traditional ionic radii, such as those given by Goldschmidt or Pauling,[25] are unsatisfactory for this purpose (for which, after all, they were never intended). The Pauling radii for Na^+ and Cl^- are 0.95 and 1.81 Å, respectively, but the minimum in the electron-density profile in NaCl occurs[26] 1.17 Å from Na^+ and 1.64 Å from Cl^-. Integration of the experimental electron density up to these radii leads to net charges of +0.95 e for Na^+ and −0.70 e for Cl^- with respect to neutral atoms, indicating that the usual ionic description is appropriate. The revised radii for Na^+ and Cl^- can be used in conjunction with observed interatomic distances in other alkali halides with the rocksalt structure to yield ionic radii for other alkali and halide ions,[27] which deviate markedly from the traditional values:

	Li^+	Na^+	K^+	Rb^+	Cs^+	F^-	Cl^-	Br^-	I^-
Morris:	0.93	1.17	1.49	1.64	1.83	1.16	1.64	1.80	2.04 Å
Pauling:	0.60	0.95	1.33	1.48	1.69	1.36	1.81	1.95	2.16 Å
Goldschmidt:	0.78	0.98	1.33	1.49	1.65	1.33	1.81	1.96	2.20 Å

The corresponding integrations of the experimental electron density in KBr[28] out to these radii yield net charges of +0.67 e for K^+ and −0.88 e for Br^-. Another estimate,[29] based on somewhat different assumptions, is +0.45 e for K^+ and −1.15 e for Br^-.

[25] L. Pauling, *The Nature of the Chemical Bond,* 2nd Ed., Cornell University Press, Ithaca, N.Y., 1940, Chap. X.

[26] H. Witte and E. Wölfel, *Z. Physik. Chem. (Frankfurt)* **3,** 296 (1955).

[27] D. F. C. Morris, *Struct. Bonding* **4,** 63 (1968).

[28] V. Meisalo and O. Inkinen, *Acta Crystallogr.* **22,** 58 (1967).

[29] K. Kurki-Suoni and P. Salmo, *Ann. Acad. Sci. Fenn., Ser. A. VI* **369** (1971) (cited in Ref. 5).

Some indeterminancy in the net charge is inevitable when the electron density does not fall to zero along the internuclear line. When adjacent atoms in a crystal are so close that appreciable overlap occurs between their high-lying outer orbitals, any attribution of electron density in the intermediate region to one atom or the other becomes arbitrary. In theoretical work on molecules, it is customary to identify net atomic charges with Mulliken populations (Eq. 8.5). The implicit assumption is that overlap density between adjacent atoms is equally divided between them, which seems questionable when the overlap density is not symmetrically distributed. Bonding density peaks seen in experimental deformation maps are sometimes markedly unsymmetrical.

For molecules the measurement of net atomic charges by integration is not very satisfactory because of the uncertainty in the volume to be assigned to the individual atoms. An alternative method is to refine the occupancy factors or population parameters (defined in Eq. 8.2) of atomic valence-shell density functions, centered on the atomic positions, by least-squares analysis. This approach is similar in principle to the analysis of deformation density by least-squares analysis, as described earlier. The core electrons may be regarded as being unaffected by bonding, at least within the detection limits of the X-ray diffraction experiment. Also, from X-ray data alone it is almost impossible to distinguish whether a nonspherical density is caused by preferential population of a certain orbital product or by anisotropic vibration of the atom. This ambiguity is eliminated if the vibration tensor can be determined independently by neutron diffraction; otherwise, only spherical-density functions can be used. The occupancy factors are then refined either in a separate least-squares cycle after positional and thermal parameters are fixed (*L*-shell projection method)[30] or simultaneously with these parameters (extended *L*-shell projection method).[31]

The net charges obtained by adding the corresponding occupancy factors are not the same as those derived from gross Mulliken populations. In the latter, populations of one- and two-center density functions are included, whereas here only one-center products are involved. However, density functions centered on adjacent atoms interpenetrate to some extent, which means that any density in the overlap region will be divided among the atoms concerned. This helps to ensure that the net charges resulting from the refinement are not very sensitive to the details of the atomic orbitals used; the effects of a contraction (or expan-

[30] R. F. Stewart, *J. Chem. Phys.* **53**, 205 (1970).
[31] P. Coppens, D. Pautler, and J. F. Griffin, *J. Am. Chem. Soc.* **93**, 1051 (1971).

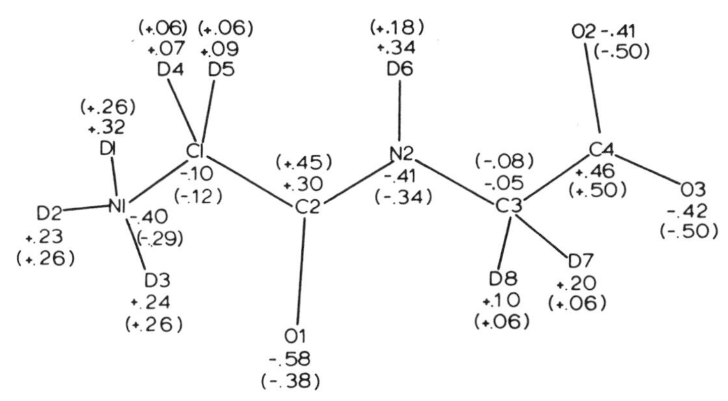

Figure 8.10. Experimental atomic charges in glycylglycine. Values in parentheses are from theoretical calculations. From Griffin and Coppens, *J. Am. Chem. Soc.* **97,** 3496 (1975). Reprinted by permission.

sion) of the atomic orbitals on both centers will tend to cancel.[32] In various tests, density functions based on Slater-type atomic orbitals with standard molecular exponents are found to be the most appropriate. With good quality data, the orbital exponents as well as the occupancy factors can be adjusted by least-squares analysis.

One of the most interesting applications has been to glycylglycine, for which accurate X-ray and neutron-diffraction data are available (Table 8.2, footnote *s*). Net atomic charges, derived from a valence-shell refinement in which positional and vibrational parameters were held at values determined by the neutron-diffraction analysis, are shown in Fig. 8.10. Although the actual numbers should not be taken too seriously, the qualitative features are in full agreement with what could be expected from chemical experience and intuition: negative charges on oxygen and nitrogen, more-or-less neutral aliphatic carbon, positive charge on the carbonyl carbon, the carboxyl carbon, and the ammonium and peptide hydrogens. The agreement with recent theoretical calculations[33] is no less pleasing. We can conclude that despite the serious conceptual difficulties, *approximate* net charges for atoms in molecules can be derived by analysis of X-ray diffraction data.

Summary and Outlook

Pictures of orbitals and of electron-charge distributions produced by theoretical calculations are beautifully symmetric and have considerable

[32]P. Coppens, *Trans. Am. Cryst. Assoc.* **8,** 93 (1972).
[33]F. A. Momany, L. M. Carruthers, and H. A. Scheraga, *J. Phys. Chem.* **78,** 1621 (1974).

aesthetic appeal, apart from their intrinsic scientific interest. The deformation densities produced by analysis of X-ray diffraction data are usually not as attractive and often contain much spurious detail—a polite term for rubbish. Nevertheless, as we have seen, the main features can often be convincingly interpreted in terms of chemically significant concepts such as bonding and lone-pair densities. Other quantities, such as net atomic charges, that can be derived from the experimental data with the help of arbitrary but reasonable assumptions, agree quite well with values derived from other sources.

Both theoretical and experimental studies of the electron density in molecules are necessary for a better understanding of molecular structure in its static and dynamic aspects. Both approaches seem to have reached a stage where the results are of borderline significance. At this stage, it is hard to resist the temptation to filter the results through the sieve of our preconceived notions about how things ought to be; we tend to select those features that seem intuitively plausible (what a splendid method we have at our disposal) and dismiss the others as artifacts (what can you expect from such poor experimental data or from a theory that contains so many approximations). In this kind of situation, the theoretical and experimental approaches to the study of electron density seem nicely complementary since the errors and approximations inherent in the one are virtually independent of those in the other. Thus one can be used to check the other. The present limitation of accurate theoretical studies to small molecules and of experimental studies to larger ones does not seem to be too much of a barrier since both approaches are likely to be extended in the required direction soon. Further, there is every reason to believe that many features of the charge density of molecular fragments can be transferred from one molecular environment to another with little change.[34] For example, the bonding and lone-pair densities of the cyano groups in tetracyanoethylene and tetracyanoethylene oxide (Table 8.2, footnotes *g*, *u*) are remarkably similar.

Perturbations of charge densities caused by intermolecular forces (nonbonding or weakly bonding interactions) have hardly begun to be studied, and most effects probably lie beyond the present limits of detectability in experimental work. However, as mentioned earlier, there are indications that the lone-pair distribution of carbonyl oxygen atoms varies with the nature and arrangement of the neighboring atoms in adjacent molecules, hydrogen bonding being the extreme case of such

[34]R. F. W. Bader, *Accounts Chem. Res.* **8**, 34 (1975).

interactions. Similarly, it has been noted that the bonding density in the amide carbonyl group of glycylglycine seems to be slightly distended towards the —NH_3^+ group. This is one of those borderline effects, whose significance cannot be assessed on the basis of a single observation, and it should be checked by further experimental studies on other peptides.[35]

[35]An excellent, up-to-date account of the present state of knowledge in the area covered by this chapter is provided by a recent special issue of *Israel J. Chem.* **16,** 87–229 (1978), entitled "Electron density mapping in molecules and crystals" (F. L. Hirshfeld, Ed., 19 articles and two appendices).

9. Geometric Constraints in Cyclic Molecules

"You couldn't deny that, even if you tried with both hands."
"I don't deny things with my hands," Alice objected.
"Nobody said you did," said the Red Queen. "I said you couldn't if you tried."

Geometric Constraints

In this chapter we consider some geometric aspects of molecular structure—in particular, how certain structural parameters of cyclic molecules are fixed or constrained to lie within narrow limits by geometric necessity. An obvious and trivial example is that the sum of the ring angles in a three-membered ring is exactly 180°. Any attempt to explain such a result in terms of a detailed theory of chemical bonding would be ridiculous. A less obvious example is provided by the structure of pentameric arsenomethane $[As(CH_3)]_5$, where the As atoms are linked into a puckered five-membered ring.[1,2] The As—As bond lengths are equal within experimental error (2.43 ± 0.01 Å), but the As—As—As bond angles vary over several degrees (100.4°, 100.0°, 105.6°, 105.4°, 97.5°; estimated S.D. ~ 0.3°).[2] Why are the angles so different while the lengths are equal? The answer is that in an equilateral five-membered ring with average angle of 100°, the individual angles *must* be different, as Waser and Schomaker were quick to recognize in their account of the earlier electron-diffraction study of arsenomethane in the vapor phase.[1] An equilateral, isogonal pentagon is only possible for a planar ring with an angle of 108°; this result cannot be said to be obvious, otherwise it would not have attracted so much attention in the mathematical literature.[3,4]

[1] J. Waser and V. Schomaker, *J. Am. Chem. Soc.* **67,** 2014 (1945).
[2] J. H. Burns and J. Waser, *J. Am. Chem. Soc.* **79,** 859 (1957).
[3] B. L. van der Waerden, *El. Math.* **25,** 73 (1970); **27,** 63 (1972); W. Lüssy and E. Trost, *ibid.,* **25,** 82 (1970); H. Irminger, *ibid.,* **25,** 135 (1970); S. Smakal, *ibid.,* **27,** 62 (1972). According to Smakal the problem was also posed by V. I. Arnold, *Proswjechtschenie (USSR)* **2,** 268 (1957) and solved by A. P. Garber, V. I. Garvackij, and V. Ja. Jarmolenko, *Proswjechtschenie (USSR)* **6,** 345 (1961). As far as I am aware, this unique property of the pentagon was first stated by J. Waser (Proposition 10, Ph.D. Dissertation, California Institute of Technology, Pasadena, 1944): "Of all n-gons ($n \geq 4$) the pentagon is the only

In this respect the pentagon is unique (apart from the trivial case of the triangle) since equilateral, isogonal tetragons, hexagons, heptagons, and so forth can be constructed for a whole range of possible angles and are not constrained to lie in a plane. Neglect or ignorance of this geometric result could lead to a profitless inquiry into the origin of the observed bond-angle differences in nonplanar five-ring molecules.

The structure of a molecule can be defined in any number of ways—e.g., by giving the coordinates of the atoms in some arbitrary coordinate system or by specifying a sufficient number of internal parameters, such as interatomic distances and angles. For most purposes, the position and orientation of the molecule can be regarded as irrelevant, in which case only the relative positions of the atoms are required. These are specified by $3N-6$ independent parameters for a system of N atoms. In terms of a Cartesian coordinate system, the origin can always be chosen to coincide with one of these positions, say atom 1, the x-axis can always be chosen along the direction of the vector from 1 to 2, and the y-axis can be chosen in the plane of atoms 1, 2, and 3. Hence the coordinates can always be listed to include six zeros (nonadjustable coordinates):

$$
\begin{array}{cccc}
i=1 & 0 & 0 & 0 \\
2 & x_2 & 0 & 0 \\
3 & x_3 & y_3 & 0 \\
4 & x_4 & y_4 & z_4 \\
 & & & \\
N & x_N & y_N & z_N
\end{array}
$$

The relative positions of the atoms are then defined by the remaining $3N-6$ adjustable coordinates (degrees of freedom). Alternatively, we could define the structure in terms of $3N-6$ internal parameters if these are independent. To define the structure of a planar molecule, only $2N-3$ independent parameters are needed since, for example, all the z coordinates can be chosen as zero.

The relative positions of N atoms arranged in a chain are completely fixed by assigning definite values to the $N-1$ bond distances, $N-2$ bond angles, and $N-3$ torsion angles. These parameters are independent, and

---

one for which the following is true: The construction of an equilateral, equiangular pentagon is possible for only two values of the angle. For all other n-gons (excepting the trivial case of the triangle) there is a whole range of angles for which an analogous construction is possible. The pentagon under consideration is planar, the possible angles are 108° and 36°." The 36° angle occurs in the nonconvex star pentagon, which is of little interest as far as molecular structures are concerned.

[4]See J. D. Dunitz and J. Waser, *El. Math* **27,** 25 (1972) for a collection of proofs provided by E. Ruch and L. J. Oosterhoff as well as by the authors.

there are $3N-6$ in all. For a branched chain of N atoms it is also possible to choose the $3N-6$ internal parameters in the same way, as $N-1$ bond lengths, $N-2$ bond angles, and $N-3$ torsion angles, but it is now necessary to ensure that the parameters chosen are truly independent. At each nonterminal atom of ligancy m, there are $m(m-1)/2$ distinct bond angles, but only $2m-3$ of these are independent:

Ligancy of central atom	2	3	4	5	6
Distinct bond angles	1	3	6	10	15
Independent bond angles	1	3	5	7	9

Similarly, about each nonterminal bond between atoms in ligancy p and q, there are $(p-1)(q-1)$ distinct torsion angles, of which only one is independent once the bond angles have been specified. Additional torsional degrees of freedom may be introduced at the cost of reducing the number of independent bond angles.

In systems that are supposed to show symmetry, the number of degrees of freedom is correspondingly reduced. For example, a regular tetrahedral arrangement of four atoms about a central atom ($N = 5$, $3N - 6 = 9$) is described by a single parameter, which is essentially a scale factor—e.g., the bond length. The bond angles are not adjustable; they are fixed at 109°28′. Thus the T_d symmetry leads to the loss of eight of the nine independently variable parameters of the five-atom system.

As the T_d symmetry is relaxed, the number of independent parameters increases, depending on the residual symmetry:

Point group	Independent bond lengths	Independent bond angles	Total
T_d	1	0	1
D_{2d}	1	1	2
D_2	1	2	3
C_{3v}	2	1	3
C_{2v}	2	2	4
C_2	2	3	5
C_s	3	3	6
1	4	5	9

For each of these symmetries, the number of distinct nonequivalent bond angles is one more than the number of independent bond angles—i.e., the nonequivalent angles are related by an equation of constraint.

In general, the algebraic relationships between dependent and independent parameters can be quite complicated, even for structures that appear geometrically simple. For example, the six bond angles subtended by the four bonds at a tetrahedral center are related by the determinantal equation:

$$\begin{vmatrix} 1 & C_{12} & C_{13} & C_{14} \\ C_{12} & 1 & C_{23} & C_{24} \\ C_{13} & C_{23} & 1 & C_{34} \\ C_{14} & C_{24} & C_{34} & 1 \end{vmatrix} = 0 \qquad (9.1)$$

where $C_{12} = \cos \theta_{12}$, etc. The left side can be regarded as the square of the four-dimensional volume of the hyperparallelepiped formed by the four unit vectors that emanate from the tetrahedral center. If these four vectors are constrained to lie in three-dimensional space—i.e., if they are not linearly independent—then this volume must be zero. Eq. 9.1 is actually the same as Eq. 5.15, which was derived in a different way. When symmetry restrictions are introduced, the relationships among the bond angles become simpler:

$$T_d(\theta_{12} = \theta_{13} = \theta_{14} = \theta_{23} = \theta_{24} = \theta_{34}) : C_{ij} = -1/3$$
$$D_{2d}(\theta_{12} = \theta_{34};\ \theta_{13} = \theta_{14} = \theta_{23} = \theta_{24}) : C_{12} + 2C_{13} = -1$$
$$D_2(\theta_{12} = \theta_{34};\ \theta_{13} = \theta_{24};\ \theta_{14} = \theta_{23}) : C_{12} + C_{13} + C_{14} = -1$$
$$C_{3v}(\theta_{12} = \theta_{13} = \theta_{14};\ \theta_{23} = \theta_{24} = \theta_{34}) : 3C_{12}^2 = 1 + 2C_{23}$$
$$C_{2v}(\theta_{12}, \theta_{34}, \theta_{13} = \theta_{14} = \theta_{23} = \theta_{24}) : 4C_{13}^2 = (1 + C_{12})(1 + C_{34})$$
$$C_2(\theta_{12}, \theta_{34}, \theta_{13} = \theta_{24}, \theta_{23} = \theta_{14}) : (C_{13} + C_{23})^2 = (1 + C_{12})(1 + C_{34})$$
$$C_s(\theta_{12}, \theta_{34}, \theta_{13} = \theta_{14}, \theta_{23} = \theta_{24}) : 2(C_{13}^2 + C_{23}^2 - 2C_{12}C_{13}C_{23}) = S_{12}^2(1 + C_{34})$$

In a ring of N atoms, the N bond distances, N bond angles, and N torsion angles must be related by six equations of constraint (ring-closure conditions). In polycyclic systems of N atoms there are even more equations of constraint since each additional ring involves at least one additional bond distance, four additional bond angles, and eight additional torsion angles, but the number of degrees of freedom remains $3N-6$. Once the bond distances in such systems are fixed, the bond angles and torsion angles may be severely constrained.

Given the constitution of a molecule, the bond distances can usually be regarded as being fixed within fairly narrow limits at standard values that are characteristic of the various bond types. Likewise, bond angles do not vary much from characteristic values unless forced to do so by ring constraints (e.g., angles in small rings). The various shapes (conformations) that a cyclic molecule may adopt therefore depend mainly on changes in the torsion angles. Indeed, as most chemists are aware, useful information about the possible shapes can often be obtained by examining the properties of models—e.g., Dreiding models, in which bond distances and bond angles are mechanically held at fixed values but rotation about single bonds is completely free. The mechanical models overemphasize ease of rotation about bonds, they underestimate variations in bond distances and angles, and they neglect nonbonded interactions altogether; nevertheless, they are extremely useful for visualizing the atomic arrangements in the various conformations that a molecule can adopt.

In our discussion of geometric constraints in cyclic molecules we shall usually regard bond distances and bond angles as being fixed. If there are no residual degrees of freedom, the ring is rigid (it can be deformed only if bond distances or bond angles are changed), whereas the presence of at least one torsional degree of freedom confers flexibility on the ring.

In three-dimensional space the basic unit of structure consists of four points (a tetrahedron), whose relative positions are defined by six parameters which can be chosen as the six edges of the tetrahedron or in several other ways. For example, the four points (atoms) can be regarded as forming a chain 1234 defined by three bond lengths, d_{12}, d_{23}, d_{34}, two bond angles, θ_2, θ_3, and one torsion angle ω_{23} (Fig. 9.1). The three lengths correspond to edges of the tetrahedron, and the lengths of the remaining three edges can be calculated from the given parameters. We shall make frequent use of the relationship involving the length of the missing edge d_{14} between the terminal atoms and the torsion angle ω_{23} about the central bond:

$$d_{14}^2 = d_{12}^2 + d_{23}^2 + d_{34}^2 - 2d_{12}d_{23}\cos\theta_2 - 2d_{23}d_{34}\cos\theta_3 + 2d_{12}d_{34}(\cos\theta_2\cos\theta_3 - \sin\theta_2\sin\theta_3\cos\omega_{23}) \quad (9.2)$$

For equal bond lengths,

$$d_{14}^2 = d^2(3-2\cos\theta_2 - 2\cos\theta_3 + 2\cos\theta_2\cos\theta_3 - 2\sin\theta_2\sin\theta_3\cos\omega_{23}) \quad (9.3)$$

and for equal bond lengths and bond angles,

$$d_{14}^2 = d^2(3-4\cos\theta + 2\cos^2\theta - 2\sin^2\theta\cos\omega) \quad (9.4)$$

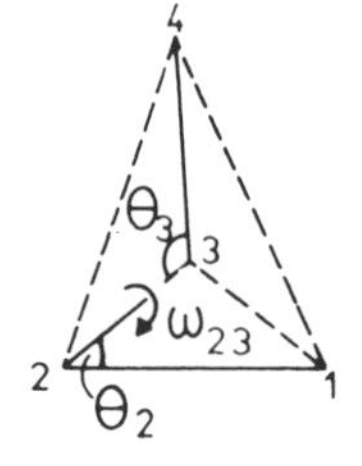

Figure 9.1. The relative positions of four points in three-dimensional space can be defined by specifying (a) the six edges of a tetrahedron or (b) the three marked edges, the two included angles, and a torsion angle, or in several other ways.

Four-Membered Rings

A four-membered ring is just a more-or-less flattened tetrahedron, and its geometry can be specified by the lengths of its four sides and two diagonals or any other six parameters from which these lengths may be calculated.

For an equilateral four-membered ring, the number of degrees of freedom is reduced. The symmetry of such a ring must be at least C_{2v}, so that only three Cartesian coordinates need to be specified to define the relative atomic positions (Fig. 9.2). These may be expressed in terms of three independent internal parameters: one bond length (taken as unity) and two bond angles:

$$x^2 + y^2 + z^2 = 1$$

$$x = \sin(\theta_2/2)$$

$$y = \sin(\theta_1/2)$$

$$z = [(1 - \sin^2(\theta_1/2) - \sin^2(\theta_2/2)]^{\frac{1}{2}}$$

From the torsion angle formula (Eq. 9.3),

$$1 = 3 - 2\cos\theta_1 - 2\cos\theta_2 + 2(\cos\theta_1\cos\theta_2 - \sin\theta_1\sin\theta_2\cos\omega)$$

or
$$\cos\omega = \frac{1-\cos\theta_1-\cos\theta_2+\cos\theta_1\cos\theta_2}{\sin\theta_1\sin\theta_2} = \frac{(1-\cos\theta_1)(1-\cos\theta_2)}{\sin\theta_1\sin\theta_2} \quad (9.5)$$

The four torsion angles are equal in magnitude, and they alternate in sign (Fig. 9.2).

Raising the symmetry to D_{2d} makes $\theta_1 = \theta_2$ so that

$$\cos\omega = \frac{(1-\cos\theta)^2}{\sin^2\theta} = \frac{1-\cos\theta}{1+\cos\theta} \quad (9.6)$$

The dependence of the torsion angle on the bond angle is shown in Fig. 9.3. When $\theta = 90°$, $\omega = 0$, and the ring is planar (point group D_{4h}), but

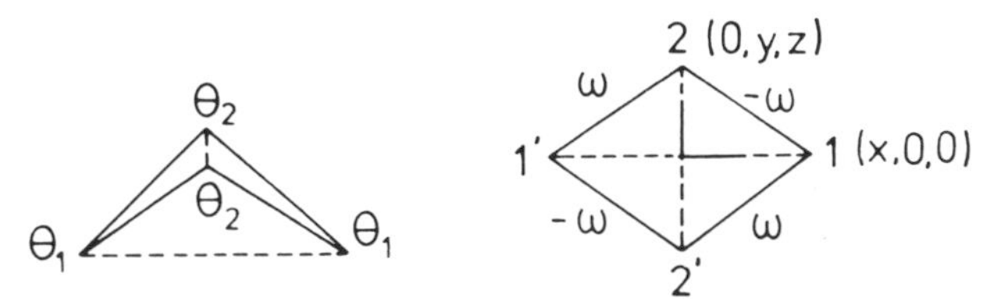

Figure 9.2. A four-membered ring with equal bond lengths has at least C_{2v} symmetry and is defined by three Cartesian coordinates. Torsion angles related by the dyad axis are equal; those related by mirror planes are equal in magnitude but opposite in sign.

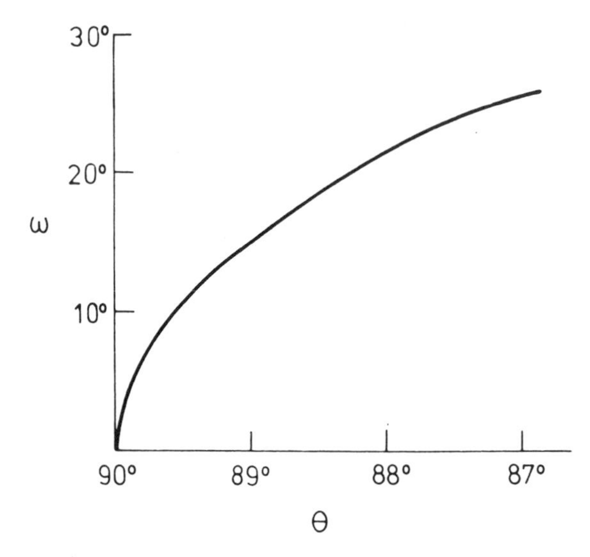

Figure 9.3. Variation of torsion angle ω with bond angle θ in a four-membered ring with D_{2d} symmetry (equal bond lengths and bond angles).

appreciable nonplanarity is obtained by decreasing the bond angle only slightly. An approximate relationship, valid for small deviations $\delta\theta$ from 90°, is

$$\omega = 15.1\ (\delta\theta)^{\frac{1}{2}}$$

where both angles are expressed in degrees. Thus a 1° change in θ corresponds to a torsion angle of about 15°. This geometric result helps to explain the observed nonplanarity of the carbon skeleton of cyclobutane,[5,6] which might otherwise seem puzzling since the out-of-plane distortion must reduce the already highly strained ring angle. This angle is indeed reduced below 90° but only by a small amount, which is sufficient to achieve a substantial deviation from zero torsion angle.

The nonplanarity of cyclobutane derivatives is sometimes expressed in terms of the dihedral angle Ψ between planes that share a common diagonal. For the D_{2d} structure, the relationship between Ψ and the bond angle θ is easily shown to be

$$\sin(\Psi/2) = \tan(\theta/2) \tag{9.7}$$

A change of 1° in θ (from 90° to 89°) corresponds to a change of 22° in Ψ. The change in Ψ is approximately $\sqrt{2}$ times as large as that in ω.

[5] J. D. Dunitz and V. Schomaker, *J. Chem. Phys.* **20,** 1703 (1952).
[6] A. Almenningen, O. Bastiansen, and P. N. Skancke, *Acta Chem. Scand.* **15,** 711 (1961).

Five-Membered Rings

A set of five points contains $5 \times 4/2 = 10$ distances, of which $3 \times 5 - 6 = 9$ are independent. However, fixing nine distances (or five distances and four angles, which amounts to the same thing) does not fix the remaining distance uniquely but leaves two possible values open. This is easily seen by consideration of Fig. 9.4, where the nine marked distances define the two tetrahedra ABCE and BCDE that share a common triangle BCE. Vertices A and D can lie on the same side of the plane defined by BCE, or they can lie on opposite sides; the distance AD will be different for these two possibilities. Thus the geometry of a five-membered ring is not uniquely defined by specifying five bond lengths and four angles. There are still two possible values for the remaining angle.

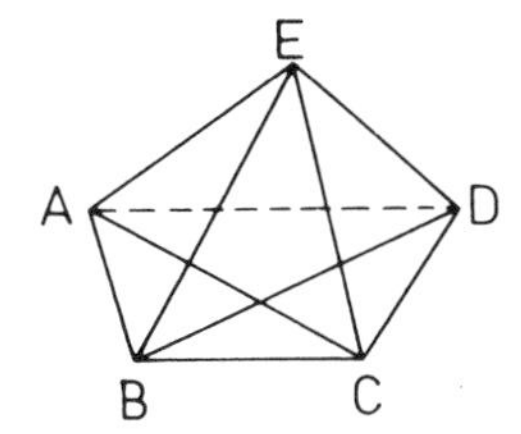

Figure 9.4. Five points (not in a plane). By fixing the nine distances (solid lines) two possible arrangements in space are left. Points A and D can be on the same side of the plane defined by the triangle BCE, or on opposite sides.

We can always take as the nine independent parameters those that would define a chain of five atoms—viz., four bond distances, the three included bond angles, and the two included torsion angles (Fig. 9.5). This selection fixes the values of the dependent parameters uniquely, but the calculations involve very tiresome arithmetic—e.g., the missing bond distance is a function of all nine variables. The best plan is probably to use the given independent parameters to derive Cartesian coordinates for the five atoms (a procedure is described in Appendix II) and to calculate values of dependent parameters from there.

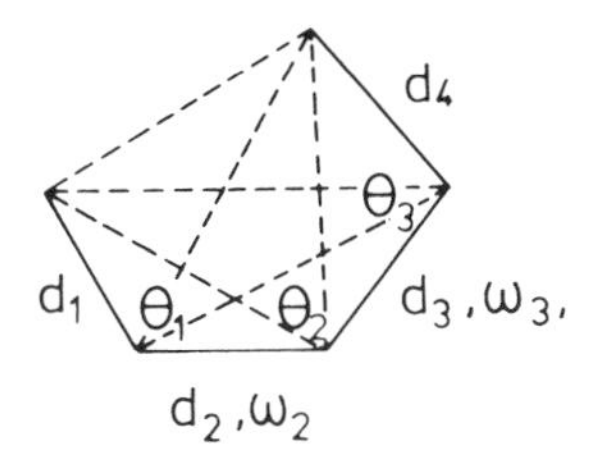

Figure 9.5. Five-membered ring regarded as a chain. Dashed distances can be calculated as functions of the nine given parameters d_1, d_2, d_3, d_4, θ_1, θ_2, θ_3, ω_1, ω_2.

The equilateral five-membered ring (of side unity) has four degrees of freedom that can be chosen as indicated in Fig. 9.6. Fixing the two adjacent bond angles θ_B and θ_C and the torsion angle ω (ABCD) determines the distance AD and hence the angle θ_E. Point E must therefore lie on a circle centered on the midpoint of AD with radius $\cos(\theta_E/2)$ normal to

AD. The position of E on this circle, together with the three given parameters, defines the ring completely.

If $\theta_B = \theta_C$, the grouping ABCD has a twofold rotation axis that passes through the midpoint of BC. Point E will not generally lie on this axis, but if it does, then $\theta_A = \theta_D$. Conversely, if $\theta_A = \theta_D$, then E lies on the axis. If $\theta_B = \theta_C$ and ω (ABCD) = 0, then the grouping ABCD has C_{2v} symmetry, and the ring must have a mirror plane regardless of the position of E on the circle. Even for these special cases of equilateral five-membered rings with C_2 or C_s symmetry (only two independent parameters), calculation of the values of the dependent parameters is quite tedious. The relevant relationships are:

C_s (atom 5 on symmetry plane):

$$4 \cos \theta_2(\cos \theta_2 - 1) + 2 \cos \theta_5 = 1$$

$$\cos \omega_{12} = \frac{1 - 2 \cos \theta_2 + 2 \cos \theta_1 \cos \theta_2}{2 \sin \theta_1 \sin \theta_2}$$

$$\cos \omega_{15} = \frac{1 + 2 \cos \theta_2 - 2 \cos \theta_1 - 2 \cos \theta_5 + 2 \cos \theta_1 \cos \theta_5}{2 \sin \theta_1 \sin \theta_2}$$

$$\cos \omega_{23} = 1$$

C_2 (atom 5 on symmetry axis):

$$16 \cos^2 \theta_1 + 4 \cos^2 \theta_2 - 16 \cos \theta_1 \cos \theta_2 + 16 \cos \theta_1 \cos \theta_5 - 8 \cos \theta_1 + 12 \cos \theta_2 - 14 \cos \theta_5 = 5$$

$$\cos \omega_{23} = \frac{2 \cos^2 \theta_2 - 4 \cos \theta_2 + 2 \cos \theta_5 + 1}{2 \sin^2 \theta_2}$$

$\cos \omega_{12}$ and $\cos \omega_{15}$ as above

The derivation of these equations is not difficult, but it is tiresome and can be recommended as an exercise only to readers who really like this sort of thing.

As mentioned earlier, an equilateral, isogonal five-membered ring is

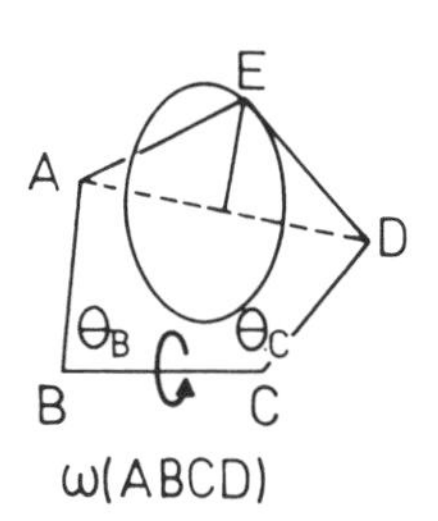

Figure 9.6. The four degrees of freedom in an equilateral five-membered ring. For given θ_B, θ_C, and ω(ABCD), line AD is fixed in magnitude and direction. Triangle ADE is still free to turn around the hinge AD.

possible only for an angle of 108°; in other words, such a ring is necessarily planar. For example, if the five angles in Fig. 9.6 are equal ($\theta_A = \theta_B = \theta_C = \theta_D = \theta_E$), then from the arguments about the symmetry of the grouping ABCD, the vertex E must lie on a dyad axis or on a mirror plane of the ring. On passing round the ring, an exactly analogous argument can be applied for every grouping of four vertices. The remaining vertex must always lie on a dyad axis or mirror plane, and the resulting system of intersecting dyad axes and/or mirror planes must possess at least D_5 symmetry. Hence, the ring must be planar. This is essentially the proof originally given by Waser,[4] but a number of others, at various levels of elegance and sophistication, are available.[3,4]

Many five-membered rings in organic molecules are at least approximately equilateral but nonplanar. The most obvious measure of the degree of ring puckering is provided by the sum of squares of the deviations of the atoms from the mean plane, and the quantity

$$q = \left(\sum z_j^2\right)^{\frac{1}{2}}$$

is known as the puckering amplitude. Computation of q from internal coordinates involves a transformation to Cartesian coordinates, but a good estimate can be obtained directly from a knowledge of the bond angles or torsion angles. For a puckered pentagon of side unity, the relationships

$$\sum \delta_j = 540 - \sum \theta_j \text{ (degrees)} = 240\, q^2$$

$$\sum \omega_{jk}^2 \text{ (degrees}^2\text{)} = 6.0 \times 10^4\, q^2$$

$$\sum \omega_{jk}^2 \text{ (degrees}^2\text{)} = 250 \sum \delta_j \text{ (degrees)}$$

are valid for infinitesimal puckering amplitudes and hold approximately even for quite puckered rings.[7] For example, the exact value of q for cyclopentane[8] (normalized to unit bond lengths) is 0.281, whereas from the bond angles:

103.19°, 105.71°, 106.37°, 104.71°, 102.31°

and torsion angles

30.57°, 7.40°, 18.67°, 37.43°, 41.89°

we obtain $\Sigma\delta = 17.7°$, $\Sigma\omega^2 = 4494$ degrees^2, corresponding to the estimates $q(\delta) = 0.272$, $q(\omega^2) = 0.274$.

[7] J. D. Dunitz, *Tetrahedron* **28,** 5459 (1972).
[8] W. J. Adams, H. J. Geise, and L. S. Bartell, *J. Am. Chem. Soc.* **92,** 5013 (1970).

Five-membered rings that are mirror symmetric (C_s or envelope forms) or those that have a dyad axis (C_2 or twist forms) can be identified from their torsion-angle patterns (Fig. 9.7). Most rings observed in organic molecules are neither envelope nor twist forms but lie somewhere in between. These intermediate forms are often described in the literature as "envelope" forms because the approximate planarity of a four-atom grouping seems to spring to the eye more easily than the presence of an approximate dyad axis.

A complete, unambiguous specification of the out-of-plane distortion of a five-membered ring can be made in terms of two parameters (since $2N-3 = 7$ coordinates can be chosen in an arbitrary plane, the mean plane of the ring). These two parameters can be chosen as the puckering amplitude q and a phase angle ϕ.[9] The individual displacements from the mean plane are then:

$$z_j = (\tfrac{2}{5})^{\frac{1}{2}}\, q \cos(4\pi j/5 + \phi) \qquad j = 1, 2, 3, 4, 5 \tag{9.8}$$

This follows since the two out-of-plane symmetry displacements of a regular pentagon transform as the E_2'' irreducible representation of D_{5h}; the right side is just the arbitrary linear combination of the two orthogonal components behaving as $\cos 4\pi j/5$ and $\sin 4\pi j/5$, respectively. As Cremer and Pople have shown,[10] the equation is also valid for rings that are not equilateral. Deformations described by ϕ values of 0°, 36°, 72°, . . . correspond to envelope forms; those described by values of 18°, 54°, 90°, . . . correspond to twist forms.

Alternatively, the out-of-plane deformation coordinates can be taken as the torsion angles ω_k, which transform in just the same way as the out-of-plane displacements z_j. The two orthogonal deformations can then be combined to give

$$\omega_k = \omega_o \cos\,(4\pi k/5 + \phi') \qquad k = 1, 2, 3, 4, 5 \tag{9.9}$$

Now the envelope forms are described by $\phi' = 18°, 54°, 90°, \ldots$, and the twist forms are described by 0°, 36°, 72°, ... Whereas Eq. 9.8 is strictly valid only for infinitesimal out-of-plane deviations, Eq. 9.9 holds very well, even for highly puckered rings.[11,12] Although the three conditions that reduce the five ω or z values to only two independent out-of-plane coordinates are of the same mathematical form:

[9] J. E. Kilpatrick, K. S. Pitzer, and R. Spitzer, *J. Am. Chem. Soc.* **69**, 2483 (1947).
[10] D. Cremer and J. A. Pople, *J. Am. Chem. Soc.* **97**, 1354 (1975).
[11] C. Altona, H. J. Geise, and C. Romers, *Tetrahedron* **24**, 13 (1968).
[12] S. Lifson and A. Warshel, *J. Chem. Phys.* **49**, 5116 (1968).

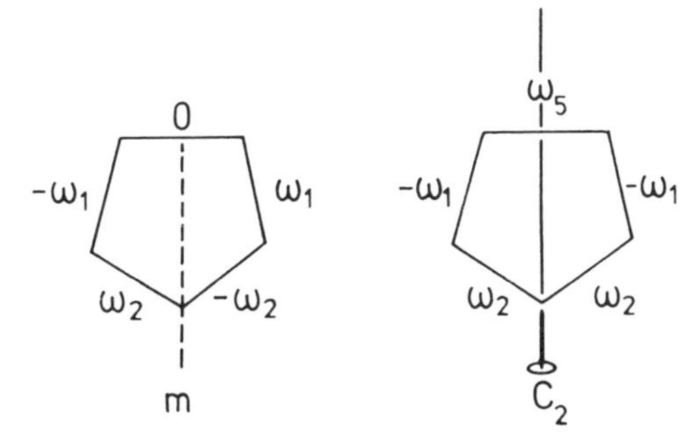

Figure 9.7. Torsion-angle patterns in C_s (left) and C_2 forms (right) of non-planar five-membered rings. For approximately equilateral rings the torsion angles ω_5, ω_1, and ω_2 are roughly proportional to 0, sin 36°, and sin 72°, respectively, for C_s symmetry and to 1, cos 36°, and cos 72°, respectively, for C_2 symmetry.

$$\sum p_j = 0$$
$$\sum p_j \cos 2\pi j/5 = 0 \qquad (9.10)$$
$$\sum p_j \sin 2\pi j/5 = 0$$

the geometric meaning of the constraint equations is different for the ω and the z values. For the z values (Cartesian coordinates) they are the conditions of no net translation or rotation; for the ω values (internal coordinates) they are the ring-closure conditions for an equilateral pentagon. For any given set of torsion angles ω_j, the values of ω_o and ϕ can be calculated from

$$\tan \phi' = \frac{-\omega_1 + \omega_2 - \omega_3 + \omega_4}{2\omega_5 (\sin 36° + \sin 72°)} = \frac{-\omega_1 + \omega_2 - \omega_3 + \omega_4}{3.0777\ \omega_5} \qquad (9.11)$$

$$\omega_o = \frac{\omega_5}{\cos \phi'} \qquad (9.12)$$

The value of ϕ' depends on how the bonds are numbered; thus, for a ring where the atoms are chemically nonequivalent (e.g., furanose), some suitable numbering convention must be adopted.[13] If the atoms are chemically equivalent (e.g., as in cyclopentane), ϕ' can be expressed as modulo 36°.

The out-of-plane deformation of the cyclopentane molecule has played an important role in the development of conformational analysis because it provided the first example of pseudorotation.[9,14] As a result of the degeneracy of the two orthogonal components of the out-of-plane deformation, the potential energy of the molecule must be independent of the phase angle ϕ, at least for infinitesimal displacements from planarity. In other words, the cyclic coordinate ϕ should run along an

[13]C. Altona and M. Sundaralingam, *J. Am. Chem. Soc.* **94,** 8205 (1972); **95,** 2333 (1973).
[14]K. S. Pitzer and W. E. Donath, *J. Am. Chem. Soc.* **81,** 3213 (1959).

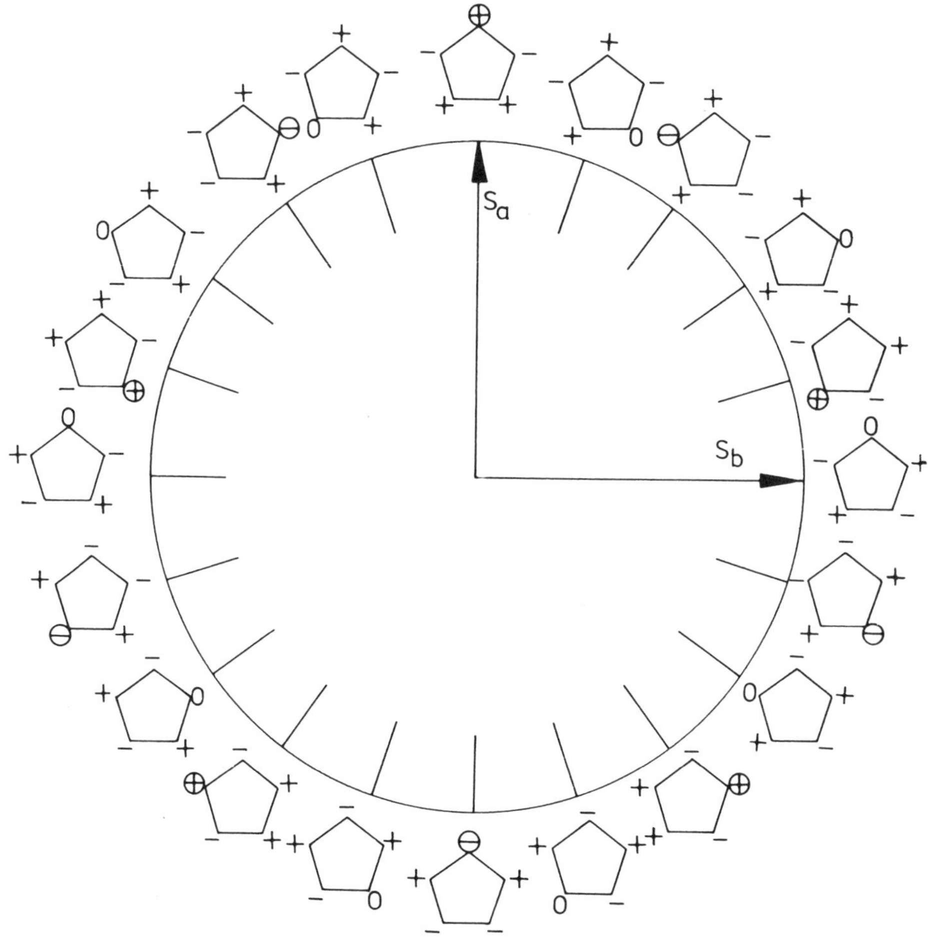

Figure 9.8. Behavior of out-of-plane displacements of the atoms as the phase angle coordinate ϕ goes around a complete cycle. Only forms with C_2 or C_s symmetry are shown; they alternate every 18°. For intermediate angles the corresponding forms have no symmetry.

essentially equienergetic path, and as it does, the molecule carries out a motion that has certain properties in common with a rotation about an axis normal to its mean plane. For example, the result of a displacement of 72° along the ϕ coordinate is the same as the result of rotating the initial conformation by −144°. The complete itinerary along the cyclic ϕ coordinate is shown in Fig. 9.8, and the corresponding pattern of changes in the torsion angles is shown in Fig. 9.9. Energy calculations[12,14,15] based on significantly different force fields yield somewhat

[15] J. B. Hendrickson, *J. Am. Chem. Soc.* **83,** 4537 (1961).

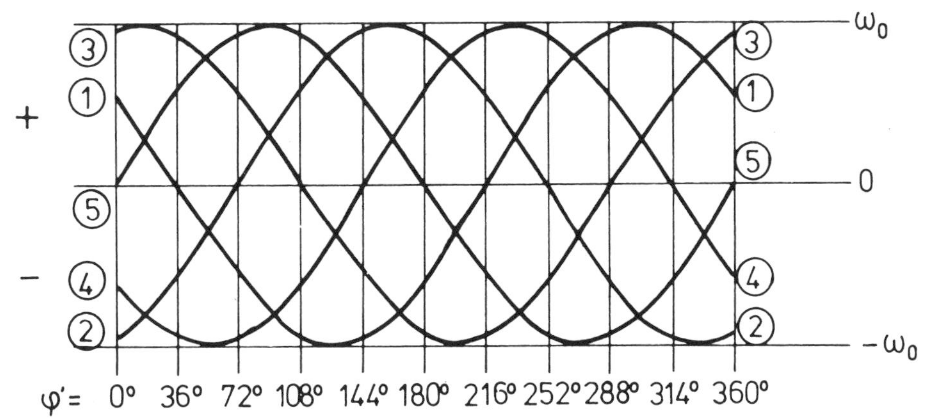

Figure 9.9. Relative values of individual torsion angles in an equilateral five-membered ring as the phase angle coordinate ϕ' passes through a complete cycle. The absolute values depend on the puckering amplitude.

different values for the puckering amplitude q (in the range 0.43–0.49 Å), but they all agree that even in this range, where q is by no means infinitesimal, the potential energy is essentially independent of ϕ. Indeed, in the calculations of Lifson and Warshel,[12] the energy along the pseudorotational path was found to be constant within 0.005 kcal/mol, and these authors comment that "the existence of an equipotential path is more due to the geometry of cyclopentane than to the exact nature of the intramolecular forces." As mentioned earlier, for the equilateral ring the sum of the bond-angle deviations from 108° and the sum of squares of the torsion angles depend only on the puckering amplitude—i.e., they are independent of ϕ, and this is also nearly true for the sum of squares of bond-angle deviations from the tetrahedral angle.

Six-Membered Rings

In addition to the rigid "chair" form of cyclohexane there is a family of flexible forms that can be changed continuously, one into another, by altering the torsion angles. The flexible forms include the "boat" form (point group C_{2v}) and the "symmetrical twist" form (point group D_2) as special cases. The rigidity of the chair form is no surprise; with six bond lengths and six bond angles fixed, the $12 = 3 \times 6 - 6$ degrees of freedom of the six-membered ring would seem to be exhausted, in which case there is no torsional degree of freedom left. What is surprising is the existence of the family of flexible forms. Models certainly suggest that the changes in torsion angles occur without concomitant changes in bond lengths or bond angles. With all six bond lengths and bond angles fixed, where does the extra degree of freedom come from? By inspecting mechanical models it is difficult to be sure that the passage from one flexible form to another does not involve a small amount of bond-angle deformation. Some such deformation must occur, for example, in a five-membered ring built from tetrahedral units, since the sum of the internal angles has to be 540° or less (at any rate less than $5 \times 109°28'$). Nevertheless, a five-membered ring can be constructed from the usual units without difficulty, and the small angle deformations that must occur here are easily overlooked.

In what follows we use the concepts of rigidity and flexibility in a mathematical rather than a mechanical sense. By rigidity we imply the presence of some functional relationship between torsion angles ω_i and the other internal parameters p_j that are kept constant; that is, in a rigid molecule we cannot alter any ω_i without changing at least one p_j, or in other words, there is at least one nonzero derivative $\partial p_j/\partial \omega_i$. By flexibility we imply that all such derivatives are zero (in a finite range of the ω_i). Even with these definitions, the family of mechanically flexible forms stays flexible.

We can obtain some insight into the problem by considering the symmetry properties of the out-of-plane distortions of a regular hexagon. The individual z coordinates (normal to the plane of the hexagon) transform like $p(z)$ atomic orbitals, and the six linear combinations that constitute an orthogonal set of symmetry coordinates have exactly the same form as the Hückel molecular orbitals of benzene.[16] This saves us some work since the MO patterns are familiar to most chemists (Fig. 9.10).

The first three sets of displacements (the bonding orbitals) do not correspond to deformations of the hexagon but to rigid-body motions; the first (A_{1u}) is a translation normal to the plane, and the next two (the degenerate E_{1g} pair) are rotations about axes that lie in the plane. For infinitesimal displacements of a regular polygon, the conditions of no net translation or rotation lead to the relationships:

$$\sum z_j = 0$$
$$\sum z_j \cos 2\pi j/n = 0 \qquad j = 1, 2, \ldots n \qquad (9.13)$$
$$\sum z_j \sin 2\pi j/n = 0$$

[16]It seems bizarre to begin a discussion of the symmetry properties of nonplanar hexagons by appealing to an analogy with the symmetry properties of molecular orbitals. Of course, it ought to be the other way around, but the MO patterns usually enter into chemistry teaching at an early stage and become familiar, while their underlying geometric basis remains relatively unknown.

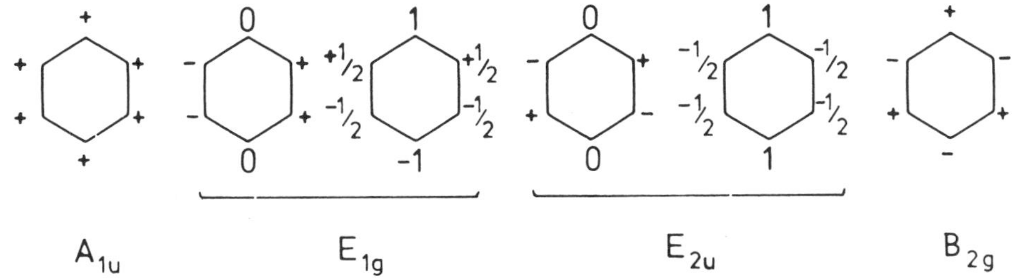

Figure 9.10. Hückel molecular orbitals of benzene. The same figures represent a set of six orthogonal linear combinations of out-of-plane displacements of the vertices of a regular hexagon. The first three combinations correspond to rigid-body movements (translation or rotation), and the second three correspond to out-of-plane deformations.

among the z coordinates (cf. Eq. 9.10 for the pentagon). More generally, these relationships fix the mean plane of the polygon whatever its shape (regular or irregular) and irrespective of which vertex is assigned index 1 in the cyclic numbering system.

The second three sets of displacements (the antibonding orbitals) correspond to actual deformations. The nondegenerate deformation (B_{2g}) can be recognized as the chair form of the hexagon with point group D_{3d}, whereas the pair of degenerate deformations (E_{2u}) shown in Fig. 9.10 correspond to the symmetrical twist (D_2) and the boat (C_{2v}) forms. The only symmetry element these two forms have in common is a dyad axis normal to the mean plane. Since this axis does not pass through any atoms or any bonds, we refer to it as a nonintersecting axis of symmetry. Now, just as in the case of the five-membered ring, any linear combination of the two E_{2u} coordinates is also a symmetry-adapted deformation coordinate transforming as E_{2u}. The two E_{2u} deformations depicted in Fig. 9.10 can be described by $z_j = \delta \sin(4\pi j/6)$ and $z_j = \delta \cos(4\pi j/6)$, respectively, an arbitrary linear combination by

$$z_j = \delta \cos(4\pi j/6 + \alpha) \qquad j = 1, 2, \ldots 6 \tag{9.14}$$

whence the extra degree of freedom would appear to be identified with the phase angle α. Values of $\alpha = 0°, 60°, 120°, \ldots$ yield the forms with C_{2v} symmetry, values of $\alpha = 30°, 90°, 150°, \ldots$ yield the forms with D_2 symmetry, and intermediate values yield forms with only C_2 symmetry. Since the torsion angles transform in exactly the same way as the out-of-plane displacements, the same considerations apply. The nondegenerate deformation transforming as B_{2g} (chair form) is not associated with any phase angle and is therefore rigid, while the presence of such a phase angle in the description of the degenerate pair of E_{2u} deformations leads to pseudorotation at least for infinitesimal deviations from planarity. For finite deviations the bond angle changes are at most second order with respect to changes in torsion angles. As in the five-membered ring, the analogous relationship

$$\omega_k = \omega_o \cos(4\pi k/6+\alpha') \tag{9.15}$$

has been shown to hold rather well even for large puckering amplitudes.[17] The pattern of torsion angle changes is shown in Fig. 9.11 as a function of the phase angle coordinate. Of course, the actual values of the torsion angles depend on the puckering amplitude, but so do the bond angles or rather the deviation of their sum from 720°. What we have not yet shown is that for finite deviations from planarity, the bond angles can be kept strictly invariant as the torsion angles change.

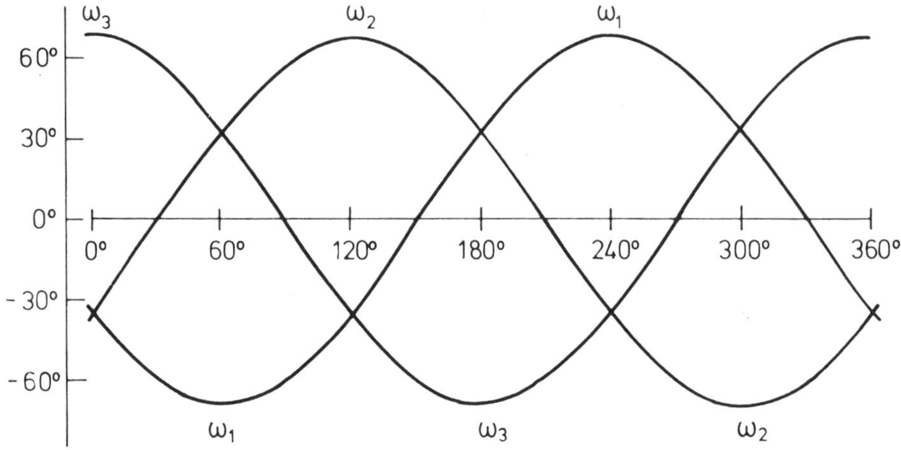

Figure 9.11. Change in individual torsion angles in an equilateral, equiangular hexagon as the phase angle coordinate α' passes through a complete cycle. The amplitude depends on the bond angle (here assumed to be 109.5°).

The exact relationship between bond angle and torsion angle in the chair form of cyclohexane is easy to derive. The projection on the mean plane is a regular hexagon, specified by a single parameter. Specification of the three-dimensional structure requires one further coordinate—the distance by which successive atoms are displaced alternately above and below the plane. Expressing the internal parameters in terms of Cartesian coordinates (Fig. 9.12), we have

$$d_{12}^2 = x^2 + 4z^2 = 1 \text{ (say)}$$

$$d_{13}^2 = 3x^2 = 2 - 2\cos\theta$$

$$d_{14}^2 = 4x^2 + 4z^2 = 3 - 2\cos\theta$$

[17]H. R. Buys and H. J. Geise, *Tetrahedron Lett.* No. 54, **1968,** 5619.

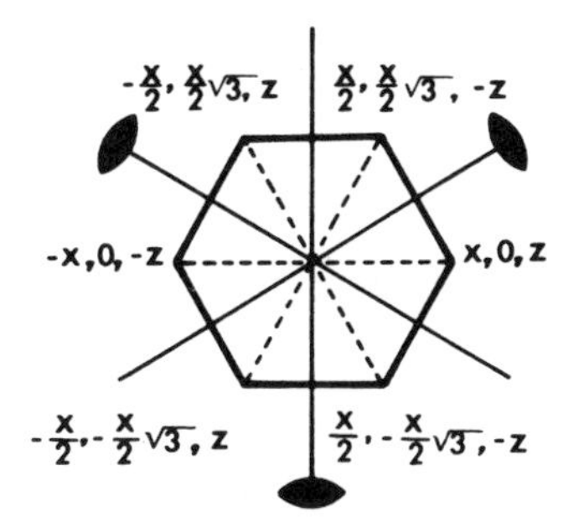

Figure 9.12. Cartesian coordinates for a six-membered ring with D_{3d} symmetry.

and from the torsion angle formula (Eq. 9.4),

$$d_{14}^2 = 3 - 4 \cos \theta + 2 \cos^2 \theta - 2 \sin^2 \theta \cos \omega$$

whence

$$\cos \omega = \frac{\cos^2 \theta - \cos \theta}{\sin^2 \theta} = \frac{-\cos \theta}{1 + \cos \theta} \tag{9.16a}$$

or in a more symmetrical form,

$$\cos \omega + \cos \omega \cos \theta + \cos \theta = 0 \tag{9.16b}$$

Hence ω cannot be changed without changing θ at the same time. The chair form is rigid; it can only be deformed by a change in the bond lengths or bond angles. An exact analysis of this kind for the members of the "flexible" family leads to an unmanageably cumbersome mess of trigonometric equations. Only for the special cases of equilateral equiangular rings with C_{2v} or D_2 symmetry are relatively simple relationships obtained among torsion angles and bond angles; however, these rings may change their torsion angles by pseudorotation, which is not possible for the chair form with its D_{3d} symmetry.

Thus far our discussion of nonplanar hexagons has been limited to the equilateral, equiangular case. What about six-membered rings where the bond lengths and bond angles are fixed but not necessarily equal? The discussion of a special case may illustrate how certain problems of this kind can be treated.[18] Consider a six-membered ring in which opposite distances and angles are equal in pairs; the three independent bond distances d_1, d_2, d_3 and the three independent angles θ_1, θ_2, θ_3 can be arranged as in Fig. 9.13. From the torsion angle formula (Eq. 9.1) it is clear that $\cos \omega_{BC} = \cos \omega_{EF}$, $\cos \omega_{CD} = \cos \omega_{FA}$, and $\cos \omega_{AB} = \cos \omega_{DE}$. Thus opposite pairs of torsion angles have the same magnitude. With opposite distances and angles equal in pairs and opposite torsion angles

[18] J. D. Dunitz and J. Waser, *J. Am. Chem. Soc.* **94**, 5645 (1972).

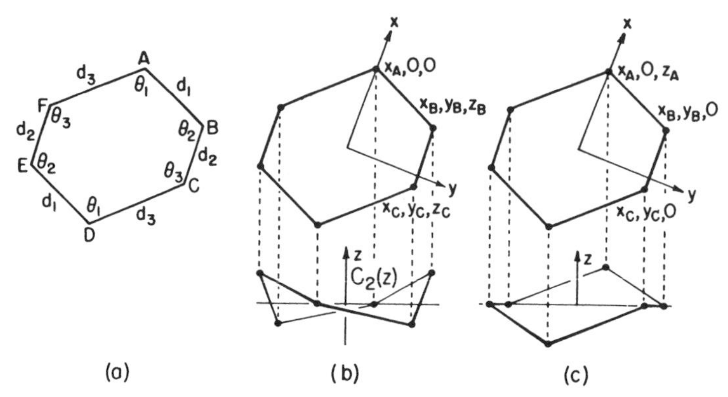

Figure 9.13. The six-membered ring with bond distances d_1, d_2, and d_3 and angles θ_1, θ_2, and θ_3 as shown on left (a) must have either a nonintersecting twofold axis $C_2(z)$ (b) or a center of symmetry (c). The former needs seven Cartesian coordinates, and the latter needs only six since the plane defined by the center and two atoms (say B and C) also passes through the two symmetry-related atoms—i.e., BCEF are coplanar with the center, defining the xy plane. The xz plane can be passed through A. From Dunitz and Waser, *J. Am. Chem. Soc.* **94,** 5645 (1972). Reprinted by permission.

of equal magnitude, the ring must have either a nonintersecting twofold rotation axis (signs of related ω angles equal) or a center of inversion (signs of related ω angles opposite). In the former case (C_2 symmetry) seven Cartesian coordinates have to be fixed to define the relative atomic positions, in the latter case only six (Fig. 9.13). With six internal parameters to be fixed, the first solution leaves one internal parameter that can be assigned an arbitrary value within certain ranges and thus corresponds to a flexible ring. In the second solution, there is no free parameter left once the bond distances and angles are fixed, so the ring is rigid. These conclusions are obviously valid for the equilateral, equiangular ring ($d_1 = d_2 = d_3$, $\theta_1 = \theta_2 = \theta_3$) except that the symmetry of the rigid form is raised from C_i to D_{3d}; the symmetry of the flexible form is only raised (from C_2 to C_{2v} or D_2) for special values of the freely adjustable parameter.

Six-Membered Rings and Octahedra

A theorem due to Cauchy states that if all external faces of a convex polyhedron are rigid then the polyhedron itself is rigid.[19] For deltahedra (polyhedra with exclusively triangular faces and the only kind we are concerned with here) a simple derivation can be made from the well-

[19]See A. D. Alexandrow, *Konvexe Polyeder,* Akademie Verlag, Berlin, 1958, pp. 112–113; L. A. Lynsternik, *Convex Figures and Polyhedra,* Dover, New York, 1963, pp. 60–66.

known Euler relationship between the number of vertices, edges, and faces of a convex polyhedron:

$$V - E + F = 2$$

or

$$F = E - V + 2$$

Each triangular face has three edges, and each edge is common to two faces. Thus if all faces are triangular, their number is

$$F = 2E/3$$

Equating the two expressions for F, we obtain

$$E = 3V - 6$$

which is just the number of degrees of freedom required to fix the relative positions of the V vertices. Hence if the lengths of the edges of a convex deltahedron are fixed, so are the relative positions of its vertices.

Instead of describing the six-membered ring in terms of its bond lengths and bond angles, we can replace the latter by distances between next nearest neighbors. The resulting arrangement of triangles framed by 1,2 and 1,3 distances is an octahedron that is convex if the ring is in the chair form (Fig. 9.14). By Cauchy's theorem, the ring is rigid once the edges are fixed. Hence, the chair forms of all six-membered rings are rigid, regardless of the presence or absence of symmetry, if the bond lengths and bond angles are fixed.

If the ring is not in the chair form, the octahedron formed by the twelve 1,2 and 1,3 distances is not convex (Fig. 9.14)—i.e., it does not entirely contain all segments that connect points on its boundary. Cauchy's theorem leaves the question of the rigidity or nonrigidity of nonconvex polyhedra open. Bricard showed that the nonconvex octahedron with fixed faces is rigid in general but that there are certain special types that are flexible.[20] In particular, he showed that a nonconvex octahedron in which opposite edges are related by a nonintersecting twofold axis is flexible. We have already derived the flexibility of this kind of figure by counting degrees of freedom.

A model of a ring with an approximate twofold axis would appear to be flexible even if it does not strictly satisfy the symmetry criterion. The derivatives $\partial\theta_i/\partial\omega_j$ would not be zero, but they would be so small

[20] R. Bricard, *J. Math. Pure Appl.* (5) **3,** 113 (1897); see also G. T. Bennett, *Proc. London Math Soc.* (2) **10,** 309 (1911).

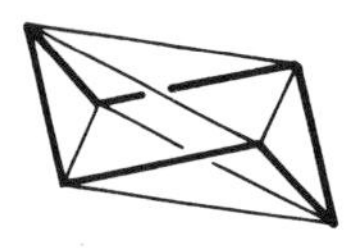

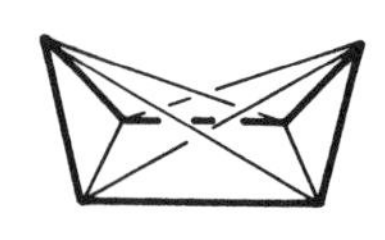

Figure 9.14. Convex (left) and nonconvex (right) octahedra. All edges of the triangular faces are 1,2 or 1,3 distances of nonplanar six-membered rings. From Dunitz and Waser, *J. Am. Chem. Soc.* **94,** 5645 (1972). Reprinted by permission.

that they would be virtually undetectable; this illustrates the distinction referred to earlier between mathematical and mechanical flexibility.

Other Approaches to the Flexibility Problem

Several other methods of dealing with the problem of the flexibility of six-membered rings with fixed sides and angles have been discussed in the literature. A mathematical analysis of the rigidity of the chair form of cyclohexane and the flexibility of the boat-twist family was given by Sachse,[21] who was the first to recognize that the cyclohexane molecule could exist in these nonplanar forms. His ideas were regarded as heretical at the time since they went against Baeyer's edict that the carbon skeletons of all cycloalkanes were planar.[22] However, despite the originality and correctness of Sachse's views, few modern readers will have the patience to work through the details of his somewhat tedious mathematical analysis. The problem was only taken up again many years later, when Hazebroek and Oosterhoff analyzed the equilateral, equiangular hexagon by vector algebraic methods.[23] Oosterhoff[24] also treated the ring with one pair of opposite angles equal but unequal to the other four—e.g., cyclohexane-1,4-dione—and suggested that this molecule probably occurs at least partly in the flexible form to account for its nonzero dipole moment. Lauwerier[25] proposed that the necessary and sufficient condition for the existence of a flexible form of a nonplanar hexagon with 12 fixed distances is the equality of opposite elements—that is, he rediscovered some of Bricard's results[20] but did not recognize that the criterion, although sufficient, is not necessary. There are other kinds of nonconvex octahedra that do not possess a nonintersecting twofold axis but are flexible, as shown by Bricard[20] and by Bottema[26]

[21] H. Sachse, *Ber.* **23,** 1363 (1890); *Z. Physik. Chem.* **10,** 203 (1892). The second paper contains the detailed mathematical analysis.

[22] A. Baeyer, *Ber.* **18,** 2268 (1885).

[23] P. Hazebroek and L. J. Oosterhoff, *Discuss. Faraday Soc.* **10,** 87 (1951).

[24] L. J. Oosterhoff, Doctoral Dissertation, Leiden, 1949.

[25] H. A. Lauwerier, *Proc. Konikl. Ned. Akad. Weten.* **69,** 330 (1966).

[26] O. Bottema, *Proc. Konikl. Ned. Akad. Weten.* **70,** 151 (1967).

in a critique of Lauwerier's paper. These papers are essentially algebraic. In most recent work on the subject, the emphasis is on energy calculations rather than the purely geometric aspects. However, as mentioned earlier, the two are not unconnected, and the existence of families of isoenergetic conformations can often be inferred from geometric arguments. The symmetry aspects of this kind of problem are dealt with by Pickett and Strauss.[27]

Larger Rings

If only the bond distances and bond angles are fixed, all rings with seven or more members are flexible, as shown by counting degrees of freedom (N bonds, N angles, and $N-6$ remaining degrees of freedom). If one torsion angle in a seven-membered ring is fixed as well as the bond distances and angles, then the ring is rigid. The remaining torsion angles are functions of the 15 assigned parameters though not single-valued functions. It is interesting that the polyhedron with edges formed by 1,2 and 1,3 distances of an odd-membered ring is bounded by a Möbius strip. This is shown for the seven-membered ring in Fig. 9.15.

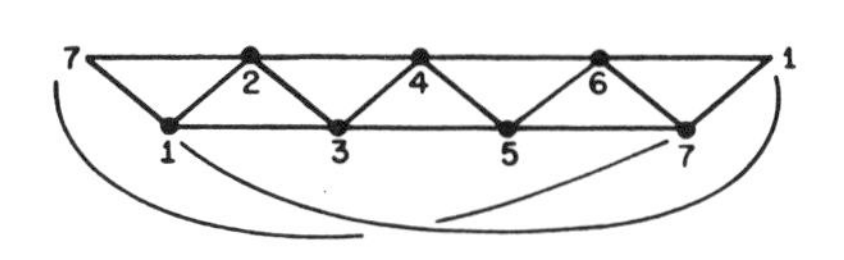

Figure 9.15. Joining the ends of the ribbon of triangles to give a seven-membered ring produces a Möbius strip. The drawing illustrates the case of equal distances and equal angles. From Dunitz and Waser, *J. Am. Chem. Soc.* **94,** 5645 (1972). Reprinted by permission.

The flexibility of the eight-membered ring with given bond distances and angles is generally frozen by assigning fixed values to two torsion angles, but there are special cases where such rings are flexible. For example, of the two conformations of cycloocta-1,5-diene shown in Fig. 9.16, the one on the left is rigid while the one on the right is flexible (it is assumed that the torsion angles about the double bonds are fixed at zero). This result is derived most easily from the known flexibility properties of six-membered rings. The planar quadrilateral ABCH can be replaced conceptually by the triangle PCH; if the edges of the quadrilateral are fixed, so are those of the triangle and similarly for the other quadrilateral GDEF. The figure PCDQGH is a hexagon with fixed sides and angles, and in one case it is in the chair form and hence rigid.

[27]H. M. Pickett and H. L. Strauss, *J. Chem. Phys.* **55,** 324 (1971).

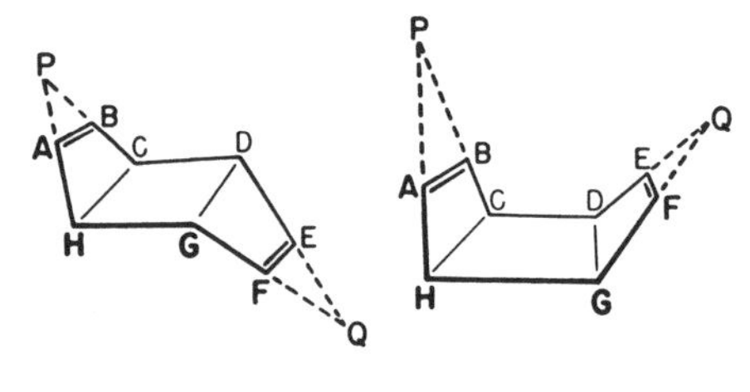

Figure 9.16. Two forms of eight-membered rings with two opposite torsion angles fixed at zero. From Dunitz and Waser, *J. Am. Chem. Soc.* **94,** 5645 (1972). Reprinted by permission.

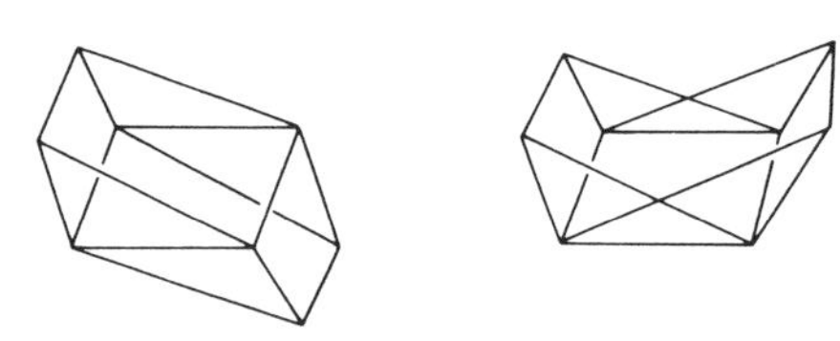

Figure 9.17. Convex and nonconvex polyhedra corresponding to rigid and flexible forms of eight-membered rings with fixed bond lengths and bond angles and with two opposite torsion angles fixed at zero.

However, if the octagon on the right has a nonintersecting twofold axis, so does the resulting hexagon, which is therefore flexible, in which case the octagon must also be flexible. These properties are readily seen by examining Dreiding models of cycloocta-1,5-diene. Note also that if the quadrilateral faces of ABCH and EFGD are regarded as fixed, then the rigid form corresponds to a convex polyhedron (an octahedron, $V = 8$, $E = 14$, and $F = 8$)[28] with fixed faces, while the flexible form again gives a nonconvex polyhedron (Fig. 9.17).

Bicyclic Systems

The bicyclo[1.1.1.]pentane skeleton provides a striking example of the effect of constraints imposed by bond lengths on the overall geometry of a small bicyclic molecule. The six chemically equivalent bonds can be assumed to be equal, and if chemically equivalent angles are also taken as equal, then the skeleton has D_{3h} symmetry, and the relative atomic positions are defined by only two parameters. We choose these as a and b (Fig. 9.18), to make the bond distance equal to $(a^2 + b^2)^{\frac{1}{2}}$; the nonbonded distances are then equal to $2b$ and $\sqrt{3}a$, the former occurring once and the latter three times. Fixing the bond distance thus introduces an equation of constraint between the two kinds of nonbonded distances:

$$4d^2\,(\mathrm{A}\cdots\mathrm{A}) + 3d^2\,(\mathrm{B}\cdots\mathrm{B}) = 12d^2\,(\mathrm{AB})$$

A part of this function is shown in Fig. 9.18 for $d(\mathrm{AB}) = 1.56$ Å. With the bond distance fixed at this value it is not possible for both kinds of

[28]There are 257 convex polyhedra with eight vertices. The one in question is number 236 in the catalogue given by D. Britton and J. D. Dunitz, *Acta Crystallogr.* **A29,** 362 (1973).

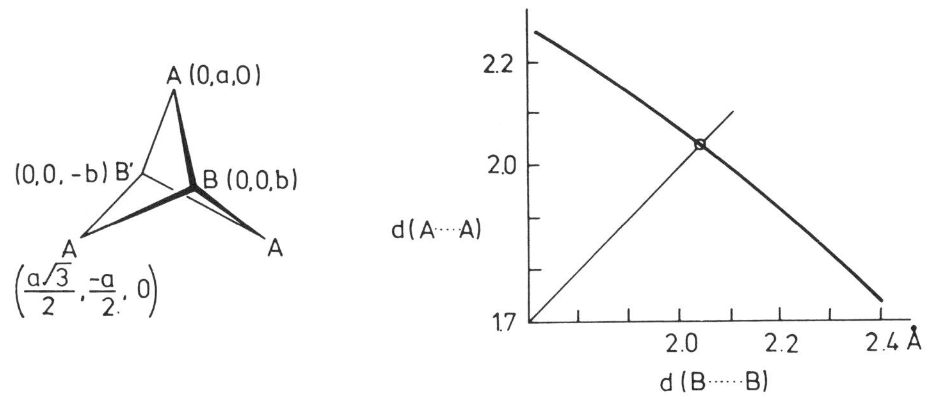

Figure 9.18. Bicyclo[1.1.1]pentane skeleton. Left: choice of coordinates. Right: relationship between nonbonded distances $d(\mathrm{A}\cdots\mathrm{A})$ and $d(\mathrm{B}\cdots\mathrm{B})$ for bond distance $d(\mathrm{AB}) = 1.56$ Å.

nonbonded distances to be greater than 2.043 Å. If one distance is greater than this value, the other is smaller. This is as far as we can go with geometric arguments alone, ignoring energy considerations altogether. According to a crystal structure analysis of a bicyclopentane derivative,[29] $d(\mathrm{B}\cdots\mathrm{B})$ is 1.89 Å, which is probably the shortest distance ever found between nonbonded carbon atoms; the A···A distances between carbons of methylene groups are about 2.15Å, slightly less than the value in cyclobutane itself. Obviously any relief of steric strain produced by increasing $d(\mathrm{B}\cdots\mathrm{B})$ beyond 1.89 Å must be balanced by the additional steric repulsion between the methylene groups.

The bicyclo[2.2.2]octane (BCO) molecule was a subject of controversy for several years. Some authors claimed that the D_{3h} conformation must correspond to a potential energy minimum; others claimed that a lowering of the symmetry to D_3 to avoid eclipsing of the three —CH_2CH_2— bonds must be associated with an energy stabilization, in which case the D_{3h} conformation would correspond to the transition state between the two D_3 enantiomers. A geometric analysis, together with some simple energy considerations, shows that neither the D_{3h} nor the D_3 conformation can correspond to a very pronounced energy minimum; rather, the energy profile associated with the proposed twisting motion of the molecule must be quite flat for considerable excursions from D_{3h} symmetry.[30,31]

The twisted D_3 skeleton is completely defined by fixing the values of

[29] A. Padwa, E. Shefter, and E. Alexander, *J. Am. Chem. Soc.* **90,** 3717 (1968).
[30] O. Ermer and J. D. Dunitz, *Helv. Chim. Acta* **52,** 1861 (1969); this paper includes an account of the controversy mentioned above.
[31] A. Yokozeki, K. Kuchitsu, and Y. Morino, *Bull. Chem. Soc. Japan* **43,** 2017 (1970).

four geometric parameters. These may be chosen as the Cartesian coordinates indicated in Fig. 9.19 or as a set of four internal parameters—e.g., the two symmetry-independent bond lengths plus one bond angle and one torsion angle. Since we are only interested in the twisting motion, the two bond lengths can be treated as constants, and for simplicity they can be regarded as equal. The geometry of the D_3 skeleton then depends on only two parameters; these are reduced to one if the symmetry is raised to D_{3h}.

With the bond lengths set to unity, the following relationships between the Cartesian coordinates and the internal parameters are obvious from Fig. 9.19:

$$x_2^2 + y_2^2 + (z_2 - z_1)^2 = 1$$

$$4y_2^2 + 4z_2^2 = 1$$

$$x_2^2 + y_2^2 + (z_1 + z_2)^2 = 2(1 - \cos\theta_2)$$

$$z_1^2 = \frac{1}{4}(3 - 4\cos\theta_2 + 2\cos^2\theta_2 - 2\sin^2\theta_2\cos\omega_2)$$

Given the values of θ_2 and ω_2, z_1 can be calculated immediately, and the other Cartesian coordinates are then obtained by substitution:

$$z_2 = (1 - 2\cos\theta_2)/4z_1$$

$$y_2 = \frac{1}{4}[4z_1^2 - (1 - 2\cos\theta_2)^2]^{\frac{1}{2}}$$

$$x_2 = \frac{\sqrt{2}}{2}\sin\theta_2\,(1 + \cos\omega_2)^{\frac{1}{2}}$$

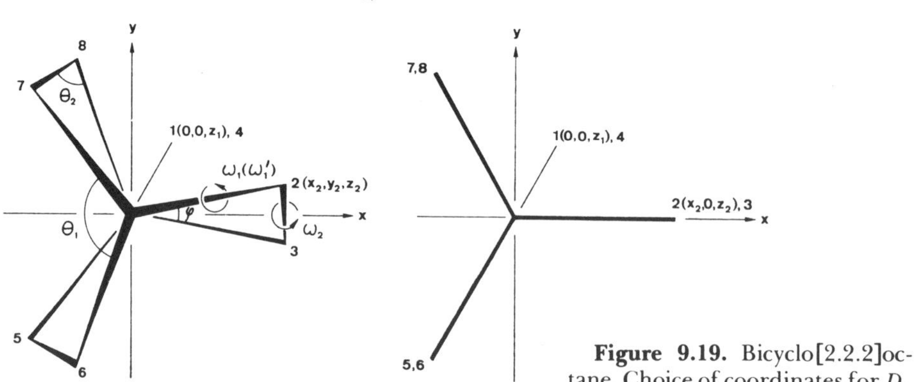

Figure 9.19. Bicyclo[2.2.2]octane. Choice of coordinates for D_3 (left) and D_{3h} (right) conformations. From Ermer and Dunitz, *Helv. Chim. Acta* **52,** 1861 (1969).

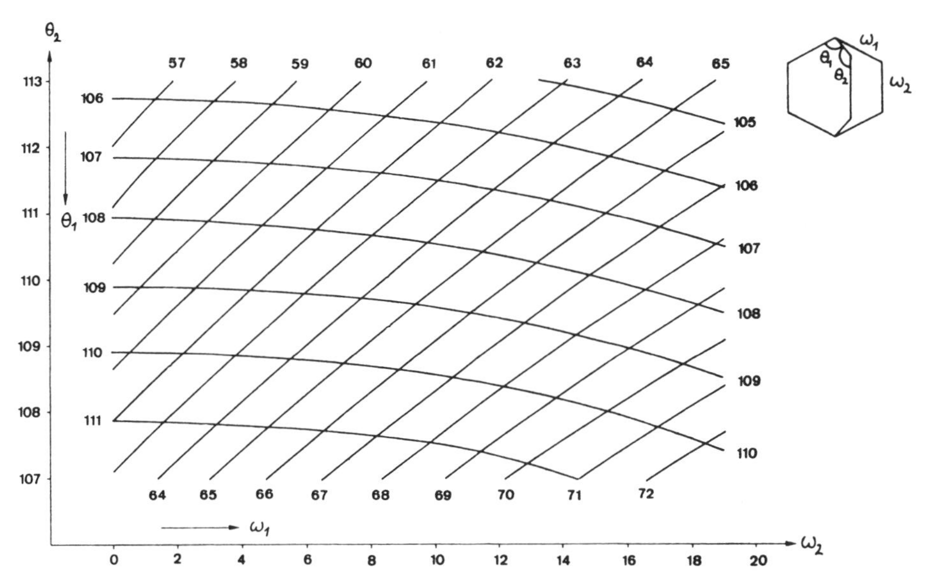

Figure 9.20. Interdependence of the four parameters θ_1, θ_2, ω_1, and ω_2. From Ermer and Dunitz, *Helv. Chim. Acta.* **52,** 1861 (1969).

Fig. 9.20 shows how the four internal parameters ω_1, ω_2, θ_1 and θ_2 are interrelated. For the D_{3h} structure with $\theta_1 = \theta_2 = 109.5°$, $\omega_1 = 60°$, $\omega_2 = 0°$. For the corresponding D_3 structure, ω_2 differs from zero, but also ω_1 differs from 60°. Since the torsion angles are not independent, any energy stabilization arising from relief of eclipsing one set of C—C bonds must be offset against an energy increase caused by less perfect staggering of the other set of bonds. The relevant part of a calculated two-dimensional potential energy surface, based on a simple force field,[30] is shown in Fig. 9.21. It turns out that the main feature, the long, flat valley extending along the ω_2 coordinate, is relatively insensitive to the details of the force field used; very similar results were obtained with several force fields.

Fig. 9.21 shows a long, narrow valley with a shallow minimum at $\omega_2 = 12.5°$; the energy barrier at the D_{3h} conformation is only about 120 cal/mol (42 cm^{-1}). Instead of the torsion angle ω_2, the twist angle ϕ (Fig. 9.19) is often used as a measure of the deviation from D_{3h} symmetry; for approximately tetrahedral bond angles, $\phi \sim \frac{3}{5}\,\omega_2$. An electron-diffraction analysis of gaseous BCO[31] yields a quasi-D_{3h} molecule with a root-mean-square twisting amplitude $\langle\phi^2\rangle^{\frac{1}{2}}$ of about 7° and an estimated barrier height of 35 cm^{-1}. Similar values are obtained for the 1-aza and

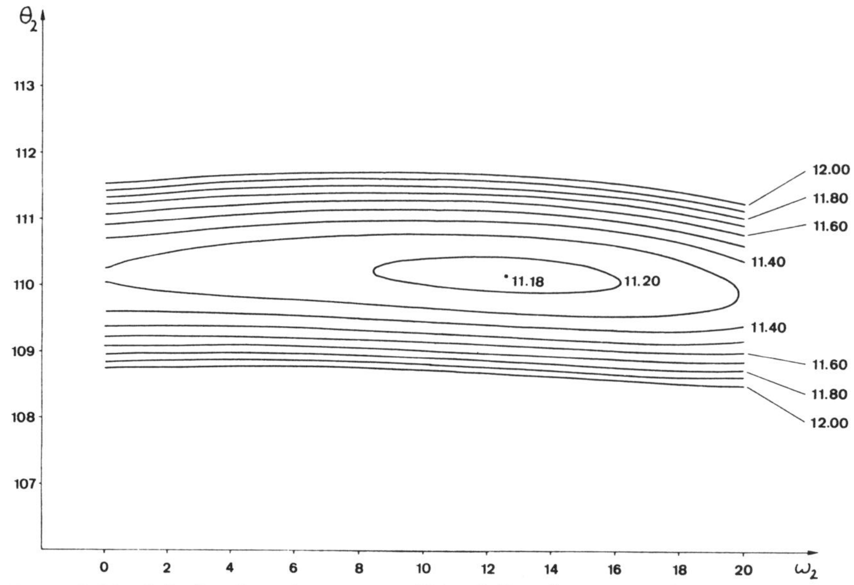

Figure 9.21. Calculated strain energy of bicyclo[2.2.2]octane as function of θ_2 and ω_2. From Ermer and Dunitz, *Helv. Chim. Acta* **52,** 1861 (1969).

1,4-diaza derivatives.[32,33] We can therefore conclude that the flat valley in the energy function is more due to the geometry of the bicyclo-[2.2.2]octane skeleton than to the exact nature of the intermolecular forces.

Polycyclic Molecules with High Symmetry

Let us consider the constraints that operate in two highly symmetrical polycyclic molecules. As in the previous examples, we regard the bond distances as fixed, and for simplicity we set them all equal to unity. This is enough to fix the torsion angles (i.e., the molecules are rigid in the sense defined earlier) and to impose very severe constraints on the bond angles.

For our first example we choose the adamantane molecule (Fig. 9.22), where the four methine carbons form a regular tetrahedron and the six methylene carbons form a regular octahedron. The relative positions of the ten carbon atoms are then fixed by only two parameters, x and u (Fig. 9.22). If the bond angles at the methine and methylene carbons are θ

[32]E. Hirota and S. Suenaga, *J. Mol. Spectrosc.* **42,** 127 (1972).
[33]A. Yokozeki and K. Kuchitsu, *Bull. Chem. Soc. Japan.* **44,** 72 (1971).

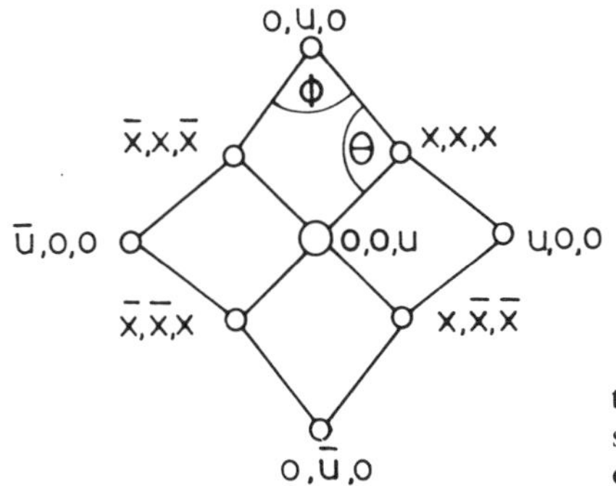

Figure 9.22. Three representations of the adamantane skeleton, which can be described in terms of two independent parameters.

and ϕ, respectively, we have

$$2u^2 = 2 - 2\cos\theta \qquad u = \sqrt{2}\sin(\theta/2)$$

$$8x^2 = 2 - 2\cos\theta \qquad x = \frac{1}{\sqrt{2}}\sin(\phi/2)$$

but setting the bond distances equal to unity means that

$$(x - u)^2 + 2x^2 = 1$$

or

$$3x^2 + u^2 - 2ux = 1$$

Thus the bond angles are related by

$$3\sin^2(\phi/2) + 4\sin^2(\theta/2) - 4\sin(\phi/2)\sin(\theta/2) = 2$$

If one bond angle is tetrahedral ($\cos\theta = -1/3$), so is the other; if one is larger, the other is smaller.

Our second example concerns the tetraasterane molecule (Fig. 9.23), where the twelve carbon atoms form a frame with D_{4h} symmetry. With

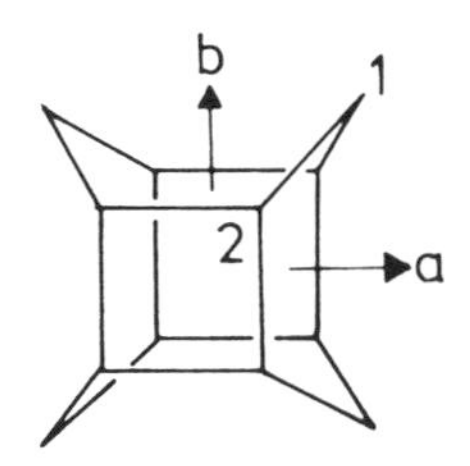

Figure 9.23. The tetraasterane skeleton can be described in terms of three independent parameters if D_{4h} symmetry is assumed —e.g., atom 1 at x, x, 0, etc., and atom 2 at u, u, z, etc.

respect to the coordinate system shown, the four atoms of type 1 occupy the special positions (x, x, 0, etc.), and the eight atoms of type 2 occupy the special positions (u, u, z, etc.), so that we need only three parameters to define the 12-atom frame. If the two symmetry-independent bond distances, $d(1-2)$ and $d(2-2')$, are set equal to unity, then

$$2(x-u)^2 + z^2 = 1$$

$$u = \frac{1}{2}$$

hence

$$2x^2 - 2x + z^2 = \frac{1}{2}$$

There are two symmetry-independent bond angles, but they must be related by an equation of constraint, as in the previous example. It is easily seen that

$$z = \sin(\theta_1/2)$$

$$x = \frac{1}{2} - \cos\theta_2$$

leading to

$$\sqrt{2}\cos\theta_2 = -\cos(\theta_1/2)$$

The two angles become equal for $\cos\theta = (1 - \sqrt{17})/8$ or $\theta = 112.98°$. The observed angles in tetramethyltetraasterane[34] are $\theta_1 = 111.5°$ and $\theta_2 = 113.4°$, which satisfy the above equation. Thus the angle widening with respect to the tetrahedral angle (109.47°) is a geometric property of the polycyclic molecule and needs no special explanation in terms of specific intramolecular forces. The tetramethyltetraasterane molecule was mentioned in Chapter 5 as one where observed vibrational parameters are in almost perfect agreement with values calculated from a rigid-body model.

[34] J. P. Chesick, J. D. Dunitz, U. v. Gizycki, and H. Musso, *Chem. Ber.* **106,** 150 (1973).

10. Conformational Maps and Space Groups

"Somehow it seems to fill my head with ideas—only I don't exactly know what they are! However, somebody killed something: that's clear, at any rate."

One-Dimensional Torsional Potential Functions and Line Groups

Fig. 10.1 represents the potential energy of the ethane molecule as a function of the torsion angle about the C—C bond, and it must be well known to nearly all chemists. The exact nature of the potential is irrelevant for our present purpose. Here we are interested only in its symmetry properties. The abscissa shows the torsion angle ω; as ω runs from 0° to 360°, corresponding to a complete revolution of one methyl group with respect to the other, a given potential energy $V(\omega)$ occurs six times at the following torsion angles: ω, $120° \pm \omega$, $240° \pm \omega$, and $360° - \omega$. Because of the C_{3v} symmetry of the methyl groups, these six torsion angles correspond to isometric, isoergonic structures.

It is useful to think of the infinite set of structures generated by varying the cyclic parameter ω in terms of a one-dimensional periodic

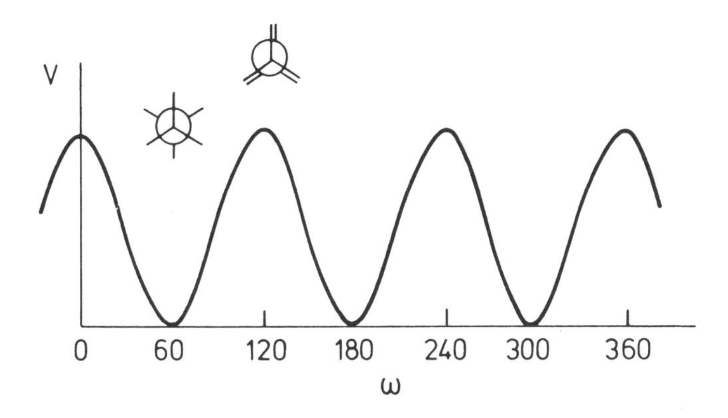

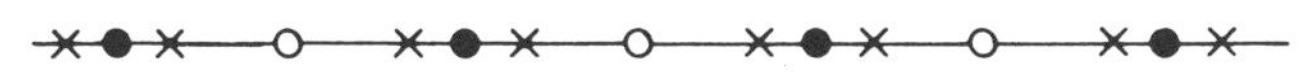

Figure 10.1. Potential energy V as function of torsion angle ω in ethane. The underlying lattice and line group ($\mathfrak{p}m$) are indicated below.

group. There is only a single one-dimensional lattice consisting of a row of equally spaced points, and there are only two one-dimensional space groups or line groups (designated $\mathfrak{p}1$ and $\mathfrak{p}m$)—the pattern may be either unsymmetrical or symmetrical about the lattice points. The pattern shown in Fig. 10.1 clearly belongs to $\mathfrak{p}m$ with lattice translation $a = 120°$. The unit cell contains two equivalent "general positions" at $\pm\,\omega$ (modulo a) as well as two nonequivalent "special positions" at the symmetry points $\omega = 0$ (modulo a) and 60° (modulo a). The asymmetric unit of the one-dimensional parameter space consists of the region $0 \leq \omega \leq 60°$, and all points within this region represent distinct (i.e., nonisometric) structures. There is an obvious correspondence between the special positions of the line group and the special symmetry of the structures they represent; a molecule with an arbitrary torsion angle has D_3 symmetry, and molecules represented by special positions have the higher symmetry D_{3h} ($\omega = 0$) or D_{3d} ($\omega = 60°$).

A similar analysis can be made for ethyl chloride (CH_3—CH_2Cl) or 1,1-dichloroethane (CH_3—$CHCl_2$) containing one group with C_{3v} symmetry and the other with C_s symmetry. The unit cell still has the lattice translation $a = 120°$, the line group is again $\mathfrak{p}m$ but the molecular symmetries corresponding to general and special positions of the line group are lower than for ethane. Now, molecules represented by the general positions have no symmetry, while those represented by the special positions $\omega = 0°$ and $\omega = 60°$ have mirror symmetry (Fig. 10.2). We may think of the mirror-symmetric CH_2Cl unit as a fixed frame bearing an axis about which the CH_3 group can be rotated to give six isometric arrangements within a complete cycle. For each of these arrangements, there are two possible orientations of the —CH_2Cl frame, related by reflection across the local mirror plane, giving $6 \times 2 = 12$ isometric arrangements. The symmetry group of the molecule,[1] its internal isometric group[2] (the group of mappings of the molecule on to itself), is thus of order 12. In the same way the isometric group of ethane is of order $6 \times 6 = 36$ since either of the two —CH_3 groups (C_{3v} symmetry, six isometric arrangements) can be regarded as the rigid frame on which the other rotates.

For a molecule of the type CH_3—$CHXY$, the 120° periodicity in the cyclic coordinate ω is retained, but the line group is reduced to $\mathfrak{p}1$, with just one general position and no special positions. Structures represented by ω, $\omega + 120°$ and $\omega + 240°$ are isometric but have no special

[1] H. C. Longuet-Higgins, *Mol. Phys.* **6,** 445 (1963).
[2] A. Bauder, R. Meyer, and H. H. Günthard, *Mol. Phys.* **28,** 1305 (1974).

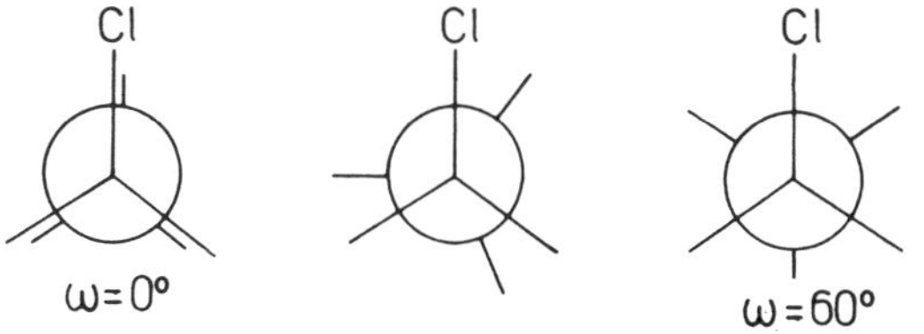

Figure 10.2. Ethyl chloride. The structures with a torsion angle of 0 (modulo 120°) and 60° (modulo 120°) have C_s symmetry; other structures have no symmetry.

symmetry. Consequently, the torsional potential function is no longer mirror symmetric about the positions $\omega = 0$ (modulo 120°) and 60° (modulo 120°). This is shown in Fig. 10.3. Apart from its 120° periodicity, the shape of the function shown has no special significance; this applies also to the positions of its turning points (maxima and minima), which are no longer determined by symmetry as in the previous examples. For any specific molecule of this type, the actual positions of the turning points would have to be determined, by theoretical calculation or by some suitable experiment, although this would not be easy. The corresponding potential for the enantiomer would then be obtained by reflection at $\omega = 0$, i.e., by reversing the sign of the torsion angle ω. The order of the internal isometric group is thus 6.

As a further example of molecules with a single torsional angle parameter we consider 1,2-dichloroethane. Fig. 10.4 shows the type of potential function involved (the exact details are irrelevant). The unit cell extends now over a complete cycle in ω, and the line group is $\mathfrak{p}m$ with general positions $\pm\, \omega$ (modulo 360°) and special positions at $\omega = 0$ (modulo 360°) and 180° (modulo 360°). Molecules represented by general positions have C_2 symmetry, while those represented by the special positions have the higher symmetry C_{2v} ($\omega = 0$) or C_{2h} ($\omega = 180°$). As in

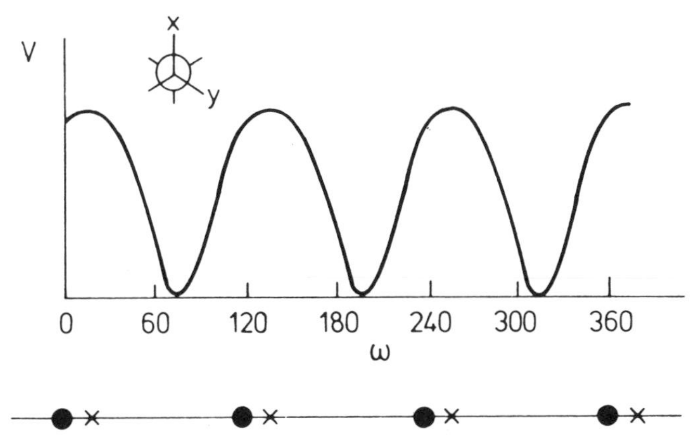

Figure 10.3. Potential energy V as function of torsion angle ω in CH_3—CHXY. The underlying lattice and line group are indicated below.

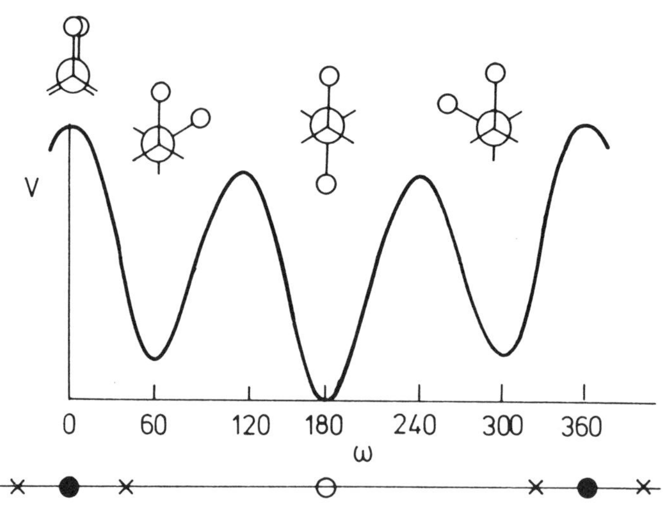

Figure 10.4. Potential energy V as function of torsion angle ω in 1,2-dichloroethane. The underlying lattice and line group ($\mathfrak{p}m$) are indicated below.

the ethane case, the two types of special position are not equivalent. Of course, the results obtained so far are rather trivial. However, the extension of these ideas to two and more dimensions leads to more interesting consequences.

Functions of the kind shown in Fig. 10.4 are sometimes used to describe the potential energy associated with torsion about the central C—C bond of molecules like *n*-butane or ethylene glycol. This is an oversimplification because it ignores the potential energy variation associated with changes in the other torsion angles. A molecule like *n*-butane possesses three independent torsional degrees of freedom, and its potential energy is a function of all three. As we shall see, the symmetry properties of the potential energy hypersurface can be described very conveniently and vividly in terms of a three-dimensional space group whose general positions correspond to a set of isometric structures, and whose special positions correspond to structures with higher symmetry. Before proceeding to problems involving three or more independent parameters, the main concepts are introduced with reference to two-dimensional plane groups.

Two Torsional Degrees of Freedom

An Example: diphenylmethane

Diphenylmethane is our first example of a molecule with two torsional degrees of freedom. Structure **1** of Fig. 10.5 shows an arbitrary confor-

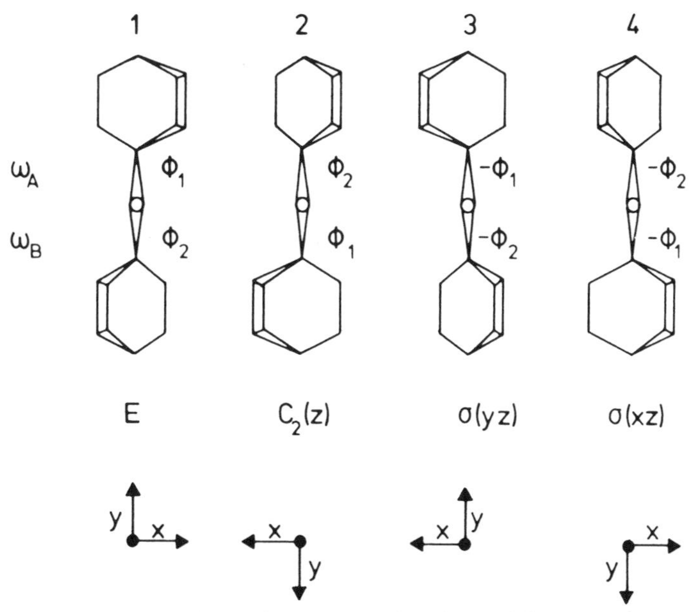

Figure 10.5. Isometric conformations **2**, **3**, and **4** obtained from **1** by appropriate rotations of the phenyl groups or by applying the symmetry operations that leave the central frame (point group C_{2v}) invariant.

mation of this molecule, designated by two torsion angles $\omega_A = \phi_1$, $\omega_B = \phi_2$ (in what follows, A always refers to the upper phenyl group and B to the lower one, and ϕ_1, ϕ_2 are the numerical values of the torsion angles). It is convenient to regard the three central atoms as a rigid frame (having C_{2v} symmetry) on which the phenyl groups may be rotated. Zero torsion angle is taken when the phenyl group in question is coplanar with the central frame, and increasing ω corresponds to a clockwise rotation of the phenyl group, viewed along the bond from the central atom to the group. Although an arbitrary conformation of the molecule has no symmetry, it is useful to imagine the C_{2v} symmetry central frame to be locked in a coordinate system with appropriately chosen axes—e.g., z along the twofold axis of the frame and x normal to its plane.

The conformations **2**, **3**, and **4** (Fig. 10.5) can be obtained from **1** by appropriate rotations of the phenyl groups. The corresponding torsion angles are:

	1	**2**	**3**	**4**
ω_A:	ϕ_1	ϕ_2	$-\phi_1$	$-\phi_2$
ω_B:	ϕ_2	ϕ_1	$-\phi_2$	$-\phi_1$

Alternatively, transformation **1** → **2** can be regarded as a rotation of the entire molecule about the z-axis, but it is not necessary to think of this rotation as a physical operation that is actually performed on the molecule—it is no more than a change of axes: $x' = -x$, $y' = -y$, and $z' = z$. In other words, conformation **2**, designated by the torsion angles (ϕ_2, ϕ_1) is entirely equivalent to conformation **1** with torsion angles (ϕ_1, ϕ_2) only the directions of the x- and y-axes are reversed.

Similarly, transformation **1** → **3** corresponds to a reflection of the entire molecule across the vertical mirror plane of the frame—i.e., to the change of axes; $x' = -x$, $y' = y$, and $z' = z$. Conformation **3** is thus the enantiomorph of **1**. Finally, transformation **1** → **4** corresponds to a reflection of **1** across the horizontal mirror plane of the frame—i.e., to the change of axes: $x' = x$, $y' = -y$, and $z' = z$. Thus, each symmetry element of the frame, operating on an arbitrary conformation **1**, leads to a torsion angle transformation:

E	C_2	$\sigma(yz)$	$\sigma(xz)$
ϕ_1	ϕ_2	$-\phi_1$	$-\phi_2$
ϕ_2	ϕ_1	$-\phi_2$	$-\phi_1$

It follows that the conformations **1**, **2**, **3**, **4** (Fig. 10.5) are isometric. Conformations corresponding to certain special values of ϕ_1 and ϕ_2 or to certain special relationships between ϕ_1 and ϕ_2 are symmetric, the possible symmetries being subgroups of the frame group $\mathbf{F} = C_{2v}$:

$$
\begin{array}{lll}
C_{2v} & \text{if} & \phi_1 = \phi_2 = 0 \quad \text{or} \quad \phi_1 = \phi_2 = 90^\circ \\
C_2 & \text{if} & \phi_1 = \phi_2 \\
C_s(yz) & \text{if} & \phi_1 = 0, \phi_2 = 90^\circ \text{ or } \phi_1 = 90^\circ, \phi_2 = 0 \\
C_s(xz) & \text{if} & \phi_1 = -\phi_2
\end{array}
$$

There is another kind of operation that transforms any conformation of diphenylmethane into an isometric conformation—a rotation of one or both phenyl rings by 180°, often referred to as a flip.[3] Fig. 10.6 shows the four isometric conformations obtained from **1** by flipping none, one (two ways), or both phenyl groups of **1**. These operations are *not* geometric symmetry operations in the usual sense since they do not correspond to any conceivable change of axes; they correspond to what Altmann[4] has called "isodynamical operations" and can be performed

[3]This term was apparently introduced by R. J. Kurland, I. I. Schuster, and A. K. Colter [J. Am. Chem. Soc. **87,** 2279 (1965)] in discussing triphenylcarbenium ions. The role of flip mechanisms in isomerization processes has been extensively discussed in a series of papers by Mislow and his collaborators; see M. G. Hutchings, J. G. Nourse, and K. Mislow, *Tetrahedron* **30,** 1535 (1974) and references therein.

[4]S. L. Altmann, *Proc. Roy. Soc., London* **A298,** 184 (1967).

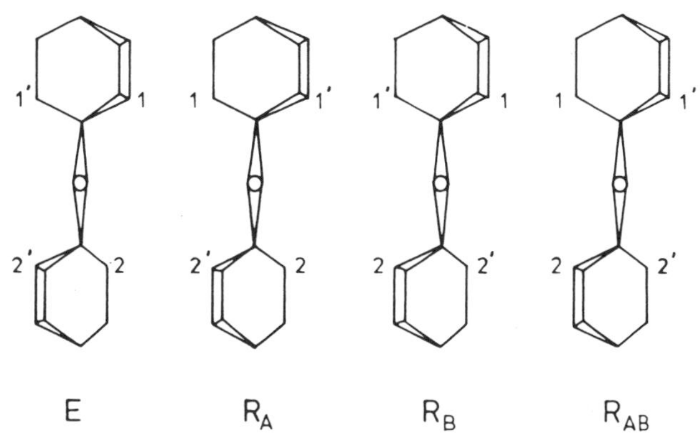

Figure 10.6. Isometric conformations obtained from **1** by 180° flips of the phenyl groups.

only by physical rotations of the phenyl groups. These rotations, or flips, form a group, the rotor group **R** with multiplication table:

E	R_A	R_B	R_{AB}
R_A	E	R_{AB}	R_B
R_B	R_{AB}	E	R_A
R_{AB}	R_B	R_A	E

that is isomorphic with C_{2v} (or more generally with $\mathbf{C_2} \times \mathbf{C_2}$).[5] Here R_A, R_B, and R_{AB} are the operations of flipping ring A alone, ring B alone, and both rings simultaneously. The corresponding transformations of the torsion angles are:

E	R_A	R_B	R_{AB}
ϕ_1	$\phi_1 + \pi$	ϕ_1	$\phi_1 + \pi$
ϕ_2	ϕ_2	$\phi_2 + \pi$	$\phi_2 + \pi$

A complete list of conformations isometric with **1** is produced by applying the rotor group operations E, R_A, R_B, and R_{AB} in turn to each of the conformations **1**, **2**, **3**, and **4**—i.e., by combining the rotor operations with the frame symmetry operations. There are 16 such combined operations, listed in Table 10.1 with the corresponding torsion angle transformations. These operations form a group **H**, isomorphic with D_{4h} (or, more generally, with $\mathbf{D_4} \times \mathbf{C_2}$) as can be checked by noting that it contains four elements (R_AC_2, R_BC_2, $R_A\sigma(xz)$, and $R_B\sigma(xz)$) of order 4 and eleven of order 2. The group **H**, the total symmetry group of the

[5]For this example the rotor group and the frame group happen to be isomorphic.

Table 10.1. Group of 16 operations (the **H** group) obtained by combining elements of frame group **F**{E, C_2, $\sigma(yz)$, $\sigma(xz)$} and rotor group **R**{E, R_A, R_B, R_{AB}}

Operation	ω_A	ω_B	Operation	ω_A	ω_B
E	ϕ_1	ϕ_2	$\sigma(yz)$	$-\phi_1$	$-\phi_2$
R_A	$\phi_1+\pi$	ϕ_2	$R_A\sigma(yz)$	$-\phi_1+\pi$	$-\phi_2$
R_B	ϕ_1	$\phi_2+\pi$	$R_B\sigma(yz)$	$-\phi_1$	$-\phi_2+\pi$
R_{AB}	$\phi_1+\pi$	$\phi_2+\pi$	$R_{AB}\sigma(yz)$	$-\phi_1+\pi$	$-\phi_2+\pi$
C_2	ϕ_2	ϕ_1	$\sigma(xz)$	$-\phi_2$	$-\phi_1$
R_AC_2	$\phi_2+\pi$	ϕ_1	$R_A\sigma(xz)$	$-\phi_2+\pi$	$-\phi_1$
R_BC_2	ϕ_2	$\phi_1+\pi$	$R_B\sigma(xz)$	$-\phi_2$	$-\phi_1+\pi$
$R_{AB}C_2$	$\phi_2+\pi$	$\phi_1+\pi$	$R_{AB}\sigma(xz)$	$-\phi_2+\pi$	$-\phi_1+\pi$

molecule, is the semi-direct product[6] of **R**, the rotor group, with **F**, the frame group,

$$\mathbf{H} = \mathbf{R} \wedge \mathbf{F}$$

The order of multiplication is important here since, for example, $C_2R_A = R_BC_2 \neq R_AC_2$.

We observed earlier that although the elements of the frame group **F** are genuine point-symmetry operations, the elements of the rotor group **R** are not, and neither are the other elements of **H** that involve products with elements of **R**. There is an analogy here between the elements of **R** and the elements of crystallographic space groups that involve pure translations—i.e., lattice translations and centerings. This analogy is best seen by looking at some of the torsion angle transformations that apply to the operations in question.

The transformation corresponding to R_A (Table 10.1) can be regarded as a *translation* of a representative point $P(\phi_1, \phi_2)$ by an amount π along the ω_A coordinate. Repetition of this operation produces a translation of 2π, a complete revolution of the torsion angle, equivalent to the identity operation. Similarly, the operation R_B corresponds to translation by π along the other coordinate ω_B, and R_{AB} corresponds to translations by π along ω_A and ω_B. To complete the analogy, the transformations that correspond to the genuine point-symmetry operations (the elements of the frame group) act on the representative point $P(\phi_1, \phi_2)$ to produce a set of general equivalent positions which, acted on by the above translations, produce additional equivalent positions. The set of 16 equivalent

[6]For definitions, proofs, and discussions of the role of the semi-direct group product in this kind of problem see J. S. Lomont, *Applications of Finite Groups,* Academic Press, New York and London, 1959, p. 29; S. L. Altmann, *Proc. Roy. Soc., London* **A298,** 184 (1967); S. L. Altmann, *Mol. Phys.* **21,** 587 (1971); M. G. Hutchings, J. G. Nourse, and K. Mislow, *Tetrahedron* **30,** 1535 (1974) and references therein.

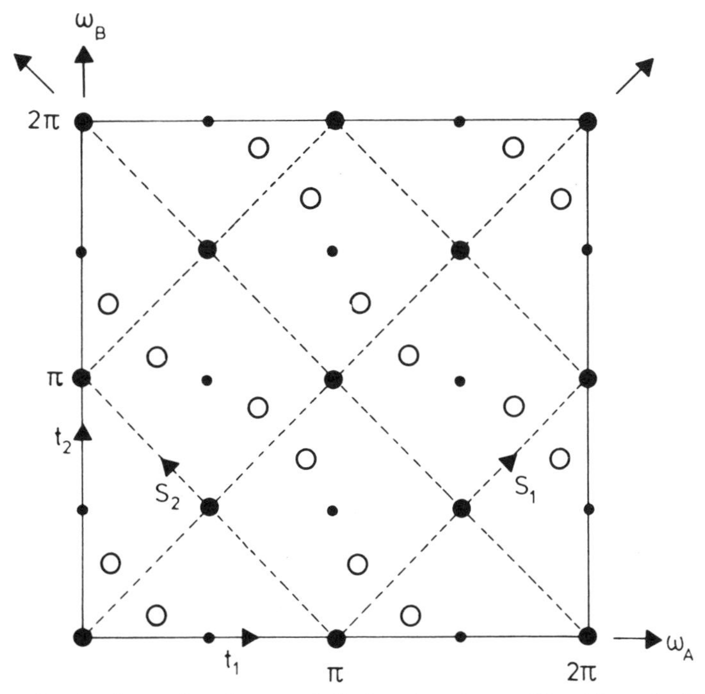

Figure 10.7. Conformational map of diphenylmethane. The 16 equivalent positions (open circles) are images of the 16 isometric conformations with different values of the torsion angles ω_A and ω_B (Table 10.1). The unit cell shown is nonprimitive, the primitive lattice having translations $\mathbf{t}_1 = \pi$ along ω_A and $\mathbf{t}_2 = \pi$ along ω_B. The plane group is *cmm*, with translations $\mathbf{S}_1 = \mathbf{t}_1 + \mathbf{t}_2$, $\mathbf{S}_2 = -\mathbf{t}_1 + \mathbf{t}_2$. The general positions of this plane group are images of arbitrary conformations; the special positions are images of conformations with nontrivial symmetry. In terms of the $\mathbf{S}_1$, $\mathbf{S}_2$ coordinate system, the special positions (and the symmetries of the corresponding conformations) are: (0, 0) etc. (C_{2v}); ($\frac{1}{4}$, $\frac{1}{4}$) etc. (C_s); (x, 0) etc. (C_2); (0, y) etc. (C_s). (See Tables 10.2 and 10.3.)

positions (modulo 2π) obtained in this way for the 16 elements of the molecular symmetry group is shown in Fig. 10.7. The figure obviously constitutes a "unit cell" that can be repeated indefinitely in two dimensions to give a periodic pattern. We can say that the elements of the molecular symmetry group **H** are mapped onto this unit cell: the pure translation operations are images of the elements of the rotor group **R**, and the point-symmetry operations are images of the frame group **F**.

There are many ways to choose a unit cell in this periodic pattern, but three possibilities stand out as especially convenient from different points of view. One is the unit cell described above, with unit translations of 2π along the ω_A and ω_B directions. The advantage of this unit cell is that the unit cell group[7] is isomorphic with the molecular symmetry

[7]In crystallography one speaks of the unit cell group **U**, the group of all elements of the space group **S** reduced modulo the unit translations. In group theoretical language, **U**

group; its disadvantage is that the unit translations are not primitive translations of the lattice. Coordinates of general and special positions of this unit cell are listed in Table 10.2.

A second possibility is to choose a unit cell based on primitive translations of the lattice, translations of π along ω_A and ω_B. These translations are labeled $\mathbf{t}_1$ and $\mathbf{t}_2$ in Fig. 10.7. The primitive unit cell contains only the four elements that correspond to the point-symmetry operations of the frame group. The general and special positions are the same as those given in Table 10.2 except that they are all to be read modulo π instead of 2π and their number is to be divided by four. The advantage of this unit cell is that it offers the most economical description of the periodic pattern, but this is gained at the price of neglecting the rotor group operations altogether; the distinction between physically different operations, rotations of individual rings through 0, π, or 2π, is lost.

Neither of the two unit cells described so far is oriented to show the standard space-group description of the periodic pattern. The mirror lines of the pattern run along the cell diagonals rather than parallel and perpendicular to the orthogonal axes, as required in the conventional space-group description. To obtain a standard orientation of these symmetry elements with respect to the unit translations, the axis transformation

$$\mathbf{S}_1 = \mathbf{t}_1 + \mathbf{t}_2$$

$$\mathbf{S}_2 = -\mathbf{t}_1 + \mathbf{t}_2$$

has to be applied. Here $\mathbf{t}_1$ and $\mathbf{t}_2$ are the primitive translations of the lattice, and $\mathbf{S}_1$ and $\mathbf{S}_2$ are the unit translations of the transformed unit cell ($|\mathbf{S}_1| = |\mathbf{S}_2| = \pi\sqrt{2}$). The coordinates must also be transformed:

$$x = (\phi_1 + \phi_2)/2$$

$$y = (-\phi_1 + \phi_2)/2$$

The lattice translation $(0, \pi)$ of the primitive cell is transformed to $(\pi/2, \pi/2)$ so that the new cell is centered (as evident by inspection of Fig. 10.7), giving a total of eight equivalent general positions, those of the plane group *cmm*. The advantage of referring the pattern to a conventional space group is that the general and special equivalent positions of

---

is the factor group of $\mathbf{S}$ with respect to the translational subgroup $\mathbf{T}$ of $\mathbf{S}$. The unit cell need not be primitive. Its order is equal to the number of elements it contains, whereas the order of $\mathbf{S}$ is infinite.

Table 10.2. General and special positions of conformational map for diphenylmethane, referred to nonprimitive unit cell defined by translations of 2π among ω_A and ω_B[a]

Number	Point symmetry	Coordinates of equivalent positions (modulo 2π)
		$(0,0;\ \pi,0;\ 0,\pi;\ \pi,\pi)\ +$
16	1	$\phi_1,\phi_2;\ \phi_2,\ \phi_1;\ -\phi_1,-\phi_2;-\phi_2,-\phi_1$
8	*m*	$\phi,-\phi;\ -\phi,\phi$
8	*m*	$\phi,\phi;\ -\phi,-\phi$
8	2	$\pi/2,\ 0;\ \ 0,\ \pi/2$
4	*mm*	$\pi/2,\ \pi/2$
4	*mm*	0, 0

[a]See Figure 10.7. The point symmetries listed refer to the map itself, not to the corresponding molecular conformations.

all two- and three-dimensional space groups are tabulated in standard works.[8] For the present example, this advantage may not amount to much since the required information can be derived by inspection of Fig. 10.7, but in more complex cases it will be extremely useful. The information is given for the plane group *cmm* in Table 10.3. Another point concerns the relationship between the coordinates x, y in the transformed conventional unit cell and the original coordinates ϕ_1, ϕ_2. The new coordinates are those linear combinations of the original ones that transform according to the irreducible representations of the frame symmetry group $\mathbf{F}(C_{2v})$—they are symmetry coordinates. It is easy to show that $x = (\phi_1 + \phi_2)/2$ transforms as the A_2, and $y = (-\phi_1 + \phi_2)/2$ transforms as the B_1 irreducible representation of C_{2v}.[9] Recall that the original coordinates ϕ_1, ϕ_2 are transformed into one another by some of the frame operations—e.g., C_2 transforms ϕ_1, ϕ_2 into ϕ_2, ϕ_1, but it transforms $x = (\phi_1 + \phi_2)/2$ into itself and $y = (-\phi_1 + \phi_2)/2$ into $-y$.

Fig. 10.7 may be useful for visualizing possible paths between different conformations, but its real value is that it provides a vivid picture of the symmetry properties of the two-dimensional torsional potential energy surface $V(\phi_1, \phi_2)$ of the diphenylmethane molecule. There is a one-to-one correspondence between the points in the plane (ϕ_1, ϕ_2) and the molecular conformations described by torsion angle pairs (ϕ_1, ϕ_2). In

[8]*International Tables for X-ray Crystallography,* Vol. 1, *Symmetry Groups.* Kynoch Press, Birmingham. 1st ed., 1952; 2nd ed., 1965; 3rd ed., 1969.

[9]Readers unfamiliar with the elementary theory of group representations may wish to consult a suitable introductory text, e.g., F. A. Cotton, *Chemical Applications of Group Theory,* Interscience, New York, 1963.

Table 10.3. Plane group *cmm*

Number	Point symmetry	Coordinates[a] of equivalent positions $(0,0;\ \frac{1}{2},\frac{1}{2})$ +
8	1	$x,y;\ \bar{x},\bar{y};\ \bar{x},y;\ x,\bar{y}$
4	m	$0,y;\ 0,\bar{y}$
4	m	$x,0;\ \bar{x},0$
4	2	$\frac{1}{4},\frac{1}{4};\ \frac{1}{4},\frac{3}{4}$
2	mm	$0,\frac{1}{2}$
2	mm	$0,0$

[a]The x and y coordinates are related to the coordinates of Table 10.2 by: $x = (\phi_1+\phi_2)/2$ (modulo π) and $y = (-\phi_1+\phi_2)/2$ (modulo π).

fact, Fig. 10.7 may be called a conformational map.[10] The general and special equivalent positions in terms of these coordinates are listed in Table 10.2. If the potential energy is assumed to depend only on the torsion angles, then for general values of these a given potential energy $V(\phi_1, \phi_2)$ must occur 16 times within the nonprimitive $2\pi \times 2\pi$ unit cell. Similarly, $V(\phi, \phi)$ occurs eight times, and so do $V(-\phi, \phi)$ and $V(0, \pi/2)$. Finally, $V(0, 0)$ and $V(\pi/2, \pi/2)$ each occur four times or only once in the primitive cell; the corresponding molecular conformations show the full frame symmetry, in this case the highest point-group symmetry the molecule can attain. The other symmetry labels listed in Table 10.2 give the symmetries of particular points or lines in the conformational map which may or may not be the same as the corresponding molecular symmetries. For example, molecules described by the torsion angle pair (ϕ, ϕ) have a twofold rotation axis, but the corresponding line in the two-dimensional map is a line of mirror symmetry. This is because the frame-symmetry element $C_2(\phi_1, \phi_2 \rightarrow \phi_2, \phi_1)$ is mapped onto a mirror line $m(x, 0)$ in the conformational map (Fig. 10.7).

In summary, there is a close connection between the total symmetry group **H** of a nonrigid molecule containing two torsional degrees of freedom ω_A, ω_B and a nonprimitive two-dimensional unit cell group with translations $\omega_A = 2\pi$, $\omega_B = 2\pi$. The equivalent positions of this unit cell correspond to isometric conformations, and its symmetry is that of the two-dimensional potential energy surface $V(\phi_1, \phi_2)$ of the molecule. The periodic pattern derived from this unit cell can be referred to a standard

[10]The term "conformational map" has been applied to two-dimensional energy diagrams or scatter plots describing the expected or actual distributions of torsion angles $\phi(C'NC^{\alpha}C')$ and $\psi(NC^{\alpha}C'N)$ in polypeptide chains. See G. N. Ramachandran and V. Sasisekharan, *Adv. Protein Chem.* **23**, 284 (1968).

space group whose unit translations lie along symmetry coordinates involving linear combinations of ϕ_1 and ϕ_2.

One troublesome point is that although certain elements of **H** (e.g., R_AC_2) are of order 4, the space group *cmm* contains only elements of order 2. Is there not a discrepancy here? Yes. To get around it, we note that although all symmetry elements of the *cmm* unit cell group are of order 2 (they are their own inverses), this is not the case for the larger unit cell with nonprimitive translations. For example, the element R_AC_2 corresponds to the torsion angle transformation $\phi_1, \phi_2 \rightarrow \phi_2 + \pi, \phi_1$ (Table 10.1). In geometric terms this means: reflect the point (ϕ_1, ϕ_2) across the diagonal mirror line $\phi_1 = \phi_2$ and displace the resultant by π along the ω_A-axis. Repetition of this operation leads to $(\phi_1 + \pi, \phi_2 + \pi)$, which is identical to (ϕ_1, ϕ_2) modulo π but not modulo 2π. However, our unit cell group is defined modulo 2π, so $(R_AC_2)^2 \neq E$. In fact, it is easy to check that if M is the transformation in question, then:

$$
\begin{aligned}
M\,(\phi_1, \phi_2) &= \phi_2 + \pi, \phi_1 \\
M^2(\phi_1, \phi_2) &= \phi_1 + \pi, \phi_2 + \pi \\
M^3(\phi_1, \phi_2) &= \phi_2, \phi_1 + \pi \\
M^4(\phi_1, \phi_2) &= \phi_1, \phi_2
\end{aligned}
$$

Symmetry operations of this kind do not appear in crystallographic discussions because the unit cell is conventionally chosen to give the smallest possible repeating unit compatible with the symmetry. In the present case, this smallest possible cell has *cmm* symmetry and is of order 8. However, if we wish to describe the molecular symmetry group in terms of a unit cell group, we need one of order 16. Hence, the usual criterion of smallest possible cell has to be discarded. The space group *cmm* describes the symmetry of the repeating pattern correctly, but if we wish a one-to-one correspondence of symmetry elements of **H** and equivalent positions we have to take the larger cell with nonprimitive translations.

Alternatively, we can regard the molecular symmetry group merely as a stepping stone to the symmetry of the potential energy surface. In this case, nothing is lost by using the unit cell based on primitive translations, whose lengths are integral submultiples of 2π that are determined by the rotor group **R**. For diphenylmethane this primitive cell has translations $\mathbf{t}_1 = \mathbf{t}_2 = \pi$ (Fig. 10.7). Now a one-to-one correspondence exists between elements of **R** and those of the translation group **T** of the periodic pattern, and a one-to-one correspondence also exists between the elements of the frame group **F** and those of the unit-cell group. The order h of the molecular symmetry group **H** is given by $f \times r$ (for m equivalent n-fold rotors, $r = n^m$).

Experimental information about preferred conformations of diarylmethanes is rather sparse. A search of the Cambridge Crystal Structure Data File (summer 1977) revealed only two crystal structures containing diarylmethane molecules lacking ortho substitutents (we can assume that substitution at other positions in the phenyl groups will not greatly alter the torsional potential)—4,4′-diaminodiphenylmethane[11] and 3,3′-dichloro-4,4′-dihydroxydiphenylmethane.[12] Both molecules have C_2 symmetry with torsion angles ϕ in the range 40°–50°.

Other Examples: The Frame Symmetry Group

The application of this approach to other nonrigid molecules with two or more torsional degrees of freedom is straightforward except perhaps for the problem of deciding what the frame symmetry group is. In the previous example it was C_{2v}. This was the symmetry of the central C—C—C frame and it was also the highest symmetry attainable by the molecule as a whole (in the two special conformations: $\phi_1 = \phi_2 = 0$, $\phi_1 = \phi_2 = \pi/2$). Our discussion of the previous example began by stressing the C_{2v} symmetry of the central C—C—C frame on which the phenyl groups could be rotated, but this was done only for illustration. If we consider, say, *p*-chlorodiphenylmethane, with the same central frame but with nonequivalent substituents, it is clear that conformations described by (ϕ_1, ϕ_2) and (ϕ_2, ϕ_1) are not isometric. Indeed, the only torsion angle transformation that corresponds to a possible frame symmetry operation is $(\phi_1, \phi_2) \rightarrow (-\phi_1, -\phi_2)$, corresponding to reflection of the molecule across the yz plane. Special conformations for which the torsion angles are 0 or $\pi/2$ are left invariant by this transformation. Such conformations have C_s symmetry, the highest point symmetry attainable by this molecule and the same as the frame symmetry.

In general, any torsion angle transformation whose effect can be reproduced (or undone) by a change of axes corresponds to a frame operation, and the frame group **F** consists of the set of all such operations. In our mapping procedure, the elements of **F** are mapped on the general equivalent positions of the *primitive* unit cell. The number of general positions equals the order of **F**. If these general positions are interrelated exclusively by *point-symmetry* operations—i.e., if every operation of **F** is imaged by a point-symmetry operation (as in all the exam-

[11] J. A. J. Jarvis and P. G. Owston, *J. Chem. Soc. (D)* **1971,** 1403; J. W. Swardstrom, L. A. Duvall, and D. P. Miller, *Acta Crystallogr.* **B28,** 2510 (1972).

[12] E. J. W. Whittaker, *Acta Crystallogr.* **6,** 714 (1953).

ples discussed so far)—then the order of the special positions of highest site symmetry is just the order of the unit cell group, which equals the order of **F**. The site symmetries of these special positions are isomorphic with the frame symmetry. These special positions also represent conformations with the highest point-group symmetry attainable by the molecule. Thus, in such cases, the frame symmetry can be chosen as the highest point-group symmetry attainable by the molecule. Some molecules can claim two different, isomorphic point groups to display their highest symmetry—e.g., ethane (D_{3h} and D_{3d}) and dibenzyl (C_{2h} and C_{2v}). These alternative frame symmetries correspond merely to different, nonequivalent special positions of the same order. We return to this point later.

However, the general equivalent positions of the primitive unit cell may be interrelated by operations that are combinations of point-symmetry and translation operations—e.g., glide planes and screw axes. In such a case the order of the special positions of highest site symmetry is lower than the order of the primitive unit-cell group and hence lower than the order of **F**, but these special positions still represent conformations with the highest point-group symmetry attainable by the molecule. Hence, if our mapping procedure leads to a set of general positions of this kind, the highest point-group symmetry attainable by the molecule is only a subgroup of the frame symmetry. We shall encounter an example of this later.

In general our list of frame operations will contain at least one that converts a chiral conformation into an enantiomeric one—i.e., reverses the signs of all torsion angles and possibly interchanges their sequence as well. Whether or not these particular transformations are feasible will depend, of course, on the details of the potential energy surface. However, if the molecule contains chiral substituents, these will not be inverted by changes in the torsion angles alone, so the transformations in question must lead to diastereomeric (i.e., nonisometric) conformations. They must therefore be excluded from the frame group.

Let us summarize results for a few other two-dimensional examples. The plane groups listed are the conventional ones (analogous to *cmm* for the example discussed in the section on diphenylmethane), and the corresponding conformational maps are shown in Fig. 10.8.

p-Chlorodiphenylmethane
Rotor group $\mathbf{R}\{E, R_A, R_B, R_{AB}\}$: $(0, 0)$, $(\pi, 0)$, $(0, \pi)$, (π, π) +
Frame group C_s: ϕ_1, ϕ_2; $-\phi_1, -\phi_2$
Unit-cell group of order 8
Plane group $p2$: $\mathbf{S}_1 = \mathbf{t}_1/2$, $\mathbf{S}_2 = \mathbf{t}_2/2$

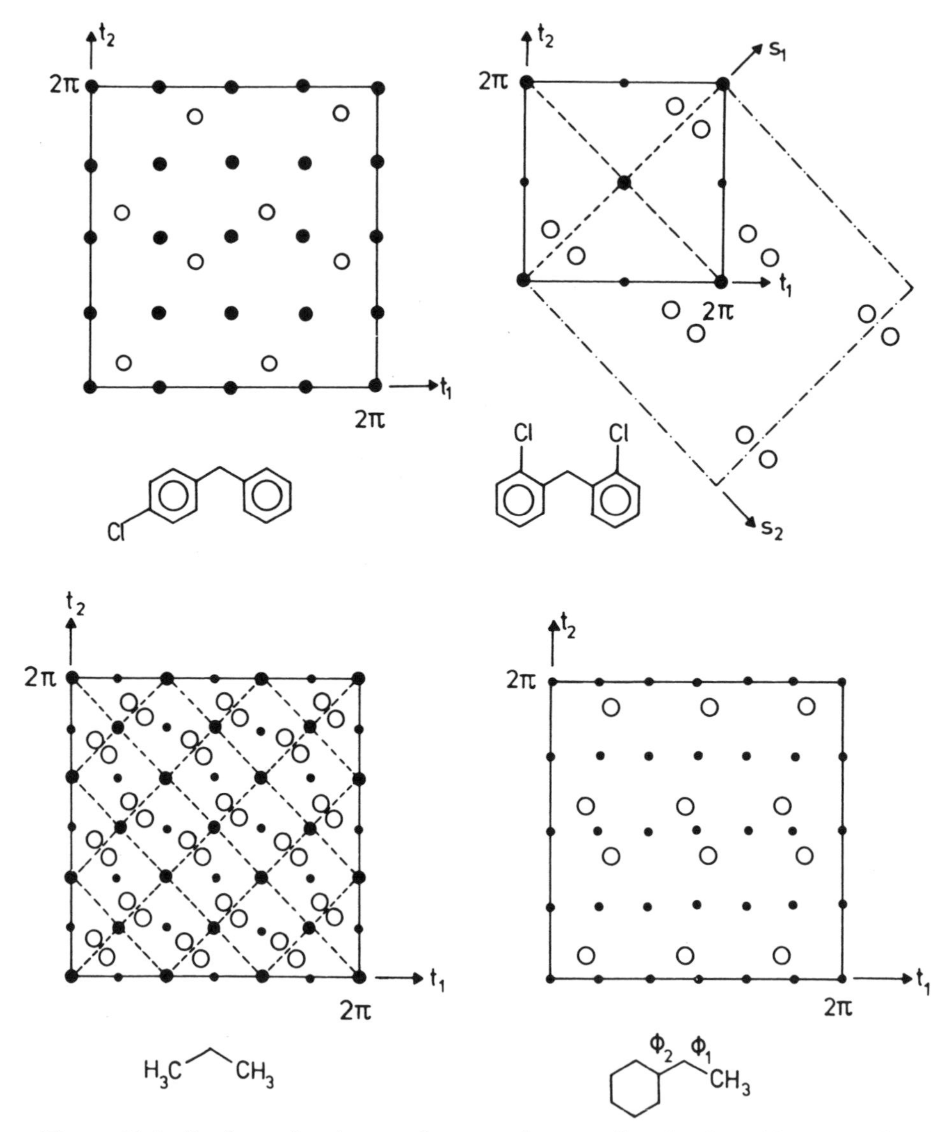

Figure 10.8. Conformational maps for several types of molecules with two torsional degrees of freedom. The equivalent positions (open circles) are images of isometric conformations with arbitrary values of the torsion angles. Special positions (filled circles and dashed lines) are images of conformations with nontrivial symmetry.

o,o'-Dichlorodiphenylmethane
Rotor group $\mathbf{R}\{E\}$: (0, 0) +
Frame group C_{2v}: $\phi_1, \phi_2; \phi_2, \phi_1; -\phi_1, -\phi_2; -\phi_2, -\phi_1$
Unit-cell group of order 4
Plane group *cmm*: $\mathbf{S}_1 = \mathbf{t}_1 + \mathbf{t}_2$, $\mathbf{S}_2 = \mathbf{t}_1 - \mathbf{t}_2$

Propane
Rotor group: (0,0), $(2\pi/3, 0)$, $(4\pi/3, 0)$
$(0, 2\pi/3)$, $(2\pi/3, 2\pi/3)$, $(4\pi/3, 2\pi/3)$
$(0, 4\pi/3)$, $(2\pi/3, 4\pi/3)$, $(4\pi/3, 4\pi/3)$ +
Frame group C_{2v}: $\phi_1, \phi_2; \phi_2, \phi_1; -\phi_1, -\phi_2; -\phi_2, -\phi_1$
Unit-cell group of order 36
Plane group *cmm*, $\mathbf{S}_1 = (\mathbf{t}_1 + \mathbf{t}_2)/3$, $\mathbf{S}_2 = (\mathbf{t}_1 - \mathbf{t}_2)/3$

Ethylbenzene
Rotor group: (0, 0), $(2\pi/3, 0)$, $(4\pi/3, 0)$
$(0, \pi)$, $(2\pi/3, \pi)$, $(4\pi/3, \pi)$
Frame group C_s: $\phi_1, \phi_2; -\phi_1, -\phi_2$
Unit-cell group of order 12
Plane group *p*2, $\mathbf{S}_1 = \mathbf{t}_1/3$, $\mathbf{S}_2 = \mathbf{t}_2/2$

Ligand Permutations

The symmetry properties of nonrigid molecules can also be derived by considering the feasible permutations of equivalent atoms that are regarded as ligands on a rigid skeleton. For this purpose, the molecule is pictured in a conformation of highest possible point symmetry, a set of symmetry-equivalent atoms is labeled arbitrarily, and the group of all feasible permutations of the labels is worked out. A feasible permutation corresponds to a ligand interchange produced by a point-symmetry operation of the molecule (change of axes) or by a low-energy dynamic process, e.g., a torsional motion of a set of ligands. For n equivalent ligands, the group of feasible permutations must be a subgroup of the complete permutational group of order n.

As an example we use the diphenylmethane molecule again. The highest point-group symmetry attainable—the frame symmetry—is C_{2v},

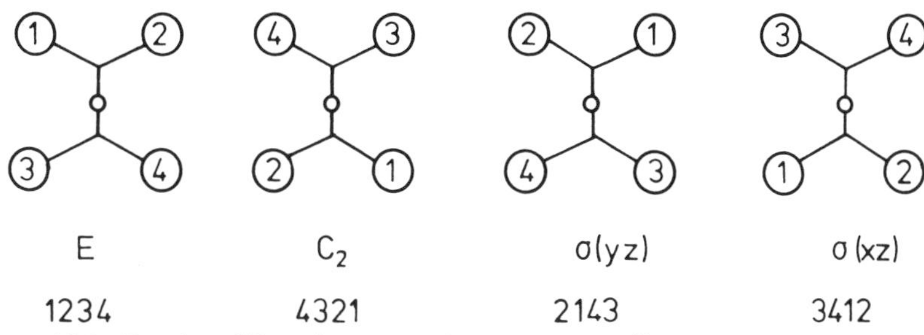

Figure 10.9. Results of ligand permutations corresponding to symmetry operations of frame symmetry group C_{2v}.

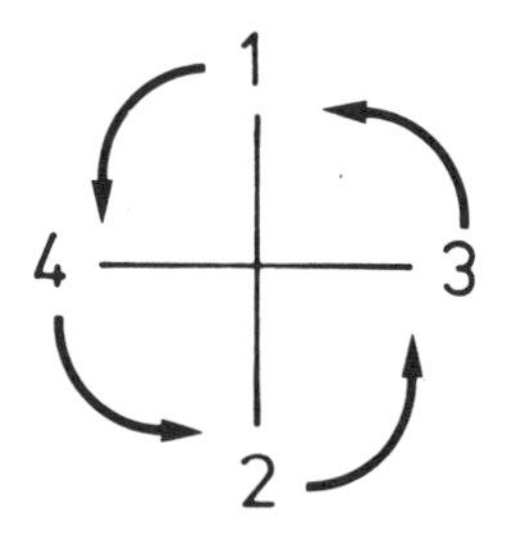

Figure 10.10. The cyclic interchange of labels (1324) means: put 3 in place of 1, 2 in place of 3, and 4 in place of 2. The result of the operation is written $\begin{Bmatrix} 1234 \\ 3421 \end{Bmatrix}$ in full, or 3421 in short, the natural order of the upper set of numbers being assumed.

and the ligand permutations that correspond to the individual elements of this group are shown in Fig. 10.9. The result of the C_2 permutation can be written 4321, meaning that ligand 4 occupies the position initially occupied by 1, ligand 3 occupies the position initially occuped by 2, and so on. Alternatively, we can say that the C_2 operation corresponds to an interchange of ligands 1 and 4, and of 2 and 3, the combined process being written as (14)(23). The operation of flipping just one ring, say the upper one, is then (12), the result 2134. The permutation R_AC_2 results in 3421. The operation is one of order 4, as indicated by the sequence:

$$\begin{array}{cc} E & 1234 \\ R_AC_2 & 3421 \\ (R_AC_2)^2 & 2143 \\ (R_AC_2)^3 & 4312 \\ (R_AC_2)^4 & 1234 \end{array}$$

or by raising the transformation matrix

$$\mathbf{M} = \begin{bmatrix} 0 & 0 & 1 & 0 \\ 0 & 0 & 0 & 1 \\ 0 & 1 & 0 & 0 \\ 1 & 0 & 0 & 0 \end{bmatrix}$$

to higher powers, leading ultimately to $\mathbf{M}^4 = \mathbf{I}$. Alternatively we can say that the R_AC_2 operation corresponds to a cyclic interchange (1324), meaning: put ligand 3 in place of 1, 2 in place of 3, 4 in place of 2, and the remaining 1 in place of 4.[13] Translated into geometric terms, this is equivalent to a rotation of the ligand labels around a fourfold axis—an operation of period 4 (Fig. 10.10). Of course, the same result was reached in our previous discussion in terms of torsion-angle transformations.

The complete list of permutations corresponding to the frame and

[13]Some authors write the cyclic interchange in question as (1423), meaning: put 1 in place of 4, 4 in place of 2, and 2 in place of 3. When reading literature on permutations it is as well to check which convention is being used.

Table 10.4. Permutation operations corresponding to frame and rotor operations with ligands labeled as in Fig. 10.9

Operation	Result	Operation	Result
E	1234	$\sigma(yz) = (12)(34)$	2143
$R_A=(12)$	2134	$R_A\sigma(yz)=(34)$	1243
$R_B=(34)$	1243	$R_B\sigma(yz)=(12)$	2134
$R_{AB}=(12)(34)$	2143	$R_{AB}\sigma(yz)=E$	1234
$C_2=(14)(23)$	4321	$\sigma(xz)=(13)(24)$	3412
$R_AC_2=(1324)$	3421	$R_A\sigma(xz)=(1423)$	4312
$R_BC_2=(1423)$	4312	$R_B\sigma(xz)=(1324)$	3421
$R_{AB}C_2=(13)(24)$	3412	$R_{AB}\sigma(xz)=(14)(23)$	4321

flip operations is given in Table 10.4. The permutations in the right column are the same as those in the left column but in a different order. For example, the operation $R_{AB}\sigma(yz)$ appears to be equivalent to the identity operation. There are thus only eight distinct permutations involved, forming a group isomorphic with D_4 of order 8.

However, in our previous analysis in terms of torsion-angle transformations $R_{AB}\sigma(yz)$ was quite distinct from the identity operation, and the group obtained was of order 16, isomorphic with D_{4h}. What has gone wrong? The answer is that we have ignored the fact that the operations in the right column all involve reflections—symmetry operations of the second kind (or improper rotations)—whereas those in the left column involve exclusively proper rotations. This situation can be remedied by adding an asterisk, say, to all the permutation operations in the right column to indicate that they correspond to geometrically distinct operations from those in the left column. The combination rule for obtaining the higher order group is then obvious.[14]

In contrast to the geometric approach involving torsion-angle transformations, the permutation approach offers no insight into the possible molecular motions that convert a given conformation into an isometric one, nor does it lead to any conclusions about the symmetry properties of the potential energy surface of the molecule—limitations that have been explicitly recognized by some of its most ardent protagonists.[15] It does not provide a map.

[14]Consider the water molecule with its hydrogen atoms labeled 1 and 2. The symmetry operations and their corresponding permutations are then: $E = (1)(2)$, $C_2 = (12)$, $\sigma_h = (1)^*(2)^*$, $\sigma_v = (12)^*$, forming a group of order 4. The pure permutations (ignoring the asterisks) only give the group of order 2.

[15]"A particular permutation . . . . emphatically says nothing about the mechanism of the isomerization it describes: no information pertaining to energetics or geometries of inter-

Three Torsional Degrees of Freedom

Triphenylmethane

The analysis in terms of torsion-angle transformation follows much the same lines as in the two-dimensional examples already discussed. An arbitrary conformation of the molecule is specified by three torsion angles $\omega_A = \phi_1$, $\omega_B = \phi_2$, and $\omega_C = \phi_3$, where the labels A, B, C each refer to a given phenyl group. For definiteness, we adopt the clockwise sequence of the labels shown in Figure 10.11. Zero value of each torsion angle can be defined as corresponding to a position where the phenyl group in question eclipses the C—H (methine) bond along the threefold axis of the central frame, which is considered to be rigid. The frame symmetry is C_{3v}, and the torsion-angle transformations produced by the frame symmetry operations are (Fig. 10.11):

	E	C_3	C_3^2	σ_a	σ_b	σ_c
ω_A	ϕ_1	ϕ_3	ϕ_2	$-\phi_1$	$-\phi_3$	$-\phi_2$
ω_B	ϕ_2	ϕ_1	ϕ_3	$-\phi_3$	$-\phi_2$	$-\phi_1$
ω_C	ϕ_3	ϕ_2	ϕ_1	$-\phi_2$	$-\phi_1$	$-\phi_3$

As in the diphenylmethane example (discussed earlier), these transformations can be regarded as the results of actual rotations of the individual phenyl groups or as symmetry operations, i.e., changes of axes. To see that they correspond to changes of axes, it is convenient to refer the frame to a nonorthogonal coordinate system with axes **a, b, c** along the three central bonds. The C_3 operation can then be interpreted as a cyclic interchange of these axes, which preserves the signs of torsion angles, a σ operation as a noncyclic interchange, which reverses signs of torsion angles. According to our definitions, the transformation $\phi_1, \phi_2, \phi_3 \rightarrow -\phi_1, -\phi_2, -\phi_3$ does not correspond to a symmetry operation and it does not, in general, lead to an isometric structure.

The set of frame transformations thus obtained can immediately be identified with the general equivalent positions (modulo 2π) of the three-dimensional space group $R32$ (see Table 10.5).

The rotor group for this molecule consists of eight elements, the corresponding torsion angle transformations involving translations of

mediate states traversed during an isomerization is implied in, or may be inferred from, a permutation" M. G. Hutchings, J. G. Nourse, and K. Mislow, *Tetrahedron* **30,** 1535 (1974).

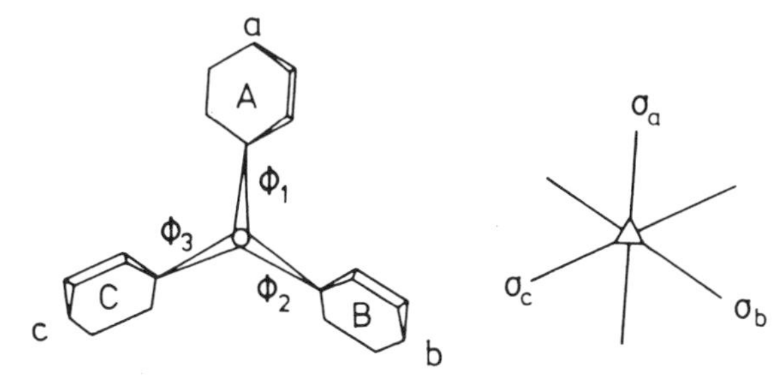

Figure 10.11. An arbitrary conformation of the triphenylmethane molecule characterized by torsion angles ϕ_1, ϕ_2, and ϕ_3. The symmetry elements of the frame, together with a convenient coordinate system, are indicated.

the representative point $P(\phi_1, \phi_2, \phi_3)$:

	E	R_A	R_B	R_C	R_{AB}	R_{BC}	R_{AC}	R_{ABC}
ω_A	ϕ_1	$\phi_1+\pi$	ϕ_1	ϕ_1	$\phi_1+\pi$	ϕ_1	$\phi_1+\pi$	$\phi_1+\pi$
ω_B	ϕ_2	ϕ_2	$\phi_2+\pi$	ϕ_2	$\phi_2+\pi$	$\phi_2+\pi$	ϕ_2	$\phi_2+\pi$
ω_C	ϕ_3	ϕ_3	ϕ_3	$\phi_3+\pi$	ϕ_3	$\phi_3+\pi$	$\phi_3+\pi$	$\phi_3+\pi$

The complete set of conformations isometric with the original conformation is obtained by combining the eight rotor group operations with the six frame symmetry operations to obtain a molecular symmetry group **H** containing 48 operations. This must be isomorphic with the octahedral group O_h since it comprises: eight operations ($6R_iC_3$, $2R_{ijk}C_3$) of order 6; 12 operations ($6R_{ij}\sigma_i$, $6R_i\sigma_j$) of order 4; eight operations ($2C_3$, $6R_{ij}C_3$) of order 3; 19 operations ($3R_i$, $3R_{ij}$, R_{ijk}, $3R_i\sigma_i$, $3R_{ijk}\sigma_i$, $3R_{ij}\sigma_k$, $3\sigma_i$) of order 2. To show that the operation $M = R_AC_3$, for example, has order 6 we merely repeat the transformation in question:

	E	ϕ_1	ϕ_2	ϕ_3
R_AC_3	M	$\phi_3+\pi$	ϕ_1	ϕ_2
$R_{AB}C_3^2$	M^2	$\phi_2+\pi$	$\phi_3+\pi$	ϕ_1
R_{ABC}	M^3	$\phi_1+\pi$	$\phi_2+\pi$	$\phi_3+\pi$
$R_{BC}C_3$	M^4	ϕ_3	$\phi_1+\pi$	$\phi_2+\pi$
$R_CC_3^2$	M^5	ϕ_2	ϕ_3	$\phi_1+\pi$
E	M^6	ϕ_1	ϕ_2	ϕ_3

The complete set of coordinates of equivalent positions (modulo 2π) is listed in Table 10.5: the translation operations correspond to the rotor group, and the point-symmetry operations correspond to the frame group. As in previous examples, the point symmetries listed refer to the symmetry of the conformational map; they are homomorphic mappings of frame symmetry elements. In particular, the σ elements of the frame are mapped onto twofold rotation axes of the conformational map. The conformations with nontrivial symmetry are shown in Fig. 10.12, along with the corresponding point symmetries of the map.

Table 10.5. General and special positions of conformational map for triphenylmethane, referred to a nonprimitive unit cell defined by translations of 2π along ω_A, ω_B, ω_C[a]

Number	Point symmetry	Coordinates of equivalent positions (modulo 2π) $(0,0,0;\ \pi,0,0;\ 0,\pi,0;\ 0,0,\pi;\ 0,\pi,\pi;\ \pi,0,\pi;\ \pi,\pi,0;\ \pi,\pi,\pi)+$
48	1	$\phi_1,\phi_2,\phi_3;\ \phi_3,\phi_1,\phi_2;\ \phi_2,\phi_3,\phi_1;\ -\phi_1,-\phi_3,-\phi_2;$ $-\phi_3,-\phi_2,-\phi_1;\ -\phi_2,-\phi_1,-\phi_3$
24	2	$\pi/2,\phi,-\phi;\ -\phi,\pi/2,\phi;\ \phi,-\phi,\pi/2$
24	2	$0,\phi,-\phi;\ -\phi,0,\phi;\ \phi,-\phi,0$
16	3	$\phi,\phi,\phi;\ -\phi,-\phi,-\phi$
8	32	$\pi/2,\pi/2,\pi/2$
8	32	$0,0,0$

[a]See Fig. 10.11. The space group of the pattern is *R*32. The point symmetries refer to the conformational map, not to the corresponding molecular conformations.

Instead of using the eightfold centered cell with unit translations of 2π along ω_A, ω_B, and ω_C, we may choose the unit cell based on the primitive translations of π in these directions. This reduced unit cell contains only the six equivalent positions that correspond to the point-symmetry operations of the frame group. The general and special positions of Table 10.5 still apply, but they are to be read modulo π instead of modulo 2π and their multiplicities are to be divided by 8. However, there is still the difficulty of visualizing the arrangement of equivalent positions and symmetry elements in a rhombohedral coordinate system. In this connection, it is more convenient to refer the pattern to a hexagonal coordinate system, as shown in Fig. 10.13. The transforma-

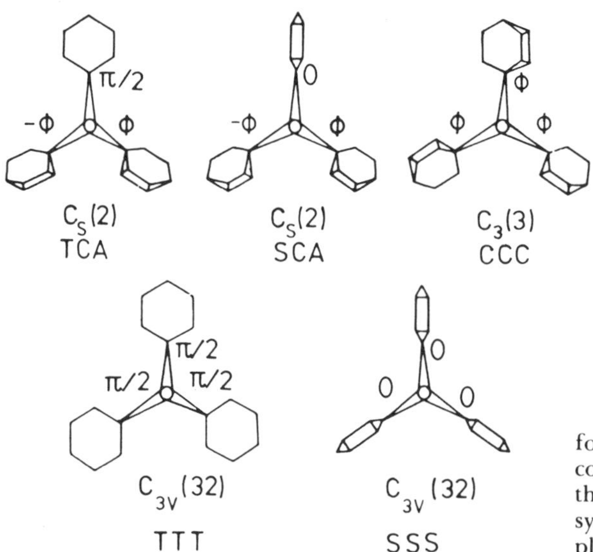

Figure 10.12. Symmetric conformations of triphenylmethane correspond to special positions of the conformational map (point symmetries in brackets). For explanation of letter code see text.

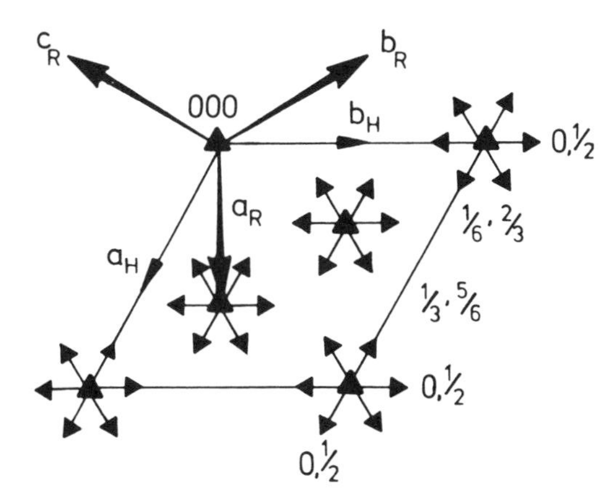

Figure 10.13. Primitive rhombohedral cell (space group $R32$) in hexagonal coordinate system: $\mathbf{a}_H = \mathbf{a}_R - \mathbf{b}_R$, $\mathbf{b}_H = \mathbf{b}_R - \mathbf{c}_R$, $\mathbf{c}_H = \mathbf{a}_R + \mathbf{b}_R + \mathbf{c}_R$. The hexagonal cell has lattice points at 0, 0, 0; $\frac{2}{3}, \frac{1}{3}, \frac{1}{3}$; $\frac{1}{3}, \frac{2}{3}, \frac{2}{3}$.

tion from rhombohedral to hexagonal axes is:

$$\mathbf{a}_H = \mathbf{a}_R - \mathbf{b}_R$$

$$\mathbf{b}_H = \mathbf{b}_R - \mathbf{c}_R$$

$$\mathbf{c}_H = \mathbf{a}_R + \mathbf{b}_R + \mathbf{c}_R$$

with a corresponding transformation of coordinates (vector components):

$$x_H = (2x_R - y_R - z_R)/3$$

$$y_H = (x_R + y_R - 2z_R)/3$$

$$z_H = (x_R + y_R + z_R)/3$$

In terms of torsion angles, $x_R = \phi_1$, $y_R = \phi_2$, $z_R = \phi_3$, are all defined modulo π; the rhombohedral cell in question is the primitive one, while the hexagonal cell has lattice points at 0, 0, 0; $2\pi/3$, $\pi/3$, $\pi/3$; $\pi/3$, $2\pi/3$, $2\pi/3$. The general and special positions referred to this cell are listed in Table 10.6. One advantage of the new coordinates x_H, y_H, and z_H is that

Table 10.6. General and special positions (modulo π) of conformational map for triphenylmethane, referred to the hexagonal cell shown in Fig. 10.13[a]

Number	Point symmetry	Coordinates of equivalent positions (modulo π) $(0,0,0;\ 2\pi/3,\pi/3,\pi/3;\ \pi/3,2\pi/3,2\pi/3)+$
18	1	$x,y,z;\ \bar{y},x-y,z;\ y-x,\bar{x},z;\ y,x,\bar{z};\ \bar{x},y-x,\bar{z};\ x-y,\bar{y},\bar{z}$
9	2	$x,0,\pi/2;\ 0,x,\pi/2;\ \bar{x},\bar{x},\pi/2$
9	2	$x,0,0;\ 0,x,0;\ \bar{x},\bar{x},0$
6	3	$0,0,z;\ 0,0,\bar{z}$
3	32	$0,0,\pi/2$
3	32	$0,0,0$

[a]Hexagonal coordinates are (in terms of torsion angles): $x_H = (2\phi_1 - \phi_2 - \phi_3)/3$; $y_H = (\phi_1 + \phi_2 - 2\phi_3)/3$; $z_H = (\phi_1 + \phi_2 + \phi_3)/3$.

they are symmetry coordinates. The z_H coordinate transforms as the A_2 irreducible representation of C_{3v} while x_H and y_H transform as the doubly degenerate E representation.

To visualize the way in which different sets of isometric conformations are obtained by varying the torsion angles, several sections of the con-

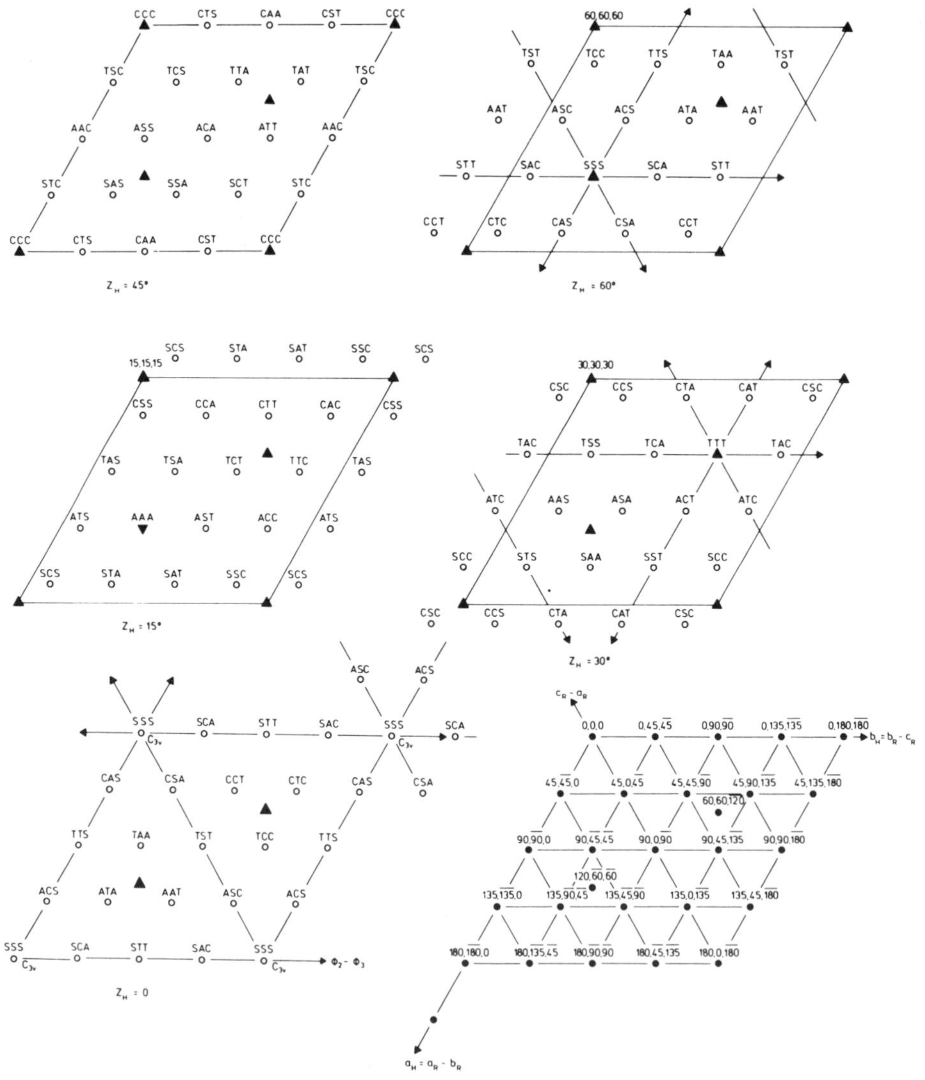

Figure 10.14. Various sections (at $z_H = 0, 15°, 30°, 45°, 60°$) through the conformational map of triphenylmethane. The map is referred to a hexagonal coordinate system. Conformations are represented by a letter code, isometric conformations having the same letters in different order, e.g., CCT, CTC, TCC (see text for details). The zero section is given both in letter code (left) and with numerical values of the torsion angles ϕ_1, ϕ_2, and ϕ_3 in degrees (right).

formational map perpendicular to the threefold rotation axis are shown in Fig. 10.14. Each conformation is coded by three letters that signify the orientations of the individual phenyl groups; S means that a given phenyl group has torsion angle 0 (or π), T means that it has torsion angle $\pm\pi/2$, C means that the phenyl group is rotated clockwise by $\pi/4$ from the S position looking along the bond from the central atom, and A means that it is rotated anticlockwise by $\pi/4$. These are the letters used in Fig. 10.12 to designate the different conformations shown there; the phenyls are to be read in clockwise order. The two bottom diagrams in Fig. 10.14 show the zero section, $z_H = 0$, of the conformational map—that on the left in letter code, that on the right with numerical values of the torsion angles ϕ_1, ϕ_2, and ϕ_3 in degrees. The origin (0, 0, 0) corresponds to SSS or $\phi_1 = \phi_2 = \phi_3 = 0$, and there are three equivalent axes in the plane along the directions $\mathbf{a}_R - \mathbf{b}_R$, $\mathbf{b}_R - \mathbf{c}_R$, $\mathbf{c}_R - \mathbf{a}_R$. Note that the mirror-symmetric conformations (SCA, STT, etc.) are mapped on the twofold rotation axes of the pattern. The next four diagrams show sections at constant values of $\phi_1 + \phi_2 + \phi_3 = 45°, 90°, 135°$, and $180°$, corresponding to $z_H = 15°, 30°, 45°$, and $60°$. The section at 30° again contains twofold axes, but they do not pass through the origin (cf. Fig. 10.13). Their intersection is the special position that maps the TTT conformation (90°, 90°, −90° referred to rhombohedral axes, or 60°, 120°, 30° referred to hexagonal axes) with the full frame symmetry. The section at 60° is a repeat of the zero section, with origin displaced by a primitive translation of the rhombohedral cell. The sections that follow merely repeat the sequence with appropriate origin shifts, the 12th section ($\phi_1 + \phi_2 + \phi_3 = 540°$) being again in register with the zero section:

z_H	Section	contains	on triad
0°	0	SSS	0, 0
15°	1	AAA	$\frac{2}{3}, \frac{1}{3}$
30°	2	TTT	$\frac{1}{3}, \frac{2}{3}$
45°	3	CCC	0, 0
60°	4	SSS	$\frac{2}{3}, \frac{1}{3}$
75°	5	AAA	$\frac{1}{3}, \frac{2}{3}$
90°	6	TTT	0, 0
105°	7	CCC	$\frac{2}{3}, \frac{1}{3}$
120°	8	SSS	$\frac{1}{3}, \frac{2}{3}$
135°	9	AAA	0, 0
150°	10	TTT	$\frac{2}{3}, \frac{1}{3}$
165°	11	CCC	$\frac{1}{3}, \frac{2}{3}$
180°	12	SSS	0, 0

Molecules of this type (C_{3v} frame carrying twofold rotors) generally occur in a propeller-like conformation with C_3 symmetry or close to it. Thus, triphenylmethane itself has C_3 symmetry ($\phi \sim 45°$) in the gas phase,[16] and the same is true for trinitromethane ($\phi \sim 26°$) and trinitrochloromethane ($\phi \sim 49°$).[17] In crystals, such molecules often deviate from C_3 symmetry but not by much—the rotors are always turned in the same sense—e.g., triphenylphosphine[18] (24°, 29°, 56°) and triphenylphosphine oxide[19] (21°, 25°, 59°). According to empirical force-field calculations,[20] the lowest energy conformation of trimesitylmethane has C_3 symmetry with $\phi = 41°$. Andose and Mislow[20] calculated minimum energy paths for the stereoisomerization of this conformation, paths connecting CCC and AAA in our terminology. The calculations were done by carrying out a series of constrained energy minimizations for various fixed incremental values of one of the torsion angles; two paths, with approximately equal energy barriers of about 20 kcal/mol, were found. Starting at the CCC conformation $\phi_1 = \phi_2 = \phi_3 = +40°$ (or $x_H = y_H = 0$, $z_H = 40°$ in hexagonal coordinates) the transition states and isometric AAA end points for paths A and B are:

	ϕ_1	ϕ_2	ϕ_3	x_H	y_H	z_H	
CCC	+ 40°	+40°	+40°	0°	0°	+40°	
transition state	+105	+22	−10	+ 66	+49	+39	Path A
AAA	+140	−40	−40	+120	+60	+20	
CCC	+ 40°	+40°	+40°	0°	0°	+40°	
transition state	+ 88	+38	−35	+ 58	+65	+30	Path B
AAA	+140	−40	−40	+120	+60	+20	

It may be of interest to see how these paths and the various symmetry-related ones look on the three-dimensional conformational map. They are shown in Fig. 10.15, where, for visualization, the hexagonal coordinate system is used. The map shows the section of the hexagonal cell bounded by $z_H \sim +45°$ and $z_H \sim +20°$. The calculated paths are indicated by the curves: solid lines for the portions with $z_H > 30°$, dashed lines for $z_H < 30°$. The twofold rotation axes of the repeating pattern (space group $R32$) occur at $z_H = 30°$ ($\pi/6$), and are indicated by arrowhead lines.

[16]P. Andersen, *Acta Chem. Scand.* **19,** 622 (1965).
[17]N. I. Sadova, N. I. Popik, L. V. Vilkov, Ju. A. Pankrushev, and V. A. Shlyapochnikov, *Chem. Commun.* **1973,** 708.
[18]J. J. Daly, *J. Chem. Soc.* **1964,** 3799.
[19]G. Bandoli, G. Bortolozzo, D. A. Clemente, U. Croatto, and C. Panattoni, *J. Chem. Soc., A,* **1970,** 2778.
[20]J. D. Andose and K. Mislow, *J. Am. Chem. Soc.* **96,** 2168 (1974).

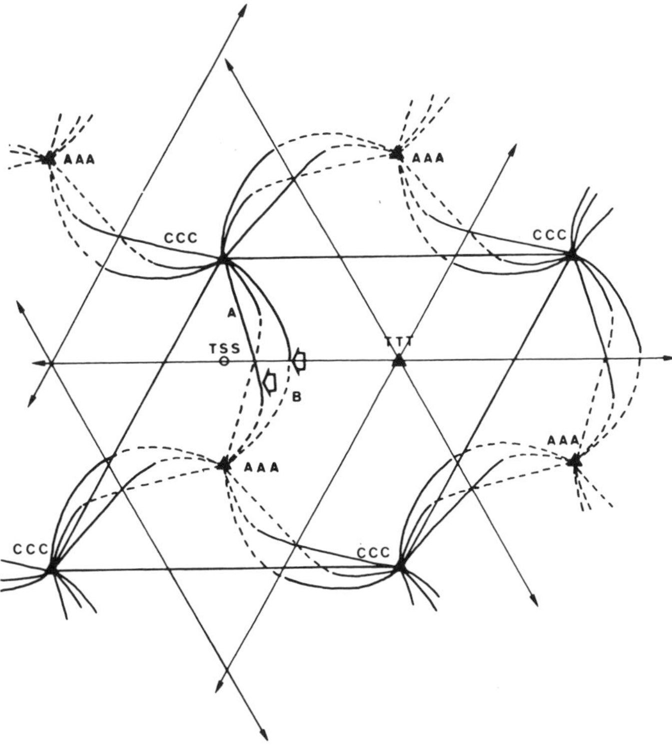

Figure 10.15. Minimum energy paths for the stereoisomerization CCC ⇌ AAA of trimesitylmethane, based on calculations by Andose and Mislow.[20] The hexagonal cell has unit translations of 180°. All paths from CCC (0, 0, 40°) to AAA (120°, 60°, 20°) or (−60°, −120°, 20°) must pass through $z_H = 30°$, containing the twofold rotation axes of the space group, indicated by arrowhead lines. Path A has its transition state (66°, 49°, 39°) about midway between TSS and TCS; path B has its transition state (−65°, −58°, 30°) close to TCA. Portions of the paths with $z_H > 30°$ are solid lines; portions with $z_H < 30°$ are dashed lines.

Several features of the map deserve comment. The CCC conformation and its enantiomer AAA are represented by points on the threefold axes of the map. At each of these points 12 paths should, in general, converge—an outgoing and incoming A, together with an outgoing and incoming B, each in triplicate. Clearly 12 energy valleys emanating from a point cannot be very well defined in the immediate neighbourhood of the point. Both types of path avoid the TSS conformation at $z_H = 30°$ that would lie on a minimum motion path, the shortest path from CCC to AAA. Path A has its transition state at a point with $z_H \sim 40°$, well away from any twofold axis, but for path B the estimated transition state lies within a few degrees (probably within the error of the calculation) of a twofold axis. If path B actually passes through this axis, then by symmetry the reaction coordinate at the crossing point must be perpendicular to the axis. The calculated path does satisfy this condition at least approximately. Moreover, if path B crosses the axis, the second half of the path must obviously follow by symmetry from the first half. According to the calculations, it appears to do so except near the C_3 axes where it cannot be very well defined in any case. If the path B is exactly symmetrical, then outgoing and incoming B paths coincide, so that only nine rather than 12 distinct paths converge at CCC and AAA conformations, as shown in Fig. 10.15.

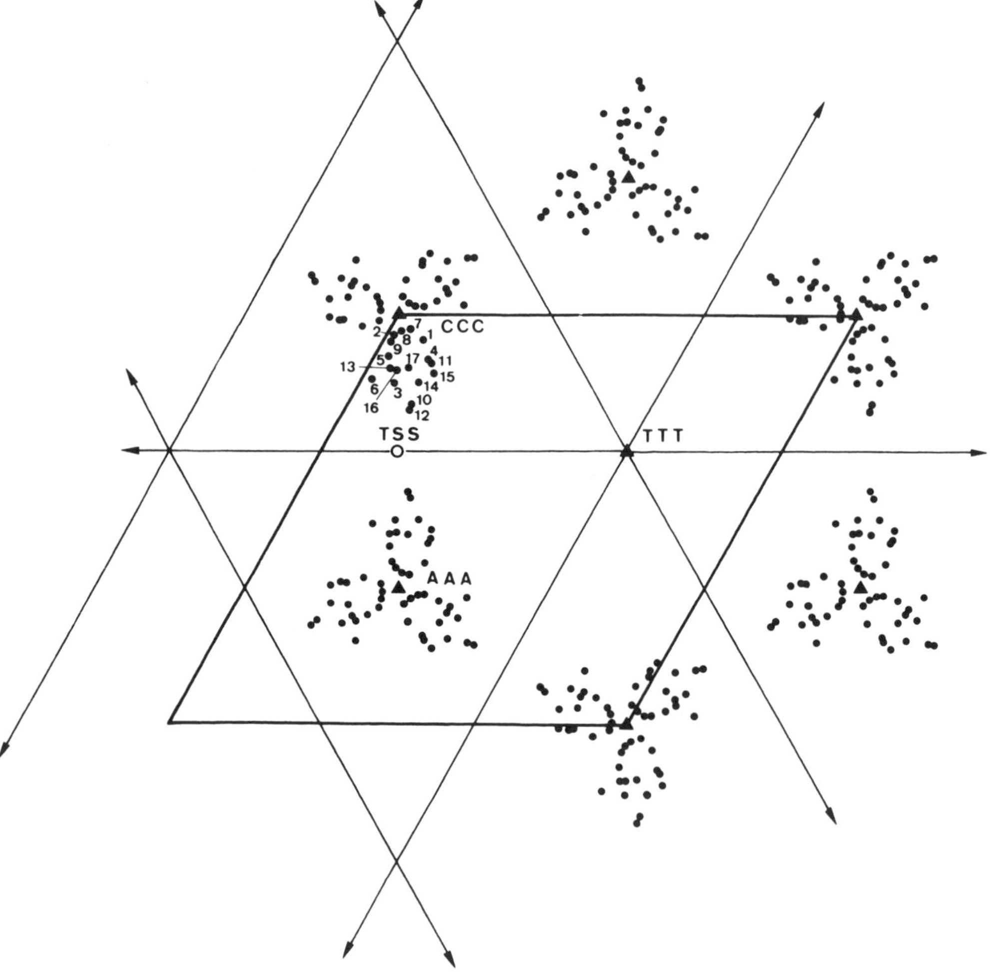

Figure 10.16. Scatter plot of 17 observed conformations of triphenylphosphine oxide molecules in various crystal structures. The ϕ_1, ϕ_2, ϕ_3 torsion angles have been converted to hexagonal coordinates x_H, y_H, z_H. Numbered points have z_H values between 30° and 45°; the others are related by the symmetry operations of the space group $R32$. Compare this with the calculated reaction path for the stereoisomerization of trimesitylmethane (Fig. 10.15).

If enough crystal structure data for triarylmethane molecules in different environments were available, we could hope to map the low energy regions of the conformational map from scatter plots (as described in Chapter 7 for several other systems). Unfortunately, not enough information for this class of molecules is available. However, as far as the symmetry arguments are concerned, we could just as well have chosen triphenylphosphine or any other molecule that contains three equivalent twofold rotors as our example. Many crystal structures containing metal complexes of triphenylphosphine are available. Depending on the metal and on the other ligands, the C—P—C bond angles vary over quite a large range, and this variation will also influence the preferred ranges of torsion angle. For triphenylphosphine oxide, the C—P—C bond angles do not vary so much from one crystal structure to another. One might therefore expect the torsion angles observed for this molecule in various environments to yield a distribution of sample points (ϕ_1, ϕ_2, $\phi_3 \rightarrow x_H$, y_H, z_H) from which the low energy regions of the corresponding conformational map might be recognizable.

Fig. 10.16 shows results[21] for 17 examples of triphenylphosphine oxide extracted from the Cambridge Crystal Structure Data File. The ϕ_1, ϕ_2, ϕ_3 values were calculated from the published atomic positions and converted to hexagonal coordinates, x_H, y_H, z_H. On the assumption that the sample points tend to lie in low-energy regions of the map, the observed distribution provides a diffuse but nevertheless recognizable reaction path for the CCC $\leftrightarrow$ AAA stereoisomerization process of this molecule. The experimental path is quite similar to the calculated paths for trimesitylenemethane[20] (Fig. 10.15)—an exact correspondence cannot be expected in view of the structural differences between these molecules.

Dibenzyl

To begin we postpone the question of the frame symmetry. An arbitrary conformation of the molecule is specified by three torsion angles, denoted in Fig. 10.17, and it is not difficult to convince oneself that, apart from ring flips, the only torsion angle transformations that lead to isometric conformations are:

	E	F_1	F_2	F_3
ω_1	ϕ_1	ϕ_3	$-\phi_1$	$-\phi_3$
ω_2	ϕ_2	ϕ_2	$-\phi_2$	$-\phi_2$
ω_3	ϕ_3	ϕ_1	$-\phi_3$	$-\phi_1$

[21] E. Bye, unpublished results.

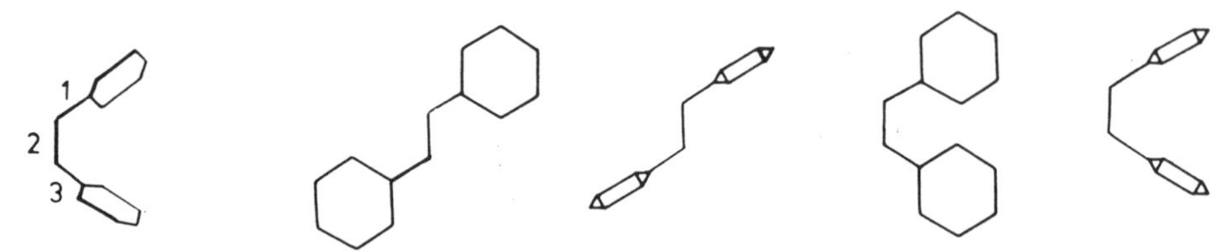

Figure 10.17. An arbitrary conformation of dibenzyl can be specified by the torsion angles around bonds 1, 2, and 3. There are four special conformations with highest possible symmetry (C_{2h} or C_{2v}); the first two have $\omega_2 = \pi$, and the second two have $\omega_2 = 0$.

Without loss of generality we can imagine the central bond to lie in the plane of the paper. The transformation F_1 then corresponds to a 180° rotation about an axis perpendicular to the paper plane (or to a corresponding change of axes). Transformation F_2 corresponds to reflection of the molecule across the plane of the paper, the transformation F_3 corresponds to the combination of these operations—i.e., inversion of the molecule through a point. According to our earlier definition, the frame symmetry **F** is then C_{2h} $\{E, C_2, \sigma_h, i\}$ of order 4, and this is also the order of the primitive unit-cell group of the conformational map. The point symmetry of the unit-cell group (Fig. 10.18) is also $C_{2h}(2/m)$, but the frame element C_2 is mapped on the unit-cell element m, σ_h on $\bar{1}$, and i on 2.[22]

The rotor group presents no difficulties. For two phenyl substituents it consists (as in previous examples) of the four elements, E, R_1, R_3, R_{13}, that are mapped by translations: 0, 0, 0; π, 0, 0; 0, 0, π; π, 0, π. The total molecular symmetry group $\mathbf{H} = \mathbf{R} \wedge \mathbf{F}$ then consists of 16 elements, and the same is true for the corresponding nonprimitive unit cell group based on 2π translations along ω_1, ω_2, and ω_3. The general and equivalent positions of the unit cell group are given in Table 10.7.

Dibenzyl is an example of a molecule whose highest attainable symmetry is expressed in terms of two point groups, C_{2h} and C_{2v}, depending on whether the torsion angle around the central bond (ω_2) is π or 0. Here we adopt the usual rule of taking zero torsion angle when bonds 1 and 3 are eclipsed (as shown in the two right diagrams in Fig. 10.17), but this is only an arbitrary convention, and we could equally well have taken this position as the one corresponding to torsion angle π if Klyne and Prelog[23] had so decreed. According to the usual convention, the origin

[22]We use the Schoenfliess symbols, C_2, σ_h, i, etc., to describe frame symmetry elements and Hermann-Mauguin symbols, 2, m, $\bar{1}$, etc., to describe symmetry elements of three-dimensional space groups.

[23]W. Klyne and V. Prelog, *Experientia* **16**, 521 (1960). The Klyne-Prelog rule is consistent with the rules of vector multiplication, but these are also based on arbitrary conventions.

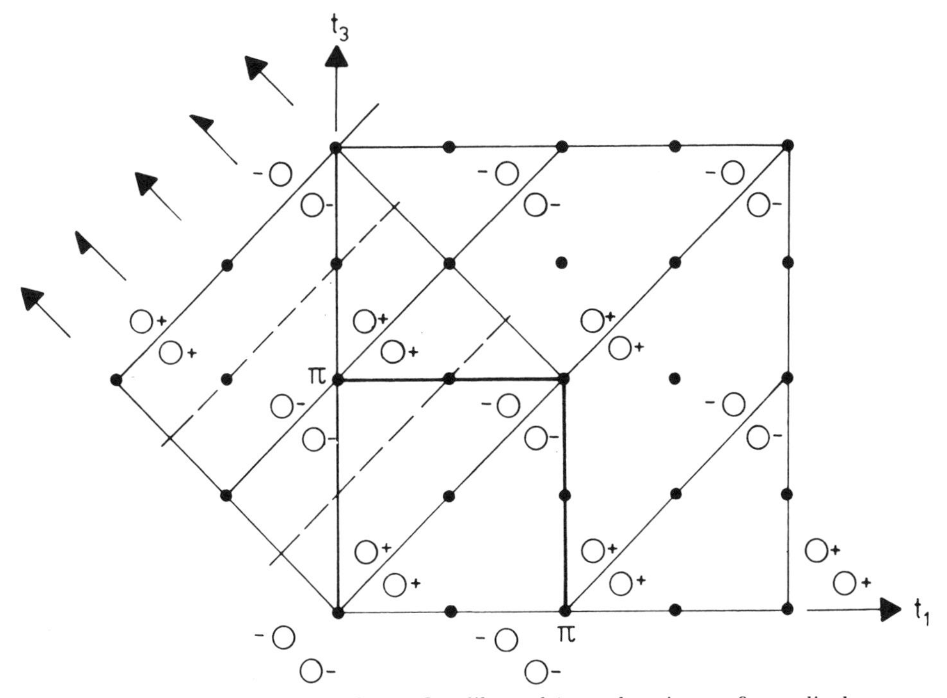

Figure 10.18. Conformational map for dibenzyl (+ and − signs refer to displacements (modulo 2π) along the coordinate axis ω_2 (perpendicular to the plane of the paper). The equivalent positions (open circles) are images of isometric conformations with arbitrary values of torsion angles. Special positions are marked. The primitive cell has translations $\mathbf{t}_1$, $\mathbf{t}_3$ of length π and $\mathbf{t}_2$ of length 2π. The space group of the map is $C2/m(C_{2h}^3)$ with unit translations $\mathbf{a} = \mathbf{t}_1 + \mathbf{t}_3$, $\mathbf{b} = -\mathbf{t}_1 + \mathbf{t}_3$, $\mathbf{c} = \mathbf{t}_2$.

Table 10.7. General and special positions (modulo 2π) of conformational map for dibenzyl, referred to nonprimitive unit cell defined by translations of 2π along ω_1, ω_2, ω_3[a]

Number	Point symmetry	Coordinates of equivalent positions $(0,0,0;\ \pi,0,0;\ 0,0,\pi;\ \pi,0,\pi)+$
16	$1(1)$	$\phi_1,\phi_2,\phi_3;\ \phi_3,\phi_2,\phi_1;\ -\phi_1,-\phi_2,-\phi_3;\ -\phi_3,-\phi_2,-\phi_1$
8	$m(C_2)$	$\phi_1,\phi_2,\phi_1;\ -\phi_1,-\phi_2,-\phi_1$
8	$2(i)$	$\phi,0,-\phi;\ -\phi,0,\phi$
8	$2(i)$	$\phi,\pi,-\phi;\ -\phi,\pi,\phi$
8	$\bar{1}(\sigma_h)$	$0,0,\pi/2;\ \pi/2,0,0$
8	$\bar{1}(\sigma_h)$	$0,\pi,\pi/2;\ \pi/2,\pi,0$
4	$2/m(C_{2h})$	$\pi/2,\pi,\pi/2$
4	$2/m(C_{2h})$	$\pi/2,0,\pi/2$
4	$2/m(C_{2h})$	$0,\pi,0$
4	$2/m(C_{2h})$	$0,0,0$

[a]Point symmetries are given both for the map and for the corresponding molecular conformations. The conventional space group is $C2/m(C_{2h}^3)$.

(0, 0, 0) of our conformational map corresponds to a molecular conformation with C_{2v} symmetry, and the conformation with C_{2h} symmetry occurs at (0, π, 0). In the alternative convention, the identification of particular conformations with these nonequivalent special positions having the same site symmetry would simply be reversed. All that is involved is a change of origin by π along the ω_2 torsion-angle coordinate. Similarly, conformations represented by the two kinds of special positions of site symmetry 2 (frame operation i), ϕ, 0, $-\phi$ and ϕ, π, $-\phi$ have molecular symmetry C_i or C_s, depending on how the origin of the ω_2 torsion angle is defined.

The extension to other molecules of the type R—X—X—R leaves the frame symmmetry (the primitive unit-cell group) unaltered. Only the lattice translations, corresponding to the rotor group, have to be changed, depending on the nature of the terminal R groups. For butane, for example, with threefold rotors, the group of translations consists of nine elements:

$$\begin{array}{lll} 0, 0, 0; & 2\pi/3, 0, 0; & 4\pi/3, 0, 0 \\ 0, 0, 2\pi/3; & 2\pi/3, 0, 2\pi/3; & 4\pi/3, 0, 2\pi/3 \\ 0, 0, 4\pi/3; & 2\pi/3, 0, 4\pi/3; & 4\pi/3, 0, 4\pi/3 \end{array}$$

The unit cell group is now of order 36. For ethylene glycol, HO—CH_2—CH_2—OH, the rotor group consists only of the identity element, and the order of the unit-cell group is reduced to 4, the order of the frame group. The general and special positions of the conformational map are the same as those in Table 10.7, but their numbers must be divided by 4 because the translation operations disappear.

From ab initio SCF calculations for ethylene glycol, two potential energy minima have been located—one corresponding to the conformation (60°, 60°, −30°), the other, approximately 2.1 kcal/mol higher in energy, corresponding to (−45°, 60°, −160°).[24]

Trimethylboron

As a final three-dimensional example, we discuss briefly some features of the conformational map of molecules assumed to have a D_{3h} frame (e.g., trimethylboron, $B(CH_3)_3$; orthoboric acid $B(OH)_3$). The torsion angle transformations are:

[24]T. K. Ha, H. Frei, R. Meyer, and H. H. Günthard, *Theor. Chim. Acta (Berlin)* **34,** 277 (1974). This paper contains a more formal derivation of the isometric group of a semirigid model of the ethylene glycol molecule with three internal degrees of freedom. The results of the SCF calculations are depicted in the form of a section (ω_2 = 60°) of a conformational map of the type discussed in this chapter.

	E	C_3	C_3^2	σ_a	σ_b	σ_c
ω_A	ϕ_1	ϕ_3	ϕ_2	$-\phi_1$	$-\phi_3$	$-\phi_2$
ω_B	ϕ_2	ϕ_1	ϕ_3	$-\phi_3$	$-\phi_2$	$-\phi_1$
ω_C	ϕ_3	ϕ_2	ϕ_1	$-\phi_2$	$-\phi_1$	$-\phi_3$

	σ_h	S_3	S_3^2	$C_2(a)$	$C_2(b)$	$C_2(c)$
ω_A	$-\phi_1+\frac{1}{2}$	$-\phi_3+\frac{1}{2}$	$-\phi_2+\frac{1}{2}$	$\phi_1+\frac{1}{2}$	$\phi_3+\frac{1}{2}$	$\phi_2+\frac{1}{2}$
ω_B	$-\phi_2+\frac{1}{2}$	$-\phi_1+\frac{1}{2}$	$-\phi_3+\frac{1}{2}$	$\phi_3+\frac{1}{2}$	$\phi_2+\frac{1}{2}$	$\phi_1+\frac{1}{2}$
ω_C	$-\phi_3+\frac{1}{2}$	$-\phi_2+\frac{1}{2}$	$-\phi_1+\frac{1}{2}$	$\phi_2+\frac{1}{2}$	$\phi_1+\frac{1}{2}$	$\phi_3+\frac{1}{2}$

where, for convenience, the angles are now expressed in cycles (in units of 2π). As before, we have taken $\phi = 0$ when a given atom of each rotor is oriented *syn*-planar to the threefold axis (Fig. 10.19). The entries in

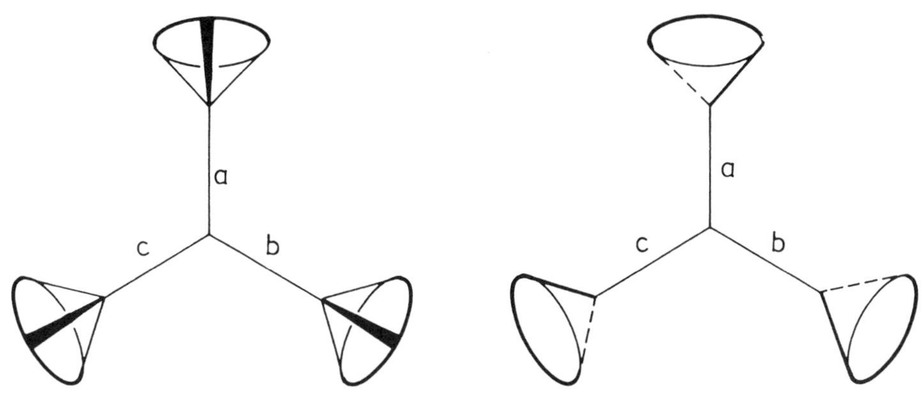

Figure 10.19. Different choices of zero torsion angle for rotors on D_{3h} frame. The choice on the right is the one that leads to torsion angle transformations that match the general positions of space group $R\bar{3}c$ with its origin on a center of symmetry.

the upper part of the list above are those derived earlier for a C_{3v} frame (triphenylmethane); those in the lower part, which extend the order of the frame group to 12, show a new feature—they involve combinations of point symmetry and translation by π along each of the coordinate axes. Any attempt to find the above list of general equivalent positions in tables of space groups would be in vain, however. Because of an unsuitable choice of origin, the coordinates are not in a form that corresponds to a standard description of a space group. Subtract ($\pi/2$, $\pi/2$, $\pi/2$) or, in cycles ($\frac{1}{4}$, $\frac{1}{4}$, $\frac{1}{4}$), from each of the above positions and write

$$x = \phi_1 - \tfrac{1}{4} \qquad y = \phi_2 - \tfrac{1}{4} \qquad z = \phi_3 - \tfrac{1}{4}$$

to obtain a revised list:

	E	C_3	C_3^2	σ_a	σ_b	σ_c
ω_A	x	z	y	$-x+\frac{1}{2}$	$-z+\frac{1}{2}$	$-y+\frac{1}{2}$
ω_B	y	x	z	$-z+\frac{1}{2}$	$-y+\frac{1}{2}$	$-x+\frac{1}{2}$
ω_C	z	y	x	$-y+\frac{1}{2}$	$-x+\frac{1}{2}$	$-z+\frac{1}{2}$
	σ_h	S_3	S_3^2	$C_2(a)$	$C_2(b)$	$C_3(c)$
ω_A	$-x$	$-z$	$-y$	$x+\frac{1}{2}$	$z+\frac{1}{2}$	$y+\frac{1}{2}$
ω_B	$-y$	$-x$	$-z$	$z+\frac{1}{2}$	$y+\frac{1}{2}$	$x+\frac{1}{2}$
ω_C	$-z$	$-y$	$-x$	$y+\frac{1}{2}$	$x+\frac{1}{2}$	$z+\frac{1}{2}$

which can be recognized as the general equivalent positions of the space group $R\bar{3}c$ with origin at a center of symmetry. With some practice an inconvenient choice of origin can be detected fairly quickly, but even if it is not, no great harm is done. The standard choice merely makes it easier to read off the special positions:

Number	Symmetry	Coordinates (in units of 2π)
6	$2(\sigma_v)$	$x, \frac{1}{2}-x, \frac{1}{4}$; $\frac{1}{2}-x, \frac{1}{4}, x$; $\frac{1}{4}, x, \frac{1}{2}-x$; $\bar{x}, \frac{1}{2}+x, \frac{3}{4}$; $\frac{1}{2}+x, \frac{3}{4}, \bar{x}$; $\frac{3}{4}, \bar{x}, \frac{1}{2}+x$
6	$\bar{1}(\sigma_h)$	$\frac{1}{2}, 0, 0$; $0, \frac{1}{2}, 0$; $0, 0, \frac{1}{2}$; $0, \frac{1}{2}, \frac{1}{2}$; $\frac{1}{2}, 0, \frac{1}{2}$; $\frac{1}{2}, \frac{1}{2}, 0$
4	$3(C_3)$	x, x, x; $\bar{x}, \bar{x}, \bar{x}$; $\frac{1}{2}+x, \frac{1}{2}+x, \frac{1}{2}+x$; $\frac{1}{2}-x, \frac{1}{2}-x, \frac{1}{2}-x$
2	$\bar{3}(C_{3h})$	$0, 0, 0$; $\frac{1}{2}, \frac{1}{2}, \frac{1}{2}$
2	$32(C_{3v})$	$\frac{1}{4}, \frac{1}{4}, \frac{1}{4}$; $\frac{3}{4}, \frac{3}{4}, \frac{3}{4}$

The unbracketed symmetry symbols refer to the conformational map, and the bracketed ones refer to the molecular conformations that are mapped on these positions. Note that the special positions of highest site symmetry still represent conformations with the highest point-group symmetries attainable by the molecule, C_{3h} or C_{3v}, but these are now subgroups of D_{3h}, the frame symmetry.

This completes the analysis of $B(OH)_3$ where the rotor group consists only of the identity element, but for $B(CH_3)_3$ with three threefold rotors, each general position would be repeated by the translation operations of the rotor group to produce additional lattice points at intervals of $\frac{1}{3}$ along each coordinate axis. With a 27-fold centering and 12 elements in the frame group, the molecular symmetry group **H** is then of order 324.

For three twofold rotors, as in triphenylboron or the triphenylcarbonium ion, the general positions are all repeated at intervals of $\frac{1}{2}$ along each coordinate axis to produce an eight-fold centered cell of order 96. Moreover a higher space-group symmetry is now attained since the translation $(\frac{1}{2}, \frac{1}{2}, \frac{1}{2})$ converts general positions of type $x+\frac{1}{2}, z+\frac{1}{2}, y+\frac{1}{2}$ into x, z, y. The new set of 12 general positions is then:

$$\pm(x,y,z;\ z,x,y;\ y,z,x;\ y,x,z;\ z,y,x;\ x,z,y)$$

making the space group $R\bar{3}m$. Now all general positions of the primitive cell (modulo $\frac{1}{2}$) correspond to point-group operations—there are none that have translational components. Special positions (and corresponding molecular symmetries) are:

Number	Symmetry	Coordinates (in units of π)
6	$m(C_2)$	x, x, z, etc.
6	$2(C_s)$	$x, \bar{x}, \frac{1}{2}$, etc.
6	$2(C_s)$	$x, \bar{x}, 0$, etc.
3	$2/m(C_{2v})$	$0, \frac{1}{2}, \frac{1}{2}$, etc.
3	$2/m(C_{2v})$	$\frac{1}{2}, 0, 0$, etc.
2	$3m(D_3)$	x, x, x, etc.
1	$\bar{3}m(D_{3h})$	$\frac{1}{2}, \frac{1}{2}, \frac{1}{2}$
1	$\bar{3}m(D_{3h})$	$0, 0, 0$

For the eight-fold centered cell of order 96, these numbers are to be multiplied by 8.

Four-Dimensional Space Groups

Thinking about objects or patterns in spaces of more than three dimensions strains our capacity for geometric visualization. This difficulty is not too serious as long as we do not have to worry about symmetry properties. We can always construct a generalized n-dimensional vector space spanned by a set of n basis vectors

$$\mathbf{a}_1, \mathbf{a}_2, \mathbf{a}_3 \ldots \mathbf{a}_n$$

and describe the points in this space by a vector $\mathbf{V}$ with n components:

$$\mathbf{V} = \sum_k x_k \mathbf{a}_k \qquad k = 1, 2, \ldots n$$

Lengths of vectors and angles between vectors can then be expressed in terms of scalar products, just as in two or three dimensions. It is much more difficult to think about crystallographic symmetry operations in hyperspaces because so many of the rules that apply in two- and three-dimensional space are no longer valid. For example, although rotation axes of two- and three-dimensional lattices are restricted to orders 2, 3, 4, and 6, a four-dimensional lattice may contain rotation axes of orders 5, 8, and 12 as well. In six-dimensional space, seven-fold rotation axes are possible. Hermann has discussed ways to derive the possible symmetry operations of lattices in spaces of arbitrary dimensionality.[25]

An additional difficulty is that there are no generally agreed names for most of the symmetry operations in four-dimensional space. The only exception, perhaps, is inversion through a point, corresponding to the transformation: $w, x, y, z \rightarrow -w, -x, -y, -z$. However, whereas inversion is an improper operation in three-dimensional space, it is a proper operation in four dimensions (as in two or any even number of dimensions)—i.e., it preserves chirality. But how should we describe the effect of the transformation: $w, x, y, z \rightarrow -w, x, y, z$? Reflection across a three-dimensional space? For that is what it is, just as reflection across a plane can be described as reflection across a two-dimensional space. In general, it seems best to avoid words that have special geometric connotations in three-dimensional space and to describe the symmetry operations simply as coordinate transformations. The order of the operation can then be found by repeating the transformation the necessary number of times to produce the identity transformation. Thus, in three dimensions, the transformation:

$$M(x,y,z) = z,x,y$$

$$M^2(x,y,z) = y,z,x$$

$$M^3(x,y,z) = x,y,z$$

corresponds to an operation of order 3 (actually a rotation of $2\pi/3$ about a line equally inclined to the three coordinate axes). It is easily confirmed that in four-dimensional space the transformation $a, b, c, d \rightarrow c, a, d, b$ corresponds to an operation of order 4:

$$M(a,b,c,d) = c,a,d,b$$

$$M^2(a,b,c,d) = d,c,b,a$$

$$M^3(a,b,c,d) = b,d,a,c$$

$$M^4(a,b,c,d) = a,b,c,d$$

but it is not as easy to imagine the "geometric meaning" of this operation. Since the determinant

$$\begin{vmatrix} 0 & 0 & 1 & 0 \\ 1 & 0 & 0 & 0 \\ 0 & 0 & 0 & 1 \\ 0 & 1 & 0 & 0 \end{vmatrix} = +1$$

[25]C. Hermann, *Acta Crystallogr.* **2,** 139 (1949).

it must correspond to a proper operation at any rate. With these preliminary remarks, we can begin to discuss some four-dimensional conformational maps.

1,1,1-Trichloropentane

Apart from rotations of the two terminal groups (rotor operations), the only torsion angle transformation that converts an arbitrary conformation of 1,1,1-trichloropentane (Fig. 10.20) into an isometric one is reversal of sign of all four torsion angles:

	a	b	c	d
E	ϕ_1	ϕ_2	ϕ_3	ϕ_4
F_1	$-\phi_1$	$-\phi_2$	$-\phi_3$	$-\phi_4$

corresponding to reversal of the chirality sense of the original, arbitrary conformation. The primitive unit cell of the conformational map thus contains only two general positions, related by inversion across the origin. The space group is accordingly $P_4\bar{1}$. As mentioned in the previous section, inversion across the origin is a proper rotation in four dimensions; the determinant of the transformation matrix is +1. (We are speaking here about the symmetry properties of the conformational map, not about the symmetry operation that converts the three-dimensional molecule into its enantiomer.)

The rotor group consists of nine elements that are mapped on a set of nine translations:

$$\begin{array}{lll} 0, 0, 0, 0; & \tfrac{1}{3}, 0, 0, 0; & \tfrac{2}{3}, 0, 0, 0; \\ 0, 0, 0, \tfrac{1}{3}; & \tfrac{1}{3}, 0, 0, \tfrac{1}{3}; & \tfrac{2}{3}, 0, 0, \tfrac{1}{3}; \\ 0, 0, 0, \tfrac{2}{3}; & \tfrac{1}{3}, 0, 0, \tfrac{2}{3}; & \tfrac{2}{3}, 0, 0, \tfrac{2}{3}; \end{array}$$

referred to the unit translations of 2π. The unit cell group (total molecular symmetry group) is then of order 18. Its $2^4 \times 3^2 = 144$ special positions, all with point symmetry $\bar{1}$ occur at

$$\begin{array}{llll} 0, 0, 0, 0; & 0, 0, \tfrac{1}{2}, 0; & 0, 0, \tfrac{1}{2}, \tfrac{1}{6}; & 0, 0, 0, \tfrac{1}{6}; \\ 0, \tfrac{1}{2}, 0, 0; & 0, \tfrac{1}{2}, \tfrac{1}{2}, 0; & 0, \tfrac{1}{2}, \tfrac{1}{2}, \tfrac{1}{6}; & 0, \tfrac{1}{2}, 0, \tfrac{1}{6}; \\ \tfrac{1}{6}, 0, 0, 0; & \tfrac{1}{6}, \tfrac{1}{2}, 0, 0; & \tfrac{1}{6}, \tfrac{1}{2}, \tfrac{1}{2}, 0; & \tfrac{1}{6}, \tfrac{1}{2}, \tfrac{1}{2}, \tfrac{1}{6}; \\ \tfrac{1}{6}, \tfrac{1}{2}, 0, \tfrac{1}{6}; & \tfrac{1}{6}, 0, \tfrac{1}{2}, 0; & \tfrac{1}{6}, 0, \tfrac{1}{2}, \tfrac{1}{6}; & \tfrac{1}{6}, 0, 0, \tfrac{1}{6}; \end{array}$$

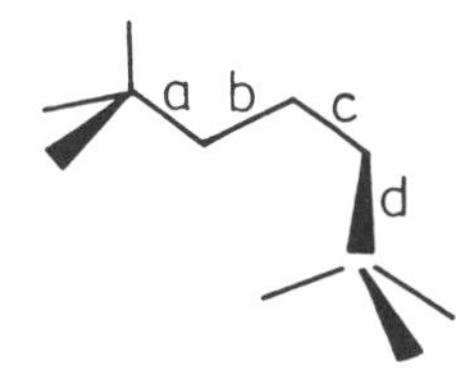

Figure 10.20. Arbitrary conformation of 1,1,1-trichloropentane.

referred to each of the nine lattice points in the $(2\pi)^4$ unit cell, and they are images of conformations with C_s symmetry. The fact that large regions of the map would correspond to impossible conformations—i.e., ones of extremely high energy—is irrelevant in the discussion of its symmetry properties.

Tetraphenylmethane and Related Molecules

In contrast to the previous example, the symmetry properties of this molecule are not straightforward.[26] The frame consists of a skeleton that has T_d symmetry, but the highest point symmetry attainable by the molecule is D_{2d}, with all four rings lying in the mirror planes or perpendicular to them (Fig. 10.21). If all four rings are rotated in the same sense and by equal amounts from one of these orientations, the molecular symmetry is lowered to D_2. However, when the common rotation angle reaches 60°, the D_{2d} symmetry is recovered and the S_4 axis reappears in a new direction at right angles to the original direction. At first sight this property seems somewhat mysterious (rather like a conjuring trick), and readers with more than a casual interest in the problem should construct a molecular model to convince themselves about its authenticity.

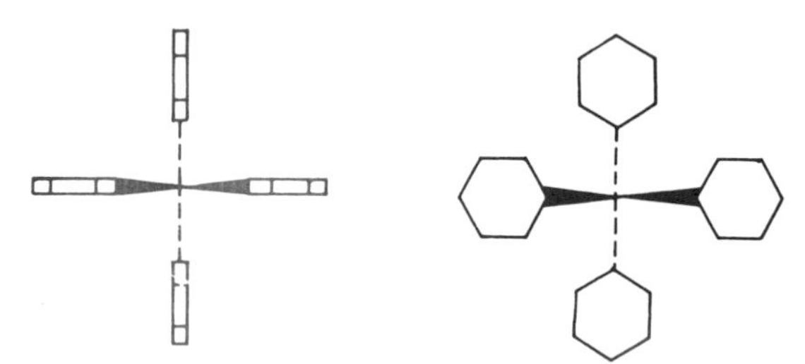

Figure 10.21. Conformations of tetraphenylmethane showing highest attainable symmetry, D_{2d}.

It is convenient to start by considering the slightly simpler case of tetrahydroxymethane (Fig. 10.22), where the rotor group has been reduced to a single element. The first problem is that of defining the zero positions of the torsion angles. With four equivalent directions of the T_d frame, this task requires special care because the numerical value of the torsion angle about any one of the axes depends on which of the other three axes is taken as a reference. For example, the torsion angle about bond a in Fig. 10.22 could be taken as −90°, +30°, or +150° with respect to c, b, or d. One choice is as good as another, but once we make a choice, we must stick to it. Say we choose $\phi_a = -90°$ to describe

[26]See M. G. Hutchings, J. G. Nourse, and K. Mislow, *Tetrahedron* **30**, 1535 (1974); J. G. Nourse and K. Mislow, *J. Am. Chem. Soc.* **97**, 4571 (1975); and F. Strohbusch, *Tetrahedron* **30**, 1261 (1974) for extensive discussions from a permutational approach.

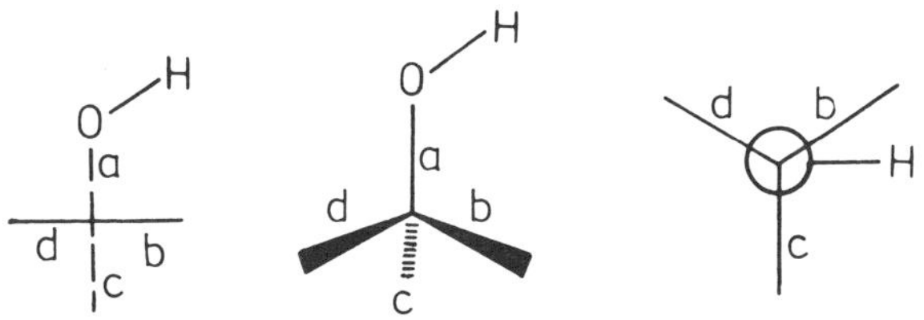

Figure 10.22. Torsion angle about bond a in $C(OH)_4$.

the orientation in question. This implies that ϕ_a is measured against c. If we now change axes to put c in place of b or d, we have to add or to subtract 120° to obtain the transformed torsion angle. With the axes labeled as in Fig. 10.22 the following relationships hold:

$$\begin{aligned}
\phi_a &= \phi_a(c) = \phi_a(b) - 120° = \phi_a(d) + 120° \\
\phi_b &= \phi_b(d) = \phi_b(c) + 120° = \phi_b(a) - 120° \\
\phi_c &= \phi_c(a) = \phi_c(d) - 120° = \phi_c(b) + 120° \\
\phi_d &= \phi_d(b) = \phi_d(a) + 120° = \phi_d(c) - 120°
\end{aligned}$$

We can now define $\phi_a = \phi_b = \phi_c = \phi_d = 0$ as corresponding to the situation shown in Fig. 10.23 where H_a is *syn*-planar to bond c, H_b to bond d, H_c to bond a, and H_d to bond b. This situation corresponds to the highest symmetry attainable by the molecule, which is D_{2d}, of order 8. It will be mapped on a special position of order 8. However, the order of the T_d frame group and of the corresponding unit cell group is 24, so that the special position of highest point symmetry must occur three times in the primitive unit cell. We therefore expect that of the 24 general positions only eight are related by pure point-symmetry operations; the remainder are related to the representative point $P(\phi_1, \phi_2, \phi_3, \phi_4)$ by operations that include translations. This is indeed the case; the operations that include translations are the images of frame operations that add or subtract 120° from the torsion angles as defined above.

A complete list of torsion-angle transformations (general positions) corresponding to the 24 possible interchanges of the axes (see Fig. 10.24

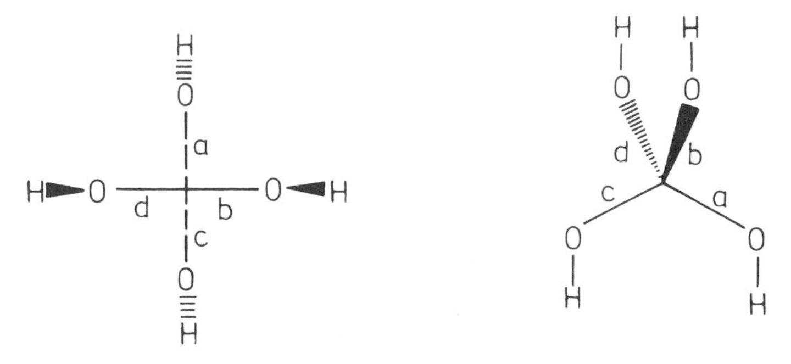

Figure 10.23. Definition of torsion angle zero for $C(OH)_4$.

for examples) of the T_d frame is given in Table 10.8. The special positions are listed in Table 10.9. For $C(OH)_4$ the unit translations are 2π for all four coordinates.

For tetraphenylmethane an additional set of 16 pure translation operations has to be included to map the group of rotor operations that characterize this molecule. These translations are, in units of 2π:

$0, 0, 0, 0;$	$\frac{1}{2}, \frac{1}{2}, \frac{1}{2}, \frac{1}{2};$		
$0, 0, 0, \frac{1}{2};$	$0, 0, \frac{1}{2}, 0;$	$0, \frac{1}{2}, 0, 0;$	$\frac{1}{2}, 0, 0, 0;$
$0, 0, \frac{1}{2}, \frac{1}{2};$	$0, \frac{1}{2}, 0, \frac{1}{2};$	$0, \frac{1}{2}, \frac{1}{2}, 0;$	
$\frac{1}{2}, 0, 0, \frac{1}{2};$	$\frac{1}{2}, 0, \frac{1}{2}, 0;$	$\frac{1}{2}, \frac{1}{2}, 0, 0;$	
$\frac{1}{2}, \frac{1}{2}, \frac{1}{2}, 0;$	$\frac{1}{2}, \frac{1}{2}, 0, \frac{1}{2};$	$\frac{1}{2}, 0, \frac{1}{2}, \frac{1}{2};$	$0, \frac{1}{2}, \frac{1}{2}, \frac{1}{2}.$

The order of the molecular symmetry group then becomes $16 \times 24 = 384$. The general and special positions listed in Tables 10.8 and 10.9 still apply, except that they are now measured in units of π rather than 2π since the primitive cell is halved along each of the four axes. Thus, for tetraphenylmethane the special position $(\frac{1}{6}, \frac{1}{6}, \frac{1}{6}, \frac{1}{6})$ is (30°, 30°, 30°, 30°) in terms of torsion angles.

For neopentane, with four threefold rotors, the rotor operations are mapped on a group of $3^4 = 81$ translation operations, giving a molecular symmetry group of order $24 \times 81 = 1944$. With the additional lattice points at intervals of $2\pi/3$ along the coordinate axes, the general positions of Table 10.8 can be written modulo $t = 2\pi/3$ instead of modulo 2π, so that their translation components disappear. This leads to new types of special positions, e.g., $(\phi_1, \phi_2, \phi_2, \phi_2)$, images of conformations with C_3 symmetry. The special positions ϕ, ϕ, ϕ, ϕ (see Table 10.9) then image conformations of point symmetry $D_2 \times C_3 = T$, and those of type (0, 0, 0, 0) and $(\frac{1}{2}, \frac{1}{2}, \frac{1}{2}, \frac{1}{2})$ (now in $2\pi/3$ units) image conformations with the full frame symmetry T_d.

It may be convenient for certain purposes to refer the conformational map to a different coordinate system that provides an alternative por-

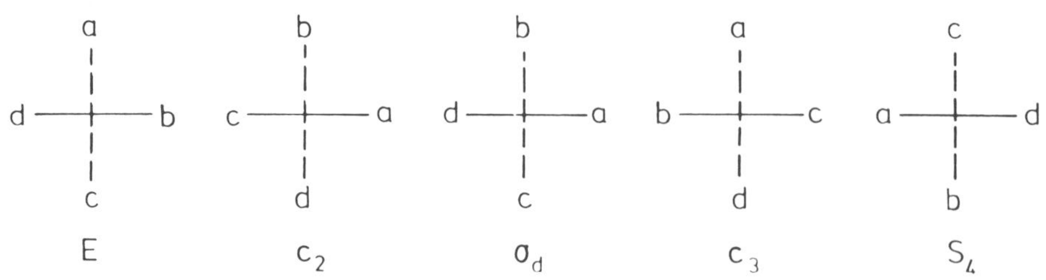

Figure 10.24. Some symmetry operations of the frame expressed as changes of axes. For the operations illustrated, we take C_2 and S_4 along the bisector of *ab*, C_3 along *a*, and σ_d in the *cd* plane, with respect to the original axial frame.

Table 10.8. Torsion-angle transformations (general positions) corresponding to the 24 frame symmetry operations of CR_4[a]

Frame symmetry element	Axis arrangement	Cycle	Torsion-angle coordinates ω_A	ω_B	ω_C	ω_D
E	*abcd*		ϕ_1	ϕ_2	ϕ_3	ϕ_4
C_2	*badc*	(*ab*)(*cd*)	ϕ_2	ϕ_1	ϕ_4	ϕ_3
	cdab	(*ac*)(*bd*)	ϕ_3	ϕ_4	ϕ_1	ϕ_2
	dcba	(*ad*)(*bc*)	ϕ_4	ϕ_3	ϕ_2	ϕ_1
σ_d	*cbad*	(*ac*)	$-\phi_3$	$-\phi_2$	$-\phi_1$	$-\phi_4$
	adcb	(*bd*)	$-\phi_1$	$-\phi_4$	$-\phi_3$	$-\phi_2$
	bacd	(*ab*)	$-\phi_2+t$	$-\phi_1+t$	$-\phi_3+t$	$-\phi_4+t$
	dbca	(*ad*)	$-\phi_4-t$	$-\phi_2-t$	$-\phi_3-t$	$-\phi_1-t$
	acbd	(*bc*)	$-\phi_1-t$	$-\phi_3-t$	$-\phi_2-t$	$-\phi_4-t$
	abdc	(*cd*)	$-\phi_1+t$	$-\phi_2+t$	$-\phi_4+t$	$-\phi_3+t$
C_3	*acdb*	(*bcd*)	ϕ_1+t	ϕ_4+t	ϕ_2+t	ϕ_3+t
	adbc	(*bdc*)	ϕ_1-t	ϕ_3-t	ϕ_4-t	ϕ_2-t
	cbda	(*acd*)	ϕ_4-t	ϕ_2-t	ϕ_1-t	ϕ_3-t
	dbac	(*adc*)	ϕ_3+t	ϕ_2+t	ϕ_4+t	ϕ_1+t
	bdca	(*abd*)	ϕ_4+t	ϕ_1+t	ϕ_3+t	ϕ_2+t
	dacb	(*adb*)	ϕ_2-t	ϕ_4-t	ϕ_3-t	ϕ_1-t
	bcad	(*abc*)	ϕ_3-t	ϕ_1-t	ϕ_2-t	ϕ_4-t
	cabd	(*acb*)	ϕ_2+t	ϕ_3+t	ϕ_1+t	ϕ_4+t
S_4	*bcda*	(*abcd*)	$-\phi_4$	$-\phi_1$	$-\phi_2$	$-\phi_3$
	dabc	(*adcb*)	$-\phi_2$	$-\phi_3$	$-\phi_4$	$-\phi_1$
	bdac	(*abdc*)	$-\phi_3-t$	$-\phi_1-t$	$-\phi_4-t$	$-\phi_2-t$
	cadb	(*acdb*)	$-\phi_2-t$	$-\phi_4-t$	$-\phi_1-t$	$-\phi_3-t$
	cdba	(*acbd*)	$-\phi_4+t$	$-\phi_3+t$	$-\phi_1+t$	$-\phi_2+t$
	dcab	(*adbc*)	$-\phi_3+t$	$-\phi_4+t$	$-\phi_2+t$	$-\phi_1+t$

[a]The translation component $t = 2\pi/3$.

trayal of its symmetry properties (cf. the transformation from a rhombohedral to a hexagonal coordinate system for triphenylmethane). For molecules with a T_d frame, one choice is to take:

$$\begin{aligned}\mathbf{a}' &= \mathbf{a} + \mathbf{b} - \mathbf{c} - \mathbf{d}\\ \mathbf{b}' &= \mathbf{a} - \mathbf{b} + \mathbf{c} - \mathbf{d}\\ \mathbf{c}' &= \mathbf{a} - \mathbf{b} - \mathbf{c} + \mathbf{d}\\ \mathbf{d}' &= \mathbf{a} + \mathbf{b} + \mathbf{c} + \mathbf{d}\end{aligned}$$

where **a, b, c, d** define the primitive cell with unit translations of 2π, π, or $2\pi/3$, depending on the nature of the rotors. The value of the determinant of the transformation matrix

$$\begin{vmatrix} 1 & 1 & -1 & -1 \\ 1 & -1 & 1 & -1 \\ 1 & -1 & -1 & 1 \\ 1 & 1 & 1 & 1 \end{vmatrix}$$

Table 10.9. Special positions of primitive unit-cell group for CR_4 (in units of 2π; $t = 2\pi/3$)

Number	Point symmetry[a]	Coordinates of equivalent positions	
24	1	General positions, see Table 10.8	
12	C_2	$\phi_1,\phi_2,\phi_1,\phi_2;$	$-\phi_1,-\phi_2,-\phi_1,-\phi_2;$
		$\phi_2,\phi_1,\phi_2,\phi_1;$	$-\phi_2,-\phi_1,-\phi_2,-\phi_1;$
		$\phi_1-t,\phi_1-t,\phi_2-t,\phi_2-t;$	$-\phi_1-t,-\phi_1-t,-\phi_2-t,-\phi_2-t;$
		$\phi_2-t,\phi_2-t,\phi_1-t,\phi_1-t;$	$-\phi_2-t,-\phi_2-t,-\phi_1-t,-\phi_1-t;$
		$\phi_1+t,\phi_2+t,\phi_2+t,\phi_1+t;$	$-\phi_1+t,-\phi_2+t,-\phi_2+t,-\phi_1+t;$
		$\phi_2+t,\phi_1+t,\phi_1+t,\phi_2+t;$	$-\phi_2+t,-\phi_1+t,-\phi_1+t,-\phi_2+t;$
12	C_s	$\phi,0,-\phi,0;$	$-\phi,0,\phi,0;$
		$0,\phi,0,-\phi;$	$0,-\phi,0,\phi;$
		$-t,-t,\phi-t,-\phi-t;$	$-t,-t,-\phi-t,\phi-t;$
		$\phi-t,-\phi-t,-t,-t;$	$-\phi-t,\phi-t,-t,-t;$
		$t,\phi+t,-\phi+t,t;$	$t,-\phi+t,\phi+t,t;$
		$\phi+t,t,t,-\phi+t;$	$-\phi+t,t,t,\phi+t;$
		Three other nonequivalent sets at:	
12	C_s	$\phi,0,-\phi,\frac{1}{2};$ etc.,	
12	C_s	$\phi,\frac{1}{2},-\phi,0;$ etc.,	
12	C_s	$\phi,\frac{1}{2},-\phi,\frac{1}{2};$ etc.,	
6	C_{2v}	$0,\frac{1}{2},0,\frac{1}{2};$	$\frac{1}{2},0,\frac{1}{2},0;$
		$\frac{2}{3}\,\frac{2}{3},\frac{1}{6},\frac{1}{6};$	$\frac{1}{6},\frac{1}{6},\frac{2}{3},\frac{2}{3};$
		$\frac{1}{3},\frac{5}{6},\frac{5}{6},\frac{1}{3};$	$\frac{5}{6},\frac{1}{3},\frac{1}{3},\frac{5}{6}$
6	S_4	$\phi,-\phi,\phi,-\phi;$	$-\phi,\phi,-\phi,\phi;$
		$\phi-t,\phi-t,-\phi-t,-\phi-t;$	$-\phi-t,-\phi-t,\phi-t,\phi-t;$
		$\phi+t,-\phi+t,-\phi+t,\phi+t;$	$-\phi+t,\phi+t,\phi+t,-\phi+t$
6	D_2	$\phi,\phi,\phi,\phi;$	$-\phi,-\phi,-\phi,-\phi;$
		$\phi+t,\phi+t,\phi+t,\phi+t;$	$-\phi+t,-\phi+t,-\phi+t,-\phi+t;$
		$\phi-t,\phi-t,\phi-t,\phi-t;$	$-\phi-t,-\phi-t,-\phi-t,-\phi-t$
3	D_{2d}	$0,0,0,0;$ $\quad t,t,t,t;$	$-t,-t,-t,-t$
3	D_{2d}	$\frac{1}{2},\frac{1}{2},\frac{1}{2},\frac{1}{2};$ $\quad \frac{5}{6},\frac{5}{6},\frac{5}{6},\frac{5}{6};$	$\frac{1}{6},\frac{1}{6},\frac{1}{6},\frac{1}{6}$

[a]Symmetry label refers to the point symmetry of the conformation that is mapped on the special positions of the four-dimensional conformational map.

equals 16, so the primed cell is 16 times as large as the primitive cell and contains 16 lattice points: its unit translations are each twice as long as those of the primitive cell.

The corresponding transformation of the coordinates is:

$$
\begin{aligned}
x_1 &= (\phi_a + \phi_b - \phi_c - \phi_d)/4 \\
x_2 &= (\phi_a - \phi_b + \phi_c - \phi_d)/4 \\
x_3 &= (\phi_a - \phi_b - \phi_c + \phi_d)/4 \\
x_4 &= (\phi_a + \phi_b + \phi_c + \phi_d)/4
\end{aligned}
$$

so that the coordinates of the 16 lattice points are (in terms of the primed axes):

$$
\begin{array}{llll}
0, 0, 0, 0; & \tfrac{1}{2}, \tfrac{1}{2}, 0, 0; & \tfrac{1}{2}, 0, \tfrac{1}{2}, 0; & \tfrac{1}{2}, 0, 0, \tfrac{1}{2}; \\
0, \tfrac{1}{2}, \tfrac{1}{2}, 0; & 0, \tfrac{1}{2}, 0, \tfrac{1}{2}; & 0, 0, \tfrac{1}{2}, \tfrac{1}{2}; & \tfrac{1}{2}, \tfrac{1}{2}, \tfrac{1}{2}, \tfrac{1}{2}; \\
\tfrac{1}{4}, \tfrac{1}{4}, \tfrac{1}{4}, \tfrac{1}{4}; & \tfrac{3}{4}, \tfrac{3}{4}, \tfrac{1}{4}, \tfrac{1}{4}; & \tfrac{3}{4}, \tfrac{1}{4}, \tfrac{3}{4}, \tfrac{1}{4}; & \tfrac{3}{4}, \tfrac{1}{4}, \tfrac{1}{4}, \tfrac{3}{4}; \\
\tfrac{1}{4}, \tfrac{3}{4}, \tfrac{3}{4}, \tfrac{1}{4}; & \tfrac{1}{4}, \tfrac{3}{4}, \tfrac{1}{4}, \tfrac{3}{4}; & \tfrac{1}{4}, \tfrac{1}{4}, \tfrac{3}{4}, \tfrac{3}{4}; & \tfrac{3}{4}, \tfrac{3}{4}, \tfrac{3}{4}, \tfrac{3}{4};
\end{array}
$$

The lattice so formed is one of the 64 Bravais lattices possible in four-dimensional space $\mathcal{R}_4$ (14 in $\mathcal{R}_3$, Chapter 2) and is described as a cubic orthogonal KU-centered (from the German description: Kombiniert überall seitenflächenzentriert) lattice.[27] Table 10.10 gives the general and special positions of the CR_4 map, referred to this KU-centered cell.

Both kinds of coordinate systems are useful. The actual molecular conformations are better described in terms of the actual torsion angles (ϕ_1, ϕ_2, ϕ_3, ϕ_4), but to apprehend approximate symmetry relations, the transformed coordinates [x_1, x_2, x_3, x_4] are more useful because some of the special positions take on especially simple forms. For example, (ϕ, ϕ, ϕ, ϕ) which images a conformation with D_2 symmetry, becomes [0, 0, 0, ϕ], and (ϕ, $-\phi$, ϕ, $-\phi$), which images one of S_4 symmetry, becomes [0, ϕ, 0, 0]. The new coordinates are symmetry coordinates that are transformed by the symmetry operations of T_d in the same way as the irreducible representations. Thus x_4 transforms as the A_2 representation, and the other three transform as the triply degenerate T_1 representation.

Some readers may like to imagine the four-dimensional space group in terms of three-dimensional sections, one for each value of the fourth coordinate.[28] The symmetry coordinates are useful for this purpose since x_4 transforms differently from the other three. It is easily shown that the three-dimensional space group involving x_1, x_2, x_3 is $P222$ for an arbitrary value of x_4. For special values of x_4, the space group becomes $P422$, the direction of the unique axis depending on the value of x_4. For example, for $x_4 = 0$, t, or $-t$, the unique axis runs along x_2, x_3, x_1, respectively, and points lying on these axes image conformations with S_4 symmetry (cf. Table 10.10). The three-dimensional space group obtained by projecting the four-dimensional one down x_4 is $P432$.

Tetraphenylmethane[29] and structurally related molecules (tetraphenylsilane, tetraphenylgermanium, tetraphenyltin, and the tetraphenylphosphonium ion)[30] nearly all adopt conformations with S_4 symmetry in

[27] H. Wondratschek, R. Bülow, and J. Neubüser, *Acta Crystallogr.* **A27,** 523 (1971).

[28] This may be a help in visualization, just as the two-dimensional sections shown in Fig. 10.14 help in visualizing the symmetry properties of the three-dimensional space group $R32$.

[29] H. T. Sumsion and D. McLachlan, *Acta Crystallogr.* **3,** 217 (1950).

[30] C. Glidewell and G. M. Sheldrick, *J. Chem. Soc., A,* **1971,** 3127; P. C. Chieh, *J. Chem.*

Table 10.10. General and special positions (modulo 1) of transformed KU-centered cell[a]

Number	Point symmetry			
16 × 24	1	x_1,x_2,x_3,x_4;	$x_3,-x_2,x_1,-x_4$;	
		$-x_1,x_2,-x_3,x_4$;	$-x_3,-x_2,-x_1,-x_4$;	
		$x_1,-x_2,-x_3,x_4$;	$x_3,x_2,-x_1,-x_4$;	
		$-x_1,-x_2,x_3,x_4$;	$-x_3,x_2,x_1,-x_4$;	
		x_2,x_3,x_1,x_4-t;	$-x_2,x_1,x_3,-x_4-t$;	
		$x_2,-x_3,-x_1,x_4-t$;	$-x_2,-x_1,-x_3,-x_4-t$;	
		$-x_2,-x_3,x_1,x_4-t$;	$x_2,-x_1,x_3,-x_4-t$;	
		$-x_2,x_3,-x_1,x_4-t$;	$x_2,x_1,-x_3,-x_4-t$;	
		x_3,x_1,x_2,x_4+t;	$x_1,x_3,-x_2,-x_4+t$;	
		$-x_3,-x_1,x_2,x_4+t$;	$-x_1,-x_3,-x_2,-x_4+t$;	
		$x_3,-x_1,-x_2,x_4+t$;	$x_1,-x_3,x_2,-x_4+t$;	
		$-x_3,x_1,-x_2,x_4+t$;	$-x_1,x_3,x_2,-x_4+t$;	
16 × 12	C_2	$0,x,0,y$;	$0,-x,0,-y$;	
		$0,-x,0,y$;	$0,x,0,-y$;	
		$x,0,0,y-t$;	$-x,0,0,-y-t$;	
		$-x,0,0,y-t$;	$x,0,0,-y-t$;	
		$0,0,x,y+t$;	$0,0,-x,-y+t$;	
		$0,0,-x,y+t$;	$0,0,x,-y+t$;	
16 × 12	C_s	$x,0,x,0$;	$0,x,x,t$;	$x,x,0,-t$;
		$-x,0,-x,0$;	$0,-x,-x,t$;	$-x,-x,0,-t$;
		$x,0,-x,0$;	$0,-x,x,t$;	$x,-x,0,-t$;
		$-x,0,x,0$;	$0,x,-x,t$;	$-x,x,0,-t$;
		Three other nonequivalent sets at:		
16 × 12	C_s	$x,\frac{1}{2},x,0$; etc.		
16 × 12	C_s	$x,0,x,\frac{1}{2}$; etc.		
16 × 12	C_s	$x,\frac{3}{4},x,\frac{1}{4}$; etc.		
16 × 6	C_{2v}	$0,\frac{1}{4},0,\frac{1}{4}$;	$\frac{1}{4},0,0,\frac{5}{12}$;	$0,0,\frac{1}{4},\frac{7}{12}$;
		$0,\frac{3}{4},0,\frac{1}{4}$;	$\frac{3}{4},0,0,\frac{5}{12}$;	$0,0,\frac{3}{4},\frac{7}{12}$;
16 × 6	S_4	$0,x,0,0$;	$0,0,x,t$;	$x,0,0,-t$;
		$0,-x,0,0$;	$0,0,-x,t$;	$-x,0,0,-t$;
16 × 6	D_2	$0,0,0,y$;	$0,0,0,y+t$;	$0,0,0,y-t$;
		$0,0,0,-y$;	$0,0,0,-y+t$;	$0,0,0,-y-t$;
16 × 3	D_{2d}	$0,0,0,0$;	$0,0,0,\frac{1}{3}$;	$0,0,0,\frac{2}{3}$;
16 × 3	D_{2d}	$0,0,0,\frac{1}{2}$;	$0,0,0,\frac{5}{6}$;	$0,0,0,\frac{1}{6}$

[a]See text for coordinates of the 16 lattice points that are to be added to the listed positions. The point symmetries refer to molecular conformations. The translation component $t = \frac{1}{3}$.

the crystalline state although the tetraphenylborate anion has D_{2d} symmetry ($\phi = 90°$) in its K^+ and $(CH_3)_4N^+$ salts.[31] The relative energies of the S_4 and D_{2d} conformations depend on the size of the central atom, D_{2d} being more favored for smaller atoms. According to force-field

Soc., A, **1971,** 3243; P. C. Chieh and J. Trotter, *J. Chem. Soc., A,* **1970,** 911; P. Goldstein, K. Seff, and K. N. Trueblood, *Acta Crystallogr.* **B24,** 778 (1968).

[31]K. Hoffmann and E. Weiss, *J. Organomet. Chem.* **67,** 221 (1974).

calculations by Hutchings, Andose, and Mislow,[32] the open D_{2d} conformation ($^oD_{2d}$) of tetraphenylmethane (Fig. 10.21, right) is more stable than the S_4 conformation, whereas for tetraphenylsilane the latter is more stable.[33]

A portion of the reaction path for the topomerization of tetraphenylmethane from one $^oD_{2d}$ conformation to another with the S_4 axis in a different orientation has been computed by Hutchings, Andose, and Mislow[32] by empirical force-field calculations; the results of a series of constrained energy minimizations with the torsion angle of one ring locked at various incremental values are shown in Table 10.11.

The topomerization itinerary is described by the authors as follows:

> Starting with the $^oD_{2d}$ conformation, as ring 1 is driven in a clockwise direction (viewed towards the center of the molecule), the other three rings rotate into an approximate S_4 conformation, but at all times remain in a more open orientation than the driven ring. As the transition state is approached, rings 2 and 3 begin to reverse their direction of rotation. The transition state is reached when the driven ring has been rotated 96°, that is, just past the closed orientation.... Further ring driving opens the possibility for ring 3 to rotate past the obstruction originally caused by the ortho-hydrogen of ring 1.... Synchronously, ring 4 rotates markedly to a less open orientation.... At this juncture ... a new conformation of approximate S_4 symmetry has been generated. Removal of constraints at this stage and subsequent full relaxation of this structure gives TPM in the $^oD_{2d}$ conformation, but with a different pairing of phenyl rings; isomerization (topomerization) has taken place.... To attain the transition state conformation, ring 1 is forced to rotate 36° further, i.e., by 96°, than required for its net overall rotation of 60°. In other words, once the transition state has been passed, ring 1 rotates back by 36°.[32]

This implies that the resultant $^oD_{2d}$ conformation is the one described by the torsion angles (30°, 30°, 30°, 210°) in our nomenclature (Table 10.11).

It is difficult to visualize paths in a four-dimensional space. Some of the complexity engendered by the present example is suggested by Fig. 10.25, where the eight symmetry-equivalent paths emanating from each $^oD_{2d}$ conformation are indicated. Although the diagram is only a two-dimensional projection (ω_a, ω_b) of the four-dimensional (ω_a, ω_b, ω_c, ω_d) space, the torsion-angle symmetry transformations (Table 10.8) ensure that all four coordinates (ϕ_1, ϕ_2, ϕ_3, ϕ_4) of any given point are projected

[32]M. G. Hutchings, J. D. Andose, and K. Mislow, *J. Am. Chem. Soc.* **97**, 4553, 4562 (1975).

[33]For tetraphenylsilane, the S_4 conformation is also favored by calculations of N. A. Ahmed, A. I. Kitaigorodsky, and K. V. Mirskaya, *Acta Crystallogr.* **B27**, 867 (1971).

Table 10.11. Strain energy[a] of tetraphenylmethane as function of torsion angle of driven ring 1

E_T (kcal/mol)	1 a	3 b	2 c	4[b] d[c]	x_1	x_2	x_3	x_4
15.9	88.8°	91.2°	88.5°	90.8°	0.1°	−1.2°	0.0°	89.9°
16.8	73.6	100.2	81.6	100.9	−2.2	−11.5	−1.8	89.1
19.1	58.6	110.2	74.3	110.1	−3.9	−21.8	−3.9	88.3
22.2	43.6	118.1	68.9	118.6	−6.5	−31.0	−6.2	87.3
24.4	33.6	122.8	66.1	123.1	−8.2	−36.5	−8.1	86.4
26.4	23.6	126.4	64.4	128.2	−10.6	−41.6	−9.8	85.6
28.3	13.6	128.1	64.9	131.3	−13.6	−45.2	−12.0	84.5
29.8	3.6	126.6	67.6	133.9	−17.8	−47.3	−14.2	82.9
30.6	−3.9	121.7	71.1	136.7	−22.5	−47.8	−15.0	81.4
30.8[d]	−6.4	120.9	70.8	137.6	−23.5	−48.5	−15.1	80.7
30.7	−8.9	116.4	73.0	139.3	−26.2	−47.9	−14.8	80.0
27.0	−16.4	88.7	79.5	151.2	−39.6	−44.2	−8.4	75.8
15.9	30	30	30	210	−45	−45	45	75

[a]From Hutchings, Andose, and Mislow [*J. Am. Chem. Soc.* **97,** 4562 (1975)]. The itinerary begins at the $^{0}D_{2d}$ conformation (90°,90°,90°,90°) = [0,0,0,90°] and ends at (30°, 30°,30°, 210°) = [−45°, −45°,45°,75°] or $[\frac{3}{4},\frac{3}{4},\frac{1}{4},\frac{5}{12}]$.
[b]Numbering of rings in original publication.
[c]Axis sequence as in Fig. 10.23.
[d]Transition state.

on the $(\omega_a\ \omega_b)$ subspace. In fact, the diagram shows not only the (ϕ_1, ϕ_2) projection of the path but also the (ϕ_3, ϕ_4), (ϕ_2, ϕ_3), and (ϕ_4, ϕ_1) projections.

According to the description cited above, the calculated path emanating from the (90°, 90°, 90°, 90°) $^{0}D_{2d}$ conformation ends at the equivalent conformation (30°, 30°, 30°, 210°). This is the endpoint reached by relaxation of the last calculated point on the path (Table 10.11), and it involves reversal of the sense of rotation of ring 1. However, on applying the torsion angle transformations to the last point L_1 of the calculated path, it is evident that this point lies fairly close to the line [−45°, −45°, x_3, 75°] which images a series of conformations with S_4 symmetry. The final point L_1 is thus fairly close to the final points L_2, L_3, L_4 of three symmetry-related paths:

	ω_a	ω_b	ω_c	ω_d
$L_1(abcd)$	−16°	89°	80°	151°
$L_2(bdac)$	−20	76	89	151
$L_3(dcba)$	−29	80	89	164
$L_4(cadb)$	−29	89	76	160

The four isometric conformations L_1, L_2, L_3, L_4 are interconverted by the

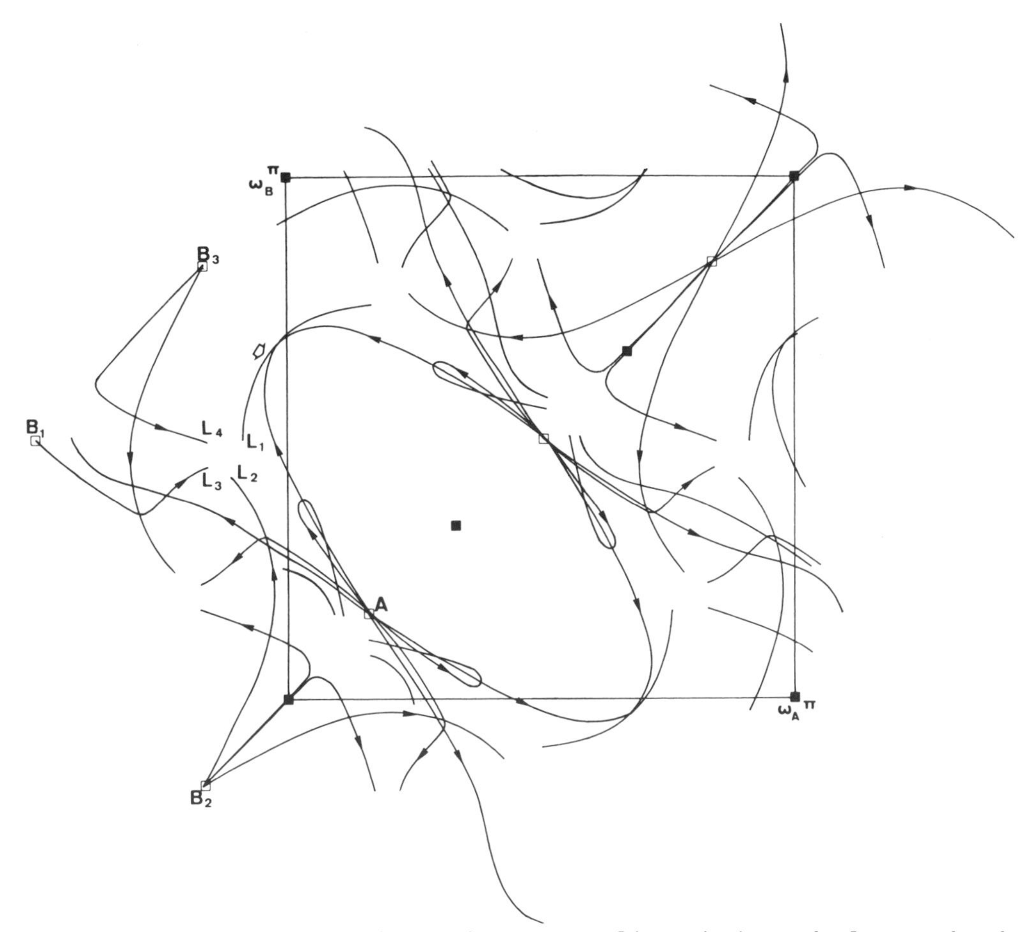

Figure 10.25. Two-dimensional projection (ω_a, ω_b) of isomerization paths for tetraphenylmethane in four-dimensional (ω_a, ω_b, ω_c, ω_d) space. Open squares are images of open ${}^oD_{2d}$ conformations (torsion angles 90°, 90°, 90°, 90° or equivalent conformations—e.g., A is 30°, 30°, 30°, 210°); filled squares are images of closed ${}^cD_{2d}$ conformations (torsion angles 0, 0, 0, 0 or equivalent conformations). The point L_1 maps the last point of the path (transition state indicated by broad arrow, upper left) calculated by Hutchings, Andose, and Mislow (Table 10.11), and L_2, L_3, L_4 map corresponding points of symmetry-equivalent paths that converge on a point with S_4 symmetry. Because of the symmetry transformations all four coordinates (ϕ_1, ϕ_2, ϕ_3, ϕ_4) of any given point are projected onto the (ω_a, ω_b) subspace. Even if the meaning of this diagram is not apparent at first, the symmetry of the pattern may be enjoyed at once.

symmetry operations $E(abcd)$, $S_4(bdac)$, $S_4^2 = C_2(dcba)$ and $S_4^3(cadb)$ or by applying small changes to the individual torsion angles (maximum change in any torsion angle is 13°). The averaged conformation (−24°, 84°, 84°, 156°) has exact S_4 symmetry and must correspond to a turning point in the potential energy, presumably an energy minimum since, according to Table 10.11, the energy slope along the path is negative at the final calculated point L_1. However, since the potential energy at L_1 is calculated to be only about 3 kcal/mol lower than the transition-state energy, the local energy minimum at the neighboring S_4 conformation cannot be very deep. In other words, the potential energy surface near the four symmetry-related points must be rather flat.

This implies that if the isomerization were treated as a dynamic process, in terms of the classical trajectory of a particle on the potential energy surface, then a particle proceeding along the calculated path as far as point L_1 could continue along any of the three symmetry-related paths. Continuation along the path through L_3 leads back to the same ${}^0D_{2d}$ structure as the initial one, with the S_4 axis in the same orientation, but continuation via L_2 or L_4 leads to the end points (−30°, −30°, 150°, 150°) or (−30°, 150°, 150°, 150°) with the S_4 axis in a different orientation from the initial one (also different from the one implied by the previous description).

Clearly the full symmetry in four dimensions must be considered if any attempt is made to locate the low-energy regions of the conformational map from scatter plots, as was done for triphenylphosphine oxide (Fig. 10.16). Many crystal structures containing tetraphenylborate, tetraphenylphosphonium and tetraphenylarsonium ions have been described, and the preparation of the corresponding scatter plots is in progress. The preliminary results[34] seem to agree quite well with expectations based on the calculation by Hutchings, Andose, and Mislow.[32]

[34]E. Bye, unpublished work.

Appendix I. A BASIC Computer Program for Calculating Interatomic Distances and Angles and Torsion Angles from Crystal Coordinates

The program transforms the crystal coordinates to Cartesian coordinates by applying Eqs. 5.30 and finds the torsion angle $\omega(ABCD) = \omega(IJKL)$ by applying Eqs. 5.35 and 5.36. The input consists of the unit-cell dimensions (line 20) and the crystal coordinates of the atoms, which are assigned code numbers $I = 1$ to N (line 46). The input is terminated by writing $I = 0$. Each torsion angle required has to be requested separately by specifying the code numbers of the four atoms concerned: interatomic distances and angles are then printed in addition to the torsion angle. If torsion angles are not required, the code numbers of only three atoms need to be specified; if only a distance is required then code numbers of only two atoms suffice. A sample output follows.

App. I Program

```
4  PRINT
5  PRINT "MOLECULAR GEOMETRY"
10  DIM X[60],Y[60],Z[60]
15  PRINT "A,B,C,ALPHA,BETA,GAMMA"
20  INPUT A1,A2,A3,W1,W2,W3
21  PRINT
25  LET P=3.14159/180
30  LET C1= COS (W1*P)
31  LET C2= COS (W2*P)
32  LET C3= COS (W3*P)
33  LET S3= SIN (W3*P)
35  LET M6=(C1-C3*C2)/S3
37  LET V= SQR (1-C1*C1-C2*C2-C3*C3+2*C1*C2*C3)
39  LET M9=V/S3
45  PRINT "I,X(I),Y(I),Z(I), I=0 FOR LAST ATOM"
46  INPUT I,X[I],Y[I],Z[I]
47  PRINT
48  IF I= 0 GOTO  75
49  LET U=A1*X[I]+A2*Y[I]*C3+A3*Z[I]*C2
50  LET V=A2*Y[I]*S3+A3*Z[I]*M6
51  LET W=A3*Z[I]*M9
52  LET X[I]=U
55  LET Y[I]=V
57  LET Z[I]=W
58  GOTO  46
75  PRINT "BOND LENGTH: I1,I2,0,0"
76  PRINT "BOND ANGLE : I1,I2,I3,0"
77  PRINT "TORSION ANGLE: I1,I2,I3,I4"
85  INPUT T1,T2,T3,T4
86  PRINT
```

```
100  LET E[1]=X[T2]-X[T1]
105  LET E[2]=Y[T2]-Y[T1]
110  LET E[3]=Z[T2]-Z[T1]
115  LET F=1
120  LET D1= SQR (E[1]*E[1]+E[2]*E[2]+E[3]*E[3])
121  LET D=D1
125  GOSUB  325
133  IF T3= 0 GOTO  499
135  LET E[4]=X[T3]-X[T2]
140  LET E[5]=Y[T3]-Y[T2]
145  LET E[6]=Z[T3]-Z[T2]
150  LET F=4
155  LET D2= SQR (E[4]*E[4]+E[5]*E[5]+E[6]*E[6])
156  LET D=D2
160  GOSUB  325
165  LET C4=-(E[1]*E[4]+E[2]*E[5]+E[3]*E[6])
170  LET S4= SQR (1-C4*C4)
175  LET A4= ATN (S4/C4)/P
176  IF A4> 0 GOTO  180
177  LET A4=A4+180
180  IF T4= 0 GOTO  497
185  LET E[7]=X[T4]-X[T3]
190  LET E[8]=Y[T4]-Y[T3]
195  LET E[9]=Z[T4]-Z[T3]
200  LET F=7
205  LET D3= SQR (E[7]*E[7]+E[8]*E[8]+E[9]*E[9])
210  LET D=D3
215  GOSUB  325
220  LET C5=-(E[4]*E[7]+E[5]*E[8]+E[6]*E[9])
225  LET S5= SQR (1-C5*C5)
230  LET A5= ATN (S5/C5)/P
231  IF A5> 0 GOTO  250
232  LET A5=A5+180
250  LET U1=(E[2]*E[6])-(E[3]*E[5])
252  LET U2=(E[3]*E[4])-(E[1]*E[6])
254  LET U3=(E[1]*E[5])-(E[2]*E[4])
256  LET V1=(E[5]*E[9])-(E[6]*E[8])
258  LET V2=(E[6]*E[7])-(E[4]*E[9])
260  LET V3=(E[4]*E[8])-(E[5]*E[7])
265  LET C6=(U1*V1+U2*V2+U3*V3)/(S4*S5)
270  LET S6=(E[1]*V1+E[2]*V2+E[3]*V3)/(S4*S5)
275  LET A6= ATN (S6/C6)/P
280  IF A6> 0 GOTO  290
281  IF S6< 0 GOTO  300
282  LET A6=A6+180
284  GOTO  300
290  IF C6> 0 GOTO  300
292  LET A6=A6-180
299  PRINT
300  PRINT " T("T1;T2;T3;T4;") ="A6
301  PRINT " W("T2;T3;T4;")     ="A5
302  PRINT " D("T3;T4;")        ="D3
305  GOTO  497
325  FOR J=F TO F+2
327    LET E[J]=E[J]/D
329  NEXT J
331  RETURN
496  PRINT
497  PRINT " W("T1;T2;T3;")     ="A4
498  PRINT " D("T2;T3;")        ="D2
499  PRINT " D("T1;T2;")        ="D1
510  PRINT
```

```
512  GOTO  85
515  END
```

App. I   Sample input and output

```
BASIC?
*READY
RUN

MOLECULAR GEOMETRY
A,B,C,ALPHA,BETA,GAMMA
? 7.877? 7.210? 7.891? 105.56? 116.25? 79.84
I,X(I),Y(I),Z(I), I=0 FOR LAST ATOM
? 1? -.1361? .1572? -.0717
? 2? -.0963? .1157? .1184
? 3? .0920? .0109? .2083
? 4? -.1955? .3344? -.1057
? 5? -.2272? .4046? -.2806
? 6? -.2182? .1553? .2003
? 7? -.4167? .2445? .1235
? 8? .2174? .0656? .3903
? 9? .2034? .2392? .5396
? 10? .1361? -.1572? .0717
? 11? .0963? -.1157? -.1184
? 12? -.0920? -.0109? -.2083
? 0? 0? 0? 0
BOND LENGTH: I1,I2,0,0
BOND ANGLE : I1,I2,I3,0
TORSION ANGLE: I1,I2,I3,I4
? 12? 1? 2? 3
 T( 12  1  2  3 ) = 46.2811
 W( 1  2  3 )     = 114.158
 D( 2  3 )        = 1.49519
 W( 12  1  2 )    = 114.083
 D( 1  2 )        = 1.49324
 D( 12  1 )       = 1.49442

? 1? 2? 3? 10
 T( 1  2  3  10 ) =-46.3004
 W( 2  3  10 )    = 114.124
 D( 3  10 )       = 1.49442
 W( 1  2  3 )     = 114.158
 D( 2  3 )        = 1.49519
 D( 1  2 )        = 1.49324

? 2? 3? 10? 11
 T( 2  3  10  11 ) = 46.2653
 W( 3  10  11 )    = 114.083
 D( 10  11 )       = 1.49324
 W( 2  3  10 )     = 114.124
 D( 3  10 )        = 1.49442
 D( 2  3 )         = 1.49519
```

Appendix II. A BASIC Computer Program for Calculating Cartesian Coordinates from Internal Coordinates

Bond distances, bond angles, and torsion angles are the internal coordinates used. Care is required to ensure that the internal coordinates chosen are geometrically independent and that they fix the structure of the molecular skeleton completely. This is done by regarding the molecule (N atoms) as a main chain with an arbitrary number of branches. Its structure is then completely fixed by specifying $3N-6$ internal coordinates consisting of $N-1$ bond distances, $N-2$ bond angles, and $N-3$ torsion angles.

The program assigns sequential code numbers $I = 1$ to 100 to the atoms in the order they are entered in the input list. For each atom entered, the input consists of the code number J of the atom to which it is directly linked (the "next atom" but always such that $J < I$), together with the values of the bond angle $W(I,J,J-1)$ for $I \geq 3$, the signed torsion angle $T(I,J,J-1,J-2)$ for $I \geq 4$, and the bond distance $R(I,J)$ for $I \geq 2$, in that order. For $I = 1, J = 0$; the six undefined quantities involving this nonexistent atom—e.g., $R(1,0)$, $W(2,1,0)$, $T(3,2,1,0)$, etc., are set equal to zero (see sample input and output). Angles are in degrees.

The origin and axial orientation of the Cartesian frame are tied to the first three atoms entered ($I = 1,2,3$). The origin is at atom 1, the X-axis is along the direction from atom 2 to atom 1, and the Y-axis is in the plane of atoms 1, 2, and 3. Thus, the Cartesian coordinates of these three atoms are:

	x	y	z
1	0	0	0
2	$-r_{2,1}$	0	0
3	$-r_{2,1}+r_{3,2} \cos \omega_{3,2,1}$	$r_{3,2} \sin \omega_{3,2,1}$	0

For each subsequent atom entered, a local Cartesian frame (X_I,Y_I,Z_I) is set up in a similar way, with its origin at the new atom I, X_I along the direction from J to I, and Y_I in the plane of $I,J,J-1$. The coordinates of a point in the Ith system are related to the coordinates in the Jth system by a rotational transformation plus a translation[1]

[1] H. B. Thompson, *J. Chem. Phys.* **47**, 3407 (1967).

$$\begin{bmatrix} x_J \\ y_J \\ z_J \\ 1 \end{bmatrix} = \begin{bmatrix} -\cos W & -\sin W & 0 & -R\cos W \\ \sin W\cos T & -\cos W\cos T & -\sin T & R\sin W\cos T \\ \sin W\sin T & -\cos W\sin T & \cos T & R\sin W\sin T \\ 0 & 0 & 0 & 1 \end{bmatrix} \begin{bmatrix} x_I \\ y_I \\ z_I \\ 1 \end{bmatrix}$$

$$\text{or } \mathbf{x}_J = \mathbf{B}_I \cdot \mathbf{x}_I$$

writing $\mathbf{x}_I$ for the expanded vector and $\mathbf{B}_I$ for the matrix. Transformation back to the initial frame is then performed by matrix multiplication:

$$\mathbf{x}_1 = \mathbf{B}_2 \cdot \mathbf{B}_3 \cdot \mathbf{B}_4 \ldots . \mathbf{B}_I \cdot \mathbf{x}_I$$

To ensure self-consistency, some of the elements of the first two matrices $\mathbf{B}_2$ and $\mathbf{B}_3$ are fixed by the program, but all other matrix elements depend on the input.

App. II Program

```
9   DIM A[8,100],Q[2,100]
10  PRINT
11  PRINT
15  PRINT "TRANSFORMATION OF INTERNAL TO CARTESIAN COORDINATES"
20  PRINT "INPUT SEQUENCE IS: NEXT,BOND ANGLE, TORSION,DISTANCE"
21  PRINT "INPUT NEXT=999 TO STOP"
22  PRINT "NUMBERING OF ATOMS HAS TO START WITH ZERO"
23  PRINT
24  PRINT
25  LET I= 0
26  LET F=1.74533E-2
30  FOR S=1 TO 7
35    LET A[S, 0]= 0
40  NEXT S
45  LET A[ 0, 0]=1
50  LET A[4, 0]=1
55  LET A[8, 0]=1
60  LET Q[ 0, 0]= 0
65  LET Q[1, 0]= 0
70  LET Q[2, 0]= 0
75  PRINT "NEXT ATOM";
76  INPUT N
77  IF N=999 GOTO  299
78  INPUT W,T,R
79  IF I= 0 GOTO  261
95  IF I>2 GOTO  140
100  IF I<2 GOTO  120
105  LET T= 0
110  LET M=1
115  GOTO  155
120  LET W= 0
125  LET T=180-T
130  LET M= 0
135  GOTO  155
140  LET M=N-1
155  LET N=M
160  LET B0=- COS (W*F)
165  LET B1=- SIN (W*F)
170  LET B5=- SIN (T*F)
175  LET B8= COS (T*F)
```

```
180  LET B3=-B1*B8
185  LET B4=B0*B8
190  LET B6=B1*B5
195  LET B7=-B0*B5
200  LET B2= 0
205  LET A[ 0,I]=A[ 0,M]*B0+A[1,M]*B3+A[2,M]*B6
210  LET A[1,I]=A[ 0,M]*B1+A[1,M]*B4+A[2,M]*B7
215  LET A[2,I]=A[ 0,M]*B2+A[1,M]*B5+A[2,M]*B8
220  LET A[3,I]=A[3,M]*B0+A[4,M]*B3+A[5,M]*B6
225  LET A[4,I]=A[3,M]*B1+A[4,M]*B4+A[5,M]*B7
230  LET A[5,I]=A[3,M]*B2+A[4,M]*B5+A[5,M]*B8
235  LET A[6,I]=A[6,M]*B0+A[7,M]*B3+A[8,M]*B6
240  LET A[7,I]=A[6,M]*B1+A[7,M]*B4+A[8,M]*B7
245  LET A[8,I]=A[6,M]*B2+A[7,M]*B5+A[8,M]*B8
250  LET Q[ 0,I]=Q[ 0,M]+A[ 0,I]*R
255  LET Q[1,I]=Q[1,M]+A[3,I]*R
260  LET Q[2,I]=Q[2,M]+A[6,I]*R
261  PRINT
262  LET I1=I+1
265  PRINT I1,Q[ 0,I],Q[1,I],Q[2,I]
266  PRINT
270  LET I=I+1
275  GOTO  75
299  PRINT
300  PRINT "TYPE 1 IF NEW INPUT SET"
305  PRINT "TYPE 0 IF END";
310  INPUT Z
315  IF Z=1 GOTO  10
320  END
```

App. II Sample input and output

```
TRANSFORMATION OF INTERNAL TO CARTESIAN COORDINATES
INPUT SEQUENCE IS: NEXT,BOND ANGLE, TORSION,DISTANCE
INPUT NEXT=999 TO STOP
NUMBERING OF ATOMS HAS TO START WITH ZERO


NEXT ATOM? 0? 0? 0? 0
 1             0              0              0

NEXT ATOM? 1? 0? 0? 1.525
 2            -1.525          0              0

NEXT ATOM? 2? 107.12? 0? 1.531
 3            -1.97568        1.46316        0

NEXT ATOM? 3? 104.08? 28.5? 1.518
 4            -.847761        2.19699       -.702565

NEXT ATOM? 4? 100.50? -33.7? 1.542
 5             .362545        1.36111       -.239745

NEXT ATOM? 4? 109.71? 91.6? 1.535
 6            -1.14158        2.27922       -2.20693

NEXT ATOM? 4? 112.82? -148.5? 1.529
 7            -.659971        3.62536       -.190399
```

```
NEXT ATOM? 999
TYPE 1 IF NEW INPUT SET
TYPE Ø IF END? Ø

*READY
```

The sample calculation refers to the skeleton

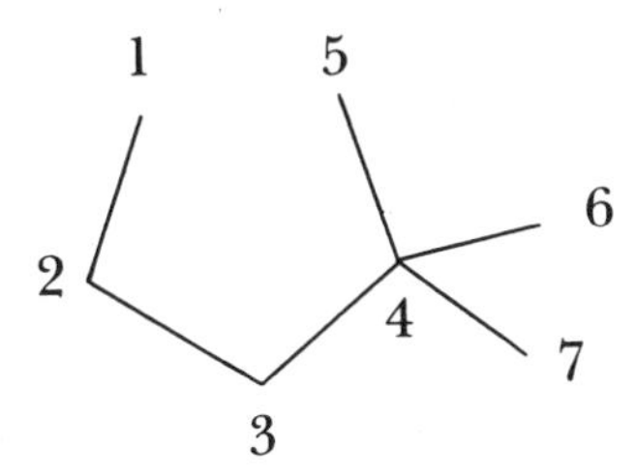

with the following input parameters:

I	J	W	T	R
2	1	0	0	1.525
3	2	107.12	0	1.531
4	3	104.08	+28.5	1.518
5	4	100.50	−33.7	1.542
6	4	109.71	+91.6	1.535
7	4	112.82	−148.5	1.529

From the output, the Cartesian coordinates are:

I	x	y	z
1	0	0	0
2	−1.525	0	0
3	−1.976	1.463	0
4	−0.848	2.197	−0.703
5	0.363	1.361	−0.240
6	−1.142	2.279	−2.207
7	−0.660	3.625	−0.190

from which the dependent bond lengths and angles may readily be calculated (e.g., using the program of Appendix I).

Index of Names

Index of Names

Index of Names

Index of Subjects

Index of Subjects

Index of Subjects